Science of Heat and Thermophysical Studies

A Generalized Approach to Thermal Analysis

Science of Heat and Thermophysical Studies

A Generalized Approach to Thermal Analysis

By
Jaroslav Šesták

Institute of Physics of the Academy of Sciences
of the Czech Republic, Division of Solid-State Physics,
Cukrovarnická 10, CZ - 162 53 Praha 6

ZÁPADOČESKÁ
UNIVERZITA
V PLZNI

University of Western Bohemia, Faculty of Applied
Sciences, Univerzitní 8, CZ - 306 14 Pilsen,
both Czech Republic

ELSEVIER
AMSTERDAM . BOSTON . HEIDELBERG . LONDON . NEW YORK . OXFORD
PARIS . SAN DIEGO . SAN FRANCISCO . SINGAPORE . SYDNEY   TOKYO

ELSEVIER B.V.
Radarweg 29
P.O. Box 211, 1000 AE Amsterdam
The Netherlands

ELSEVIER Inc.
525 B Street, Suite 1900
San Diego, CA 92101-4495
USA

ELSEVIER Ltd
The Boulevard, Langford Lane
Kidlington, Oxford OX5 1GB
UK

ELSEVIER Ltd
84 Theobalds Road
London WC1X 8RR
UK

First edition 2005

Library of Congress Cataloging in Publication Data
A catalog record is available from the Library of Congress.

British Library Cataloguing in Publication Data
A catalogue record is available from the British Library.

ISBN–13: 978-0-444-51954-2
ISBN–10: 0-444-51954-8

♾ The paper used in this publication meets the requirements of ANSI/NISO Z39.48-1992 (Permanence of Paper).

Working together to grow
libraries in developing countries

www.elsevier.com | www.bookaid.org | www.sabre.org

ELSEVIER   BOOK AID International   Sabre Foundation

Transferred to digital printing in 2007.

"Heat is disordered energy.
So with two words the nature of heat is explained.
The rest of this article will be an attempt to explaine the explanation."
Freeman J. Dyson (1954)

vi

The book is published on the occasion of the 2005 ***International Year of Physics***, which was herewith devoted to the anniversary of Kepler's (1600-1612), Fürth's (1915-1939) and Einstein's (1905-1907) stays in Prague

*Kepler Johannes (*1571 - ⁺1630)*
*Einstein Albert (*1879 - ⁺1955)*
*Fürth Reinhold (*1893* - ⁺1979)*

We also remember the anniversary dates of some distinguished Bohemian scholars and alchemists of the Middle Age:

Agricola Georgius (Georg Bauer) (⁺1555)
Hájek Tadeáš (from Hájků) (⁺1600)
*Marcus Marci Jan (from Kronland) (*1595)*
*Stolcius Daniel (*1600)*
Komenský Jan Amos (Comenius) (⁺1670)
Diviš Prokop (Procopius) (⁺1765)

as well as the birthdays of some early forward-looking Czech thermodynamicists:

*Strouhal Čeněk (*1850)*
*Wald František (*1860)*
*Regner Albert (*1905)*

The book is also dedicated to the forty years anniversary of the foundation of the ***International Confederation for Thermal Analysis (ICTA)*** (in London and Aberdeen 1965) by the following pioneers

Bárta Rudolf (1897–1985)
Berg Lev Germanovič (1896–1974)
Mackenzie Robert Cameron (1920–2000)
Murphy Cornelius Bernard (1918–1994)
Wendlandt William Wesley (1927–2000)

PREFACE

At the beginning of the 1980s, I accomplished my long-standing ambition [1] to publish an extended treatise dealing with the theoretical aspects of thermal analysis in relation to the general subject of thermophysical properties of solids. The pioneering Czech version appeared first in 1982 [2], successively followed by English [3] and Russian [4] translations. I am gratified to remark that the Russian version became a bestseller on the 1988 USSR market and 2500 books were sold out within one week. The other versions also disappeared from the bookshops in a few years leaving behind a rather pleasing index of abundant citation responses (almost 500 out of my total record of 2500).

Recently I was asked to think over the preparation of an English revision of my book. Although there has been a lapse of twenty years, after a careful reading of the book again, I was satisfied that the text could be more or less reiterated as before, with a need for some corrections and updating. The content had not lost its contemporary value, innovative approach, or mathematical impact and can still be considered as being competitive with similarly focused books published even much later, and thus a mere revision did not really make sense.

In the intervening years, I have devoted myself to a more general comprehension of *thermal analysis,* and the associated field of overlaying thermophysical studies, to gain a better understanding of the science of *heat* or, if you like, *thermal physics* or *science of heat.* It can be seen to consist of the two main areas of generalized interest: the *force fields* (that is, particularly, the *temperature* expressing the motional state of constituent particles) and the *arrangement deformations* (that is *entropy,* which is specific for ordering the constituent particles and the associated 'information' value). This led me to see supplementary links to neighboring subjects that are, so far, not common in the specialized books dealing with the classically understood field of thermo-physical studies or, if you like, forming the generalized and applied domain of *thermal analysis.*

This comprehension needed a gradual development. In 1991, I co-edited and co-authored other theoretical books [5, 6] attentive to problems of non-equilibrium phase transitions dealing with the non-stationary processes of nucleation and crystal growth and their impact on modern technologies [7], and later applied to glasses [8]. A more comprehensive and updated Czech book was published recently [9]. I have also become occupied in extensive lecturing, beside the short courses mostly read on thermodynamics and thermal analysis (among others, in Italy, the USA, Norway, India, Germany, Argentina, Chile and Taiwan), I enjoyed giving complete full-term courses at the Czech University of Pardubice (1988–1999, "Modern materials"), the University of Kyoto (1996 and 2004, "Energy science"), Charles University in Prague (1997–

2001, "Thermo-dynamics and society" and "On the borderland of science and philosophy of nature"), and at the US University of New York, the division in Prague (1999-, "Scientific world"). I was also proud to be given the challenge of supervising an associated cooperation project with the University of Kyoto (2001-2004) and was honored to be a founding member of its new faculty on energy science (1996). It gave me space to be in contact with enquiring students and made it easier for me to think about thermal science within a wider context [10] including philosophy, history, ecology, sociology, environmental anthropology, informatics, energetics and an assortment of applied sciences. It helped me to include completely new allocations to this new edition, e.g., Greek philosophical views and their impact on the development of contemporary ideas, understanding caloric as heat to be a manufacturing tool, instrumental probe and scholarly characteristic, early concepts of temperature and its gradients, non-equilibrium and mesocopic (quantum) thermo-dynamics, negentropy as information logic, generalized authority of power laws and impact of fractal geometry, physics applied to economy (econophysics) or submicroscopic scales (quantum diffusion) and, last but not least, the importance of energy science and its influence on society and an inquiring role into sustainable environments.

A number of Czech-Slovak scientists participated in the discovery of specialized thermophysical techniques [3], such as dielectric (*Bergstein*), emanation (*Balek*), hydrothermal (*Šatava*), periodic (*Proks*), photometric (*Chromý*) or permeability (*Komrska*) methods of thermal analysis including the modified technique of accelerated DTA (*Vaniš*). There are also early manuscripts on the first national thermodynamic-like treaties dealing with wider aspects of heat and temperature that are worth indicating. They are good examples of creative and illustrious discourse, which served as good quality precedents for me. They were published by the world-famous Bohemian teacher and high-spirited Czech thinker *Jan Amos Komenský (Comenius)*, the renowned author of various educational books among which the treaty on *"The nature of heat and cold, whose true knowledge will be a key to open many secrets of nature"* (available as early as 1678 [11]). Among recent example belongs the excellent book `*Thermal phenomena*` 1905 [12] authored by *Čeněk Strouhal*, a Czech educator and early builder of modern thermal physics.

My aim has been to popularize the role of heat from the micro- up to the macro-world, and have endeavored to explain that the type of processes involved are almost the same – only differing in the scale dimension of inherent heat fluxes. A set of books, popularizing the science of physics through the freely available understandability (from *Prigogine* [13] to *Barrow* [14]), served me here as examples. I am hopeful (and curious too) that this book will be accepted as positively as my previous, more methodological and narrowly-focused publications where I simply concentrated on theoretical aspects of thermal analysis, its methods, applications and instrumentation.

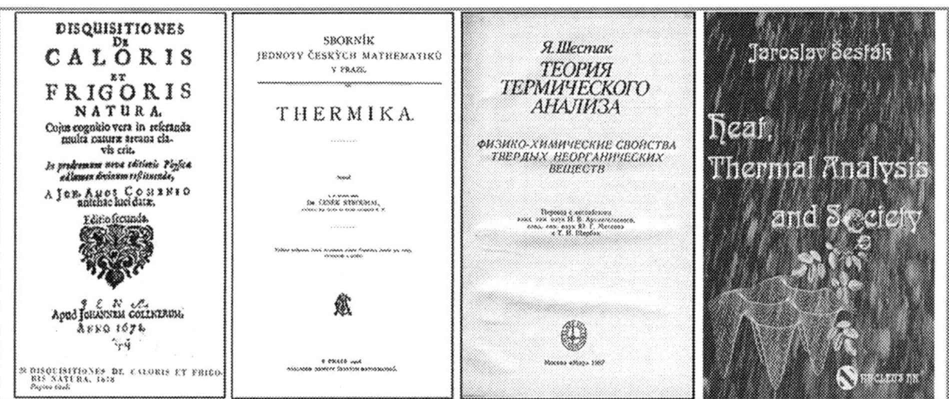

Fig. 1. – The symbolic manifestation of the title pages of four selected books related to the scientific domain of heat; left to right: J.A. Comenius 1678, Č. Strouhal 1906 and J. Šesták 1987 and 2004. *It is worth noting that the latter publication (the revealed cover of which was designed by the author) preceded the present updated and broadly reconstructed version in your hands. The previous book contents [9] was purposefully intended to assign a yet unusual amalgamation between the author's scientific and artistic efforts and ambitions so that the book included 60 of the art (full-page) photos printed on the coated paper, which were used not only as the frontispieces of each chapter but also helped to compose book's extended appendix. Though unconventional, such anticipated interdisciplinary challenge hopefully refreshed the scientific comeliness and gave a specific charisma of this previous book, which was aimed to present and seek deeper interconnections between the science and philosophy of nature (cf. www.nucleus.cz and info@nucleus.cz).*

By no means did I want to follow the habitual trends of many other thermoanalytical books (jointly cited in [3,9]), which tried for a more effective rendition but still merely rearranged, for almost 40 years, more or less unvarying information [15] with little innovativeness that would help the reader's deeper edification. Therefore, all those who are awaiting a clear guidance to the instrumental basis of thermal analysis or to the clearer theory of thermodynamics are likely to be disappointed whereas the book should please those who would care to contemplate yet unseen corners of thermal science and who are willing to see further perspectives.

However, this book, which you now have in your hands and which I regard as a somehow more inventive and across-the-board approach to the science of heat, did not leave much space to work out a thorough description of any theoretical background. The interested readers are regretfully referred to the more detailed mathematics presented in my original books [1–9] and review articles cited in the text. It is understandable that I have based the book's new content mostly on the papers that I have published during the past twenty years. I also intentionally reduced the number of citations to a minimum (collected at the end) so that, for a more detailed list of references, the readers are kindly advised to find the original literature or yet turn their attention to my previously published books [1–9] or my papers (a survey of which was published in ref. [16].

I have tried to make the contents as compact as possible while respect-ing the boundary of an acceptably readable book (not exceeding 500 pages) but, hopefully, still apposite. In terms of my written English in this version of my book, its thorough understandability may be questionable but excusable as it is written by an author whose mother tongue is fundamentally and grammatically very different – the Slavic language of the Czechs.

I would also like to enhance herewith my discourse on heat by accentuating that the complete realization of an absolute zero is impossible by any finite processes of supercooling. In such an unachievable and unique 'nil' state, all motion of atoms would cease and the ever present fluctuations ('errors'), as an internal driving force for any change, would fade away. The system would attain the distinctly perfect state, which is deficient in defects. We know that such a perfect state is impossible as, e.g., no practical virtuous single crystal can truthfully exist without the disordering effect of admixtures, impurities, vacancies, dislocations, tensions and so forth. This is an objective reason to state here that no manuscript could ever be written faultlessly. In this context, any presentation of ideas, items specification, mathematical description, citation evidence, etc., are always associated with unprompted mistakes. As mentioned in the forthcoming text, any errors (i.e., a standard state of fluctuations) can play the most important roles in any positive development of the state of matter and/or society itself, and without such 'faults' there would be neither evolution nor life and even no fun in any scientific progress.

Therefore, please, regard any misprints, errors and concept distortion that you will surely find in many places in this book, in a more courteous way of 'incontrovertibly enhanced proficiency'. Do not criticize without appreciating how much labor and time has been involved in completing this somewhat inquisitive but excessively wide-ranging and, thus, unavoidably dispersed idea-mixing approach, and think rather in what way it could be improved or where it may be further applied or made serviceable.

In conclusion I would like to note that the manuscript was written under the attentive support of the following institutions: Institute of Physics of the Academy of Sciences of Czech Republic(AV0210100521); Faculty of Applied Science, the West Bohemian University in Pilsen (4977751303); enterprise NETZSCh Gerätebau GmbH., Selb (Germany) as well as the Municipal Government of Prague 5 and the Grant Agency of Czech Republic (522/04/0384).

Jaroslav Šesták
Prague, 2005

CONTENTS

Petite characteristics of some important scholars and scientists of the past
and a selection of recent thermoanalysts related to the book contents.

1. SOME PHILOSOPHICAL ASPECTS OF SCIENTIFIC RESEARCH

1.1. Exploring the environment and scale dimensions

The understanding of nature and its description are not given *a priori* but developed over time, according to how they were gradually encountered and assimilated into man's existing practices. Our picture of nature has been conditioned by the development of modes of perceiving (sensors) and their interconnections during mankind's manufacturing and conceptual activities. The evaluation of sensations required the definition of measuring values, i.e., the discretion of what is available (experience, awareness, inheritance). A sensation must be classified according the given state of the organism, i.e., connected to the environment in which the evaluation is made. Everyday occurrences have been incorporated, resulting in the outgrowth of so-called *custom states*. This is, however, somewhat subjective because for the sake of objectivity we must develop measures independent of individual sensation, i.e., scales for identifying the conceptual dimensions of our surroundings (territorial and/or force-field parameters such as *remoteness* (distance) *or warmth* (temperature), having mutually quite irreconcilable characteristics).

Our educational experience causes most of us to feel like inhabitants of a certain geographical (three-dimensional) continuum in which our actual position or location is not necessarily indispensable. A similar view may be also applied to the other areas such as knowledge, durability, warmth etc. If we were, for example, to traverse an arbitrary (assumingly 2-D) landscape we would realize that some areas are more relevant than others. Indeed, the relative significance of acknowledged objects depends on their separated distance – which can be described as their '*nearness*' [17]. It can be visualized as a function, the value of which proportionally decreases with the distance it is away from us, ultimately diminishing entirely at the '*horizon*' (and the space beyond). The horizon, as a limit, exists in many of our attributes (knowledge, experience, capability).

When the wanderer strolls from place to place, his 'here', 'there', his horizon as well as his *field of relevance* gradually shifts whilst the implicit form of nearness remains unchanged. If the various past fields of relevance are superimposed, a new field of relevance emerges, no longer containing a central position of 'here'. This new field may be called the *cognitive map,* as coined by *Havel* [17] for our positional terrain (as well as for our knowledge extent). Individual cognitive maps are shaped more by memories (experience, learning) of the past than by immediate visual or *kinestatic* encounters.

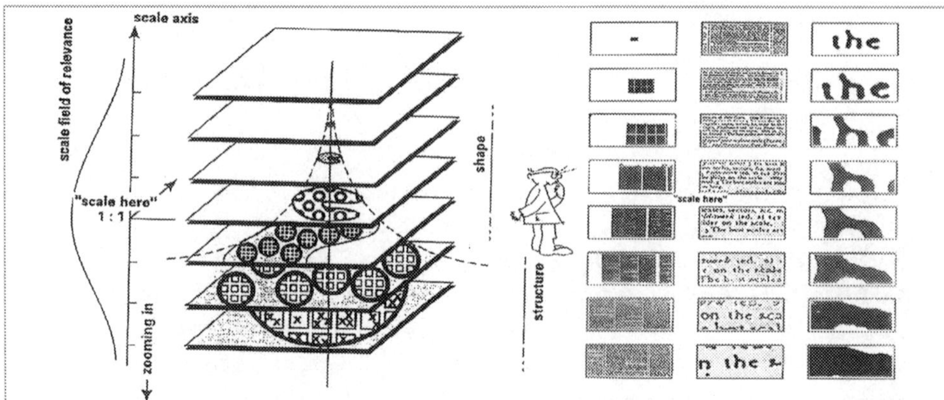

Fig. 2. – Illustrative zoom as a shift in the scale dimension and right a symbolic communication user who exploits a limited range of scales to its explication [9,17]. Courtesy of *Ivan M. Havel*, Prague, Czech Republic

It is not difficult to imagine a multitude of cognitive maps of some aggregates to form, e.g., a collective cognition map of community, field, etc., thus available for a wider public use. However, to match up individual maps we need to be sure of the application of adequately rational levels, called *scales* [18]. Returning to the above geographical illustrations we may see them as the superimposition of large-scale maps on top of another smaller-scale maps, which, together, yield a more generalized dimension, called the '*spatial-scale axis*'. A movement in the upward direction along this scale axis resembles *zooming out* using a camera (objects shrink and patterns become denser) while the opposite downward movement is similar to *zooming in* (objects are magnified and patterns become dispersed or even lost). Somewhere in the center of this region (about the direct proportions one-to-one) exists our perception of the world, a world that, to us, is readily understandable.

Our moving up and down in the scale is usually conditioned by artificial instruments (telescopes, microscopes) that are often tools of scientific research. Even when we look through a microscope we use our natural vision. We do not get closer to the observed but we take the observed closer to us, enabling new horizons to emerge and employ our imagination and experience. We may say that we import objects, from that other reality, closer to us. Only gradually have the physical sciences, on the basis of laborious experimental and theoretical investigations, extended our picture of nature to such neighboring scales.

Let us consider the simplest concept of *shape*. The most common shapes are squares or circles easily recognizable at a glance – such objects can be called '*scale-thin*' [17]. Certainly there are objects with more complex shapes such as the recently popular self-similar objects of fractals that can utilize the concept of the scale dimension quite naturally because they represent a recursive scale order. It is worth noting that the term *fractal* was derived from the Latin word *'fractus'* (meaning broken or fragmented) or *'frangere'* (break) and was coined by the Polish-born mathematician *Mandelbrot* on the basis of *Hausdorff* dimension analysis. The term *fractal dimension* reveals precisely the nuances of the shape and the complexity of a given non-Euclidean figure but as the idiom dimension does not have exactly the same meaning of the dimension responsive to Euclidean space so that it may be better seen as a property [9].

Since its introduction in 1975, it has given rise to a new system of geometry, impacting diverse fields of chemistry, biology, physiology and fluid dynamics. Fractals are capable of describing the many irregularly shaped objects or spatially non-uniform phenomena in nature that cannot be accommodated by the components of Euclidean geometry. The reiteration of such irregular details or patterns occur at progressively smaller scales and can, in the case of purely abstract entities, continue indefinitely so that each part of each part will look basically like the object as a whole. At its limit, some 'conclusive' fractal structures penetrate through arbitrary small scales, as its scale relevance function does not diminish as one zooms up and down. It reflects a decision by modern physics to give up the assumption of the scale invariance (e.g., different behavior of quantum and macroscopic particles).

Accordingly, the focus became the study of properties of various *interfaces*, which are understood as a *continuity defect* at the boundary between two entities regardless of whether it is physics (body surface, phase interfaces), concepts (classical and quantum physics, classical and nonequilibrium thermodynamics), fields of learning (thoughts, science and humanities) or human behavior (minds, physical and cultural frontiers). In our entrenched and customary visualization we portray interfaces only (as a tie line, shed, curve) often not monitoring the entities which are borderlined. Such a projection is important in conveniently picturing our image of the surroundings (models in physics, architectonical design). Interfaces entirely affect the extent of our awareness, beyond which our confusion or misapprehension often starts. As mentioned above, it brings into play an extra correlation, that is the interface between the traditional language of visible forms of the familiar Euclidean geometry and the new language used to describe complex forms often met in nature and called fractals.

The role of mathematics in this translation is important and it is not clear to what extent mathematical and other scientific concepts are really independent of our human scale location and the scale of locality. *Vopěnka* [19] in 1989 proposed a simplifying program of naturalization of certain parts of mathematics: *"we should not be able to gain any insight about the classical (geometrical) world since it is impossible to see this world at all. We see a world bordered on a horizon, which enables us to gain an insight, and these lead us to obtain nontrivial results. However, we are not seeing the classical, but the natural (geometrical) world differing in the classification of its infinity as the*

4

form of natural infinity" (alternative theory of semi-sets that are countable but infinite beyond the horizon).

One consequence is the way we fragment real-world entities into several categories [17]: *things, events* and *processes.* By things, we typically mean those entities which are separable, with identifiable shapes and size, and which persist in time. Events, on the other hand, have a relatively short duration and are composed of the interactions of several things of various sizes. Processes are, in this last property, similar to events but, like things, have a relatively long duration. However, many other entities may have a transient character such as vortices, flames, clouds, sounds, ceremonies, etc. There is an obvious difference between generic categories and particular entities because a category may be scale-thin in two different ways: generically (atoms, birds, etc.) or individually (geometrical concepts, etc.).

There is an interesting asymmetry with respect to the scale axes [18]; we have a different attitude towards examining events that occur inside things than what we consider exists on their outside. Moreover there are only a few relevant scales for a given object, occasionally separated by gaps. When considering, for example, a steam engine, the most important scale is that of macroscopic machinery while the second relevant scale is set much lower, on a smaller scale, and involves the inspection of molecules whose behavior supports the thermodynamic cycle. Whatever the scale spectrum in the designer's perspective, there is always one and only one relevant 'scale-here' range where the meaning of the object or process is located as meaningful.

In the case of complex objects, there is a close relationship between their distribution over scales and a hierarchy of their structural, functional and describable levels. We tend to assign objects of our concern into structural levels and events as well as processes into functional levels. Obvious differences of individual levels yield different descriptions, different terminology (languages) and eventually different disciplines. Two types of difficulty, however, emerge, one caused by our limited understanding of whether and how distinct levels of a system can directly interact and, the other, related to the communication (language) barriers developed over decades of specialization of scientific disciplines (providing the urgent need for cross-disciplinarity).

One of the first mathematical theories in science that dealt with inter-level interactions was Boltzmann's statistical physics, which is related to thermodynamics and the study of collective phenomena. It succeeded in eliminating the lower (microscopic) level from the macroscopic laws by decomposing the phase space to what is considered macroscopically relevant subsets and by introducing new concepts, such as the mannered entropy principle. It requested to widely adopt the function of logarithm that was already and perpetually accustomed by nature alone (physiology, psychology). In comparison, another scaled sphere of a natural process can be mentioned here where the gradual evolution of living parts has been matured and completed in the *log/log* relations, called the *allometric* dependence.

Another relevant area is the study of order/disorder phenomena, acknowledging that microscopically tiny fluctuations can be somewhat 'immediately' amplified to a macroscopic scale. What seems to be a purely random event on one level can appear to be deterministically lawful behavior on some other level. Quantum mechanics may serve as another example where the question of measurement is actually the eminent question of interpreting macroscopic images of the quantum-scale events. Factually we construct 'things' on the basis of information, which we may call information transducers.

The humanities, particularly economics, are further fascinating spheres for analysis. However, their evaluation can become more complicated as individual scale-levels may mutually and intermediately interact with each other. Namely any forecasting is disconcerted assuming that any weather prediction cannot change the weather itself while economic prediction activity displays the inevitable dependence of what is being evaluated or forecasted.

Yet another sphere of multilevel interactions is the concept of active information – another reference area worthy of mention. Besides the reciprocal interrelation to ´entropical´ disorder we can also mention the growth of a civilization's ability to store and process information, which encompasses at least two different scales. On the one hand, there is the need for a growing ability to deal with entities that become composite and more complicated. On the other hand, there is a necessity to compress information storage into smaller and smaller volumes of space. Human progress in its elaborateness is hinted at by the triangle [14] of his rivalry scales: time, t, information, I, and energy, E.

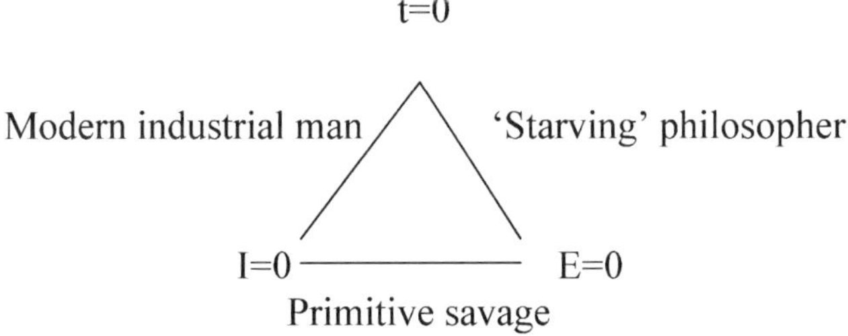

Cyberneticist *Weinberg* is worth noting as he said *"time is likely to become, increasingly, our most important resource. The value of energy and information is, ultimately, that it gives us more freedom to allocate our time"*. If we have lots of time we do not need much information because we can indulge in haphazard, slow trial-and-error search. But if time becomes expensive, then we need to know the fastest way to do things and that requires lots of information and time organization. The above treated spatial 'here' suggests an obvious analogue in the temporal 'now' (temporal field relevance), which is impossible to identify without involving 'change' (past and future).

One of the most widespread concepts in various fields (physics, society and/or mind) is the notion of 'state' [20]. In fact, there is no any exact physical definition of what is the state alone and we can only assume that a system under consideration must possess its own identity connected to a set of its properties, qualities, internal and external interactions, laws, etc. This 'identity' can then be called the *state* and the description of state is then made upon this set of chosen properties, which must be generally interchangeable when another same system is defined. Thermodynamics (cf. Chapter 6), as one of the most frequent users of the notion of state, is presented as a method for the description and study of various systems, which uses somehow a heterogeneous mixture of abstract variables occasionally defined on different scales, because the thermodynamic concept involves an energetic communication between macro- and microlevels. For example, heat is the transfer of energy to the hidden disordered molecular modes, which makes it troublesome to co-define the packaging value of internal energy when including processes at the macroscopic level, such as the mechanical work (of coordinated molecular motion as whole). Associated with this is the understanding of thermodynamic variables as averaged quantities and the assignment of such variables to individual parts that may be composed together to form other parts [9, 20].

Another important standpoint is the distinction between 'phase' as the denomination of a certain intensive state and the different forms of matter (as liquids or solids). In fact, phase keeps traditional meaning as a 'homogeneous part' and already Gibb's writings were marked in this respect with a great conciseness and precision so that he also treated the term phase from a statistical point of view, introducing the words 'extension-in-phase' to represent what is generally referred to today as 'phase space', i.e., all of the possible microstates accessible to the system under the constraints of the problem. The relation between these two ideas is as follows: at equilibrium a thermodynamic system will adopt a particular phase in its macroscopic concept, which is represented by a large number of possible microstates, which may be thought of as an occupying extension-in-phase or regions of phase space. A *phase* thus represents a statistical projection onto fewer dimensions from a region of phase space. This representation is not possible if the word phase is used to merely represent a state of matter

Any thermal process requires another scale-dependent criterion (which is often neglected) that decides whether or not any (thermally) stable state is stable enough and in what scale, viewed dimensionally, still maintains its stability. When the approached stability is of a simple natural scale this problem is more elementary but when equilibrium exists in the face of more complicated couplings between the different competing influences (forces) then the state definition of stability becomes rather more complicated. There can be existent equilibriums, characterized by special solutions of complex mathematical equations, whose stability is not obvious. Although the comprehensively developed field of thermal physics deals with *equilibrium states* it cannot fully provide a general law for all arbitrary "open" systems of stabilized disequilibria but, for example, it can help to unlock an important scientific insight for a better understanding of chaos as a curious but entire source for systems evolution.

1.2. Warmth and our thermal feeling

One of the most decisive processes of man's sensation is to understand warmth – the combined effect of *heat* and *temperature*. A stone under sunshine can be regarded as torrid, sunny, tepid, warm, hot, radiant, caloric, sizzling, fiery, blistering, burning, boiling, glowing, etc., and by merely touching it we can be mistaken by our previous feeling so that we cannot discern what is what without additional phraseology, knowledge and practice. Correspondingly, under a freezing environment we can regard our sensation as wintry, chilly, cold, frosty, freezing, icy, arctic, glacial, etc., again, too many denominations to make an optimal choice.

We, however, would feel a different effect of sensation in our hand if in contact with an iron or a wooden bar that are both of the same temperature. Here we, moreover, are unintentionally making a certain normalization of our tactility by accidentally regarding not only the entire temperature of the bar but also the heat flow between the bar and the hand. Therefore the iron bar would feel to us colder. Curiously, this is somehow similar to the artificial parameter called entropy that explicates different qualities of heat with respect to the actual temperature. Certainly, and more realistically, it should be related to the modern understanding of transient thermal property known as warm-cool feeling of fabrics (particularly applied to textiles) related to thermal absorptivity (characterizing heat flow between human skin and fabrics at given thermal conductivity and thermal capacity). The higher the level of thermal absorptivity, the cooler the feeling it represents (cf. paragraph 5.7).

Early man used to describe various occurrences by vague notions (such as warmer-cooler or better-worse) due to the lack of development of a larger spectrum of appropriate terminology. Only the Pythagorean school (~500 BC) resumed the use of numbers, which was consummated by Boole's (~19th Century) logical mathematics of strictly positive or negative solution. Our advanced life faces, however, various intricacies in making a precise description of complex processes and states by numbers only, thus falling beyond the capacity of our standard mathematical modeling. Increased complexity implies a tendency to return from computing with exact numbers to computing with causal words, i.e., via manipulation of consequently developed measurements back to the somehow original manipulation of perceptions, which is called 'fuzzy logic' [21] and which is proving its worth in modern multifaceted technologies .

The scale-concept of temperature [22,23], due to its practical significance in meteorology, medicine and technologies, is one of the most commonly used physical concepts of all. In a civilized world even small children are well acquainted with various types of thermometers giving the sign for "temperature" of either, sick children, outdoor suitability of environment, working state of a car engine or even microscopic distribution of energies. It should be noted, however, that the medical thermometer can be deemed by children to be rather a healing instrument decisive for an imperative command to stay in bed, while the outdoor thermometer decides how one has to be dressed, the position of a pointer on the dial in the car thermometer has some importance for the well-being of the engine while the absolute zero represents the limiting state of motionless order of molecules.

As a rule, there is no clear enough connection among these different scales of "temperature" given by particular instruments. For teenagers it is quite clear that all the things in the world have to be measured and compared so that it is natural that an instrument called a thermometer was devised for the determination of certain "exact" temperature – a quantity having something to do with our above-mentioned imperfect feeling of hotness and coldness. The invention of temperature was nothing but a further improvement of modern lifestyle in comparison with that of our ancestors. Eventually, all adults believe that they know what temperature is. The only persisting problem is represented by various temperature scales and degrees, i.e. *Fahrenheit*, centigrade or *Kelvin* and/or *Celsius*. The reason for their coexistence remains obscure and common meaning is that some of these degrees are probably more ´accurate´ or simply better – in close analogy with monetary meaning of dollars and euros [22].

Roughly speaking, it is true that modern *thermal physics* started to develop as the consequence of thermometer invention, thus making possible optional studies of quantitative thermal phenomena. It is clear that there were scientific theories dealing with heat effects before this date and that the discovery of the thermometer did not make transparent what temperature really is. It still took a long time before scholars were responsive enough to what they were actually doing to carry out experiments with thermometers. In this light it may be quite surprising that the essential part of ancient natural philosophy consisted just of what we now may call thermal physics. These theories and hypotheses worked out by old philosophers remained still active even after the invention of the thermometer and was it a matter of curiosity that led to the build up of predicative theory of thermal phenomena paying little attention to such an important quantity as temperature? To give an explanation it is important to say a few words about these, for us quite strange, theories, in the following chapters.

The first conscious step towards *thermal analysis* was man's realization that some materials are flammable. Despite a reference to the power of fire, that is evident from early records, man at an early stage learned how to regulate fire to provide the heat required to improve his living conditions by, *inter-alia*, cooking food, firing ceramic ware and extracting useful metals from their ores. It occurred in different regions, at different times and in different cultures, usally passing from one locality to another through migration of peoples or by transmission of articles of trade.

The forms of power (energy, in contemporary terminology) generally known to ancient peoples numbered only two; thermal and mechanical (as the extraordinarily knowledge of electric energy, documented e.g. in the Bible, should be considered exclusive). From the corresponding physical disciplines, however, only mechanics and optics were accessible to early mathematical description. Other properties of the structure of matter that include thermal, meteorological, chemical or physiological phenomena were long treated only by means of verbal arguments and logical constructions, with alchemy having here a very important role.

Purposeful application of heat as a probing agent imposes such modes of thermal measurement (general observation), which follow temperature changes in matter that are induced by absorption, or extraction of heat due to state changes. It is, fundamentally, based on the understanding of intensive (temperature) and extensive (heat, entropy) properties of matter. In an early conception of heat it, however, was widely believed that there was a cold "frigoric" radiation [24] as well as heat radiation as shown above. This gave credence to the fluid theory of *'reversibly flowable'* heat. Elements of this caloric theory can be even traced in the contemporary description of flow mathematics.

Thermal analysis reveals the thermal changes by the operation of thermophysical measurements. It often employs contact thermometers or makes use of indirect sensing of the sample surface temperature by various means (pyrometery). Therefore, the name for outer temperature scanning became *thermography*, which was even an early synonym for thermal analysis. This term is now restricted only for such thermal techniques that visualize temperature by thermal vision, i.e., the thermal screening of the body's surface. Under the caption of thermal analysis, the heating characteristics of the ancient Roman baths were recently described with respect to their architectural and archeological aspects. In detail, the heat loss from the reconstructed bath was calculated and the mass flow rate of the fuel was determined, allowing for the estimation of temperature and thermal conductivity [25]. It shows that the notion of thermal analysis should be understood in broader conjectures, a theme that is upheld as one of the goals of this book.

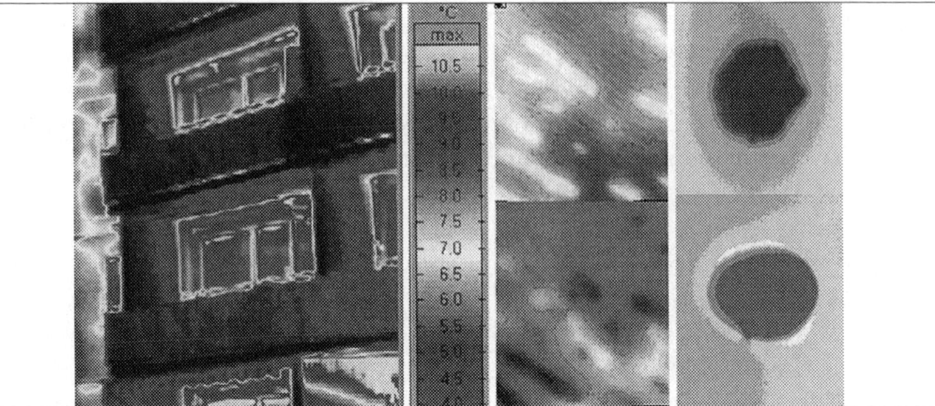

Fig. 3. – Thermography: examples of thermovision of selected objects (left, classical view of heat-loss for buildings with notable windows), which is a useful method of direct screening the temperature (intensity scale shown in the center bar). Middle: the onion-like epidermal cell formation directly scanned on the surface of a metallic sample during its freezing, which is a specially developed type of thermoanalytical technique recently introduced by *Toshimasa Hashimoto* (Kyoto, Japan). Another style of such temperature scanning (right) was applied to visualize the immediate thermal behavior of a water drop (0.5 ml) deposited on (two brand) of textiles at the initial temperatures of 25°C (and humidity of 40 % with the T-scale laying between 15 to 30 °C) newly pioneered by *Zdeněk Kus* (Liberec, Czechia).

1.3. Databases in thermal material sciences

It is clear that the main product of science is information, and this similarly applies for the section of thermally related studies, too. There is seldom anything more respectable than the resulting data bibliographic bases, which store the information gathered by generations of scientists and which put them in order. On the other hand, there are still controversial issues and open problems to be solved in order that this information (and such derived *databases*) will better serve the ultimate endeavor of science – the pursuit of discovery and truth.

Let us make some remarks related to our specific field of interest, i.e., thermal science specified as *thermal analysis* as well as the accompanying *thermal treatment* [3]. Let us mention only the two most specialized journals, *Thermal Analysis and Calorimetry* (JTAC) and *Thermochimica Acta* (TCA), which cover the entire field of thermal analysis and related thermophysical studies, and which naturally belong to a broader domain of journals concerned with material thermal science [26]. These two journals are members of a general family of about 60 000 scientific journals that publish annually about 10^6 papers on 10^7 pages. The questions then arises as to the appropriate role of such specific journals, and their place among so many presently existing scientific periodicals. The answers to these questions may be useful not only for their Editors, but also for prospective authors trying to locate their articles properly, as well as for researchers needing to identify suitable journals when the interaction between thermal specialties or disciplines pushes them beyond the borders of familiar territory.

It is generally recognized that almost three-quarters of all published articles are never cited and that a mere 1% of all published articles receives over half of the citations from the total number. These citations are also unequally distributed over individual journals. Articles written by a Nobel-prize winner (or other high-profile scientist) are cited about 50 times more frequently than an average article of unknown affiliation cited at all in a given year. About 90% of all the actual information ever referred to represents a mere 2000 scientific volumes, each volume containing roughly 25 papers. The average library also removes about 200 outdated volumes each year, because of shortages of space, and replaces them with newer issues.

What is the driving force for the production of scientific papers? Besides the need to share the latest knowledge and common interests, there is the often repeated factor of "publish-or-perish" which is worthy of serious re-thinking, particularly now in the age of resourceful computing. We have the means of safeguarding the originality of melodies, patents and even ideas, by rapid searching through a wide range of databases, but we are not yet able (or willing?) to reduce repetitions, variations and modifications of scientific ideas. Printed reports of scientific work are necessary to assure continued financial support and hence the survival of scientists and, in fact, the routine continuation of science. It would be hypothetically possible to accelerate the production of articles by applying a computer-based "Monte Carlo" method to rearrange various paragraphs of already-existing papers so as to create new papers, fitting them into (and causing no harm in) the category of "never-read" articles. Prevention or restriction of such an undesirable practice is mostly in the hands of scientific referees (of those journals that do review their articles) and their ability to be walking catalogues and databases in their specialization.

The extent of the task facing a thermal analyst is potentially enormous [27-29]. For the 10^7 compounds presently registered, the possibility of 10^{14} binary reactions exists. Because all reactions are associated with thermal changes, the elucidation of a large number of these 10^{14} reactions could become a part of the future business for thermochemistry and, in due course, the subject of possible publications in JTAC, TCA and other journals. The territory of thermal treatment and analysis could thus become the most generally enhanced aspect of reactivity studies – why? The thermal properties of samples are monitored using various instrumental means. Temperature control is one of the basic parameters of all experiments, but there are only a few alternatives for its regulation, i.e., isothermal, constant heating/cooling, oscillating and modulated, or sample determined (during quenching or explosions). Heat exchange is always part of any experiment so reliable temperature measurements and control require improved sophistication. These instruments can be considered as "information transducers", invented and developed through the skill of generations of scientists in both the laboratory and manufacturers' workshops.

The process of development is analogous to the process for obtaining useful work; where one needs to apply, not only energy, but also information, so that the applied energy must either contain information itself, or act on some organized device, such as a thermodynamic engine (understood as an energy transducer). Applied heat may be regarded as a "reagent" [3], which, however, is lacking in information content in comparison with other instrumental reagents richer in information capacity, such as various types of radiation, fields, etc. We, however, cannot change the contributed information content of individually applied reagents and we can only improve the information level of our gradually invented transducers.

This may be related to the built-in information content of each distinct "reagent-reactant", e.g., special X-rays versus universal heat, which is important for the development of the field in question. It certainly does not put a limit on the impact of combined multiple techniques in which the methods of thermal analysis can play either a crucial or a secondary role. Both interacting fields then claim superior competence (e.g., thermodiffractometry). These simultaneous methods can extend from ordinary combinations of, e.g., DSC with XRD or microscopy, up to real-time WAXS-SAXS-DSC, using synchrotron facilities. Novel combinations, such as atomic force microscopy fitted with an ultra-miniature temperature probe, are opening new perspectives for studies on materials [3,9,16], and providing unique information rewards. However, the basic scheme of inquiring process remains resembling [3].

At the end of the 20th Century, *Chemical Abstracts Service* (CAS) registered the 19,000,000-th chemical substance and since 1995, more than a million new substances have been registered annually [29]. The world's largest and most comprehensive index of chemical literature, the CAS Abstracts File, now contains more than 18 million abstracts. About a half of the one million papers, which are published annually in scholarly journals deal with chemistry that is considered as a natural part of material thermal science. The database producer *Derwent* and world's largest patent authority registers some 600,000 patents and patent-equivalents annually; 45% of which concern chemistry. One of the most extensive printed sources of physical properties and related data, *Landolt-Boernstein* Numerical Data and Functional Relationships in Science and Technology, has more than 200 volumes (occupying some 10 meters of shelf space).

In the area of enhanced electronic communications and the global development of information systems, electronic publishing and the Internet certainly offer powerful tools for the dissemination of all types of scientific information. This is now available in electronic form, not only from computerized databanks, but also from primary sources (journals, proceedings, theses and reports), greatly increasing the information flow. One of the greatest benefits of the early US space program was not specimens of Moon rocks, but the rapid advance in large and reliable real-time computer systems, necessary for the lunar-project, which now find application almost everywhere.

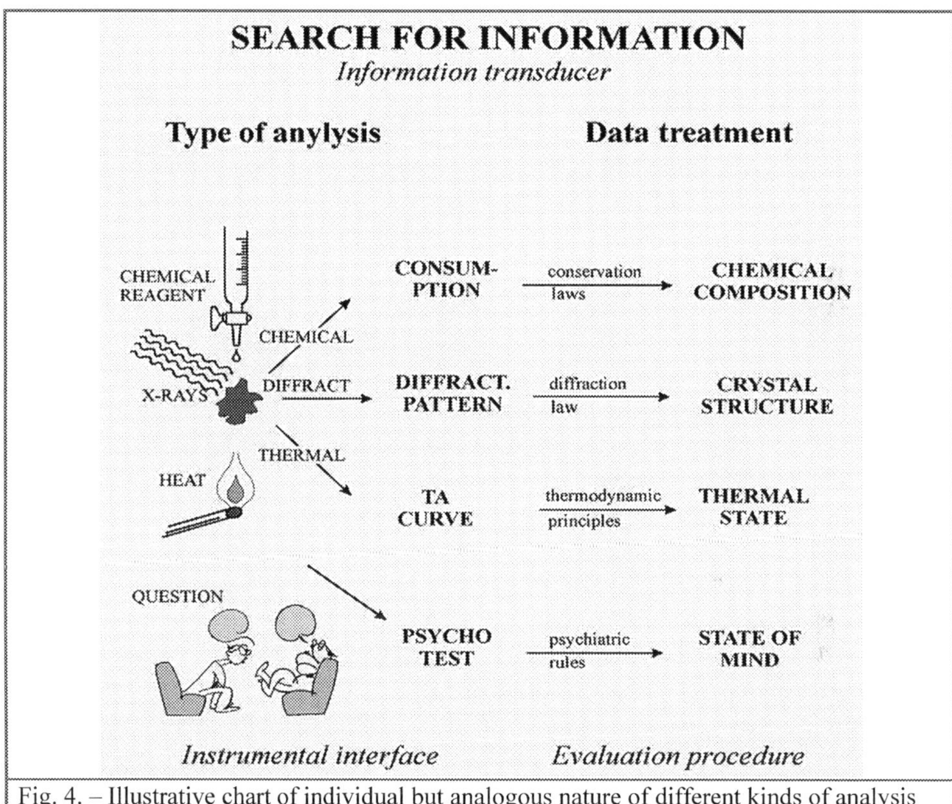

Fig. 4. – Illustrative chart of individual but analogous nature of different kinds of analysis

However, because of the multitude of existing data of interest to material thermal science and technology, and the variety of modes of presentation, computer-assisted extraction of numerical values of structural data, physico-chemical properties and kinetic characteristics from primary sources are almost as difficult as before. As a consequence, the collection of these data, the assessment of their quality in specialized data centers, the publication of handbooks and other printed or electronic secondary sources (compilations of selected data) or tertiary sources (collections of carefully evaluated and recommended data), storage in data banks, and dissemination of these data to end users (educational institutions and basic scientific and applied research centers), still remain tedious and expensive.

The total amount of knowledge, collected in databases of interest for materials science, is impressive. On the other hand, the incompleteness of this collection is also alarming. The 11 million reactions covered by the BCF&R database constitute only a negligible fraction of the total number of 200,000,000,000,000 binary reactions between 19 million already registered compounds, not even considering tertiary reactions, etc. In other words, lots of substances are known, but little is known of how these substances react with each other. We cannot even imagine how to handle such a large database containing information on 10^{14} reactions. The number of substances registered grows by more than a million compounds annually, so the incompleteness of our knowledge of individual compounds increases even more rapidly.

Materials thermal databases expand steadily, becoming more and more difficult to comprehend. Man perceives serially and the speed with which he receives information is small. It is estimated that an average researcher reads 200 papers annually. This is negligible to respect of the one million papers published in the sixty thousand scholarly journals throughout the world though a specialists needs to read a fraction. If a person could read the abstracts (about two minutes each) of almost 10^6 papers on relevant chemistry and physics processed during the last year by Chemical Abstracts Service, it would take him almost 20,000 hours in order to optimize the selection of those 200 papers. And it would take more than two years to complete!

Fortunately, there are other ways of making priority selections. One can trust the search for information to computers, which will quickly locate it by title, keywords, authors or citations, using complicated algorithms. Unfortunately, the possibility of looking for atypical papers, which may bring unusual solutions beyond the frame of the algorithms used, is, however, lost. Such papers may be very important and valuable.

During recent years, most of the great discoveries made in any domain of science impact thermal science in view of thermochemistry and thermal material science. It has developed from the previously uncommon concepts: quasicrystals, low-dimensional systems (quantum semiconductors often based on GaAs structures), optoelectronics, non-crystalline and nano-crystalline material (particularly in the field of metals), synthesis of high-temperature oxide superconductors, manganets and ferrites (exhibiting magnetocaloric effects), fast growing sphere of fullerenes, macro-defect-free cements, biocompatible ceramics and cements, as well as the rapidly moving discipline of associated nanotechnologies. It follows that Nature has provided such unusual and delicate chemical mixtures enabling us to discover its peculiarities and curiousness. There, however, is not any reason to expect these compounds to occur spontaneously in natural environments, like a planetary surface, or evolve from interstellar material.

The intellectual treasure contained in scientific papers is great and any simplification of this body of knowledge by computer searches may lead to irreplaceable losses. People, however, will rediscover, again and again, things that were already described in old and forgotten papers which they were not able to find buried in overwhelming data sources. This rediscovered knowledge will be published in new papers, which, again, will not fully succeed in passing into the hands of those who could make use of them. The unwelcome result is steadily and quickly growing databases, which might hopefully be capable of re-sorting overlapping data. We can even joke that the best way to make some data inaccessible is to file them in a large database. Curiously, large databases may be even seen to act like astronomical black holes in the information domain.

The steadily growing databases may distract a large number of scientists from their active research, but it can also give jobs to new specialists engaged in information and data assessment itself. Scientists may spend more and more time in searching ever more numerous and extensive databases, hopeful of becoming better organized. This allows them to be acquainted with the (sometimes limitless) results of the often-extensive work of other scientists. On the other hand, this consumes their time, which they could otherwise use in their own research work and, are, accordingly, prevented from making use of the results of the work of the other scientists. Gradually the flow of easily available information may impact on even youngsters and students, providing them with an effortless world of irrationality developed through games; perpetual browsing the Internet, trips to virtual reality, etc. However, let us not underestimate its significant educational aspects associated with computers (encyclopedia, languages, etc.) and their capability to revolutionize man's culture. Another, not negligible, aspects is the beauty of traditional book libraries; the bygone treasures of culture, and often a common garniture of living rooms where all books were in sight and a subject for easy access and casual contemplating. Their presence alone is one that fills me with personal contentment.

If the aim of Science is the pursuit of truth, then the computerized pursuit of information may even divert people from Science (and, thus, curiously thus from the truth, too). We may cite *"If knowing the truth makes a man free"* [John 8:32], the search for data may thus enslave him (eternally fastening his eyes to nothing more than the newborn light of never-ending information: a computer display).

What is the way out of this situation? How can we make better use of the knowledge stored in steadily growing databases? An inspirational solution to this problem was foreshadowed already by *Wells* in 1938. He described an ideal organization of scientific knowledge that he called the 'World Brain' [30]. *Wells* appreciated the immense and ever-increasing wealth of knowledge being generated during his time. While he acknowledged the efforts of librarians, bibliographers and other scientists dealing with the categorizing and earmarking of literature, he felt that indexing alone was not sufficient to fully exploit this knowledge base. The alternative he envisioned was a dynamic "clearing-house of the mind", a universal encyclopedia that would not just catalogue, but also correlate, ideas within the scientific literature.

The World Brain concept was applied in 1978 by *Garfield*, a founder of the *Institute for Scientific Information* (ISI), of the ISI citation databases and, in particular, co-citation analysis [31]. The references that researchers cite establish direct links between papers in the mass of scholarly literature. They constitute a complex network of ideas that researchers themselves have connected, associated and organized. In effect, citations symbolize how the "*collective mind*" of Science structures and organizes the literature. Co-citation analysis proved to be a unique method for studying the cognitive structure of Science. Combined with single-link clustering and multidimensional scaling techniques, co-citation analysis has been used by ISI to map the structure of specialized research areas, as well as Science as a whole [32].

Co-citation analysis involves tracking pairs of papers that are cited together in the source article indexed in the ISI's databases. When the same pairs of papers are co-cited with other papers by many authors, clusters of research begin to form. The co-cited or "core" papers in the same clusters tend to share some common theme, theoretical, or methodological, or both. By examining the titles of the citing papers that generate these clusters, we get an approximate idea of their cognitive content. That is, the citing author provides the words and phrases to describe what the current research is about. The latter is an important distinction, depending on the age of the core papers. By applying multidimensional scaling methods, the co-citation links between papers can be graphically or numerically depicted by maps indicating their connectivity, possibly to be done directly through hyperlinks in the near future. By extension, links between clusters can also be identified and mapped. This occurs when authors co-cite papers contained in the different clusters.

Thus, the co-citation structure of research areas can be mapped at successive levels of detail, from particular topics and subspecialties to less-explicit science in general. It seems to be useful to have the numerical databases of materials related to the ISI's bibliographic databases. Each paper bearing the data under consideration cites and is cited by other papers, which determine its coordinates in the (bibliographic) map of (materials) science. In this way, definite data (a definite point in data space) is related to a definite point in bibliographic space (image of these data in bibliographic space). The correlation between data (objects, points in data space) is expressed (capable of locating) as correlations between their images in bibliographic space (which is a well-approved technique developed and routinely performed by ISI).

1.4. Horizons of knowledge

The structure of the process of creative work in natural sciences is akin to that in the arts and humanities as is apparent from the success of computerization, which is itself a product of science [33]. An inspired process requires certain mind harmonization or better endowment of rhythm accordance, which is necessary in any type of communications (language, digits). Besides fractal geometry (natural scene, artistic pictures, graphical chaos, various flows, reaction kinetics), as an alternative to Euclid's strict dimensionality (regular ornaments, standard geometrical structures, models of solid-state reactions), there are no margins that science shares with art in some unique and common way of 'science-to-art'. They both retain their own subjects and methods of investigations. Even an everyday computer-based activity, such as word processing, or even computer-aided painting, has provided nothing more than a more efficient method of writing, manuscript editing, graphics, portrayal or painting (popular 'Photoshop') similarly applied even to music. It has indeed altered the way in which the authors think; instead of having a tendency to say or draw something, now they can write in order to discover if they have something to write (or even to find).

Spontaneously mutual understanding through a concord of rhythms is, as a matter of interest, a characteristic feature of traditionally long-beloved music, well familiar in various cultures [9]. The way that it forms melodies and that it combines sequences of sound brings about the swiftness of an optimum balance of surprise and predictability. Too much surprise provides non-engaging random noise while too much predictability causes our minds to be soon bored. Somewhere in between lays the happy medium, which can intuitively put us on a firmer, rhythmical footing. The spectrum of sequences of sounds is a way of gauging how the sound intensity is distributed over different frequencies. All the musical forms possess a characteristic spectral form, often called '1/f-noise' by engineers, which is the optimal balance between both, the unpredictability, giving, thus, correlations over all time intervals in the sound sequences.

So that when a musical composition is in style, i.e., that is highly constrained by its rules of composition and performance, it does not give listeners too much of new information (adventure). Conversely, if the style is free of constraints the probabilistic pattern of sounds will be hard to make it result in a less attractive style than the optimal 1/f spectral pattern. Distinguishing music from noise, thus, depends then entirely on the context and it is sometimes impossible and even undesirable to discern. It is close to the everyday task of physics, which is the separation of unwanted, but ever-presented noise away from the authentically (repeating) signal, or even ballast figures out of the true mathematical theory, which all is long-lasting challenge now effectively assisted by computers. All other creative activities, like painting, poetry, novel writing or even architecture have displayed similar trends of getting away from constraints. The picture of artistic evolution is one of diminishing returns in the face of successful exploration of each level of constrained creative expression. Diversity has to be fostered and a greater collaboration, easier connection and eavesdropping between minds, between people and organizations should be a measure of progress.

Separation of natural sciences from philosophy, and the development of specialized branches of each scientific field, led to the severance of thinking which is now tending back to re-integration. Its driving force is better mutual understanding, i.e., finding a common language to improve the comprehension of each other and the restoration of a common culture. Thus, the cross-disciplinary education, aiming to bridge natural sciences and humanities, i.e., certain 'rhythmization of collaborating minds', has become a very important duty to be successfully run through the third millennium. It should remove the mutual accusation that the severe philosopher's ideas have initiated wars and bright scientific discoveries have made these wars more disastrous.

All human experience is associated with some form of editing of the full account of reality. Our senses split the amount of facts received; recognizing and mapping the information terrain. Brains must be able to perform these abbreviations together with an analysis of complete information provided by individual senses (such as frequencies of light, sound signals, touch discerning, etc.). This certainly requires an environment that is recognizable, sufficiently simple and capable to display enough order, to make this encapsulation possible over some dimensions of time and space. In addition, our minds do not merely gather information but they edit them and seek particular types of correlation. Scientific performance is but one example of an extraordinary ability to reduce a complex mass of information into a certain pattern.

The inclination for completeness is closely associated with our linking for (and traditional childhood education towards) symmetry. Historically, in an early primitive environment, certain sensitivities enhanced the survival prospects of those that possessed symmetry with respect to those who did not. The lateral (left-right) symmetry could become a very effective discriminator between living and non-living things. The ability to tell what a creature is looking at it clearly provided the means for survival in the recognition of predators, mates and meals? Symmetry of bodily form became a very common initial indicator of human beauty. Remarkably no computer could yet manage to reproduce our various levels of visual sensitivity to patterns and particularly our sense of beauty.

Complex structures, however, seem to display thresholds of complexity, which, when crossed, give a rise to sudden jumps in new complexity. If we consider a group of people: One person can do many things but add another person and a relationship becomes possible. Gradually increasing this scenario sees the number of complex interrelations expand enormously. As well as applying nature, this also applies for the economy, traffic systems or computer networks: all exhibits sudden jumps in their properties as the number of links between their constituent parts overgrows. Cognizance and even consciousness is the most spectacular property to eventually emerge from a very high level of complexity achieved within a connected logical network, like the top neuronal organization of the brain.

Such a complex phenomenon can be explained as the outcome of a huge number of mundane processes organizing themselves, over a long period of time, into a structure which learns in the same way that a neural network does: consciousness would be like a computer 'internet' system that evolves by a microscopic version of natural selection gradually incorporating its novel hardware and software. Supercomputers can always outperform the brain in specific abilities, particularly by applying speed in performing repetitious tasks. But a high price is paid for their lack of adaptability and their inability to learn about themselves or combine with others, in yet unknown ways, enabling collaboration. In the nearest future the personal computers will be able to figure out what the owner is doing and provide him with some kind of useful service without any cognitive load on him (e.g., monitoring health).

It is worth noting that the long-lasting battle of whether the thoughtless computer can ever beat the first-class intuition of a chess world-champion has recently turned in favor of a computer; despite its perfunctory capability to merely check the millions of possibilities, it is also able to find the most appropriate move in a negligible period of time. In some way, the computer expeditiousness may thus become competitive to man's forethought and discretion. Physicists, however, believe in basic mathematical structures behind the laws of nature and, to the astonishment of biologists, they dare to introduce computational things (like quantum gravitation and intrinsic non-computability at the microscopic level) in order to explain macroscopic features of the mind as a complex computational network.

Computing, in its usual sense, is centered on the manipulation of numbers and symbols. Recently, there has arose computing with words, in which the objects of computation are propositions drawn from a natural language, e.g., small, large, light, dark, heavy, close, far, easy, difficult, etc. Such manipulation is inspired by the remarkable human capability to perform a wide variety of physical and mental tasks without any measurements and any calculations, such as parking a car or driving it in heavy traffic, performing diverse skills and sports or understanding speech and body-language. Underlying this remarkable capability is the brain-like capability to manipulate perceptions of distance, speed, color, sound, time, likelihood and other characteristics of physical and mental objects. A basic difference between perceptions and measurements is that, in general, measurements are crisp whereas perceptions are fuzzy. The fundamental aim of science is continuous progression from perceptions to measurements. But alongside the brilliant successes of these steps forward there is the conspicuous underachievement and outright failure to build computerized robots with the agility of animals or humans.

The new computational theory based on perceptions, introduced by *Zadeh* and called *fuzzy logic* [21] is based on the methodology of computing with words, where words play the role of labels of perceptions, expressed as propositions in a natural language. Among the basic types of constrains are possibilistic, veristic, probabilistic or random sets, fuzzy graphs and usualities. There are two major imperatives: (i) if the available information is too imprecise (or exceedingly multifarious) to justify the use of numbers, or (ii) when there is a tolerance for imprecision or indistinctness, which can be exploited to achieve tractability, robustness, prudence, low-solution expenditure and better rapport with reality. Although the fuzzy logic is still in the initial stages of development, it may gradually play an important role in the conception, design and utilization of informatively intelligent systems.

Recently we have also become attentive to the traditional method of learning only from our mistakes, customarily passing on information genetically from generation to generation. We can pass on information by world of mouth, over written message, over airwaves or using modern network means of the Internet. This information can influence any member of the species that hears it. The time that it takes to convey information is now very short and its influence extremely wide. The evolution of human civilization witnesses the constant search for better means of communication [34]. In 1943 the chairman of IBM has said that the world market can absorb about five super computers and just 30 years later there was the opinion that there is no reason for every individual to have a desk-computer in their homes. Earlier everyone expected that computers would just keep getting bigger and more powerful as well as expensive, the reality was opposite: Computers got smaller and cheaper, and more and more people could afford to own them. Their co-acting effectiveness developed most impressively by linking them together into huge world spanning networks. It helped the further development of intellectual capabilities of individual 'brains' not by their further evolution but by the sophisticated level of their computerized collaboration. In any complex system it is not so much the size of the components that are of primary importance but the number of interconnections between them alternating the neuron network. Creating the power of a gigantic interconnected computer, through a web of connections between an over-frowning numbers of small devices, is a pattern that has developed within the human brain, however, the Nature, undoubtedly, got there first.

It is a well-known opinion that collaboration provides exciting opportunities for research and understanding, which individuals working alone are unable to realize. But collaboration presents new issues to be addressed as we design new environments for this cooperation. The key is to identify how truly "great" feats of teamwork materialize. Nowadays, it is clear there are not as many brilliant individuals as there are brilliant collaborations. In considering collaboration, one must deal with issues involved in the inter-relationships among the individuals in a particular group activity. For example, how can the system reduce friction that may develop between two persons or groups? The value added to a project by collaboration is not only found in the replication of information for each participant, but more importantly, the type of people who use the system. It is not the task of mercantile calculus but of non-equilibrium thermodynamics.

The kind of networks we require depends on what type of experts we want to draw together. People are information processors, but if you define them as such, you end up with a warped sense of what people are really all about. As we change the quantity and quality of people with information, we change the quality of their behavior. This seems rational, but if you think about it, if information was really the most valuable and important thing, the people who run our organizations would be the smartest – that clearly is not the case. Some other variable must be making the impact – and that variable is *intelligence* – one of the few elements of modern mankind that cannot be distributed democratically. For example, people smoke, even though they know it is harmful to their health, drink alcohol and drive, even though they know they should not threaten others by their drunken outing. One of the most important design shifts is that we must structure information not for itself, but for its effect on relationships. We are slowly moving from "creative individuals" to "creative relationships" as a new source of information.

The real value of a medium lies less in the information it carries than it does in the communities it creates. The Internet is as much a medium of community as it is a medium of information retrieval. Consider *Gutenberg* in the 15th century. At the time, the new medium was movable type. The Bible was the first book ever published and became the medium of the community. During the Reformation period, alternative understanding and interpretations of the Bible developed simply because of its wider distribution through the society of the time. The results of successful collaborations in the past are many, among others: the flying airplane and the atomic bomb, quantum physics and thermal sciences, the double helix and personal computers. Even the Internet was originally a tool to help physicists to collaborate. Such cooperation consists of several aspects: *Communication* = an essential ingredient but not a synonym; bandwidth does not control success; *Participation* = a means to the end; a required attribute, but again, not a synonym and *Process* = a shared creation/discovery that the individuals could not have done it alone.

Famous biologists, the Nobel prize winners *Watson* and *Crick,* both said they could not have come up with the double helix secret of life working alone (e.g. without knowing the x-ray image of DNA observed by *Franklin*), though their instinct helped them to realize that the base pairs do not match like with like but complementary (as A-T and G-C). It means that there are individual geniuses, but the problems that they face are often bigger than they can solve if working in isolation. The value came from the interactions and the spread of technology now allows more people with more information to interact. A key element of all success is 'shared space' – it is needed to create 'shared understanding'. The properties of the environment shape the quality of the collaboration. Chalkboards have limits for example, but computers present many more possibilities. We can scale shared space by making it intelligent – perhaps we need to think of the computer as a facilitator. Intelligence is derived and applied in a myriad of ways. Sometimes we must recognize that others have more intelligence than ourselves – sometimes we must realize that we hold the key to success if we could just convince others that this is the case. Communication is an act of intelligence, but often the final and/or initial stages of the communication link become "fuzzy" and the message does not get sent or received as intended. The "quality" of the intelligence becomes muddied.

In this context, the disposition of a man-made intelligence will soon become of important attentiveness. Biological life proceeded from very complex interactions of originally simple inorganic units through the continuous process of self-organization. Its imitation complement, called "artificial life", is believed to arise cognitively from complex (logic) interactions to take part within computer (neuron) software. Both such variants, come what may, follow the original vision of *Neumann's* and *Turing's* idea of a certain structure (or organism) simulated by a cellular (digital) automaton, today electronic computer.

Many common health troubles are due to viruses, a short length of DNA/RNA wrapped in a protein coating that fits cell receptors and replicates itself using the cell's machinery. It can be anticipated as an infectious structure (organism) where its appropriately managed mission (having implementation as its driving force) is rewarded by benefiting more space to live. If implanted they often effect possible mutation, ensuing undesirable syndrome or, under the interplay character of positive circumstances may eventually help to form more stable complexity leading thus to evolution.

In a similar fashion, computer viruses can be seen as embryos during the operation (soon also creative) process, in which the virus competes for more working time (or memory space) in the computer processor, closely similar to any animals' fight for food. Viruses can multiply when finding breeding-ground, developing new strategies on how to survive or even the capabability to messenger themselves to other environments. It was already shown that an artificial creature can be formed by a certain bunch (collection) of instructions (coding) alone. They however still lack the feature of real life – mutative self-replication in order to trace evaluation adaptation to their surroundings and cohabitants.

We must regroup in our own mind and attempt to examine the issues in a fresh manner in order to see the other person's viewpoint. Collaboration is much more than sharing workspace and experiences. Collaboration, by its very nature, is a form of intelligence on its own – the process of collaboration is as important as the starting point and the end. If we can keep this in mind, then we may be able to achieve the "greatness" that comes from great collaboration. Have you ever met or worked with someone with whom you seemed able to communicate with by telepathy – someone who was able to anticipate your ideas and needs and vice-versa. It is a wonderful thing to find someone with whom you are "simpatico". It might be your spouse or a close colleague or a son, daughter or other close relative. It may be a friend who shares common values, interests, skills, etc. The basis for the closeness can be many different things. The question is [34]: *"Can we create or even, hope to create, such closeness over the Internet?"*

But if we want to predict the future and understand what is likely to be upon us, it is necessary to step back and examine the most important revolutionary technology to ever appear on Earth. What is the Internet and from where did it come? Some will talk about AOL or Microsoft as if they are the same thing as the Internet. Others will refer to pornography and the dangers inherent on the "Net" from this dark side of society. Throughout the history of humanity, there have been many significant revolutions, such as the Renaissance and the Industrial Revolution that permanently changed how people lived their lives. But none of these changes has occurred as quickly, as universally and as unceremoniously as the Internet revolution. The Internet affects every corner of our world in profound ways – at home, at school, and at work – our lives are different, not necessarily better, as we move into the Information Age. The motto of the Information Age is *"Information is Power"* and if this is true, we are the most powerful generation that has ever populated the face of the Earth.

From simple words to the latest hit-song to the ingredients required in a favorite recipe on how to make a pipe bomb, there is almost nothing you cannot find out about with a little effort and access to the World Wide Web. In the early days of the 1990s, many businesses were forced to adopt Internet-based technology. But today, companies involved in mining, manufacturing, transportation, communications, etc. are now using the Internet routinely since the benefits are so self-evident. Businesses, such as bookstores, are not necessarily located down the road anymore, they can now be found online. It is worth recognizing that an Internet-based business can market their products globally without needing an actual store – they only require good distribution channels. The Internet automates many business processes and transactions, reduces costs, opens up new markets, and empowers customers and potential customers in ways that couldn't even be imagined 10 years ago. The Internet allows small businesses to compete with large corporations, provided that a professional online presence is developed and maintained. We can communicate almost instantly with anyone in the world, whether it is with family or friends abroad or business associates. Letters and Faxes of yesterday are now the E-mails of today and who knows what may turn up in the near or distant future.

In the world of ever-increasing figures the *information theory* has become an important tool, which helps to elect a best element in a large set by employing some relevance criteria, which is particularly of growing importance in the current information-based economy. Contrary to information thirst in technology economy gathers large amounts of data as a widely affordable task, which requires, however, selection, examination, approval and judgment of such a huge information input often too plentiful to a consistent (and finite) processing capacity. The filtered result must be presented in a user-friendly manner, e.g., in an optimum ranking order when the quality of each selected item cannot be measured directly but only in an analogous element pairwise comparison. Each element is evenly put on a site of a liner chain with periodic boundary conditions and can spread over neighboring sites, thus forming a domain. The associated algorithm (searching engine) is programmed to stop when a domain occupies the whole lattice and the value attached to the domain is then favored as the best one [37].

There, however, is always a risk of over-abundant cumulating of ballast information through, e.g., unwanted distribution of unasked data, advertising and promoting rubbish or other yet unknown inflows. This danger has become particularly painful in the recent Internet, which assimilates and helps intermediate large amounts of unsolicited Email (Spam). The consequent need of unpolluted figures is thus acquired through a sort of data cleaning (data filters) in order to confine the storage load. Here we can behold certain analogy with a biological computer (our long-perfected brain), which terribly needs a sensible management to avoid brain overflow (or even its perpetual outgrowth), intricate by multiple levels of input data (rational, emotional, etc.). It is worth mentioning that recent hypothesis correlates the sleeping phase called REM (rapid eye movement) with an assortment and filtrating processes, which factually undertake survival role of the disposal of a day-package of input information to those that are useful to store or useless to disremember (trash). This allocation process depends on the antecedent state of comprehension so that the REM stage is found longer for newborn and shorter for older individuals apparently depending to the extent of experience how to analyze, sort and collect input data. Of course, this kind of self-improving process is yet outside the capability of our artificial automatons-computers.

We can imagine that the Internet makes research blind to proximity and scale. It, however, may bring some disadvantages rarely discussed. While the whole human quest for further knowledge may not crash to a stop as a result of some uncontrollable computer virus destroying everything (all important steps in the progress of civilization paid a high price such as yellow fever victims during the construction of the Panama canal). These adverse factors might well move human progress along certain unwitting tracks or, at least, frustrate progress by a reduction in the diversity of views. Unnoticed it might even forbid the types of questions that are undesired to ever appear.

Forbidden knowledge is also a controversial subject. All modern states have secrets that they hope to keep concealed from certain people for various reasons. Recently this issue has been running into controversy with the imposition of restrictions on the Internet lying beyond the means of a government's computer system to break. Like any proprietary of dangerous possessions (guns or even cars) it may be a subject to some imposed restrictions for the common goods, just as in the same way that like the PIN numbers of credit cards are secured. Powerfully widespread communication networks may become exceptionally vulnerable to the novel forms of attacks – conventional assault of society order (robbery, shooting wars) being replaced by computer and mobile-phone fraud, cyber-terrorism (hackers' attacks on the computer control systems to trigger a disaster), cyber-espionage, etc. It is not far from early installed religious taboos usually framed in order to maintain the exclusivity of certain gods. Moreover the insightful network of satellites, massive use of cellular phones and coupling all home facilities, schools and further organization with Internet, can help establishing a kind of police state of everyone being under continuous but unnoticed control (communication, opinions, and location). Such an inquisitiveness state can even match the *Orwell's* famous sci-fi of ever-watching "Big Brother" situated in the early year 1984 as an ironic picture of the past Soviet Union dictatorship.

Strict domination over communications seems to be an important, if not crucial, tactic in the recommencement of modern wars, where the instantaneous knowledge of the enemy's location, and the decoding its messages, facilitates the precise guiding controlled missilery. The blocking or jamming of your enemy's communication resources along with its broadcasting and, on the contrary, dissemination of false instructions and spreading propaganda or fear is an equally powerful weapon as the use of conventional explosives. Widely accessible information can help to dismantle autocratic regimes without uprising; just by free available news. Alternatively, too much free information can narrow the creative approach of people or better students if readily available for transcription without the individual's creative impact.The harder the competition, the greater is the pressure to gain a marginal advantage by the adoption of innovation. Progressives will have been better adapted to survive in changing environments than conservatives. Sometimes science advances by showing that existing ideas are wrong, that the past measurements were biased in some way or old hypothesis were misleading.

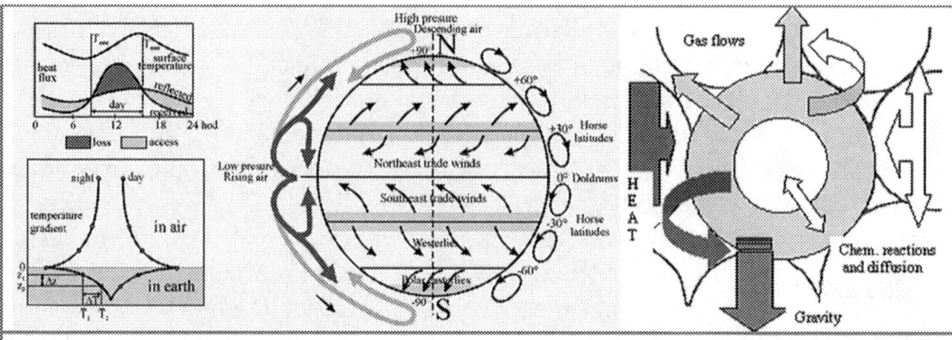

Fig. 5. - The atmospheric convection on the rotating Earth shows vast rivers of air that cover the globe from the equator to the poles. In the equator the sun's rays have the larges angle of incidence so that the surface and the air is warmed most extensively therefore the air starts to rise. On contrary, the area of poles are least warmed so that the hot air masses have tendency to move from the equator towards the poles at higher altitudes of atmosphere while the cold air from the poles moves opposite. The rotation effect, however, breaks both the Northern and the Southern Hemisphere's convection cell into three-to-three parts causing in addition circulation of air so that winds in each cell get a skewed. The winds between the equator and the thirty-degree parallel are called 'trade winds' while the winds between the thirty and the sixty-degree parallels are called 'westerlies'. A more detailed view on inherent fluxes and temperature gradients is shown left because of the repetitious changes in the heat delivery conditions (day and night), which becomes the base of changes of the Earth's weather. In fact, it is a macroscopic outlook to the generalized prospect of thermal analysis. Zooming down to a microscopic scale the globe can serve as the model of a tiny particle (right) in a reaction conglomerate where the heat is delivered by furnace and atmosphere by decomposing gas but where gravity force has unidirectional downward action. Such a portrait is apparently more complicated by accounting on the mutual interactions, various flows and diffusion paths as discussed further on in the Chapter 10.

Much of the everyday business of science involves the gradual expansion of little islands of knowledge, deepening the interconnections between the ideas and facts within their environments. Progress is made not by new discoveries but by finding new ways in which to derive known things, making them more simple or efficient.

Comfort, however, may become disincentive to further modernization. Most inhabitants of the Western democracies live luxuriously when compared with the lot of their distant ancestors. Looking forward we might wonder whether the direction in which advanced technological societies are moving will create less work, longer lives and greater leisure, and whether this might eventually remove the incentive to innovate in science and technology. On the other hand, the technological process itself can reveal a serious downside. It often creates environmental problems that outweigh the benefits that the technology was designed to alleviate. Also our own situation in the Universe and our technical capabilities have not been intended in a design view to the completion of our knowledge of the Universe, which does not exists for our convenience only.

There are limits with what we can do and know – the skill and knowledge is cumulative and we can only but build on what we already know. Existence is precarious – as the world becomes an increasingly sophisticated technological system, it is intensifying the risk from the consequences of its own headlong rush for development. It is difficult to get politicians and democracies to plan for the far future as there are enough problems for today and tomorrow. Even if a civilization would grow befittingly and would not suffer self-destruction, it will ultimately face environmental crises of cosmic proportions as the earth lacks resources, stars run out of nuclear energy and galaxies disintegrate – nothing to worry about yet in our somehow what narrow-looking and often self-foolish societies.

1.5 Measurability and knowability

It is understandable that early science possessed neither exact information nor stored or otherwise catalogued data so that it dealt predominantly with easily describable phenomena, such as motion. It was taken as a fact that there existed force acting on a thrown stone to stop it motionless on ground but it was not clear enough what forced the heated air to flee above fire. It, however, may be found quite surprising that the requisite part of ancient natural philosophy consisted just of what we now may call thermal physics and that the theories and hypotheses suggested by old philosophers were, more than one and a half centuries after the invention of thermometer, still active. How was it possible to build up the predicative theory of thermal phenomena ignoring such a field-force quantity as temperature? To give an answer to this question it is worth to say a few words about these, for us as modern scientists, quite strange theories.

Let us first make clear that such basic but opposite forces, like gravity and buoyancy, have a different effect in different arrangements. Let as take the Earth as a representation of a huge particle with its own enveloping atmosphere which layer is imprisoned by concentric effect of gravity. The Earth is unequally heated by Sun forcing macroscopic areas of air to move due to temperature caused differences in its density which, we humans, witness as weather, cf. Fig. 5. On much large scale we cross the threshold of cosmos and perceive its new peculiarities laying, however, beyond or text. On contrary, for a microscopic system of miniature grains (often found agglomerated) the gravity force is directionally straightforward and the surrounding atmosphere is usually created by chemically induced decomposition of grains. The resulting micro-flows respect again the compacting force of gravity and escaping strength of gases under the temperature gradients imposed by heated furnace. In yet smaller level we enter the special world of quantum mechanics, worth of a special noting, again.

In an often desired 'cutoff point' we can even get rid of either of these ever-present forces. For example, we can carry out our thermohysical study in a spaceship laboratory under conditions of negligible microgravity. It was shown [6, 35] that in such a case the force of gravity (keeping the melt in a terrestrial laboratory inside a crucible holder and shaping almost flat outer surface of the melt) is replaced by the predominant force of surface energy, which compresses the melt to form a ball as to achieve minimum of its outer surface (often pushing the melt out of the crucible). Another technique makes possible to diminish the effect of thermal energy (noise) when adjusting the laboratory experiment to comply with very low temperatures (usually below 4K) so that the other consequences can become dominant (quantum mechanic effects).

An important circumstance is provided by the energetically balanced state, often called *equilibrium*, the final state of time evolution at which all capacity for a system's change is spent. A basic contribution to its understanding was by early mechanics, the traditional branch of science dealing with plain motion of objects in material systems and the forces that act on them. Let us recall that the *momentum, p* [N.s = m.kg/s] of an object of *mass, m*, and *velocity, v*, is defined as the product (*mxv*). The *impulse* of a force on an object over a time interval equals to the object's change of momentum and can bee seen as a sudden driving forward, push or impetus or the effect caused by impelling force. The *force, F* in Newtons [N = m kg/s^2] is the influence on a body, which causes it to accelerate (a vector equal to the body's time rate of change of momentum) and *work, W* [in Joules = kg m^2/s^2] is the force that is exerted over times the distance over which it is exerted. The *power* [in Watts = kg m^2/s^3] is the measure of the amount of work done [J/s] or energy expended. Secularly, it has been found infeasible to give a generally strict definition to *energy, E* (as we can witness from *Poincare* to *Feynman*). We do know there is the principle of the *conservation of energy*, which simply signifies that there is something that remains unvarying, i.e., preserved absolutely constant. Nonetheless, energy is comprehended as the capacity for doing the work and work is thus the transference of energy [J/s] that occurs when a force is applied to a body that is doing some action, such as moving, changing its state (of temperature, stress, etc.).

Energy has to be completed by a degree of its accessibility, which requires even a more complicated connotation of a new term called *entropy, S* [J/K]. Practically, this quantity determines a systems capacity to evolve irreversibly in time. Its ascendant meaning follows from thermodynamics where it is understood as the function of state whose change in differential irreversible processes is equal to the heat absorbed by the system from its surroundings (*Q* in [J]) divided by the absolute temperature (*T* in [K]) of the system. Further on it has a definite meaning in statistical mechanics as the measure of complexion determined by logarithmic law, S = k log W, where W of the number of possible arrangements of microscopic states involved in the system set-up and k is the *Boltzmann* constant. In mathematical context it expresses the amounts of disorder inherent or produced and in communication it is rented as a measure of the absence of information about a situation, or videlicet, the uncertainty associated with the nature of situation.

We can find an apparent association between the *Brillouin's* relation between entropy and information and the Einstein's relation between mass and energy, which can be exposed through their transferring factors, which is, for the first case, equal to one bit, i.e., $k\,ln2 \cong 10^{-23}$ [J/K] while for the latter case it is related to the reciprocal square of the speed of light, $1/c^2 \cong 10^{-21}$ [s^2/cm^2]. They are of the similar gneseologic principle pairing the system and the observer.

By means of the connotations of contemporary science, let us define the two basic but opposing forces more exactly, previously mentioned to have the status of an unintentionally experienced omnipresent strength. The force *aiming down* to the rest is now associated with gravity, recently characterized by the universal gravitational constant, g ($= 6.672059\ 10^{-11}$ [N m^2 kg^{-2} or m^3 kg^{-1} s^{-2}]), as a natural part of the basic unit called the *Planck (quantum) length* (of the magnitude of $10^{-34} = \sqrt{(gh/c^3)}$, where h { $= 6.6260755\ 10^{-34}$ [J Hz^{-1}]}, and c { $= 299\ 792\ 458$ [m s^{-1}]} are respectively the *Planck* constant and the speed of light in vacuum}. The other force *tending upwards* to swell (symbolized by fire) is currently represented by the thermal kinetic energy of particles, $mv^2/2 = kT$, where k (= $1.380\ 658\ 10^{-23}$ [J K^{-1}]) is the *Boltzmann* constant and T [K] is temperature (as a natural part of the basic unit called the *thermal (quantum) length, $\hbar/\sqrt{(m\ k\ T)}$*). Here we should mention that there exists two connotation of Planck constant either in [J s] as $\hbar$ or in [J Hz^{-1}] as h (= $2\pi\ \hbar$).

The inherent regularity is the consequence of fact that the space and time cannot be boundlessly divided below certain limits because of their interfering fluctuations. Therefore we can distinguish elementary basic length (10^{-34} m) and time (10^{-43} s) as based on the fundamental Planck and gravitational constants together with the speed of light which signifies that the distance of the Planck length passes the light at the Planck time. When completing with the charge of an electron, e, the use of these four fundamental quantum constants broke also through the new definition of SI units in metrology where the electric voltage can be defined on basis of the *Josephson* effect observed for superconductors (i.e., Volt in terms of 2e/h) as well as the electric current through the *Quantum Hall Effect* measured for semiconductors (i.e., Ohm in terms of h/e^2).

The strength of electromagnetic force of nature and with it the whole of atomic and molecular structure, chemistry and material science, is determined by a pure number, the *final structure constant* (equal to $\mu_0\ c\ c^2/2\ h$), which numerically matches the value of about 1/137. This is one of the famous unexplained numbers that characterize the universe. Present theories of physics lead us to believe that there are surprisingly few fundamental laws of nature although there are almost endless arrays of different states and structures that those laws permit to exist.

Besides man's oversights, a particular uncertainty of different levels and weights is always inherent in any of the assemblages arising from its set-up and inherent physical properties. Especially *quantum mechanics* states the principle of microscopic uncertainty as the simultaneous determination of position (x) and momentum (p), $\Delta x \Delta p = h$ and/or time (t) and energy (E), which is written as $\Delta t \Delta E = h$. It puts on view a *specific limit* (identified, e.g., by the quantum volume of a particle/fermions, $\Delta x^3 = N (h/2\pi)^3/\Delta p^3$) abridge as a specific span, which cannot be overtaken by any finite process leading either to equilibrium ($\Delta t \to 0$, attaining the static state of classical crystal structure, $E = $ constant) or to disequilibrium (providing the dynamically static state of 'lively equilibrium' under continuing t and supplying E) often exemplified by self-organized processes. Typical are the oscillatory Zhabotinsky – Belousov reactions carried out exclusively under a continuous input and self-catalyzing feedback curiously constrained by the Planck constant [36] through both the principle of least action and the effect of thermal length, $\hbar/\sqrt{(m\ k\ T)}$. Factually, most of the macroscopically accounted diffusion-scaled processes are drawn near the dimensions appropriate to the micro-world following the Schrödinger wave equation when many such processes possess the matching mathematical formulae. It appears that the quantum world in solids is in the hands of electrons while protons realize that for aqueous environments. In addition, it also shows that the micro-space possess a fractal structure (*Nottole*), i.e., $v \sim (v_1+v_2)/(1+ v_1 v_2/c^2) \sim ln(\Delta x_1/\Delta x_2)$. It is not unreliable that even the synchronized staying power of heat (Q) and temperature (T) may have certain, similar limits, such as the product of $\Delta T \Delta Q$ obeying a certain limiting constant. It can encompass such a meaning that we cannot determine the temperature precisely enough for large heat flows and vice versa.

Through the process of better understanding of the notions 'light-fire-heat-energy-entropy' it was recognized that although energy is conserved in physical processes, it is degraded into less ordered and less useful forms as is habitually observed in a closed system that tends to became more and more disordered. Whenever entropy, literally, prevails over energy, the resulting structure is dominated by randomness (and fractal and chaos) rather than by our familiarity of a traditional 'Euclidean orderliness'. This is not a law like gravity but, instead, a principle having a statistical nature and thus turns out to be very important for consideration of what is technologically possible and practically attainable. Yet later succession was found in the interconnection between the notions of entropy and information and their gain or loss. It follows that information can be seen as a commodity and it takes effort to acquire it. So that it is possible to classify all technological enterprises in terms of the amount of information that is needed to specify the structure completely and the rate at which the information needs to be changed in order for the system to improve or arrange or organize. It needs lot of efforts, understanding and willingness to get across the boundary of narrow minds and special considerations. There is no reason to think that the Universe was constructed for our convenience.

Deeper thought about nature and its knowability was already particularized by *Majmonides* at the turn of 13th Century in his book "Handbook for scatterbrained afloaters" where he attempted to categorize knowledge to that which *is known, knowable but yet unknown* and *unknowable*. In his view God is unknowable and whether one can have faith in God or not, both are unverifiable and its testimony is up to one's conviction and it can be relocated on many other subjects of thinking. His famous example was a person who tries to move an ocean by poring it out by a pail, one by one. Clearly it is not feasible but it is also not (theoretically) unachievable! He said that it is always more difficult to answer the question *'why'* (philosophy, e.g., why a stone is moving, life begansurfacing) than to reply to the question of *'how'* (sciences, e.g., how is the stone moving in its current trajectory? How did life begin?). This became reflected in the modern way of the presentation of basic questions about nature as those, which are *answerable, yet unanswerable, ultimately unanswerable* and *senseless to ever ask*. Here we should remember *Cantor* who said that *'the art of asking the right question in mathematics is more important that the art of solving them'* as well as *Einstein's* phrase *'any physical theory cannot be fully testified but just disproved – we are often blind trying to understand our nature'*.

In a spiritual quest we can say that humankind was entitled to comprehend only what God deigned to reveal. Adam and Eve were banished from Eden for eating from the tree of knowledge – God's knowledge. *Zeus* chained Prometheus to a rock for giving fire, the secret of the gods, to a mortal man. When *Adam*, In Milton's "Paradise Lost", questioned the angel *Raphael* about celestial mechanics, Raphael offered some vague hints and then said that the rest from Man or Angel the Great Architect of the Universe did wisely to conceal. All the limitations and forbidden regions were swept aside with Newton's monumental work "The Principia" (1687). There, in precise mathematical terms, *Newton* surveyed all phenomena of the known physical world, from pendulum to springs, to comets and to grant trajectories of planets likely benefiting from the early work of the Greek philosopher *Apollonius* from *Porgy* (about the 200's BC), who calculated all possible geometrical and algebraic properties of ellipses, parabolas and/or hyperbolas. After Newton the division between the spiritual and physical was more clear and physical world became knowable by human beings. *Einstein* rejected Newton's inertial structure, going one step further to give space a relativistic frame, which obliged him once to ask whether God had a choice in how the Universe was made or whether its laws are completely fixed by some fundamental principles, citing his famous "*I shall never believe that God plays dice with the world*".

Fig. 6. - *Newton* (Sir) *Issac* (1642 – 1726) British mathematician (also physicists and yet unfamiliar alchemists), most famous and respectable scientist, founder of mechanics.
Einstein Albert (1879 – 1955) German physicist, originator of theories of relativity, quantum theory of photoelectric effect, theory of specific heats or Brownian motion.

Similarly to Newton also Einstein has his way paved by the work of *Poincaré*, who was Einstein's challenger prior to his death in 1912, which, unfortunately, happened sooner than the relativistic theory got underway. Nonetheless, Newton and Einstein were unique artists who devoted themselves to simplicity, elegance and mathematical beauty to be admired by many generations. The assessment of the basic constant of Nature gives us an exceptional opportunity to have such geniuses, the observers, who can distinguish and comprehend the laws and the inherent constants in charge. We should be responsive enough to see that if the universal constants would be slightly different there probably would be no such observers who would be able to appreciate the constants and even to have them ever recognized.

1.6. Commencement of thermal studies

Let us repeat that veritable views claim that the contemporary science of thermal physics and the interrelated field of thermal studies started to develop in the 17th Century following the invention of an instrument that enabled quantitative studies of thermal phenomena. However, this statement should not be interpreted quite speciously, or in such a way that there was no scientific theory dealing with heat effects before this date. Equally wrong would also be to believe a more widely shared understanding that after the thermometer became a recognized instrument then scholars had a clear idea of what temperature (and heat) was, and that by making experiments with thermometers they were aware of what they were actually doing.

Initially, *Rumford's* conception of heat included the belief that there is a cold *"frigoric"* radiation (consisting of cold particles called *'frigoids')* existing side by side with the heat radiation This idea was time-honored by *Poincare* who even made an authentic experiment trying to distinguish between such heat carrier particles [9,24]. The instrument consisted of two opposite parabolic mirrors aligned in a precisely collateral position, common focal point equipped by a socket to keep a thermometer for measuring temperature changes while the other focal seated replaceable holders to operate various substances. In the case of a candle insertion, the increasing temperature in the opposite focus documented that there was a heat flow between the mirrors in the direction from candle towards the thermometer. On contrary, when the candle was replaced by a piece of ice the thermometer displayed a definite temperature decrease which was explained by the opposite flow of particles of cold from the ice towards the thermometer. Our challenge is to leave the solution to readers but everybody would note that the flow is always from hotter to cooler spot even from thermometer to ice.

The forms of *potency* (*energy* in our contemporary terminology derived from Greek *'ergon'* – action and *'energia'* – activity), generally known to ancient people, numbering only two again, were explicitly mechanical and thermal (the knowledge of electric energy documented, e.g., in the Bible should be considered as extraordinarily exclusive). From the corresponding physical disciplines, however, only mechanics and optics were available for clear mathematical description. The rest dealing with the structure of matter and including thermal, meteorological, chemical or physiological phenomena were treated only by means of verbal arguments and logical constructions. The most representative theory of this type had been formulated by *Aristotle* already in the fourth century BC and survived from ancient to middle age. It was based on the famous doctrine of the four Rudiments (or conventionally Elements) and its everlasting inference even into modern views of physics acts as a joining feature in the forthcoming text.

By measuring particular temperature it becomes clear that we do not live in a thermostatic equilibrium. Instead we reside in an obvious disequilibrium. This can be also witnessed in a series of a self-organized patterns and processes lying evidently outside the traditional concept of equilibrium. It links with a delicate interplay between *chance* and *necessity, order* and *chaos* as well as amid *fluctuation* and *deterministic laws* and it is always under subtle interactions between *mass flows* and *heat flows*. The idea of equilibrium is also reflected in our gradual development of the understanding of our Universe, which has progressed from a static view up to the present complex world of novelty and diversity, the description of which is also a modest aim of this book.

Mankind would like to live in thermo-stable environments, but the Earth's ecosystem, sometimes simplified as a sort of a *Carnot* super-engine (or better myriads of tiny motors working as a team), is not stable and pleasant everywhere. The systems used by mankind to obtain what it wants is usually done so in a dissipative way by heating and cooling, and with machines when manipulating these disequilibria in micro- (reactions), ordinary- (machinery, experiments) and macro- (ecosystem) levels. Disequilibria can thus be a source of desired work but also grounds for thermo-stabilizing as well as destabilizing power. Disequilibria are also a global datum of Nature as it derives from the cosmological history of the present Universe (cosmic microwave background). On a smaller scale, the Earth (atmosphere, biosphere, etc.) experiences a non-equilibrium due to the constant influx of heat delivered from the Sun. On an even smaller scale, we have everyday experience of the variety of non-equilibrium processes visualized in the formation of, e.g., snowflakes or living cells and we can also encounter it during laboratory experiments. However, all dissipative processes take place with a time arrow. They belong to a world in which exist the past and the future, memories, culture and discoveries, including the recent invention of the laws of thermodynamics and irreversibility themselves.

History of thermal physics illustrates the everlasting dilemma of how to distinguish between the overlapping notions of heat and temperature. Though this puzzle has brought up devious but fruitful ideas about immaterial character of *'phlogiston'* and *'caloricum'* as fluids it enabled a better understanding of early views to different transformations. Scientists often challenge the dilemma of which understanding is more efficient. In this example, *Black* was already aware of the incorrectness of the material theory of caloric but, at the same time, he also realized that such a fluid representation is more convenient for the phenomenal cognoscibleness and associated mathematical description. We may even face the similar situation today when dealing with new but yet unobservable notions of dark matter and dark energy, which is helping us to solve the analogous mystery of gravity and anti-gravity forces in the Universe. *Black's* great intuitive invention was the introduction of measurable quantities such as heat capacity and latent heat. There were long-lasting struggles between famous scientists of different scientific school backing and opposing caloric, like *Laplas, Fourier* against *Bernouli*, etc.

Newton was one of the greatest scientists. He gave to us, amongst other things, the deterministic description of our physical world whilst always aware that it could be a part of a superior universe (extending from the very small to the extremely large). He intuitively associated heat conduction with temperature gradients called *'gradus caloricum'* (whereupon gradient is derived from Greek *'gradus'* which means felicitous, congenial). *Newton* even tried to make some quantitative observations by heating one end of a metallic bar and observing heat propagation by detecting the progress of melting of various substances (wax) at different distances. It helped him to formulate the law of cooling without knowing what heat actually was. At the turn of eighteenth century, *Buffon* applied Newton's cooling law in an effort to estimate the age of the Earth based upon an approximation of cooling rates of various substances. He arrived to the value of 75 000 years. *Leibnitz* in his '*Protogaea*' assumed that the outer surface of the molten proto-earth would solidify irregularly leaving the inner part still liquid. On the basis of *Fourier's* theory, *Thompson* tried to approximate how much energy could be stored in the Earth that was formed from a contracting cloud of interstellar dust that has since cooled and condensed, and he arrived to the age of about hundred million years. Impact of the additional energy received from radioactivity decay increased it by four times until radioactive chronology pushed the value up to the present estimate of $4.6\ 10^9$ years – quite a difference from the original statement of the Irish bishop *Ussher* in the year 1654 who said that the biblical "let there be light" happened just on the 23rd March of the year 4004 BC.

The scientist *Kelvin* also tried to make a more accurate attempt at estimating the age of the Earth, arriving at the value of 100 000 000 years, which was based on the amount and rate of energy lost by radiation to the Earth's surrounds from the time when it was formed from contracting interstellar dust. The most important invention of *Kelvin*, however, was to abandon the standards necessary for correct temperature measurements so that he defined temperature on the basis of the efficiency of thermal work whilst putting faith in the idea of caloricum. In fact, even the current definition of temperature is somehow awkward due to the fact that the absolute zero temperature is a limiting state, which cannot be achieved by any finite process. It would have been more convenient to define such a 'nil' point as an untouchable infinity (logarithmic proposal by *Thompson)* even under the penalty of a different structuring to thermodynamics.

Hess enunciated an important law, which is now associated with his name, that the amount of heat evolved in a reaction is the same irrespective of whether the reaction occurs in one stage or in several steps. It, however, became indispensable to make thermal observations more accessible, which turned out to be feasible by *Seebeck's* investigations who, in 1821, observed that an electric current was generated when one junction of two dissimilar metals in contact was heated, even if only by the hand. The current became easy detectable by using a magnetic needle in a spiral coil termed a galvanometer by *Ampere*, who based his work on the observation of *Oersted*. This system grew to be popular when applied to high temperatures in 1826 by *Becquerel* who used the junction of platinum and palladium. The increase in conductivity on lowering temperatures was first observed by *Davy* and later confirmed by *Ohm* during his investigation in 1826 that established the law that bears his name. Actual thermodynamic concepts became necessary, if not irreplaceable, when the determination of thermophysical properties became needful and significant.

It was initiated as early as in the middle of the 17th Century by *Boyle* and *Mariott* who related the pressure and volume of air, and by *Hooke,* who determined the simple relation between the deformation and stress. A century later it was supplemented by *Gay-Lussac's* detection that the heat expansivity of all gases is the same, which lead to the famous state equation of gases. It strongly affected the conception of heat to be associated with the internal motion and collision of molecules, first connected with their linear movement by *Kroening* in 1856. From the balance of impulses it followed that the pressure is proportional to the medium kinetic energy of molecules and is equal to 3/2kT where T is the absolute temperature and k is the *Boltzmann* constant. It acted in accordance with the *Petit* and *Dulong* observations on heat capacities showing that the medium kinetic energy of a fixed atom in the crystal network is equal to 3kT. It proved that, for each degree of freedom, for an atom to move it required a mean energy of kT/2.

Important roles in early thermal physics were played by statistics and the *Boltzmann* relation between the ratios of heat over temperature (entropy). These, as well as the probability of the arrangement of atoms, were used by *Planck* to show that the *Wien* radiation law for low frequencies and the *Rayleigh-Jeans* radiation law for high frequencies could be explained by individual atoms that emit discrete frequencies. *Planck's* work allowed him to derive his radiation law on the basis of the thermodynamic equilibrium between emitted photons and the absolutely black surroundings encountered on the condition of maximum entropy and thus providing a bridge between the micro- and macro- world of very small and very large collections. Recently these principles have been used to analyze the background structure of the entire universe, i.e., the arguable *vacuum,* previously known as the 'ether', which is now understood to be bursting with unidirectional zero-point radiation (seen as the ocean resource of energy which is borrowed on account of the creation of asymmetry and returned back on the energy re-execution).

1.7. Touching ultimate hypotheses

According to the *Heisenberg principle of uncertainty,* within its super-ultramicroscopic dimensions the Universe appears to be full of extravagant, vehement and fiery fluctuations (known as "boiling vacuum" or "Fermi Sea" or even "physical vacuum"). We can envision that virtual particle-antiparticle pairs pop out of the vacuum, travel for a short distance loop and then disappear again on Planck's timescales so fleeting that one cannot observe them directly. Nevertheless we can take notice of them as another indirect effect can be brought up to measurement. The virtual particles affect the spectrum of hydrogen in a tiny but calculable way that has been confirmed by observations. Accepting this premise we should contemplate the possibility that such virtual particles might endow empty space with some nonzero energy. The problem is, however, that all such estimates lead to absurdly large values falling beyond the known matter in the observable Universe, which, in consequence, would force the Universe to instantly fly apart.

It was not until the 1980s that a new promising and rather comprehensive approach to the ultimate theory of Universe appeared, based on the idea that all particles and forces can be described in terms of one-dimensional objects along with their interacted membranes of two-dimensions and higher. It surfaced as a suitable description of our 'elegant' Universe [38], which is close to the Pythagorean's idea of chord harmony tuned in rational ratios. This so called *string* and later *superstring* theory (*Green, Schwarts, Gamow*) is mathematically built on the thought structure of vibrating *strings* (of various shapes and a range of tension) that factually represent non-material and tiny fragments (little filaments of deep-buried dimensions) infiltrating the Universe to cooperatively and harmoniously act as a symphony of cosmic breath (often dedicated to *Ailos, the Greek goddess of wind*). The Universe is thus displayed in terms of a somewhat strange geometry of 'compactified' multi-dimensional space (visualized as a bizarre pattern often called *Calabi-Yano manifold*). We should not forget to mention another approach, known as *loop quantum gravity*, which seeks a consistent quantum interpretation of general relativity and predict the space as a patchwork of discrete pieces ("quanta") of volume and area.

The known human physical laws are since 1916 assumed to be the consequence of the laws of symmetry (*Noesther*), i.e., the conservation of a quantity (position, momentum, spin, etc.) upon the transfer of symmetry (translation, rotation, mirroring, etc.) within space and in time. The symmetry holds at that juncture when the laws of physics are unaffected when three transformations are all applied at once: interchange of particles and antiparticles (*charge conjugation*), reflection in a mirror (*parity inversion*) and reversal of time (*time inversion*). It shows that the four basic types of forces (gravity, electromagnetism, weak and strong interactions) must naturally exist in order to enable Nature to respect a given calibration accord and, thus, the original question of why there are such forces is now reduced to the question of why does Nature respect such symmetry? It required a new system of representation (e.g., *Feynman's diagrams*) that could provide a new physical framework such as *quantum flavor dynamics*, known as SU(2), and *quantum chromo dynamics*, SU(3), and a new theological visualization of God as the symmetry architect.

Historical cosmology assumed that such an energetic Universe flowed through 'ether' (*'aither'*) which was something celestial and indestructible, possibly related to *Aristotelian* primeval matter (*'plenum'*) or the recent view of omnipresent force-fields. It gave birth to a material universe in such a way that a delicate impact (*'bud'*) in such a primordial 'reactive solution' provided a dissipative structure, or fluctuation, capable to grow. This idea is also very close to the *reaction-diffusion model of space-time* [39] and challenges the traditionally established model of the `Big Bang`. It certainly is different with respect to the accepted 'mechanical' anticipation of waves resulting from the primeval big explosion that would have never occurred without reason ('spontaneously') and where the wave's transporter medium holds constant entropy (conservation of energy). On the contrary, the diffusion waves can interact with the carrier to undergo metabolic changes, enables entropy to increase and, most importantly, enabling self-organizing (and thus is non-linear in its character, assuming the Universe is an open system, like the terrestrially-scaled weather).

We may remind ourselves that according to conventional wisdom, *Poincare* failed to derive an early relativity theory mainly as a result of his stubborn adherence to the nineteenth century concept of the ether as an utter void while *Einstein*, just few years later, succeeded because he stayed behind the assertion of a new kind of ether that was the superfluous space-time substratum. His relativity postulate was called an 'absolute world'by *Minkowski*. Fifteen years later, however, *Einstein* used *Mach's* ideas on rational movement in order to cancel the action-at-a-distance and refreshed the idea of the ether because his relativity space is endowed with physical qualities. The ether should be regarded as a primary thing (similarly to the prematurely *Aristotelian* plenum) assuming that matter is derived from it [40], citing '*the metrical qualities of the continuum of space-time differ in the environment of different points of space-time being conditioned by the matter existing outside of the territory under consideration*'.

We can anticipate that gravity arises from the geometry of space and time, which combine to form 4-D spacetime. Matter tells spacetime how to curve and spacetime tells matter how to move. Any massive body leaves an imprint on the shape of the *spacetime*, governed by an equation formulated by Einstein already in 1915. The Earth's mass, for example, makes time pass slightly more rapidly for an apple near the top of a tree than for Newton's head 'just conceiving his theory of gravity in its shade'. Literarily, when the apple falls, it is actually responding to this warping of time. The curvature of spacetime keeps the Earth in its orbit around the Sun and drives distant galaxies ever further apart. Given thus the success of replacing the gravitational force with the dynamics of space and time, we may even seek a geometrical explanation for the other forces of nature and even for the spectrum of elementary particles. Whereas gravity reflects the shape of spacetime, electromagnetism may arise from the geometry of an additional fifth dimension that is too small to be seen directly.

This is a true domain of above mentioned string theory where we meet several extra dimensions, which results in many more adjustable parameters. One extra dimension can be wrapped up only in a circle. When more extra dimensions exist, the bundle can possess many different shapes (technically called *topologies*) such as a doughnut or their joins, other ridiculous spheres (six-dimensional manifold, etc.). Each doughnut loop has a length and a circumference, resulting in a huge assortment of possible geometries for such small dimensions. In addition to such `handles`, further parameters correspond to the location of `branes` (a kind of specific membrane) and the different amounts of flux wound around each loop. Vast collection of possible solutions are not mutually equal as each springs' configuration exhibits its own potential energy, flux and brane contributions and the curvature itself of the curled-up dimensions. Associated energy can, again, be seen as a kind of vacuum energy, because it is the energy of the spacetime when the large four dimensions are completely devoid of matter and fields. How such a hidden space behaves depends on the initial conditions, where the singularity that represents the start point (Big Bang) on the space-curve happened to lie in the initial vast landscape of possibilities, which inhabit the hierarchically supreme yet quizzical super-universe of curled proportions. It reminds us of a possible world of fluctuations curling on the endless sea level of wholeness predicting that our Universe might just occupy one (and for us living, appropriate, but generally random) valley out of virtually infinite landscape possibilities. This novel landscape picture seems of help in resolving the quandary of vacuum energy, but with some unsettling consequences.

In Einstein's theory, the average of the Universe is tied to its average density, so that geometry and density must be linked. The high-density universe is thus positively curved like the surface of a balloon (*Lobachevsky* hyperbolic geometry) and, on contrary, the low-density universe is negatively curved like a surface of a saddle (*Riemann* elliptic geometry), and the critical-density universe is flat. Many precise measurements of the angular size of the variation of the microwave background residual radiation were good enough for researchers to determine that the geometry of the Universe is flat (current curvature provides a value very close to one, ≈ 1.05). But many different measurements of all forms of matter, including cold dark matter, a putative sea of slowly moving particles that do not emit light but do exert attractive gravity, showed that visible matter contributes only by less than 1/4 of the critical density. Thus vacuum energy, or something very much like it, would produce precisely the desired effect, i.e., a flat universe dominated by positive vacuum energy would expand forever at an ever increasing rate, whereas one dominated by negative vacuum energy will collapse. Without vacuum energy at all, then the future impact on cosmic expansion is uncertain and in the other world, the `physics of nothingness` will determine the fate of our Universe.

1.8. Science of thermal dynamics

Looking back to novel ideas of the 19th Century we see that they assisted the formulation of an innovative and then rather productive field of physics (thermal physics), which gradually coined the name *thermodynamics* though the involved term 'dynamics' has not its appropriate position. Thermodynamics became a definite physical domain based on general laws of energy conservation and transformation, and where its consequences serve practical applications. It studies real bodies that consist of many integrating parts. The basic assumptions for such real mixtures are the same as those that are also fulfilled by pure components, but rather that all the properties of mixtures are the consequence of the constitution of the components, although not a simple summation of components as various interactions between components must be taken into account. One of the most important laws is the second law of thermodynamics, which, simply stated, says that any spontaneous process must be only such that any resulting state is more easily accessible than the state of any previous (competing) process. Thermodynamics can also provide an idea of a global (non-material) driving force showing that the process always follows the circumfluent contour lines of the straightforward access to all intermediate states on its reaction path. This, in fact, is very similar to the objective of our curved universe, where the easiest path is controlled by gravity lines of force or, in everyday experience, by the uniplanar contour line of comfort walking.

Every thermodynamic process is associated with a certain dissipation of energy, as well as an associated degree of entropy production, which is also the measure of the irreversibility of time. Inherent thermal processes are sensitive to the criterion (or perhaps our point-of-view) of what is stability, metastability and instability. There can exist equilibria characterized by special solutions of complex mathematical equations, whose stability is straightforward, however, the comprehensively developed field of thermal physics primarily deals with equilibrium states that are unable to fully provide general principles for all arbitrarily wide, fully open systems, which would encompass every unsteady (or somehow re-re-establishing) disequilibrium. It, however, can promote the unlocking of important scientific insights, which lead to a better understanding of chaos and are an inquisitive source for the evolution of systems (such as life).

Any thermodynamic system bears its own clear definition criteria, i.e., a real material body has its volume and is circumscribed by its surface, which means the defect on the exterior otherwise homogeneously defined body. A system can be any macroscopic entity, biological being or even ecological formation. A system can be imagined to be composed of certain material points or sites, which we may call subsystems, where each point, or any other fraction of a whole, cannot be merely an atom or a molecule but any elementary part of a body that is understood as a continuum. Thanks to the body's interactions with its surroundings, there are certain primitive characteristics, representing the behavior of the system, as each constituent point can move and can change its temperature. There certainly are other important quantities that play crucial roles in the general laws of balances, i.e., mass, momentum, energy and/or entropy. Energy in space, however, is directionless, which is why we often use the help of an auxiliary parameter: the directional vector of momentum across a surface.

A system's interaction can be identified with any type of reciprocation of balances. The number of parameters is often greater than the number of balance laws, so it is necessary to have other axioms and interconnections. Among the most important are causality, determinism, equipresence, objectivity, material invariance, time irreversibility, and the decisive effect of nearest neighbors, and nearby history.

When the thermodynamic parameters of a process change so slowly that their relaxation time is longer than the time of observation, these processes are called quasistatic and have important technical consequentiality. This, in fact, means that these processes are in states of both equilibria and disequilibria at the same time. Without this simplification, however, we would not be able to study most of nature's processes because only a negligible fraction of them are time independent. Generally we assume *equilibration* during which any well-isolated system achieves, within a long enough time, its stable, well balanced and further noticeably unchanging ground state (i.e., death, which is the stabilization of original 'fluctuation').

An isolated ecosystem could thus sustain inherent life for only a limited period of time, often much less than that, which is necessitated to undertake from the onset of isolation until reaching thermodynamic equilibrium. Such a well-confined situation can be compared to the thermal death of our 'one-off-universe' so that any thermodynamic equilibrium is factually a global attractor for all physical processes when separated from their surroundings. Having reached the ultimately balanced state, no gradients would exist and no further changes would be possible and in this petrified state, even time would be defeated of its meaning as its passage could not be verified by reference to any change as well as the observation of properties could not be made, only inferred, because even observation requires some kind of interaction between the measured system and an observer. We can match them up to the first category of antithesis in physics (black holes).

An important quantity is heat, which is often understood as the energy absorbed or evolved by a system as a result of a change in temperature but the words of *heat and energy are not synonymous*. We must see the concept of heat as referring to a process whereby energy is passed from one system to another by virtue of the temperature difference so that the word heat factually applies to the process. It is worth noting that an unharnessed flow of heat from a higher to a lower temperature represents in actuality a loss in work (accessible) effect and is therefore equivalent to friction in the production of irreversibility

Therefore, the most basic and also unlocking quantities in thermodynamics are *temperature, heat* and *entropy*, where the latter is an artificially introduced phenomenological (extensive) parameter expressing as the ratio of heat and temperature. Temperature and entropy characterize collective behavior of the studied corpus of constituent species. Temperature is a measurable quantity and represents the degree of the mean energy, which is inter-exchanged by species (atoms) that constitute the system. On the other hand entropy as a measure of the system displacement is not a directly measurable parameter, similarly to heat; we can observe only associated changes. From a standard phenomenological viewpoint, we can also treat entropy as 'normalized' heat in order to match its conjugate pairing with temperature, obtaining the sense that the higher the temperature the higher the 'quality' of heat [9].

Creation of order in a system must be associated with a greater flux of entropy out of the system than into the system. This implies that the system must be open or, at least, non-isolated. In this context any *structure* must be a spatial or temporal order, describable in terms of information theory and a higher-order *dissipative structure* stands for self-organization indicating that such a system dissipates energy (and produces entropy) in order to maintain its organizational order. It follows that all real systems, such *ecosystems,* are not closed to interactions with surroundings, which is a kind of sink for energy higher in entropy, and closer to equilibrium. *Schrödinger* pointed out that *"living organization is maintained by extracting order from the environment"* and *Boltzmann* proposed that *"life is struggle for the ability to perform useful work"*.

Entropy has accoutered various connotations, corresponding to the system's make up, i.e., incidentalness of the internal evolution of the arrangement of system constituents, which can be definite species or logical thoughts. It has also initiated a perception of chaos that is excessively original within habitual mathematical descriptions but has opened-up new perspectives on disorder as a vital part of our universe where we employ only practicable solutions (e.g., acceptable ever-presented vibration of aircraft wings to assure safe flying against their anomalous quivering indicative of danger).

However, there is persisting and old idea of how to manipulate entropy. It is often attached to delaying the moment of death, re-obtaining order once disorder has set in. *Maxwell* taught us that a supernatural actor, a demon, is required to be able to sit next to a hole drilled in a wall separating two parts of a vessel containing gas. Watching the molecules and opening and closing the hole, the demon can let faster molecules go on one side and slower molecules stay on the other side. Disequilibrium, or information, is obtained. Maxwell's demon is a mental artifact used to explain the concept of entropy but cannot be a real device. But the idea is fertile; we can conceive the heat flow diversion (allowed) but not its reversion (forbidden). Respecting the entropy principle, we can imagine an automaton operating that flows continually from warmer to cooler according to a purely thermodynamic logic. Instead of being a stable solution, entropy ensures only dissipative structures such as another thermodynamic demon – life.

The theory of thermal science is important for our practical applications as it teaches us that the best yields can be only obtained from infinitesimal changes, which are associated with reversible processes where all adjustments are close to zero, i.e., motion damps out and concretization would last exceptionally long. However, if some action is due to a real execution it must possess a finite driving force, even if it is negligibly small. We have always to decide whether to employ high driving forces with lower work efficiency or make it more energetically economic upon decreasing gradients but paying time for its prolonged realization. For getting an easy profit the thermodynamic laws are somehow cruel and impossible to circumambulate (even by any persuasive ecological arguing). There is an aphorism that, in the heat-to-conversion game, the first law of thermodynamics says that one cannot win; at the best we can only break even. The second law reveals that we can break even only at absolute zero and the third law adds that we can never reach the desired zero temperature.

As all other living things do, we 'Homo sapiens' – a most significant form of life, demand and would require thermal energy in the form of food and heat [41]. The manmade haste in the mining of energy may be seen by the Nature hazardous as an activity rather close to wrongdoing. The Nature may wish to have a control over human activity [42] like the goddess '*Venus*' liked to visit the God '*Volcano*' as to keep graceful control of his dubious behavior in his underground dwellings. We wordlessly witness accumulation of ash from the combustion of fossil-fuels, radioactive wastes and other manmade litters, without pushing natural recycling by the ecosystem, which is vulnerable to events it has never witnessed before. The ancient Greeks saw the ecosystem as 'Nature of the Earth' to be something agitated by divine vitality and even now we should keep remindful of its beauty. It is clear that naturally enough smart '*Goddess Earth*' is beginning to show her first signs of suffering.

"We are not just observers we are actual participants in the play of the universe." Bohr once said that actually applies to the forthcoming description of the generalized concept of fire. The fight of mankind for a better life, while striving to survive in the universe/nature and while hunting for eternal fire, should therefore be understood as a fight for lower chaos (entropy) and maximum order (information), not merely seeking for sufficient energy, which seems be, more and less, plentiful. We thus should not underestimate certain self-organization tendencies noticeable not only on the micro- but also on the macro-level of nature noting (according to *Kauffman* [43]) that *"even biospheres can maximize their average secular construction of the diversity of autonomous agents and the ways those agents can make a living to propagate further on. Biospheres and the Universe itself create novelty and diversity as fast as they can manage to absorb it without destroying the yet accumulated propagation organization which is the bases of further novelty"*,

2. MISCELLANEOUS FEATURES OF THERMAL SCIENCE

2.1. Heat as a manufacturing tool and instrumental reagent

The notion of fire (light, flames, blaze, heat, caloric and more recently even energy) is thoroughly recognized as an integrating element, rudimentary in the pathway of ordering matter and society [41]. It has a long history, passing through several unequal stages in the progress of civilization. From the chronicles of interactions of society with fire we can roughly distinguish about four periods. Perhaps the longest age can be named the period *without fire* as the first human beings were afraid of fire/blaze (like wild animals) but, nevertheless, eventually gained their first sensation of warmth and cold. The first man-made fireplace dates as being about 300.000 years old, but artifacts of ashes resting from different burned woods (apparently composed unnaturally by early man) can be even associated with prehistoric *homo erectus* dating back to one and half million years ago (*Koobi Fora, Kenya*). Another extended era was associated with the growth of the continuous experience of *using fire* which helped, in fact, to definitely detach human beings from animals (fire as weapons or as a conscious source of warmth) by substantially aiding the cooking of meat as to make it more easily digestible. The associated development of cooking also increased the range of palatable foodstuffs and allowed for more time to be spent on activities other than hunting.

A definite advancement came with the recent but short period of *making fire,* up which preceded the *exploitation of fire* that included the domestication of fire and its employment as a tool and energy source until the present use of heat as an instrumental reagent. Even the ability to use fire for many specific purposes is conditioned by the smallest flame defined by the balance between the volume of combustible material available and the surface area over which air/oxygen can fuel the combustion reaction. As the volume of combustible material gets smaller, the surface becomes too small for the flame to persist, and it dies. Small stable flames are well suited to the needs of *creatures of human size* the tallness of them is affected by the straight downward force of gravity. The opposing direction (buoyancy) of heat was viewed as a contradictory force to gravity that holds people attached on the earth forcing all bodies to finish in rest (i.e., the equivalence of motion and rest for a thrown stone). Moreover people witnessed that a small ignition (spark, impetus) could lead to catastrophic results (fires, avalanches), i.e., unique circumstances (singularity) can give way to the concealed potentiality hidden in the system.

It was found to be curious but apposite that our terrestrial atmosphere contains offensive gases like oxygen, which should easy react but instead coexist and form a mixture far from chemical equilibrium representing thus an open, non-equilibrium system characterized by a constant patchy flow of energy and matter. The proportion of free oxygen eventually stabilized at 21%, a value amply determined by its range of flammability. If dropped to below 15% organisms could not breathe and burning would become difficult while exceeding 25% combustion may become spontaneous and fires could rage around the planet. Oxygen, in fact, was a toxic by-product in the originally massive blue-green bacteria photosynthesis (while splitting water molecules into their components) and its increased atmosphere "pollution" resulted in the self-elimination of these bacteria (seen as a signified 'global catastrophe').

The earliest inventions of mankind were always connected with application of fire in providing and/or processing not only food but also in procuring natural or processed materials for making them useful or ornamentally attractive. The first man-made artifacts were hand-molded from clay known as early as about 15000 years BC and primeval ceramic objects stem 7000 years later from the Mesolithic period. The name ceramics is derived from Greek word *'keramos'*, i.e., potter's clay or vessel, but its origin may be even older, from Sanskrit, where a similar word means 'firing'. The *'potter's wheel'* is a Mesopotamian great invention from the 3rd millennium BC and was responsible for a significant improvement in ceramic technology and cultural life.

Based on the experience gained in producing glazed ceramics by smelting copper ores and in preparing mineral dye-stuffs, people discovered how to obtain enameled surfaces by fire-melting the readily available mixture of silica, sand, lime, ash, soda and potash. The first steps in this direction were made as early as some 7000 years ago by putting in service *natural glasses,* as 'tektites' (believed to be of meteoric origin), obsidians (glassy volcanic rocks), 'pumice' (foamed glass produced by gases being liberated from solution in lava) and 'lechatelierite' (fused silica in deserts by lightening striking sand or by meteorite impact). Through history, glass was always regarded to have a magical origin: *just to take plentiful sand and plant ashes and, by submitting them to the transmuting agencies of fire to produce melt which whilst cooling could be shaped into an infinite variety of forms which would solidify into a transparent material with appearance of solid water and which was smooth and cool to the touch, was and still is a magic of the glass workers art.*

Invention of man-made glass was accomplished somewhere on the eastern shores of the Mediterranean prior to 3000 BC, but no hollow glass vessel dating earlier than the second millennium BC has been found. Glass technology reached a high degree of perfection in Mesopotamian and Egypt, especially sometime during 500 BC when one of the most important brainchild, a *'blowpipe',* was invented (around 50 BC probably in Syria), which turned glass into a cheap commodity and provided the stimulus for the proliferation of glasswork throughout the Roman Empire.

Nearly as old is the technology of binding materials obtained by lime burning for the preparation of lime-gypsum mortars. These materials were already used in Phoenicia and Egypt in about 3000 BC. It was the Romans who invented concrete made of lime with hydraulic additives known as 'pozzolana' and volcanic tuff. In *Mesopotamia* the experience of firing ceramics was also applied to the melting of copper, and, later, of gold and silver that led finally to the metallurgical separation of copper from oxide ores. Copper, however, was known in *Baluchistan* as early as by 4000 BC and bronze by 3500 BC. Around 2500 BC excellent properties of alloyed copper (silver, gold, arsenic) were discovered followed by tin in 1500 BC and iron about 500 years later. About 3000 BC articles on lead also appeared and starting from 2500 BC the Indians began to monopolize the metallurgy. Whilst bronzes were mixed phases prepared intentionally, the admixture of carbon was introduced into the cast iron, and, later, steel by chance, and its unique role could not be explained until recently. Even ancient Chinese metallurgists clearly appreciated that the relative amounts of copper and tin in bronze should be varied depending on the use for which the articles were intended.

Iron could already be melted and poured by about 500 BC and steel appeared as early as by 200 BC depending on the thermal and mechanical treatment. The medieval skill belonging to the Persians in cold-hammering swords of exceptional quality was based on the ability to pour bimetallic strips of low (tough) and high (brittle) carbon-containing steel followed by their prolonged mechanical interpenetration. All such progress was intimately associated with the skill and knowledge of fire-based craftsmen, such as metal-smiths, who had had to have the experimental know-how of, for example, how intense the fire must be to make the metal malleable, how fast to quench an ingot, and how to modify the type of hearth to suit the metal involved.

Operational stuffs, called *materials,* have always played an important position in the formation and progress of civilizations [7,9]. It was always associated with the relevant level of intercourse *(communication)* and comprehension *(information)*. We can learn about their history from archeological sites as well as from recent dumps and waste-containers. The recent tremendous growth of certain groups of individualistic materials will, in time, be necessarily restricted in favor of those materials that are capable of recycling themselves thus possessing the smallest threat to nature.

Applicability of materials is traditionally based on working up their starting ('as cast') mass to shape a final and refined structure. Today's science, however, tends to imitate natural processing as closely as possible and, in this way, wastes less by positioning definite parts to appropriate places. This ability forms the basis for the development of so-called *'nano-technologies'*. It has long been a dream of scientists to construct microscopic machines, such as motors, valves or sensors on the molecular scale. They could be implanted into larger structures where they could carry out their invisible functions, perhaps monitoring internal flows or even some vital biological functions. For example, the newly made *dendrimers,* tree-shaped synthetic molecules, can possess the ability to capture smaller molecules in their cavities, making them perfect to deal with biological and chemical contaminants. It is believed that they will be eventually able to detect cancer sells, and destroy them by delivering a specific drug or gene therapy. They may become capable of penetrating the white blood cells of astronauts in space to detect early signs of radiation damage, and may even act as toxicant scanners in the battle against bio-terrorism.

Living structures are yet becoming more and more of an observable example, particularly assuming the synergy of various components in combined materials. Such composites are common not only in nature but also known within human proficiency: from Babylonian times (3000 BC), when resin impregnated papyrus or tissue was used, up until recent times when the icy airfields of Greenland used sandwich composites made from brittle ice strengthened by layers of newspaper. It appears that we must follow the models of the most efficient natural composites, such as hollow fibrils in cellulose matrix that make up wood, or collagen fibers such as in hydroxyapatite found in our teeth.

In the reproducibility of technological processes, the artificial production of glass and cement (originally volcanic and slag) and iron (primarily meteoritic) have played a decisive role in the progress of civilizations – their production recently amounting to a tremendous 10^{11} kg per year (with the total world need of energy sources close to ten gigatonnes of oil equivalent). Reproducibility of technological processes, in which ceramics and metals were worked to produce the desired wares, could be secured only by experienced knowledge (information, data-storing) as well as by the resulting accustomed measuring techniques. Its development could only take place after people learnt to think in abstract categories such as are used in mathematics whose beginning reach as far back as the fourth millennium BC. Production became feasible only after it matured to be consciously determined, keeping constant and optimal values, such as mass proportions of input raw materials. Thus, as well as the others, the Egyptians knew the double-shoulder balances as early as 3000 BC. The notion of the equality of moments as the forces acting on the ends of an equal-arm lever was surely one of the first physical laws discovered by man and applied in practice. Weighing (of pouring in water) was even used as a convenient means to measure time intervals more precisely.

Besides growth of industrialization, we should also realize what the more powerful and far-reaching the benefits of a technology are, the most serious are the by-products of technologies' misuse or failure. The more a forward-looking manufacturing process can bring about randomness, the further its products depart from thermal equilibrium, and the harder it is to reverse the process that gave rise to them. As we project a future of increasing technological progress we may face a future that is advanced but increasingly hazardous and susceptible to irreversible disaster.

In order to make technological progress, mankind had to acquire progressive techniques of producing high enough temperatures. First they applied controlled heat not only by using the closed kilns but also by using sitting kilns or smelters in regions prone to high winds. The role of air had to be understood in order to get the process of burning coordinated. It was also vital to keep certain temperatures constant; hence it was necessary to introduce some early-experimental temperature scales. The relation between input mixtures, mode of fire treatment and resulting properties were initially recognized and then, subsequently, experimentally verified during the preparation of other compounds such as drugs, dying materials and, last but not least, in culinary arts. This all led to the early recognition of three principles of early thermal treatment and analysis [44], i.e., *amount of fire (temperature) can be level-headed, applied fire affects different materials in different ways, and that materials stable at ordinary temperatures can react to give new products on firing*.

It is clear that fire has always played a significant role as an explicit *tool* either in the form of an *industrialized power* (applied for working materials by men long ago in the process of manufacturing goods), or later as *an instrumental reagent* (for modern analysis of the properties of materials). In contrast to a mechanical machine, that provides a useful work upon the consumption of potential energy received from the surrounding world, it was recognized that any practical use of heat must include changes involved in the explored system itself. In the principle sphere of dynamo-mechanics (in an ideal world detached from disturbances caused by e.g. friction) the efficiency of mutual conversions between potential and kinetic energies can approach a theoretical 100%.

Heat engines, however, cannot be only a passive machinery – just protecting two components, with different temperatures, to get in touch, or not, with each other at the right moment (similar to the restriction of two parts, moving with different speed, to come, or not, into the required mechanical contact with each other). We should remember that reversible changes furnish the capacity to effect the system and its control – the initial conditioning can restrict the dynamics of an object. Model thermodynamic behavior can be similarly specified if well-defined reversible changes are available for an equilibrium state. For that reason any irreversibility is seen negatively as it is realized by lively irrepressible changes when the system gets out of standard control. It follows that thermal systems can be controlled only partially; accidentally they may arrive at a spontaneous stage of a surprising character.

Fig. 7. - A most unusual manner of treating external wastes in a natural way is done by the holy scarab in Egypt (left). This pill-peddler is known for using the wastes of others in a rather original fashion while exercising skilful motions of its legs to form and roll a ball from the excrements of other animals. Such a biological prototype of waste managements, which Nature has evolved through a long-lasting optimization, is the best model for our waste treatment and renders the possibility to foreshadow the symbolic furtherance for a better-established process of futuristic recycling of human wastes. (Moreover the beetle's teguments are of the most tough and durable composite materials, which again is worth increased attention by material's engineers for better material tailoring and use of nanotechnologies.) On the other hand the progress of civilization has made it possible for man to expand his exploitation of fire (right) to such an extent that the Earth is being overrun by excess of waste heat and its by-product material litter. It, however, stays against the primary role of natural fires, which originally contributed in the creation of landscapes through the regular burning of localized areas due to lightening. Thanks to fires, the countryside mosaic and plant diversity have been maintained and, on the other hand, thanks to mosaic structure, the extent of the fire damage has been kept locally restricted. This natural course, however, has been disturbed by man-made woods, and mono-cultivations due to agriculture, which causes national fires to be regionally more disastrous. Even a more radical trend is the ultimate forest nonexistence, when the land is clear-cut and exploited for manufacturing, which obstructs the natural cycles merely producing man-made litter useless (and even harmful) in the view of ecosystem.

2.2. Mechanical motion and heat transfer

The incompatibility between the time-reversible processes taking place in the mechanics of body-like species (including molecules), and irreversible thermal processes responsible for the equilibration of temperatures (e.g. in an unequally heated rod) required a better understanding of the inherent phenomena involved, particularly the nature of heat. In sequences it led to the formulation of a consistent science of *thermal physics*, developing the related domain of *thermal analysis* and touching on any adjacent field of science where temperature is taken into account.

In fact, it was initially based on the early and erroneous premises of a non-material notion called *'thermogen'* (or superior term 'caloric'). It can be also seen as the application of *different* premises regarding the manner of individual approach of theoretical modeling. The caloric theory gave an obvious solution to the problem of thermal expansion and contraction, which postulated and described an opposing force to that of the attraction between particles caused by gravity. In order to stay away from a total gravitational collision, which would produce a single homogeneous mass, the opposing force was considered to be the *'self-repulsive caloric'*. Such an early 'fluid' hypothesis became important in the formulation of modern laws and was common to the way of thinking of *Archimedes, Epicureans* and later used in *Carnot's* and *Clausiu's* concept of thermodynamics [45, 46], up to present day truly dynamic theories of non-equilibrium, introduced by *Prigogine* [47, 48]. Even our everyday use of heat flow equations applied in thermal analysis bears the "caloric" philosophy.

However, the notions of heat and temperature (*temper – temperament*, first used by *Avicena* in about the 11th Century) were not distinguished until the middle of the 17th Century (*Black* [49]). It took another 150 years until the consistent field of thermal science was introduced on the basis of *Maxwell's* works [50,51] and named by *Thompson* as *'thermodynamics'* according to the Greek terms *'thermos'* – heat – and *'dynamis'* – force. The contradiction of its 'dynamic' notion against its 'static' applicability (*thermostatics*) is thus a common subject of examination. The thermodynamic account was complicated by the introduction of an artificial quantity to counterpart the intensive parameter of temperature in its sense of gradients. After the Greek *'en'* (in-internal) and *'trepo'* (turn) it was called *entropy* (meaning transformation and proposed by *Clausius* in sound analogy with the term energy). This action eliminated heat from a further mathematical framework. In effect the famous law of thermodynamics gives the *quotation of energy conservation law but only under specific conditions for heat that is not fully equivalent with other kinds of energies*. Heat cannot be converted back to mechanical energy without changes necessary to affiliate heat with entropy, *via* the second law of thermodynamics, which intuitively states that heat cannot be annihilated in any real physical process. It brought a conceptual disparity between the fully reversible trajectories of classical mechano-dynamics and the evaluator world governed by entropy. It may be felt that the domain of thermodynamics was initially positioned at a level of *esoteric* ('those within') doctrine.

Only statistical thermodynamics (*Boltzmann*) interconnected the micro-cosmos of matter with phenomenological observables under actual measurements. The famous *Boltzmann* constant gave the relation between the phenomenological value of temperature and the average kinetic energy of motional free atoms – submicroscopic species. In the evaluation of heat capacity, *Einstein* and *Debye* associated temperature with these perceptible micro-particles (array of atoms), each oscillating around lattice sites where all degrees of freedom correspond to the vibrational modes (assumed either as independent harmonic oscillators or as an elastic continuous medium with different frequencies). The vibrational nature of some forms of energy (as that of heat) was a very valuable and inspirational idea related to an old *Pythagorean* idea of chord harmony tuned in rational ratios.

The 19th Century was the period of the industrial revolution. Scientists studied the efficiency of machines of all sorts, and gradually built up an understanding of the rules that govern the conservation and utilization of energy. The laws of thermal physics were one of the results of these investigations, most famously stating that the artificially introduced parameter known as entropy of a *closed* system can never decrease. In practice, this means that even though energy can be conserved in some physical processes, it is gradually degraded into less ordered and less useful forms. It is habitually observed that the entropy of any closed system becomes steadily more disordered. This is not a kind of law as gravity might be classified, but a principle having a statistical nature and very important to consider what is technologically possible.

Later an interconnection between entropy and information (its gain or loss) was found. If we are to obtain information about the state of a system then there is always a cost. It was *Maxwell's* sorting demon that made it clear that it is not possible to create or cause a violation of the second law of thermodynamics, any more than it is possible to show a profit at roulette by always betting on all the numbers. The cost of such a long-term strategy always outweighs the possible benefits. It follows that information can be seen as a commodity that takes effort to acquire, and it is possible to classify all technological enterprises in terms of the amount of information needed to specify its structure, or what level of information would be required to introduce a change in the system (organization improvement, etc.). Information is thus expensive to acquire – it costs time and energy. There are limits to the speed at which that information can be transmitted, and limits to the accuracy with which it can be specified or retrieved. Most importantly, however, is that there are powerful limits on how much information can be processed in a reasonable period of time. Even the speed at which light travels is limited.

An ever-increasing experience with burning revealed that released heat always leads to volume increase. Cognizant employment of fire was accordingly utilized by *Savery* and *Newcomen* (1705) and improved by *Watt* (1800), as well as by *Trevithick, Cugnot* and finally by *Stephenson* (1813) while constructing a *steam heat engine* and, later, a functional *locomotive*. In that time, they were not at all aware that part of the heat is transformed to motion. It was, however, a qualitative step in making a better use of coal (as a traditional source of heat for personnel working in manufactures) and made it the main source of mechanical energy for use in industry and home, replacing other traditional sources such as animals, water and wind.

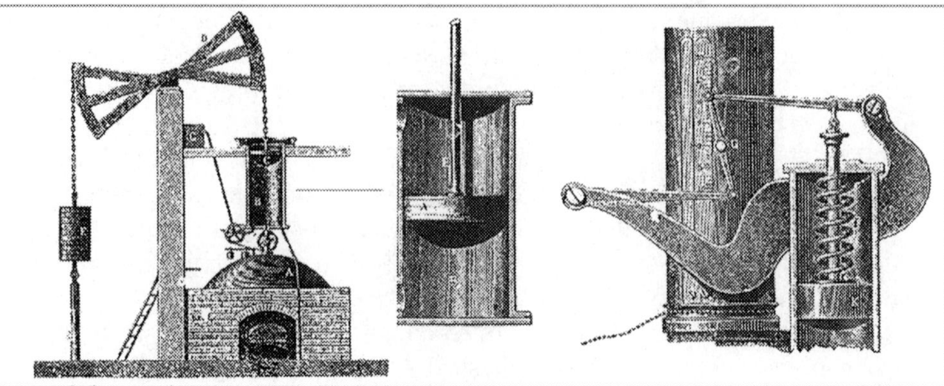

Fig. 8. – Newcomen's steam engine as the primary milestone in the novel exploitation of heat and its alternative use other than for effortless getting warmness with the original sketch of cylinder with piston (middle), which can be acquainted with the shared employment of all four elements (where the earth-based construction represents the joining 'in-form' = information transducer). Right it is added by illustrative drawing of the mechanical control torque, factually the first automaton realized through the Watt's innovative governor.

This new scheme actually interconnected the three of the early elementary forms [52,53] heating *water* by *fire* to get a thick *air* (steam) capable of providing useful work (moving piston and turning a wheel). Later on, this idea was improved by *Lenoir* (1868, gas-engine) and particularly *Ott* (1878) *and Diesel* (1892), who imprisoned the burning fire directly inside a cylinder and, thus, constructing thus a shaped *earth*. Certainly, a confident know-how was indispensable, or perhaps better to say, a particular use of the directed (in) formation. The miracle of the nineteenth century, i.e., the four stroke combustion engine, was brought to life!

Restricted by thermal laws, it gave a practical dimension to the innovative field of thermodynamics and showed the essence of a sequence process controlled by the four cycle series, which both *started and ended in the same* position. The encircled loop of given pair of the associated (intensive and extensive) parameters granted access to a convenient estimate of the energy gain (or, better, generally assumed assets, goods).

All this put an emphasis on the importance of mechanical 'know-how' to everyday life. Heating a room when it is cold and cooling it when it is hot is also a matter of know-how (an exchange of information). Various thermal flows are set up to partially modify the natural laws determining the movement of the air in the room. The ingenious control torque (governor) of a heat engine, invented by *Watt* using the principle of the excess centrifugal force, was realized experimentally by the famous parallelogram with springs, two balls and steam valves. It illustrates well the advanced information portrayal though there was nothing known about its inherent control function in the terms of modern state vectors.

However, the moment of birth for modern scientific thought probably occurred a few years earlier, in 1811, when *Fourier* was awarded the prize of the French Academy of Sciences for the material description of heat transfer in solids citing: `the heat flow is proportional to the temperature gradient`. This simple proportionality revealed a new potential that later achieved more general applicability (analogical equations applicable for diffusion, quantum-mechanics of waves, etc.). In its simplicity *Fourier's* law was considered somehow similar to the *Newton's* laws of mechano-dymamics, but its novelty led to a set of new classifications, the evaluation of which put heat and gravity at a contradiction to one another. Gravity can move a mass without causing an internal change, whilst heat puts matter in motion while reforming, at the same time, its internal properties. Similarly to *Boerhaave's* statements [54] *"heat is propagated and its distribution is equalized under all circumstances"* it supports the idea that propagation of heat is an *irreversible* process. Moreover, there was a belief that something is 'qualitatively conserved' and something is 'quantitatively converted'. In the year 1847, *Joule* introduced an equivalent for physical-chemical changes and helped to finally define energy as a joining element between physical-chemical and biological processes.

The idea of the *thermal engine* has developed into a wide variety of engines, turbines, pulse-jets and other power cycles all still governed by the same principles of thermodynamic efficiency for the external and/or internal utilization of heat/fire. Similarly to any hydroelectric turbine, the combustion engine is an open system – it consumes oxygen and fuel at higher potential energies than at which it produces heat and combustion products. Such a desired pattern of activity produces, however, disorder and thus fully complies with the second law of thermodynamics. The energy produced by the automotive open system of an engine is literally kept alive on the production of excess amount of heat depleted to its surroundings. The second law of thermodynamics tells us that we need to do work in order to acquire information.

This way of thinking is very general and allows us to quantify the cost of any computation. In any sphere of human activity it is not enough to be in possession of a procedure to solve a problem. We need also to know the cost of its implementation, either in terms of time, money, and energy or computation power. This knowledge opens to us the possibility of finding a procedure that is better, in the sense of being more cost-effective.

Foreshadowing the work of *Prigogine's* self-organization, there was introduced in 1920s the *science of structures,* as pioneered by *Bogdanov's tektology* (from the Greek *'tekton'* – builder), and the first attempt to arrive at a systematic formulation for the principles of organization, which can function in non-living and living systems. It was followed by *Bertalanffy's general system theory* [55] aimed to develop self-guiding and self-regulating machines that try to solve the problems of interactions and regulation, which lead to an entirely new field of investigation that had a major impact on the further elaboration of the system view of life.

It inspired *Wiener* to invent a special name *'cybernetics'* (derived from the Greek *'kybernetes'*, meaning steersman) as to describe the joint science of control and communication in the animal and the machine. Its importance became the theory of information developed by *Shannon* when trying to define and measure the number of signals (information) transmitted through telegraph lines. It is worth mentioning that the clockworks of the seventeenth Century were the first autonomous machines, and for the next three hundred years they were the only machines of their kind. Computers invented recently are novel and unique machines that can work autonomously, once turned on, programmed and kept ongoing by sufficient energy supply. Computers, however, do something completely new – they process information. As *Descartes* used the clock as a metaphor for the body, *Neumann* used the computer in his cybernetic believe and as a metaphor for the brain introducing there the human relevant expressions of intelligence, i.e., computer *memory* and *language*.

Any variety of useful machinery, that has been developed by means of human cognition (mind) have, however, a relatively short history, just a couple of thousand years. The evolution of life, on contrast, exhibits a history that is several orders longer, perhaps billions of years, and even a live cell can be understood as a *precisionally fine-tuned machine* that was gradually (self) constructed through the ages as an extraordinary sophisticated set of interrelated parts (molecules) that harmoniously act together in predictable ways insuring thus the progressive development, variability, and selective survival of more of its kinds and efficient use of energy. Its "know-how" is recorded in DNA memory that possesses a capacity for creative mutations of its structure (internal development of certain "cognition" [43,56]), which permits further evolution of its machinery cell into forms needed to prosper successfully upon the Earth's continuously changing face. It is clear that the space-time structures do not remain stable and that some of the subsequently mentioned principles, such as chaos and/or entropy versus order and/or information, penetrated into other fields such as humanities, sociology, economics, etc. A generalized course of cycling, which makes a process in such a reliable way as to exhibit coequality of the initial and final points, is necessary to retain the procedural capacity for its continual periodicity.

The short history of human civilization does not compare with the much longer history of life. The recently constructed mechanical engines bear a level of matchless sophistication in comparison to the mechanism of living cells, which have undergone very slow but continual development. Thus, the present-day propensity for the fabrication of macroscopic human tools tends to avoid the undesirable aspects of self-destruction by introducing, e.g., *pure technologies,* they produce ecologically harmless byproducts (*clean engines)* but are, however and unfortunately, less efficient. This downside is, however, being overcome by techniques such as *miniaturization,* which tries to simulate a more efficient level of the molecular tools from living cells. The as yet conventional approach of a "top-down" technology starts with a large clump of material, which is then formed by being cut into a desirably useful shape yet carried out mostly within the capability of our hands. Energy saving miniaturization (the earlier mentioned "nanotechnology") would, by contrast, be a "bottom-up" approach that would involve the stacking-up of individual molecules directly into a useful shape, and, moreover, that would be controlled by the "power of our minds" in a similar way to the proteins that are assembled from individual amino acids. We can think back to the vision of the "nanobots" age, which predicted the manipulation of small species to actually form miniature engines (or more imminent micro-computers).

Similarly to the processing instructions necessary for the operation of macroscopic machines (done either by an experienced engineer, by the help of a printed manual, or by the fully automatic control of a programmable computer) and realized at the one-dimensional level of a written message, we have, in necessary addition, to look for the analogy of microscopic instructions already known within the models of the three-dimensional level of DNA structures. Such an approach is also related to the nowadays message passed by the human cognition and based on the early inaugurated need how to understand fire from its visible wholeness down to its imaginable composition of internal vibrations. This has become an important domain of communication and is closely related to spheres of information (and reversed entropy) to be discussed herein, and in more detail, later on.

2.3. Special but not rare case of self-organization

By the end of the nineteenth century, there were available two different mathematical tools available to model natural phenomena – exact, deterministic equations of motion and the equations used in thermal physics, which were based on the statistical analysis of average quantities present in complex systems: Whenever any non-linearity appeared, it was put into linear regimes, i.e., those whereby double cause produces double effect and the solution is linear in the known term (classical thermodynamic laws). In these linear systems small changes produced small effects and large effects were due either to large changes or to a sum of many small changes. Moreover the imperfect measurements of an event did not prevent one from gathering its main properties, and in turn, this fact produced the property of easy experiment repetition. The description of phenomena in their full complexity, the equations that deal with small quantities, such as infinitesimal changes, at maximum, with constant changes (e.g., linear heating) took, however, time to develop.

An example of such a simple model is the linear spring where the increase of its tension is proportional to the increment whereby it is stretched. Such a spring will never snap or break. The mathematics of such linear object is particularly felicitous. As it happens, linear objects enjoy an identical, simple geometry. The simplicity of this geometry always allows a relatively easy mental image to capture the essence of a problem, with the technicality growing with the number of its parts. Basically, details are summed up until the parts became infinite in number, and it is here that the classical thermodynamic concepts of equilibration leading to the *equilibrium structures* [13,47,57] belong (i.e., crystals and other static body-like structures). In such conditions each molecule is in an equilibrium state with its own environment and with the potential and kinetic energies of neighboring molecules that are, in the overall level, mutually indifferent.

On the other hand we know the historical prejudice against non-linear problems, i.e., neither a simple nor a universal geometry usually exists. We can consider the cases of a strained spring oscillations now with a non-linear restoring force where the resonance does not happen in the same way as in the above mentioned linear case. It is possible for the tilt to become so pronounced that the amplitude is not single valued and shock jumps may occur conceivably coming to pass in wave packages. It is similar to the case of a continual inflow of energy which forces the molecules to coordinate in larger clusters. In such non-linear systems, small changes may have dramatic effects because they may be amplified repeatedly by self-reinforcing feedback being, thus, the basis of both instabilities and the sudden emergence of new forms of order that is so characteristic of *self-organization* [9,36,47,57]. Such a drastic move away from a standard configuration can lead to states displaying *spatial or temporal order*. We call these regimes *dissipative structures* to show that they can only exist in conjunction with their environment (non-equilibrium thermodynamic laws) and are granted by the continuous access to an energy source. We can reveal that such a dissipative system in its dynamic 'equilibrium' is literarily fed by *negative* entropy (i.e., producing positive entropy – plainly said, it is factually in the state that we may associate with our normal awareness of heating). If constraints (or, better, 'enforced fluctuations') are relaxed the systems return back to standard equilibrium and, consequently, the entire long-range molecular organization collapses.

During the 1960s, when *Prigogine* developed a new systematic approach to so called non-linear thermodynamics that was used to describe the curiosity of self-organization phenomena experienced outside the reaches of equilibrium. The dissipativeness of a process accounts for the irreversible dispersion of the part of energy during its transformation from one form to another, and in technical processes, it is mostly associated with decrements and losses. Dissipation of energy (due to diffusion, friction, etc.) is thus associated with wastes. However, a new kind of equilibrating disequilibrium can be assumed as a competition between energy and entropy where, at low temperatures, the energy is likely to yield ordered structures with low component entropy, while at high temperatures the entropy becomes more decisive. Mutual molecular motions turn out to be more important producing chaos and structural disorder.

Non-equilibrium thermodynamics of a completely open system also enables the formation of configuration patterns in a similar way to classical 'thermostatics' (stable crystals), but it is of a quite different 'dynamic' nature (turbulence, vortices), showing that the *dissipation can become a source of order*. Microscopic characteristics of the equilibrium distribution lie in the order of molecular distances (10^{-10} m), while in the dynamic case of the energetic super-molecular its ordering reaches clusters of 10^{-4} up to 10^{-2} m. In short we can say that the classical (equilibrium) thermodynamics was the first reaction of physics to the complexity of nature's pretentiousness towards the atrophy of dissipated energy and its conclusive forgetting of the initial conditions (in its sense of diminishing original structures).

Ordered disequilibria are customarily exemplified by the well-known *Bernard-Raleigh instabilities*, a very striking ordered pattern of honeycomb (hexagonal) cells [9] appearing under certain conditions of heating a thin layer of a liquid, in which hot liquid rises through the center of the cells, while cooler liquid descends to the bottom along the cell walls. It occurs only far from the equilibrium state (originally represented by the uniform temperature throughout the liquid) and emerges at the moment when a critical point of instability is achieved. A constant flow of energy and matter through the system is necessary condition for this self-organization to occur. It may be observed even at a macroscopic scale in the atmosphere when the ground becomes warmer than air. Physically, this non-linear pattern results from mutually co-coordinated processes because molecules are not in random motion but are interlinked through multiple feedback loops thus acting in larger cooperative aggregates, mathematically described in terms of non-linear equations. Farther away from equilibrium, the fluxes become stronger, entropy production increases, and the system no longer tends towards thermostatic equilibrium.

Upon involving two or even more variables in connection with spatially inhomogeneous systems, the higher-order non-linearity gives rise to more complex phenomena. The best examples are systems controlled by simultaneous chemical reactions and mass diffusion. From the mathematical point of view, the system becomes localized at the thermodynamic branch and the initially stable solution of the appropriate balance equation *bifurcates* and new stable solutions suddenly appear often overlapping. One such a possibility is time-symmetry breaking, associated with the merging of time-periodic solutions known as limit

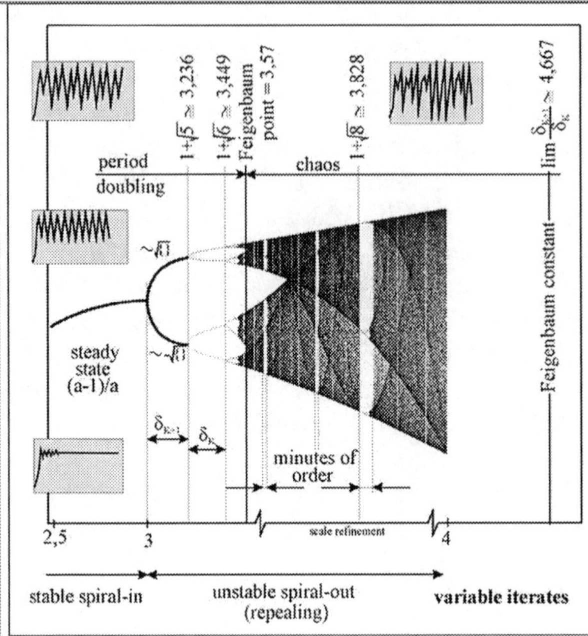

Fig. 9. – Final-state bifurcation chart (or the so called *Feingenbaum diagram*) that is one of the most important representation of chaos theory. Through the studies of the quadratic iterator, x_{n+1} = a x_n (1 − x_n) (for n gradually amounting to 1, … 3, …i), both antagonistic states of order and chaos can be ruled by a single law regarding the effect of multiplying constant, a. The essential structure subsists two branches bifurcating, and out of these branches another two branches bifurcating again, and so on, providing the period-doubling regime. The peri-ods are initially even (2, 4, 8, 16,…) followed by disordered cycles of chaos exhibiting with its increasing complexity the windows of stable periods of order characterized by odd integers. It is continued by duplicated even period (6, 12, 24, …), and so on, with incessant and infinitesimal depth. Factually, shadowed band of chaos is regularly interrupted by white window of order where the final state again collapses to only few points, corresponding to the attractive periodic orbits. Points seem to condensate at certain lines which border bands that encapsulate the chaotic dynamics ending at a = 4 with a single band spanning the whole unit interval. There are an infinite number of such windows, the one between 3.828 < $\underline{a}$ < 3.857 being the most prominent one characterized by 3 (and followed by 5, 7, 9, etc., the period-9 window already hard to find), which can be found in the reversed order (i.e., 3 right of 5, etc.). We can easily discover its inherent self-similarity as we can see smaller and smaller copies of the whole bifurcation diagram finding the complete scenario of period-doubling, chaos, order and band splitting again, however, on a much smaller scale. Literally, bifurcation moves from the stable state to the period-doubling regime where the length of bifurcating branches became relatively shorter and shorter following certain geometric law (i.e., self-similarity of the quadratic iterator, where the vertical values in the first and third magnification are reversed to reflect the inversion of previous diagram). The parameter, a, beyond which the branches of the tree could never grow marks the period-doubling regime. This threshold became known as the Feigenbaum point (the limit of rectangles sequences is reaching the value of 3.5699456..), which factually splits the bifurcation diagram into two distinct parts of (period-doubling) order and utter chaos. There is a rule that quantifies the way the period-doubling tree approaches the Feigenbaum point. Measuring the length of the two succeeding branches in the a-direction the ration turns out to be approximately 4.6692. This number is called Feigenbaum constant and its appearance in many different systems was called universality (to have the same fundamental importance of numbers like π or $\sqrt{2}$).

cycles whose period and amplitude are stable and independent of the initial conditions, cf. Fig. 9.

Their importance lies in the fact that they can constitute models of rhythmic phenomena commonly observed in nature, particularly assuming biological or chemical clocks materialized in picturesque world of various bayaderes and strips endowed with various sea shells. The way to gradual increase of complexity by a mechanism of successive transitions became opened leading either to the loss of stability of a primary branch and the subsequent evolution to a secondary solution displaying asymmetry in space similar to stable rotating waves, as observed in the classical *Belousov-Zhabotinsky* reactions. Such transitions are sometimes accompanied by some remarkable trends, e.g., certain classes of reaction-diffusion systems under zero-flux boundary conditions, may exhibit no net entropy production change when the system switches from the thermodynamic branch to a dissipative structure. It well reflects elementary consequences of universal geometry.

Let us reiterate case of a gross organization where we perceive a set of discs, the largest the main cardioids, one abutting upon the next and rapidly diminishing their radii. How rapidly they diminish their size we can derive from the fact that each one is x-times smaller that its predecessor with the parameter ratio attaining in its limit the *Faigenbaum* universality constant of 4.6692... and thus characterizing the previously mentioned universality at the transition to chaos, see Fig. 9.

Take for example the well known case of resistors carrying large electrical current, which can exhibit negative differential resistance, i.e., currents that even decreases with increasing voltage supporting oscillations, rather than steady currents. Another example are the instabilities that occur in thin wafers in certain semiconductors (GaAs). If the electrical potential across the semiconductor exceeds a critical value the steady current that is stable at lower potentials abruptly gives way to periodical changes in the current, often called *Gunn* oscillations. However, the most apparent correlation between a choice of organized structures is provided by different living organisms like butterflies, shells or even animals (skin ornamentation – zebra) and was elaborated by *Turing*, who formulated the hypothesis that such patterns are, in general, the result of reaction-diffusion processes, which are mostly applied at the early stages of the cell's growth.

Nevertheless, this self-organization does not contradict the second law of thermodynamics because the total entropy of the open system keeps increasing, but this increase is not uniform throughout disorder. In fact, such dissipative structures are the islands (fluctuations) of order in the sea (background) of disorder, maintaining (and even increasing) their order at the expanse of greater disorder of their environment.

We should remember that, particularly in the living world, order and disorder are always created simultaneously at the moment when the transfer of heat begins to play a crucial role. It does not violate *Boltzmann's* early supposition that *"entropy can be understood as a representation of increasing molecular chaos and consequent disappearance of initial non-symmetricality"*.

2.4. Stimulation responses and generality of power laws

Most of our cultural experience lies in the linearly perceived world. We are taught from early childhood to apprehend things, objects and their surroundings geometrically and to foresee them in the basic visible forms of lines (and aggravated blocks) and the associated circles (and spheres). The question arises whether such an austerely 'linear' education is satisfactorily intrinsic to our naturally sensed education (through the process of evolution, which materialized in the development of general responses of our organisms). We may inquire if some other functions (obviously exhibiting a damping character) are even more natural (particularly as organisms avoid the self-destruction processes due to incessant surfeit). Furthermore, we may benefit from sensing the loveliness provided by the structure of nature available through various ways of sensation. For example, most of us have experienced eye vision (and its brain perception) of a forest. From a distance the forest may be seen as a two-dimensional object, getting closer its three-dimensional imprint becomes increasingly obvious, but on the actual entering the woods their complex structure can be felt as a beautiful mixture of bizarre patterns and forms, which are not part of our standard cognition of linearity. We may experience the same feeling when zooming into the morphological picture of a microscopically observed structure. Its joined curiosity has become the driving force for the recent development of alternative portraying of geometry and commensurate mathematics in terms of *fractals* [9,59].

It is well known that every physiological property of sensation is related to some physical property of the corresponding stimulus. *Sensation* is the process by which our sensory systems gather information about the environment. *Perception* is the selection process of organization and interpretation of sensations. *Stimulus* is a form of energy that can excite the nervous system and the transformation of external stimulus into neural impulses is called *transduction*. The *absolute threshold* is the weakest stimulus that can be detected in a modality and the *relative threshold* is the smallest change in intensity that can be detected. In psycho-physical terms, intensity is related to the amount of stimulus energy falling on the sensory surface. If the original physical intensity, I, is low, even a very small change, Δ, is noticeable; however, if it is high, a relatively larger change is required. This principle is represented in *Weber's* Law where $\Delta I/I$ = constant. A more general principle relating the sensor intensity (response), S, to the stimulus dimension (measured in distance from the threshold), I, was introduced by *Fechner* and is presently known as *Weber-Fechner* logarithmic law, $S = const\ log\ I$. It shows the importance of a *logarithmic function* as the operation introduced by nature for the biological necessity of gearing up the sensation change to proceed more slowly, having thus a certain self-protection role against damage due to sensation overload.

An even more general psycho-physical law is the power law [60], named after *Stevans,* which states that $S = \text{const } I^N$, where N is the modality-dependent exponent (either power or root) and I is again the above-mentioned intensity of physical stimulus. The exponent N was estimated to equal 0.5 for brightness, 0.6 for loudness, 0.9 for vibration, 1.4 for heaviness and 1.6 for temperature. Some special cases of *Stevans* Law are worth noting: if the exponent N is less than unity, it complies with the above case of compressed stimulation, and when N = 1, it provides the case of the so-called 'length exception' to the *Weber-Fechner* Law, where every change in I produces an isomorphic change in S. *Stevanson's* Law, however, is applicable to all modalities, and all qualities within a modality, which takes place under any circumstances, and which is suitable to describe so-called 'operating characteristics'. This term was appropriated from the spheres of engineering, as *any* sensory transducer behaves similarly, which is true even for expanded stimulations, where N > 1 (as in the 'pain' exception).

There are many other experimental scales related to the damping role of our sensation such as *Pogson's Law* to evaluate the magnitude of stars, *Banfort's scale* of wind-strength or *Mercalli-Cancani* magnitude of earthquake. Power law has also instituted its value in such diverse areas as are models for the evaluation of syntactic communication [61] where it become an agreeable ingredient to facilitate most theories of language evolution. Its particular form is known as *Zipf's* law, which states that the frequency of a word is a power function of its rank [62]. Another sphere of operation is the partitioning course of commonalty property [58,63]. There is also a close allometric relationship between energy flow rates and organism size as introduced by *Peters* [58].

Likewise we can derive an optimal strategy for the investigator to formulate the question, and the information is thus a measure of how successful this strategy is. Namely, the information can be seen as the average number of questions asked until the right knowledge box is discovered. Clearly, as the box size, s, decreases the information must increase and its plot versus the logarithmic inverse scale reveal the traditional power law, $I_s \approx I_o + D_I \log_2 (1/s)$, where I_o is a constant and the fractal D_I characterizes the information growth. Factually it can be seen as the additional amount of information obtained when doubling the resolution and is often called the information dimension ($1.21 < D_I < 1.24$).

Important is the strength of the acoustic signal and particularly the music harmony, which was revolutionized by turning from the original *natural tuning* (traditional since *Pythagoreans,* using the ratios of chords length from 1, 9/8, 5/4, 4/3, 9/2, 5/3, 15/8 to 2) to the equidistant interval division called *tempered tuning* [64, 65] introduced by *Werckmeister* in the year 1691 and practically employed by *Bach* in his famous "48" piano composition in 1722. It reads for the frequency, $log\,f = log\,f_o + m(1/12)\,log\,2$, and the staves of sheet music are thus separated in the logarithmic scale.

It is worth mentioning that for getting the same increase in loudness as we get, e.g., between one singer and two, we have to double the size of the choir, similarly in sensing weight or light or temperature. Since any real tone is accompanied by overtones we can increase its loudness by strengthening either the fundamental or the overtones. In an organ, for example, which is pretty loud anyway, there is not much point in trying to make it even louder by bringing in more pipes to play exactly the same notes. But if, instead, we can bring in more pipes to reinforce the harmonics, we are putting our effort where it produces most effect. It is also possible to calculate the degree of overlap from any pair of notes, and to predict how much dissonance they should generate when sounded together. Clearly the traditional musical intervals ratios stand out from those around them as being particularly free of dissonance.

These insights were largely developed by *Helmholtz* and published in his book "On the Sensation of Tones" where he pointed out that, when two notes sound together, there will be many difference tones at the frequencies separating the various harmonics. These will only be heard faintly, but unless the fundamental frequencies are in simple ratios again, they will be dissonant with the primary tones. It leads to the theory of chords and understanding of the role of the fundamental bass, which gives a certain paradox about *Helmoholtz's* place in the history as seen by musician. It had further consequences in physics, for example, we can see the emitted light produced upon heating chemicals elements to be arranged in a particular set of frequencies, rather like a musical scale. Electron waves trapped inside an atom can be recognized as standing waves. So the reason why *Bohr's* orbits existed was essentially the same as the reason the different modes of oscillation exist inside a musical instrument and the radiation of light from an atom must be very like the sounding of a musical note. *Heisenberg* pointed out that a musical note should never be considered in isolation from the ear that hears it, which involves a fundamental uncertainty because of the ear's critical bandwidth; so also an electron should always be thought of in relation to the experiments which measure it, and this will involve a fundamental indeterminacy – famous principle of uncertainty.

The associated logarithmic function touches another inquisitive sphere – psychology – where an almost anonymous Czech priest *Šimerka* [66] in the year 1882 made an experimental record and analysis of confrontation and quarrels based on his confession practice. He proposed that the *strength of conviction, P,* and *number of reasons, D,* are logarithmically proportional as $P = log\,(1 + D)$ showing the table of D versus P pairs: $0 - 0$, $1 - 0.7$, $2 - 1.1$, $3 - 1.4$, $10 - 2.4$, $20 - 3.0$ and $50 - 3.9$. It follows that the most confident belief is for just one motive regardless if it is certainty or insinuation, defamation, journalese, publicity or politics while the cause multiplication does not improve the trustworthiness to a great extent – only three times for as many as twenty reasons. Therefore the most important sphere became information [9,67] which will be dealt with separately in further parts of this book.

The above discussed and naturally originated power laws are very important for the approval of validity of the laws in physics, either derived theoretically or developed on the basis of experiments. All resulting relations stand for the functional dependence f [68], between the i-quantities, x, characterizing the given phenomenon and numerical values, which depends on the selection of the j-units, k, independent to the character of law. For example, it reads that $x = f(x_1, x_2, x_3, ...,x_i)$ so that $x_1^{k1} x_2^{k2} x_3^{k3} ...x_i^{kj} = 1$, which is called the dimension analysis. It can be illustrated on the transport phenomena relating two alike processes, 1 and 2, of heat transfer applied for temperatures $T_2 = k_T T_1$, with generalized time, k_t , distance, k_x $(=x)$ and heat diffusivity, k_a. It follows that $k_a k_t/k_x^2 = 1$, which makes it directly available in the form of two dimensionless numbers, which were earlier introduced by *Fourier* and *Biot* for the heat and mass transport as $F_o = k_a t/x^2$ and $Bi = \alpha x/ \lambda$ (alternatively as $F_{om} = Dt/x^2$ and $Bi_m = \alpha_m x/D$, where D is the diffusion and α is thermal expansion coefficients). Another renowned number of that kind was devised by *Rayleigh* in the form, $R \cong \alpha g \Delta T/ (k_a v)$ [69] where symbols g, ΔT, k_a and v are respectively gravitation acceleration, thermal gradient, thermal diffusivity and kinematic viscosity. Certainly, there are many other important dimensionless numbers, among others let us mention *Reynolds*, $Re = vx/v$ or *Prandtl*, $Pr = v/ k_a$, both useful in the theory of motion of fluids.

2.5. Thermal science and energy resources

We cannot exclude a brief examination of all possible types and impacts of energy [70–76] in concluding these two introductory chapters for a book dealing with heat, which we use to employ to live and to assure us of a desirably comfortable life. It is fair to say that one of the most embarrassing questions to ponder is how much we exploit energy and what is the availability and sustainability of its reservoirs. It is well-known that energy resources can be divided into two broad categories: *energy capital sources*, i.e., those sources of energy which, once used, cannot be replaced on any time scale less than millions of years and *income energy sources*, which are more and less continuously refreshed by nature (or even by man assisting nature). For large scales we often utilize a more easily imaginable unit, coined as Q, which equals to 3×10^{14} kWh [74] (roughly representing the mount of energy required to bring the US Lake Michigan to boil – about 5 000 km^3 of water).

Over the past two millennia, the total world energy consumption has been approximately 22 Q, corresponding to an average annual use of about one hundredth of Q. During the period of the so-called industrial revolution (the past one-and-half century) some 13 Q were consumed. The total world rate of energy consumption has drastically changed, from $Q = 0.01$ in the year 1850, to 0.25 (in 1950's), and to 0.45 expended recently towards the end of 20th Century. About 20% of this value was consumed by the US itself. The important question is how efficient is our use of energy? Since the sixties the average energy-efficiency increased by a few percent to about 32 %, but most power stations are machine-based and use heat engines and the efficiencies are limited to no more than the maximum limit given by the ideal *Carnot* thermodynamic cycle, which implies a realistic upper bound limit of about 40% [76].

Oil (petroleum) is thus the most widely used energy source with the currently estimated world reserves of about 15 Q (with about another 10 Q available from additional sources such as tar sands and oceans shelves. Here, however, one has to be cautious with any estimates, which always depend on many factors, i.e., how well do we know the geology of our planet, where the source can be found elsewhere, and if a source could get naturally refilled (recent theories of methane deep-earth diffusion and sea-base sediments). Coal is our most plentiful fossil fuel energy source, having Q of about 35 and natural gas amount is approximated to about 20 Q (though it is not clear yet what is the origin of natural gas and whether it can be restocked by the outward emergence of methane that is still plentiful in the deep-rooted zones of the Earth). Roughly speaking, we have altogether some 80 Q of effectively extractable fossil fuels, promisingly one hundred.

The most important question arises: how many years will these fossil reserves last? If we double, or even triple, the above guessed numbers it would still provide, more or less, a pragmatic estimate in survival years until the reserves would be actually used up – not too long at all. It goes side by side with another question: what about the other conventionally exhaustible sources? A reasonably conservative estimate for the reserves for nuclear fission is roughly 30 Q if a light-water reactor technology (LWR) is employed and perhaps thirty times more for a scenario involving the use of breeder reactor technology. Use of geothermal energy is more questionable as the present extraction rate is very low, close to a mere 0.0005 Q. An estimate for the total recoverable energy from geothermal sources is, however, as high as about 60 Q while the time to exhaustion of the estimated geothermal energy reserves may vary within the range from several thousands to mere hundreds years. The world population has explosively increased from one to five billion and we can ask what about our future. Five potential scenarios can be foreseen [9,74]:

A) Turn out to be modest and reduce the world consumption to the limit of the year 1850 where the total Q was about 0.05 for five billion people (to range up to 0.1 Q for double population) – it would have keep going on for almost another 20.000 years.

B) Halt the energy consumption at the current level and for the present (unchanging) five billion people, i.e., at 0.45 Q level. Current technical measures could assure prolongation of our lifestyle for an extra 2 years or so.
C) Maintain the present population of five billion people but let the rest of the world achieve the living level of the average US citizen. In order to match the US current consumption per capita, the world annual energy use would increase to 1.7 Q and would allow our further persistence for only 500 years.
D) Halt the energy consumption at the current level (as in C) but allow the population to increase to 10 billion, yielding the energy use to double, i.e., Q = 3.4, which, however, shrinks the continued existence yet to half.
E) Assuming 10 billion (as in D) but letting the previous energy consumption grow by another 50% (to the tremendous value of 5.1 Q, which is 10 times as in A or 3 times as in C) and which implies survival of civilization for a mere one hundred or, optimistically, two hundred years.

Discussing sincerely any reliable energy conservation program we have to see the consequences of extremes. Any crucial change of our lifestyle, dictated by the first degree of energy conservation economy (A), would fundamentally affect the citizens of industrial nations but would be more courteous to the underdeveloped countries. Therefore, it would be impossible to bargain its anticipation because the history records reveal that any abrupt wealth changes are accompanied by major political upheavals. Perhaps we could label such a drastic economical process with suddenly enforced energy crisis, Earth disastrous catastrophes or as a result of the radical upgrading of the efficiency of energy production or due to some new yet not foreseeable discoveries. It would not be easy at all, as any courses of action to decrease energy requirements and to search for new energy production technologies that would manage to secure even scenario B have been mostly unsuccessful.

Not considering fusion, as a hopefully soon to be available optional and plentiful source, we have to account for renewable sources [9,74,78] from hydropower (<0.05; dams, tidal), biological stock (<0.015; wood, specialized plants, farm waste materials), direct temperature differences (<0.07 wind, waves) and solar-thermal-electrics (<0.07; i.e., the provision of heat for the direct heating and for variety of techniques that convert sunlight to electrical energy, however, difficult to estimate more precisely), altogether approximately 0.2 Q; prospectively 0.3 Q in yet outlying future. We should note that the use of hydropower energy is almost saturated and its long-range estimated doubling cannot be attained without an advanced employment of see drifts. On the other hand, the practically exploitable wind energy [79], which amounts to about 10^3 GW (out of the totally presented 10^6 GW), makes available further escalation of variously constructed windmills (mechanical efficiency of propeller is close to 45% being proportional to the second power of the length of blades, the angular velocity of their tops cannot, however, exceed the speed of sound). Nonetheless, such a projected spreading out may be restricted by some inevitable side-effects most severe is the offensive sound, which can seriously disrupt nature. Inherent noise can enforce animals to vacate nearby areas and even cause fish to leave shallow seas. Low amplitude sound, often difficult to perceive, can become a source of health problems thus the windmills must be situated in isolated areas and their location should not disturb the landscape, too. Unfortunately, the wind velocity is also not constant in magnitude or direction, even from the top to bottom of large rotors, which imposes cyclic loads and increases noise pollution. All this briefly illustrates how many often unseen or even neglected problems can be involved.

Burning natural biological material, the most conventional source of energy, which is best experienced by mankind, is now capable of yielding only hundredths of Q per year. We should also note that planting new trees and other vegetation to replace those we burn (often so hastily) is a true income energy source, moreover neutral with respect to the environmental balancing of the content of carbon dioxide. We, however, have to be careful to consider how "green" the so called green technologies are in reality because even biomass conversion may show an inefficient profit if all provisions are made for heat, work and chemicals consumed during its man-made production so that the true final yield may be negligible.

Besides traditionally asking for how long our resources could last, we may pose the equally valid query: what is the future carrying capacity of renewable reserves to evenly accommodate a certain number of human populations? The answer is: for scenario A about 8 billion, severely decreasing for B close to one, and being the juncture for both C and D down to 0.3 billion. For the most energy demanding scenario E, it goes critically to accommodate tenths of billions people only if providing them tolerable life. Even if our estimates are 100 % misguided they still give sufficiently alarming data to seriously think about them!

Within our deeper exploratory principles of thermodynamics [76-78] we could, certainly, manage to exploit even very small efficiencies and small thermal gradients if we have enough time to carry out the process and provide sufficient technical means to manage its sophistication. For example, if an electricity generating system would be driven by the temperature difference between the surface and the depth temperatures of the ocean, the maximum efficiency obtainable is about 3.5 %, which would reduce down to an operating efficiency of mere 2 %. If approximately 1700 Q of the solar energy, which falls upon the ocean annually, were utilized by this OTEC (Ocean Energy Thermal Energy Conversion) it would provide tremendous 35 Q; even a portion of this amount would be a nice prospect for a very high-tech future hungry for energy. Certainly it does not account for possible ecological disasters, which may result in assumingly robust alteration of world weather (including changes of sweeping streams and deep-sea whirls). Similar exploratory problems we may face when approaching a prospective exploitation of massive deposits of hydrocarbons laid on the bottom of deep oceans as it may suddenly produce an easy-initiated and thus uncontrolled escape of massive amounts of methane, which is a very effective 'green-house gas' crucially effecting Earths' climate.

Even nuclear fission is not limitless. The scenario involving the use of breeder reactor technology, which may last not more than several hundred years only, is almost nothing in the distant-time-view of future civilizations. Dependent on the capital source of energy, which will eventually become exhausted, the world inhabitants will need to undergo a drastic downward shift in the human population. It will happen due to the uncompromising revolution in the energy lifestyle, unless, of course, a radically new system of energy conversion will be revealed in order to span our traditional business governed by a low-efficiency thermal engine.

Over a normal year, however, the Earth receives over five thousand Q, and about half becomes available for such purposes as powering photosynthesis, warming the earth surface and also providing available energy for mankind – this is about ten thousand times the current energy requirements of the human race. We can imagine that covering just one percent of desert would satisfy the present world energy thirst. If the surface of US highways were made from solar cells it would exceed four times the US need of energy. All is always only 'if', because we must accustom new materials and further 'spacious' requirements, accept that the sun does not shine 24 hours a day, acknowledge that the light density varies, highways surface deteriorate, dirt must be cleaned up and sweeper labor employed, cells would fade and corrode, electricity must be circulated far away, etc.

Existing conversion units [72,77,78] and energy production units are high-priced requiring grand input energy to manufacture and have a long lifetime to pay it back in the form of after-created energy. Typical efficiency of commercially fabricated solar cells do not exceeds more than 25 % (even those inspected on the laboratory scale where they stay close to theoretical values). We can expect progress on amorphous ($\eta=13$ %) and crystalline silicon (24%), $CuInSe_2$ (18 %) and composed (multilayer) semiconductors (GaInP/GaAs/Ge at 32 %), the latter, however, signifying a danger due to the poisonous nature of arsenic in its consequent disposal. On the other hand, silicon would have no dangerous wastes and is about 10^4 more plentiful on the earth than arsenic, gallium or indium but still needs a lot of energy for its manufacturing. If solar energy is to be more widely used, it must become economically competitive – the consuming price is still above 5 US cents per kWh, at best, while conventional electrical power plants can still deliver it ten times cheaper. For real progress, photovoltaic energy production needs a strong economic impulse (earnest background and bounteous financial support) as well as a suitable profit like that which is still thriving in the coal or oil business.

So far the most successful method for generating electricity directly from incoming energy from the Sun is the solar-thermal-electric conversion (STEC) system – traditionally we heat water contained in reservoirs (pipes) with darkened surfaces. We can also let sunshine fall on a large number of mirrors, which reflect and focus rays onto a central receiver (equipped with a heat engine) or allow natural heating of a large, duly covered area furnished with the central (solar) chimney operational for a hot-air driven turbine. In a very dry and cloudless atmosphere and at sea level, the maximum power density falling on the Earth's surface is a mere 1.07 $[kW/m^2]$. When the Sun is 70° from the zenith, the power density measured by a flat collector is about 0.25 $[kW/m^2]$, which can be improved for the central receiver by adjustable mirrors positioned against the sunlight. A fluid flowing through a thermal receiver is, however, treated classically by a thermodynamic cycle with the associated engine's efficiencies. Therefore, the overall efficiency is difficult to estimate but is unfortunately always low.

Climate is among the most important variables; at best, we can assume that about 15% is converted into electrical energy assuming the storage efficiency in neighborhood of 80%. The annual electric generation would not thus exceed 4.65 kWh/m^2, so that in order to satisfy the world consumption of 0.85 Q (scenario C) some 466.000 square kilometers of land would require its covering by mirror collectors (or solar cells), which would occupy about 0.3 % of the earth surface area (e.g. 1000 km^2 is about 0.644 % and 1 % equals approximately 1554 km^2). This is technologically next to impossible nowadays, not only from the viewpoint of building costs and construction complications but also from the standpoint of intricate maintenance of collectors in distant areas deserts, ownership) as well as the other isolation-related intricacy.

Even the yet imaginary but smart modes of solar cells placed in the 'climatically undisturbed' space, either simply orbiting the Earth or even placed in the 'Lagrange location' where the gravities of the Earth and Sun are balanced (obviating the need for maintaining the position outside the shadow of the Earth) are not yet without serious doubts. The main problem is the transfer of enormous amounts of energy back to the Earth because there we have no experience of what the supposed wireless connection, by way of the huge microwave radiation beam, could cause to the Earth's aerial weather (both in the atmosphere and magnetosphere). Other yet unknown aspects of unidentified leftover liquidation from detached sources of novel origin may always bring new surprises to our ecosystem besides the immediate need of very accurate beam localization on the relatively sizeable Earth receivers (even a negligible beam misfit would destroy everything on the way and become even a possible source of terrorist threat when misused). Certainly, the other question is to use all types of alternative sources directly at the spacecrafts where solar cells are so far irreplaceable. Among others, the astronauts can easily improve their electrical economy by using oxide-based superconductors naturally operating outside the rocket at the ambient temperature (of about 72 K). The temperature difference between the illuminated and shadow surfaces of the spacecraft is another point of a straightforward exploitation even if the application of some traditional machines may be found surprising (e.g., a rather effective *Stirling* gas-operated engine [9] capable to handle minute gradients).

For a better illustration of the current state let us add some approximate data for electricity producers trying to include certain 'ecosystem' feedbacks, i.e., typical capacity of a power plant against the area exploited: (i) Power plant using biomass – output 1 MW at the mean area exploitation of 3 million m^2 per one MW, (ii) solar photovoltaic power plant – output 1 MW with about 50 thousand of m^2 needed for 1 MW, (ii) classical heat power plant fed by coal – output approximately 500 MW with 500 m^2 compulsory area for one MW and (iii) nuclear power plant – output of about 2 GW at the $80 m^2/MW$.

Certainly, there are other classification measures, e.g., production of wastes (exhaust gases – CO_2, solid wastes such as gypsum, ash – sometimes radioactive) and their elimination as well as the electricity production cost, manufacturing lifetime, cost of the plant disposal, etc. However, the most common littering productions of civilization are residual wastes pointlessly distributed to the atmosphere and even to the outer space. It may be any type of radiation beginning from heat, over light to communication, which is send to the environment by means of natural attributes of all man-made activities (heat engines, light bulbs, transmitters). This unsheltered activity is not only wasting of energy but also the amplification of polluting framework [81–84].

It should be noted, however, that the most important biophysical cycle is the production and consumption of pollutant carbon dioxide. In the plant growth phase, CO_2 is consumed and oxygen released while the reverse process takes place during combustion. Nature has invented a super-sophisticated chemical cycle to safeguard the CO_2 content, while humankind has developed its counterpart, somehow unaware to infringement to the global steady state. For example, polyp's vigorous utilization of dissolved CO_2 in sea water produces the enormous quantity of corals, thus conserving huge amount of CO_2 in solid state as well as plankton itself, whose growth was recently enhanced by sea-fertilization using iron-rich recuperative ('artificial ocean forests'). Existence of dimethylsulfide, a plentiful gas produced by microscopic phytoplankton), which can be found in large populations at surface layers of the middle ocean (investigated in the tempestuous sea region next to south Australia). Upon the reaction with air, dimethylsulfide oxidizes and forms particles containing sulfur and sulfates, which float upwards and act as condensation sites for clouds formation. The created cloud-cover increases the shadowing of the sea surface and thus reduces phytoplankton growth and, on the contrary, if a lack of sunshine causes an inadequate amount of phytoplankton growth, then the diminished production of dimethylsulfide decreases the cloud formation and allows the return of full sunshine to recuperate plant escalation.

Metabolism of a man-size living creature is in the order of a few hundreds of Watts. The development of our technological lifestyle necessitated vast burning of fossil fuels and human consumption rose to several kW. If the total mass of atmosphere is about 10^{15} t then about 10^9 t of CO_2 is sitting in it assuming that CO_2 neither sticks to one place nor reacts to withdraw. Though the biomass cycle oscillates within one year with an amplitude of about 10^4 t, it is assumed with confidence that almost 10^5 t of CO_2 have been inserted into the atmosphere in the past half decade (at the rate of about 0.4% per year) – i.e. the same value which is estimated for the period of 200 years since 1750. It is known that CO_2 may cultivate plant growth; on the other hand, it is transparent to the incoming solar radiation but opaque (with a high absorption cross-section) to the outgoing radiation transformed by Earth to longer waves. For an idealized Earth spherical shell and the so-called screening *Stefan-Boltzmann* number, δ, assumed at the first approximation to be linearly dependent on the CO_2 concentration we can write the relation $T_{rise} = T_{fontal} \, (\delta/\delta^*)^{1/4}$. Introducing the above data, we can assume a temperature increase of 1.8 K by the year 2005. This is not the only effect, the change of CO_2 may initiate driving of unpredicted bifurcations and we know that going across a bifurcation threshold means passing from one dynamic regime to another, possibly with surprising outcomes, sensitive even to negligible alteration. Perpetual increase of the CO_2-containing surface envelope, shielding the infrared re-radiation from the Earth as the wave transductor, is traditionally called the greenhouse effect (global warming) and sometimes might be overemphasized or misjudged considering the eccentricity and periodicity of global patterns.

In this view the most discussed domain is therefore the CO_2-free production of energy as occurs, e.g., in *nuclear power plants*. This advantage, however, is often degraded by the fear of radioactive contamination. It is known that the strict operating rules allow a failure of nuclear power plants to happen less than 10^{-6} per year, which means that out of about 500 functioning reactors only one can collide within 2,000 years – this is very good safety considering the possible dangers provided by other everyday activities. An associated problem is the safe storage of radioactive waste, which can be either: (i) Encapsulated into hard rock deposit sites (even beneficial for possible later reprocessing), (ii) recycled by breeding procedure inside the reactor or outside in a cyclotron or (iii) propelled to the sun by rockets. Despite the latter being the cheapest way, requiring only 10^{-4} of the produced energy, it would be yet very insecure because the cosmic carriers have a reliability of only 99% compared with a reactor's fail-safe rate of 99.9%). It certainly would also be undesirable to get rid of such precious and factually prospective material, which is suitable for reprocessing by future, yet unknown technologies. Moreover the burned nuclear fuel can even be subjected to a direct application in more progressive ranks of nuclear reactors upon the direct fission of their molten fluoride salts (actually such a prototype was used as an alternative power source of aircraft and spacecraft).

However, a wider application of nuclear technology is still considered intolerable by the inexpert public, which often has a prejudiced attitude to estimates of possible risk and profitability. Even in the short term, nuclear power would seem to be the only solution to overcome and even survive the energy crisis and assure sustainable protection of our environment – nonetheless, in the general scheme of global energy flow, it still represents a negligible portion of our need. We should not be fooled by various ecological reasons and motivations because they might involve unintentionally subjective views favoring certain aspects of given waste management or energy production or might have unseen backing by financial lobbies. We should be conscious that we are continuously facing hazards to health (and even life) as we manufacture more and more chemicals, machineries and technologies producing byproducts whose true environmental effect may often only be recognized after a delay. Moreover, we are also subjectively alarmed by the prospect of certain danger by propaganda warning us not to eat genetically modified food, not to accumulate radioactive waste or not to produce injurious gases. However, many will happily keep their bad habits of smoking cigarettes and drinking alcohol, all of which together clearly amounts to a very eminent health risk. We should remember that all we often want is to have our life of luxury protected, but prefer to have someone else to blame for inherent risks.

What level of radiotoxicity of ionizing radiation is assumed to chemically interact with living cells in the same way as chemically induced toxicity? Curiously, one of the most poisonous substances is pure oxygen that acts in living cells as a strong toxin (originating free radicals), whose effect is comparable to exposure by irradiation. Oxygen can thus be assumed as a 'pollutant' to whose toxicity we gradually adapted (breathing corresponds to the 30% radiation exposure, which is allowed as an annual dose per capita). Life, however, has adapted to withstand its coexistence with radiation, whose natural background is 100 times stronger than that so far introduced by man-made nuclear activity. Some organisms (crabs, scorpions) reveal insensitivity to irradiation. Interesting is the case of operative natural reactors found in the African state *Gabon*, which about 1.8 million years ago functioned over a period of several hundred thousand years, fed by microorganisms that delivered uranium for its continuous fission from neighboring shallow waters. Since that time radioactive wastes, surprisingly, did not enhance serious contamination of the surrounding soil beyond a tolerable level, giving thus a good indication for the feasibility of a long-term storage of nuclear wastes in natural deposits.

In the overall view and despite all setbacks, the above data support a wider employment of nuclear energy as the only consistent solution at present, as well as in the near-future. We are forced to wait for some novel solutions and, in the meantime, should not be disloyal to nuclear sourced provisions whose use is hindered by often very naive argumentation about the hazard of nuclear wastes, which is relatively more benign for environmental sustainability than the huge amounts of gaseous exhausts, pollutant ashes and other neutralization-based by-products triggered by classical power-plants, whose energy profit is concomitant with ecological damage.

However, the widespread concept of sustainability is not rigorous (time and again difficult to mathematically portray) and the popular idea of sustainable development often runs counter to the basic laws of thermodynamics, taking almost mythical shape. Since our artificial process of overproduction cannot be inexhaustibly socked by the biosphere's entropy draw-off-pump, sustainable development is possible only locally, and only at the expense of creating entropy dumps elsewhere. Thus, ongoing degradation of the environment is the only (though inevitable) way to compensate for our wellbeing – the manmade overflow of entropy.

It is comprehensible that the future evolution of society and culture should not be driven by philosophers or politicians and their stagnant ideas but by technological innovative changes and associated theories (preferably applied in economics and other humanities). We, however, have to be prepared that the bureaucrats, pseudo-ecologists, selfish-businessmen and other regulators will try to maintain control of the rate of change mentioning that *"there is nothing more useless than doing efficiently what shouldn't be done at all"*.

2.6. Thermal processing and manufacturing of advanced materials

An important area of interest is the study of behavioral properties of various *interfaces*, which are understood as a *continuity defect or disorder emergence* created at the boundary between two entities regardless if it is physics, chemistry or sociology [9]. It affects the extent of our responsiveness beyond which starts our uncertainty, misapprehension, or for want of a better word, *chaos* [9,48,59]. The word *'chaos'* is familiar in everyday speech, normally meaning a lack of order or predictability in mathematics or a disordered state of mind or society. A characteristic property of chaotic (highly dynamic) systems is pathological sensitivity to initial positions. The hereditary meaning of chaos is an empty space or not yet formed matter (*Theogony,* about 700 BC). It was probably derived from Greek *'chaskó'* – drifting apart, gape, opening and *'chasma'* – abyss, chasm, gap, and was associated with the primeval state of the Universe (also related to *'apeiron'*) see Fig 10.

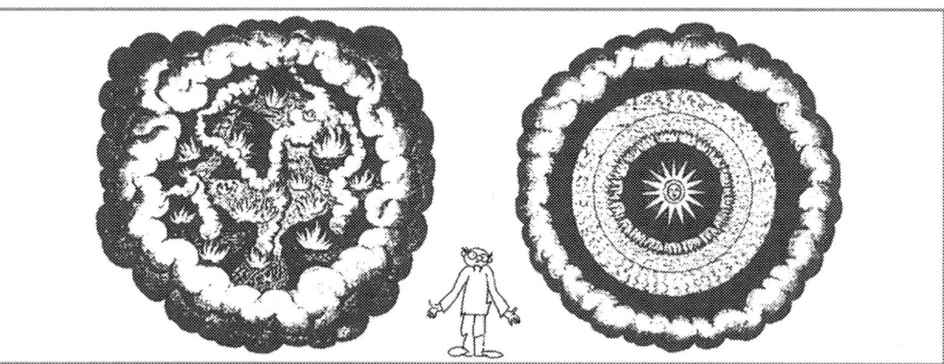

Fig. 10. – Chaos ('hyle' or 'pan') is understood as a mixture of backwash elements (left) confused by the inherent fight of heat against cold and humidity against dryness. Crucial effect of the central sun helped to disperse chaos and the four elements became organized in the concentric circles (right) around the sun in the middle, which is followed by fire (as the most powerful rudiment) and then by air, water and earth. On contrary, in the hermetic conception the central position upholds the earth with water above (as the retractile force JIN) and the outer spheres are fulfilled by air and fire (far away stars as blaze perforations in the firmament) representing the centrifugal force (JANG).

In the Chinese tradition, chaos was seen as homogeneous space that preceded the constitution of directions and/or orientation (i.e., separation of four horizons) in the sense of 'great creation'. In Egyptian cosmology, chaos (*'nun'*) represented not only the state preceding the great creation but also the present state of coexistence with the world of forms/structure, which serves also as a limitless reservoir of field forces where forms dissolve under infinitesimal time. In alchemy, chaos was linked to primeval matter (*'nigreden'*) capable of creating a 'big masterpiece'. Chaos was a symbolic representation of the internal state of alchemists who first needed to overcome their unconsciousness to become ready for transmutation. In *'Genesis'* chaos is understood as a symbol of paltriness and non-distinctiveness but also as a source of feasibility. The word chaos is sometimes taken to mean the opposite of *'kosmos'* and in that latter term has connotation of order. Arabic meaning of *'chajot'* is also closely regarded with life. For the Epicurean conception chaos was a source of a progressive transformation. Chaos thus became a definite domain of present day science showing that disorder can disclose windows of order (order through fluctuations) and that, *vice versa,* order bears inherent minutes of disorder (disorder as information noise). The theory of chaos provided the basis for various progressive specialties of numerous branches of knowledge that appear important for any further advancement of science and society.

The above mentioned boundary connecting two entities brings into play another important correlation, the interface between the traditional language of visible forms of the familiar Euclidean geometry and the new language used to describe complex forms often met in nature, previously mentioned as *fractals*. The existence of interfaces turns out also to be a focal point of modern material science assuming the two traditional limiting cases of crystalline order (long-rage arrangement) and disorder (amorphous\glassy short-range arrangement), which has recently been added by the intermediate state of mesoscopic display of nano-crystallites. In later cases the impact of interface area becomes almost comparable with that of the bulk it separates and thus may greatly affect the resulting properties. Matching surfaces, which separate two different phases must also comply with several microscopic and masoscopic requests: (i) intermediate chemical composition and activity, (ii) sooth interphase tension, (iii) harmonize structural units (network) and (iv) possibly focus on fractal selfsimilarity.

Characteristics of interfaces turn out to be important for many tailored materials such as fine-metals (subcritical magnetic domains), modulated structures (quasicrystals revealed in metallic glasses or even ice incommensurable nuclei of icosahedras and dodecahedras in liquid water), low-dimensional quantum dots and wells and many others domains concerning cryopreservation, biological pathways, therapeutic dietary supply, etc. Further it also involves the interface creation often connected with the process of new phase formation, called *nucleation* affecting rather divergent fields such as ecology (formation of smog, aerosols, dust, fog), biology (critical size of viruses), medicine (growth of cancer), pharmacology (grains` delayed dissolution), nuclear energy (wall crack formation due to irradiation) and may even converge some aspects of cosmology.

Thermodynamic studies and broader application of thermophysical methods have played an important role in the steady development of new materials and in the material headway, allowing civilization to progress [7]. Progress has simultaneously taken place on the macroscopic and microscopic scales [9], which can be exemplified by the building structure as a matter of construction fashion and urban design as well as by the building's internal equipment (standard furniture, luxury belongings, communication instruments) together with the functional and physical property of materials involved.

Until recently, scientists were doubtful whether higher complexity can produce qualitatively new behavior and whether the discovery of very complex materials has somehow broken a conceptual barrier opening active research in broader classes of multicomponent (and now even low-dimensional) systems. It touches equally the research of quantum states with their auspicious optical and scattering characteristics as well as the use of unexpected technologies like novel ultra-short-energy-pulses furnished by lasers and/or heat-convection-free experiments performed under microgravity [6,35]. Besides yet increasingly sophisticated technologies we also restore old-manufacturing practices now re-applied to modern processing, such as the calendar as rolling machine for rapid quenching [85, 86]. The goal of progress is to obtain such materials that would, on the one hand, allow the effects and the properties of materials used to date to be produced with much better parameters tomorrow and, on the other hand, grant material capability to act as ready-to-use reserves when new effects and properties are discovered and exploited in the next-generation technology.

A common drawback is the insufficient cross-disciplinary transfer of specialized skill from branches that are not directly related though they could mutually serve as inspiration. We can see that the building design, architectonic styles, applied materials, decorations, etc., have long served in the fortitude of civilization, which spontaneously included the design aspects of traditional (Euclidean) and sensational (fractal) geometry. There is also a non-negligible tendency to underestimate the importance of the search for people gifted for science and ready to undertake its missionary vocation and student's subsequent inter-disciplinary education with respect to learn both the 'feeling' of material and the 'tactility' for experimentation.

It is well known that major discoveries in materials and their engineering have driven the progress of civilization. Such a process was complex as the art of processing was passed down from generation to generation but little of the ancient knowledge survived major historical population shifts and cultural transformation. Much of the art was rediscovered in different eras and parts of the world. We, for example, can recognize with amazement the sophistication of Damascene sword steel, the maturity of 3000 years old blast furnaces discovered in Africa and the perfection and durability of Roman aqueducts, some being still in use today. Transistors, among other revolutionary changes in the second half of the 20th Century (such as the associated quantum Hall Effect, integrated circuits, optoelectronics, oxide superconductivity), upshot the growth of electronics. The transparency of silica glass improved only slowly over centuries, until studies on optical fibers were begun in earnest, and the transparency subsequently increased by orders of magnitude within just a few years. However, in the emergence and rapid ascent of new materials in the the term 'materials science' was unknown in the 1950s, and slowly accepted within universities in the 1960s becoming, however, the sign of smartness in the 1990s.

Recently it has achieved the new state of a joint approach within the scope of the so-called field of *'intelligent processing and manufacturing of materials'* [87], which tries to enfold the three necessary aspects of fruitful progress: (i) tailoring the quality functional material, (ii) constructing and producing materials under the most economical (and ecological) conditions and (iii) assuring the material's promotion, distribution and marketing as well as its enhancement for a yet wider demand (along with the attention to an acceptable impact on sustainable level of civilization). In this respect we should hark back the importance of such a global manufacturing and economic approach and remember that socialism likely collapsed because it did not allow prices to tell the economic truth. Similarly, capitalism might also disintegrate when it does not allow prices to tell the ecological truth. Novel applications of *fuzzy logic* may assist its agreeable solution.

Fig. 11. – Living objects as compositional constructions are lengthily rewarded together with the aptitude of human imagination and fractal-like design as well as the building capacity of applied materials then available. From left: Stone, fractal-like composition of the frontal portal of the gothic jewel of the St. Vitus Cathedral in Prague (Czechia) founded by *Charles IV* in the middle of fourteens century (Architect *Matyáš of Arras* replaced in 1352 by *Petr Parléř*); Cultural Centrum of J.-M. Tjibaou in Noumea (New Caledonia, 1994); Opera in Sydney (New South Wales 1973, architect *Bjorn Utzon*) and a new design proposed by *Marcus Novak* (2002, USA) typically showing liquid-like architecture ('trans-architecture') made of conjoined lamina and cross-fade linkage (P. Zellner, "Hybrid Space" Thames & Hudson, London 1999).

New products achieve success through a combination of sound technical pattern and imaginative industrial design. This amalgam creates so-called *'product character'*, which provides functionality, usability, and satisfaction of ownership. In particular, satisfaction is greatly influenced by the aesthetics, association, and perception that products carry, a combination that we can refer to as product personality. The overall character of the product is a synthesis of its loveliness, functionality, usability, personality and economy. Smart composite structures with built-in diagnostics will soon appear on the market replacing the traditional use of standard materials (metals, various polymers). Functional ceramics is expected to become significant in view of possible integration into future 'smart' systems (intelligent automata) composed of sensors (information sources), actuators (movement), batteries (power) and computers (coordination and decision).

Progressive tailoring of magnetic as well as of mechanical, ferroelectric, dielectric and other specific properties is for long the center of attention. Currently, their elaboration respects the so-called mesoscopic ordering states popularized (and terminologically often misused to attain reputation and money) under the generalized idiom of nanomaterials and nanotechnologies [88, 89]. For example, the extent of magnetic exchange interactions is effective across a given width of magnetic domain walls, and the disordered nano-crystallites of a sub-critical domain size ('*finemetals*') would thus appear as magnetically disordered in a similar way as truly non-crystalline, yet classical glasses ('*metglasses*'). Similarly it brings new dimensions to nonlinear optoelectronics where again nano-crystalline based waveguides play an important role in infrared optics. Silica glass fibers alone could cause the frequency doubling of infrared laser beams suggesting those even non-crystalline solids can have large second-order susceptibilities. Such oxide glasses can also serve as useful transparent matrices for semiconductor and metals nano-crystallites to form nano-composites (formerly known as colloids, e.g., ruby glass containing gold) with large third-order susceptibilities. Controlled uniform size distribution of such nano-particles is needed for nonlinear devices and soliton switching as well as for waveguide lasers while non-uniformity is required in the applications for optical data storage.

Order-disorder phenomena in the system with lower dimensions are another separate and well-emerging category providing new boundary problems such as nanometer range phase separation in the thin amorphous films prepared by CVD (chemical vapor deposition) as is known, e.g., for germanium. It touches non-stoichiometric semiconductors prepared via non-equilibrium MBE (molecular beam epitaxy) or MOCVD (metalorganic chemical/vapor deposition). Such a matrix can be generally understood as being a submerged disordered system of defects with nano-crystalline dimensions. When the system's characteristic dimension is comparable with the electron wavelength, the quantum electron phenomena (i.e. dimensional absence of electron resistance) become important and such derived materials are known as quantum wells, wires and/or dots. If for an appropriate thickness of a semiconductor layer the disorder of the interface is controlled by remote doping, a high mobility transistor function can thus be achieved. Quantum-sized dots can also be conventionally formed by dispersion in a suitable matrix, their optimum size to be estimated on the basis of the ratio of material permittivity over the effective mass.

Carbon spherical web structures (fullerenes), oxide multilayered complexes (clays) and organically modified silicate gels (ormosils) can serve as the host for inorganic materials and polymers serviceable from the higher electronic applications to the pollutant cleaning purposes. Submicron sized halides in composite glassy electrolytes can increase ionic conductivity, and nanometric ('pinning') centers can improve superconductivity of complex cuprates [9,90]. Nano-crystallization of porous silicon can play an important role in the better management of photoluminescence when taking into account the role of separating interfaces of silicon grains, which was recently shown to be responsible for the blue photoluminescence, which quality was ascertain with the remaining (nanosized) separating layers, as revealed by the early studies carried out on inorganic and organic silanes.

To speak about the neighboring domain of superalloys, we can cite the importance of inhibition during the sub-critical formation of nuclei upon the application of rapid changes of environmental conditions (e.g. temperature quenching [86]). Similar considerations apply to the embryo's generation in such diverse spheres as preventive medicine, which deals with the growth of viruses and commencement of diseases as well as environmental sciences, which pays attention to the formation of pollutants (or even variety of snow). Resistance of plants to the damage due to ice crystals formed upon freezing of aqueous solutions belongs to this category and is aided by the beneficial presence of low-molecular carbohydrates (sugar). Self-protection is accomplished by raising the body's liquid super-saturation to high viscosity, which is then capable to undergo vitrification. Formation of glassy (freeze-concentrated) phase may also improve the stability of frozen food and is equally important in cryopreservation of human implants in liquid nitrogen and shelf life in general. The stability can be naturally and artificially improved by the additions of 'cryoprotectors'. They avoid damage caused by inappropriate cooling as experienced at too slow rates, where cells can be killed by overly lifted salt concentration owing to intracellular water loss as a result of osmosis. At faster rates, cells can be destroyed by intracellular ice formation due to insufficient water flow-out or the undesirable phase separation of lipids and proteins in membranes (protecting their mesomorphic liquid-crystalline state capable to commence a glass transition).

As mentioned above the interfaces are a challanging sphere for various advanced exploration. It extends from nanometersized processors dispersed in a thin film, which can be spread discretionary on the suitable (often glassy) surface in the form of a skeletal layer and which network would allow their diffusely self-adjustement and the formation of mutual interlock according to the minute need of job processing and data evaluation (giving the first step towards the new generation of a self-installed computer hardware) to the biomimetic processes, called 'surface-induced mineralization', which can use self-assembled monolayers as a template for tailored nucleation. For example, this method is currently subjected to tungsten as a mechanically strong substrate material for human implants. In contrast to the traditional deposition of the affixed coating by biocompatible hydroxyapatite layers the as-cast-Ti-surface is first subjected to the acid etching (TiH-formation) and then to the NaOH treatment, followed by subsequent thermal annealing, which results in the formation of a gel layer composed of an amorphous sodium titanate, which behavior is similar to bio-apatite that is the main inorganic component of living bone tissue enabling thus a strong bond between the bone tissue and the implant material without any intermediate fibrous layer [9,91]. Worth noting is the growth of pre-treated TiO_2 layers proficient to yield not only biocompatibility but also with the capability to elicit immune and hypersensitive reactions, interactions with biological pathways, processing and sterilization on biodegradation or, most curiously, even self-sterilization effects (anti-bacteriologic action of TiO_2-coated surfaces, often deposited by sol-gel method, maintained under permanent ultraviolet light).

Adjustment of chemical composition and associated morphology requirements together with the application of suitable tailoring to achieve matching structure of appropriate fractality of contact interfaces, is still only halfway to the so called "tissue engineering" though the micro-additives of various organic molecules enabling easier mineralization (such as proteins, glycoproteins and polysaccharides). The procedures described above [92, 93] are still following the natural self-reparation of our organism, such as an implant-aided bone fracture, injured skin or missing teeth undergo repair but, even when this does occur, this usually involves nonspecific reparative tissue rather than the regeneration of the specific functional tissue that has been affected. The new philosophy of tissue engineering would involve the use of technologies of molecular and cell biology, combined with those of advanced materials science and processing, in order to produce tissue regeneration in situations where evolution has determined that adult humans no longer have innate powers of regeneration. Such tissue engineering is based on the creation of new tissues *in vitro* followed by surgical placement in the body or the stimulation of normal regeneration repair *in situ* using bioartificial constructs or implants of living cells introduced in or near the area of damage.

Understanding the mechanisms underlying the mystery of the order-disorder transitions from nonliving to living offers hope for prolonging the quality of life by, e.g., helping to design better therapeutic treatments for diseases of the skeletal system or perhaps even dietary supplements, which inhibit the onset of diseases. An important step is seen in the osteogenic properties of bioactive glass-ceramics capable to release soluble (hydrated) silicon, which activates bone cells to produce growth factors. Formation of ordered proteins (DNA-like structures) on inorganic bioactive substrates (clays) could not be solved on mere entropy concepts as it may involve a mechanistic solution based on the order-disorder paradox of life. Namely, hydrated (three-member) silica rings can be easily formed on the activated silica surfaces (fractured or otherwise bio-stimulated) creating thus penta-coordinated silica atoms in a metastable transition state due to the aminoacid interaction with trisiloxane rings. Such a Si-OH complex can act like an inorganic enzyme by providing a favorable reaction pathway for polypeptide synthesis and thus possibly foreseeing the foundation of the first secret code on the pathway of life creation [94].

Chapter 3

3. FIRE AS A PHILOSOPHICAL AND ALCHEMICAL ARCHETYPE

3.1. Sources and effects of fire

Fire [9,41,44,95-99] was believed to have both heavenly and earthly origins: it is brought from the sky by lightning, and it lives in the underworld of volcanoes. Due to its ubiquitous nature and its association with both good and evil, fire has been (and in some places still is) worshipped by many peoples throughout civilization. Because of various psychological reasons, fire is considered to be a personified, animated or living power: it is red like human blood and warm like the human body, it shines brightly in the night and may have a form of "eternal life" or by constant rekindling can be made into a "perpetual fire". Masculine fire (principle *YANG* – light) is thought to fight from the center and to have the power to decompose what nature joined before while the feminine fire (principle *YIN* – shadow) attacks from the surface, is difficult to withhold, and often disappears as smoke. Fire was believed to extend throughout the celestial spheres and even time was thought to move in cycles (*'ekpyrosis', 'conflagratio'*) involving a period of its destruction by fire during the cycle's involution and/or end. Fire has for all intents and purposes accompanied mankind's thoughts, beliefs and doings from the very beginning until toady's serious scientific treatises including the theory of chaos applied (e.g., the heat transfer and/or distribution conditioning Earth's weather) not forgetting its mystical but regarded basis, see Fig. 12.

The generation of fire, which would be unachievable without the aid of fire bores or saws, was also sometimes perceived as a sexual act that imagines male and female firewood. Corresponding views were most probably pronounced among Aborigines and such a conceptual framework consequently influenced ideas of fire in the body of humans, especially of women, also as a center of sexual life. In archaic civilizations with sacral kings, the sacred perpetual fire (the so called state fire) of the residence and temples of the royal ancestors was believed to be a phallic symbol, and was said to be sacred for virgins, who were viewed as wives of the state fire. The extinguishing and rekindling of fire at the inauguration of a prince to kinghood points to the idea of a spirit of the princes within the state fire and also to the cyclical renewal of the state in the purifying act of fire, which signifies the beginning of a new era. According to some *Hermetic* and *Gnostic* doctrines it was thought that the soul emanated from the God, fell into the body casting its internal fire, and at death returned to its former home. Thus, it was believed that during cremation the soul is helped to separate from the body and continue its journey to the heavens by fire. Fire has duly become a mandatory attribute of almost all holy places.

Fig. 12. – Partially oriented graph (left) illustrating various quantitative (circles) and qualitative (bars with arrows) connotations associated with the time-honored concept of the four-elements. In the middle, a loop consisting of two dragons (called 'Uruboros' eating each other tails) is revealed together with the symbols of the four elements in the corners. The upper dragon has wings to symbolize 'volatility' while the lower dragon has only legs to mark 'solidity'. The dragon and fish are mutually replaceable (right), and from two fishes similarly biting each other's tail the symbol of 'YIN – YANG' is created, which has had an important role in Chinese alchemy and philosophy

Burning, as a source of fire, is invisible. It is only the product of burning, i.e., *flame,* that is visible. The physical appearance of flame exhibits a surprising similarity for different substrates (fuels) and is scientifically reasoned to be a universal portrayal of conglomerated chemical reactions resulting in energy production in the form of heat and light, cf. previous Fig. 7. Flame propagation is explained by two theories: heat conduction and heat diffusion. In heat conduction, heat flows from the flame front, the area in a flame in which combustion occurs, to the inner cone, the area containing the unburned mixture of fuel and air. When the unburned mixture is heated to its ignition temperature, it combusts in the flame front, and heat from that reaction again flows to the inner cone, thus creating a cycle of self-propagation. In diffusion, a similar cycle begins when reactive molecules produced at the flame front diffuse into the inner cone and ignite the mixture. A mixture can support a flame only above some minimum and below some maximum percentage of fuel gas. These percentages are called the lower and upper limits of inflammability. Mixtures of natural gas and air, for example, will not propagate flame if the proportion of gas is less than about 5 percent or more than about 20 %.

At the beginning of science, however, flame was proclaimed to be just an optical illusion and only a visual specter that was felt not to have any substantial purpose – it illuminates and animates its surroundings, creating the illusion of liveliness. Fire can also create a vision of it being a living organism (*'agile' sive/i.e. 'ignis'*) that exhibits growth and change, and that has a need for food and air. Fire is composed of very sophisticated internal structures of flames, and shows continual instability, self-structuring and self-reproduction. Flame is the visible pattern of fire and was treated by scientific, poetical and mystical essays.

There is a Latin proverb '*Ignis mutat res*' – 'fire changes things' (which is often exemplified by a burning candle; the more the wick is flaming the more is extinguished being buried in the melted wax thus feeding back the actual fuel supply). This saying implies that fire has the power to change the properties of matter, metals become ductile, raw food-stuffs can be transformed into a meal. Fire is the kindest servant and the fiercest master; an open log fire and its fireplace ('*focus*') is a symbol of the intimate asylum of the family home unit but, at the same time, is the source of potentially devastating danger and thus a focus of destruction. Fire is a source of expansion and contraction, annihilation and purification. The light of ideas reveals the truth, the glow of fire proofs its genuineness. Everything that flares up ends in ashes. Fire is self-destructing; its process of burning turns itself into a worthless thing. Fire is a fundamental beginning with its final effect being the entire end. Fire is often associated with chaos.

It is commonly pointed out that ancient people were familiar with four types of phenomena related to the glow that they associated with the ignition of fire. These sources were thought of as discharges of power: (i) lightning, which was earlier believed to be a sort of burning vapor, (ii) the way a torpedo fish stuns its prey (known to the early Egyptians and later recorded by Greek and Roman naturalists), (iii) St. Elmo's fire, which is the pale glow sometimes seen on the tips of pointed objects during stormy weather (again described by the ancient Romans in their military camps), and (iv) the tiny sparks associated with the curious property of attraction shown between pieces of rubbed amber and known as the amber affect (among the ancient Greeks amber became to be called '*electron*', which was the name also given to the native silver-gold alloy that had a similar color as that of pale yellow sunlight). No connection between these four phenomena was made until comparatively modern times when we recognized and separated their thermal and electrical character, giving rise to the two basic scientific fields of thermal physics and electromagnetism. *Theophrastos* was known for defining the stone called '*lyncerium*' and for having observed and discussed both the amber effect and natural magnetic behavior.

Some writers give credit to *Thales*, who is known for his proposition that all things are indeed water and that its basic nature or cause is '*arche*'. Since that time onwards the early 'natural myths' were slowly replaced by the thoughts and writings of ancient Greek philosophers [9,100-102], explanations that may be termed as being 'scientific'. Particularly in cosmology, two types of such explanations developed, one of which was theological in character and was referred to sometimes as 'organismic' because of its analogy with physical phenomena and the behavior of organisms. For example, one such hypothesis concerning amber was that rubbing developed in it a certain "longing", a need that was satisfied by the amber 'catching' its "prey" by projecting arms, as do many living things (not too far from understanding the behavior of caloricum).

Another such explanation, known by the Greeks as the respective appellations of *'symphatia'* and *'antipathia',* was that all objects, both animate and inanimate, possess states of similarity or oppositeness. Thus all objects that are mutually "sympathetic" would tend to unite or interact, whereas those "antipathetic" or naturally contrary to one another would exhibit the opposing tendency.

The second type of explanation (a physical explanation) is agreeable with the notion of fire, and may be termed as "materialistic" or "mechanistic". Its hypothesis was advanced in the Century following *Plato* by the Greek philosopher *Epicurus* (and based on *Democritus*) who regarded the universe as consisting of two parts, matter and free space. He proposed that atoms move straight downward according to a natural tendency, supposing that the atoms that are falling through empty space collide by virtue of a self-determining power. This power causes atoms to swerve a little from their vertical direction of fall. This deviation was termed *'parenclisis'* and enabled philosophers to explain the existence of objective chance and, also, of free will. This logic has had a direct impact on present-day thermal physics, in which we describe a system by a set of so called phenomenological qualities (e.g., temperature, pressure) that are not directly connected with the detailed microscopic structure of matter, but are manifested by directly measured values. Any such system exhibits small spontaneous deflections from the predicted overall state, called *fluctuations,* and are caused by the particular tectonic-configuration of matter. Under standard conditions, fluctuations play a negligible role in most systems or, ate least, their effect is averaged. Only under certain circumstances do they become perceivable (often at the vicinity of bifurcations; a novel notion in the present day science) or even play a crucial function in the "spontaneity" of a system, a self-ordering commencement of the above mentioned "free will". The concept of the rotary movement of atoms led to the formation of innumerable worlds, separated from each other by empty inter-mondial spaces called *'metacosma'.* This was followed by the modern thesis of quantum uncertainty that is, in part, derived from the same roots. It, however, was rejected by some important physicists of the time who never accepted the probabilistic spirit of quantum mechanics and, instead, adhered to the *Democritean* access of the order of necessity (*'ananke'*).

The term *"free will"* can lead to a confusing paradox because the only freedom that we have concerns the possibility to do what we wish but not necessarily the subject that we wish, best interpreted by the German philosopher *Schopenhauer* citing *"if I wish I could give away my property to the poor, but I cannot wish to wish"* later reformulated by *Barrow* [14] as *"no being can predict what he will do if he will not do what he predicts he will do"*. It is related to *Chew's* formulation of the so-called *'bootstrap'* hypothesis (1970s) for a continuous dynamic transformation taking place within itself, which is mostly related to the composition and interaction of sub-nuclear particles (the existence of each particle contributes to forces between it and other particles, leading to a bound system in which each particle helps to generate other particles). It may even be said to be similar to *Marutana's* concept of *'autopiesis'* (i.e., self-making) [103] for a distinctive organization of (mostly living) systems (sharing the same Greek roots as the word *'poetry'* – creating). Autopiesis is used to analyze phase portraits or fractals within the framework of topology, which may appear similar to the action analysis of strongly interacting hadrons within the network of high-energy collisions, however, bootstrap does not form any desirable boundary as the living systems do.

3.2 Early Greek philosophical views

All nations of the world have their own mythology [100-102]. Myths, in the sense of the Greek term *'mythos'* that means speech (tale or story), bring poetic views of the world around us although customarily considered synonymous to something widely spread and not often true. For the Greek poet and philosopher *Hesiod,* the universe was a moral order close to the idea of an impersonal force controlling the universe and regulating its processes of change. The notion of *philosophy* that is joins images of fondness with knowledge (likely introduced by *Pythagoreans* and similar to the term *mathematical,* meaning conception or theorem *'mathema'*) came probably into existence when people were no longer satisfied with such supernatural and mythical explanations [104]. It proclaimed *'some are influenced by the love of wealth while others are blindly led on by the mad fever for power and domination, but the finest type of man gives himself up to discovering the meaning and purpose of life itself. He seeks to uncover the secrets of nature. This is the man I call a philosopher for although no man is completely wise in all respects, he can love wisdom as the key to nature's secrets'.*

The Greek word philosophy was actually derived from the notion of love (*'philia'),* which marked (or better explained) the attraction of different forms of matter, and of another opposing force called strife; hate (*'neikos'*) to account for separation. Love *'philia'* together with wisdom (*'sophia'*), factually composing the word philosophy, which first appeared first in the fifth Century BC and primarily concerned itself with the problem of "The One and the Many". Simply stated it involved the attempt to explain of the infinity of things we meet in the Universe (the Many) and the early Greeks believed that the single unifying thing (the One) can be some kind of a material substance, like water, stone or fire. They were concerned with finding an unchanging principle that lay behind all changes. The stable unchanging component of the Universe, which the Greeks called *'arche'* and living (and growing) nature was associated with the notion of *'physis '* (meaning nature as a procreative power).

People gradually began to suspect that there was a logical order in the universe and that humanity had the capacity to discover it. *Milesians* (birthplace of the first cosmopolitan and "philosophical" ideas that eventually made possible the leisure-pursuit called *'schole'*) introduced the approach in which a single element, that contained its own principle of action or change, lay at the foundation of all physical reality. Its founder *Thales* was the first who tried to explain all things by the reduction to one simple principle, one *'arche'*. Such a single viewpoint is generally called *monism* and a more generalized approach, to see all things being alive, is *'hylozoism'*.

This new paradigm sees the world as an integrated whole rather than a dissociated collection of parts and it may also be called as an *ecological view*, if the term ecological (from Greek *'oikos'* – household*)* is used in a much broader sense, e.g., to see an engine not only as a functional whole, composed of parts, but also to perceive how the engine is embedded in its natural and social environment including its manufacturing (raw materials) or functioning (waste management). Today's science nearly always tries to reduce the complex world to as few principles as possible and the idea of reduction to a single principle is still alive. The physicist's search for the unified theories or, better, for theories of everything (i.e., looking for a data compression to achieve a particular interwove *'final'* theory) can serve as an illustration. On the other hand the Eastern sages have emphasized that the nature of our world cannot be reduced to a number of simple principles and any reduction inevitably leads to misinterpretation. They are aware of the complex interconnections of all aspects of nature and even of the connection of these aspects with our minds

In early times, the *Milesians* tried to explain things by the reduction to one simple principle (*'arche'*) and by viewing everything from a single point (*'monism'*). *Anaximenes* introduced the important idea that differences in quality are caused by differences in quantity, citing *"when it is dilated so as to be rarer, it becomes fire; while winds, on the other hand, are condensed air. Clouds are formed from air by felting; and this, still further condensed, becomes water. Water, condensed still more, turns to earth; and when condensed as much as it can be, to stones"*. On contrary the *Pythagorean School* was a more religious society that cultivated secrecy and speculated that power could be obtained through knowledge. They developed a theory that provides a form or limit (numbers) to the "unlimited", saying that things consist of numbers.

Numbers (*'arithmos'*) were natural in character and represented bounds (*'peras'*) and their ratios were called *'logos'*. They thought that number was a unifying principle in the Universe, so that anything that could be numbered was ultimately linked to other things with the same number. Numbers had meanings apart from their relationships with other numbers. Thus, musical harmony was linked to the motion of the heavenly bodies. The discovery that there were numbers that could not be represented by fractions precipitated a crisis so deep that these numbers had to be called irrational numbers that lay beyond the arithmetic pattern of the Universe. In medicine, the *Pythagorean* saw the principle of harmony at work (body as a musical instrument). *Philolaos* began to teach and lectured about a 'central fire' in the cosmos and located the home of the chief God *Zeus* there.

It was *Xenophanes*, a celebrated teacher in the *Pythagorean* School, who took the Gods of Greek mythology and, one-by-one, reduced them to certain 'meterological' phenomena, especially to clouds. God was, in his view, an immaterial eternal being, spherical in form, like a universe, and lots of modern believing scientists often understand God in this spirit, i.e., identifying God with something very abstract, with mathematical or physical principles of the universe.

Later *Zeno of Elea* introduced a proof of contradictions[1] using the term '*aporia*' and taught that space and time were immanent in our conceptions. It is also close to the notion '*paradox*', which is a synthesis of two Greek words '*para*' and '*doxos*' meaning beyond a belief, and in its modern `counter-intuitive` findings throw often light upon something fundamental (such as the Maxwell's demon giving the insight of intimate linking for the seemingly disparate concepts of entropy and information).

The concepts of space and time are not things as they were in themselves ('*noumena*') but rather our way of looking at things ('*phenomena*' – nowadays phenomenology). *Heraclitus* again redirected his attention to the change substituting dynamic '*pyr*' – fire for the static '*arche*' of the *Milesians*. He said that water, air and even '*apeiron*' are some substances or even material objects, but fire is the process or becoming. Fire cannot be static. It is not a "thing", it is the primary form of reality. Fire itself exhibits the tension of opposites and, indeed, depends upon it. The world is an ever-living fire citing "*this world, which is the same for all, no one of the Gods or humans has made; but was ever, is now, and ever will be an ever living fire, with measures of it kindling, and*

[1*] Zeno's customary arguing showed the famous story of Achilles who could not overtake the tortoise (having a prematurely start) because he must always reach the point that tortoise has passed so that logically the tortoise would always be ahead. The involved arguments implicate paradoxes against perpetual motion as well as against the multiplicity of length (later following the idea of a formal and skeptical rationality when bisecting a line, which swill always leave us with another segment that can itself be bisected, and so on, thus never reaching a single point, cf. Chapter 10, dealing with fractals). In the current viewpoint, however, it appears obvious that any non-denumerable infinity of points in a line is much larger than any infinity Zeno could have imagined and that the sum of infinite series of numbers, like convergent geometrical progression, is known to have a finite sum. In the past, the elucidation of Zeno's paradoxes were contributed by such personalities as Newton, Leibnitz or Cantor but, recently, its apparent illogicality obtained a new meaning within the framework of quantum mechanics where the measurement itself affects the state of quantum system under observation. By the technique of a fast checking the decaying quantum system can be slowed down and even temporarily come to rest. Within this outlook, when the Heisenberg principle of limited distinguishability is applied for a finite length interval, Δx, the change of the associated impulse, Δp, may stretch to infinity due to the impact interference of measuring probe. In that case the Heisenberg product must be averaged, which means that we move from the configuration (quantum) space to the phase (mechanic) space of stochastic language, see the paragraph 7 dealing with the concept of quantum diffusion, Chapter 6.

measures going out". Neither the Gods, nor they, nor the souls of human beings could escape final destruction citing *"all things are an exchange for fire, and fire for all things, even as wares for gold and gold for wares. Fire lives the death of air, and air lives the death of fire; water lives the death of earth, earth that of water"*.

The circular process was called *'ekpyrosis'. Heraclitus* taught that all changes in the world arise from the dynamic and cyclic interplay of opposites, and saw any pair of opposites as a unity or a whole. This unity he called *'logos'* which can be similarly applied to "awake people", those who could make themselves understood (as matter of interest – those who avoided public life of their city states *'polis'* were called strangers and were rated as second-rate citizens *'idios'*). The existence of opposites depends only on the difference in direction of motion; the principle states that The Universe is in a state of flux and that all things are, at the same time, identical and non-identical assuming that the way up and down is indistinguishable. Here we can also cite the famous aphorism *"you cannot step twice into the same river"*. The process of change is a process of opposites and diversity creates the problem of identity which is not self-evident, *"it is impossible for fire to consume its nourishment without at the same time giving back what it has consumed already. This presents a process of external exchange like that of gold for wares and wares for gold"*. Fire was traditionally a part of limitless *'apeira'*, sacred and self-referenced *'apeiron'* (indifinite, boundless) primordial beginning   non-material subsistence. Fire (*'pyr' – flamma)* delivers light (~ eyesight), that is transmitted (~ hearing) by air (*'aer'- flatus)*, reflected (~ appetite) by water (*'hydor' – fluctus)* and absorbed (~ tactility) by earth (*'ge'- moles)*.

It is part of myth that *Prometheus* stole the fire from *Zeus (*thought by *Platonians* to have actually happened to the blacksmith *Hephaestus)*. It is of interest that the word *'promethean'* is derived from the Sanskrit name for drill and can thus be understood as a personification of the act of making fire. *Oastane,s* the teacher of *Demokritus,* was aware that a natural power existed a (possibly fire in the sense of energy) that can overcome all other powers and is thus capable of creating unification but also is ready to diminish it repeatedly. It, however, was not specified until speculation of some early Greek philosophers, notably *Empedokles,* who was apparently the first to name the four *basic elements* (cf. Fig.12 and 13.) that signified the substantiality from which all subsistence (or being) were composed.

88

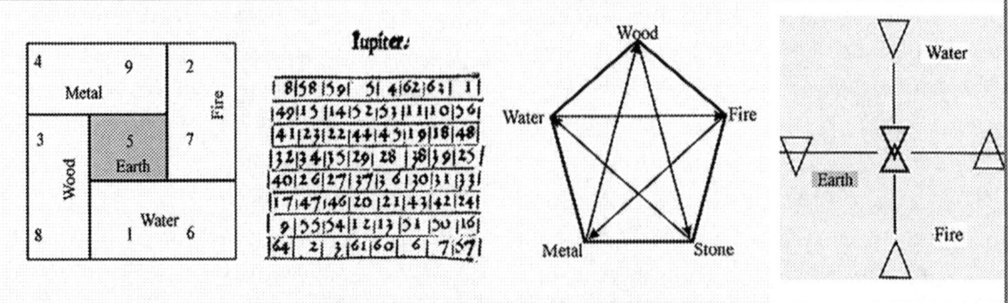

Fig. 13 – Left, the numerical symbolism, initially familiarized by the Chinese concept of figures, where the central number five represented the Earth (as the allied symbol for the great China). Though other elements bear the traditional character, the important behavior of particular number grouping was associated with the display of identical sums read for vertical, horizontal and diagonal recon. It can be extended to higher orders of numbers, and such squares became later popular in the Middle Ages as the subject of various mystical prognostic and magic periapt (middle), often specified to certain astrological terms, planets, etc. On the middle right, the early Indian concept of material elements is depicted showing clockwise (along the solid lines) the creation, which reveals along the dashed lines with arrows the destruction (symbolized, e.g., as the extinction of fire by the action of water). It is noteworthy, that this early scheme may emblematically account for the formation (water-to-wood) and annihilation (wood-to-fire) of life. Far right is the most traditional representation based on triangles where the vertical symbolizes the opposites: fire - our own self (faith) which should float up during our life, and water (our feeling, love, which should be spirited as seen from 'above'). If the spirit is treasured to achieve worship they became interpenetrated (middle overlapping triangles) and can reach attunemention (hexagon, see next Fig. 15.).

In Greek, however, the elements are termed as *'stoicheia'* (today's chemical stoichiometry) and the entire name *'elementa'* (beginning) was authentically derived from LMN the first letters of the Etruscan (Phoenic) alphabet. The *Empedokles* concept of such four patterns/roots (*'rhizómata'*) was, seventy years later, made widely known by *Aristotle* but it came together with the fifth platonian subsistence/being *'quinta essentia'* that was thought to interject a certain qualitative principle (*'arche'*). It was correspondingly conceptualized as the ether (*'aither'*) – something celestial and indestructible (derived from *'aitho'* meaning glowing, flickering) possibly related to the Aristotelian "primeval matter" (*'prote hyle'*) and interpreted as the presence of subjects. The four elements had been proposed gradually through the ideas of *Anaximenes* – air, *Xenophan* and *Parmenides* – earth and *Herakleitos* – fire, and he also emphasized that fire most completely reveals the "heavenly" reality of our universe, i.e., its order (*'kosmos'*). Sanctified fire gave a basis to the so called "Empedocles complex" where the love of fire is bound with its respect and the instinct of life with the perception of death [105,106].

Aristotle was concerned about the general forms and the cause of being (science of being as being), and discussed the notions of potentiality (*'dynamis'*) and actuality (*'entelecheia'*). He also proposed that elements determine not only the degree of warmth and moisture of a body but also their natural motion upwards and downwards according to the preponderance of air or earth. All things are in one way or another analyzable down to the basic bodies – fire tends to rise upwards and become air while water tends to fall downwards and become earth. All motions and all actions succeed in bringing into actuality what is potentially contained in the process of heat transfer such as evaporation and condensation. Mere potentiality without any actuality is the "prima materia" – existing nowhere by itself.

According to *Aristotle,* a body can only be moved if there is a mover in contact with it and if the mover communicates with the object by power involved in the movement (the first particles of air are moved first moving than other particles and finally moving the whole object). The power involved in the initial movement of a body decreases, however, in proportion to the distance moved, so that with time, the thrown body comes to a rest. This was taken for granted in almost every philosophy of nature until Newton's brilliant step to overcome this intuitive principle by the introduction of the dissipation of energy into the process of motion. *Aristotle's* consideration of the effect of heat led him to the conclusion that metals, glass and stones that melt on heating are composed of water, whereas materials that merely softened contain various amounts of earth – infusible stones are earthy. Similarly, liquids that did not solidify upon cooling were thought to contain a lot of air and those that readily solidified were supposed to compose mostly of water, and those that thicken were thought to contain more earth or more air. His written observation on "Generation and Corruption" describe flame as a burning of smoke, a claim which was retained by *Theoprastus* in his book "On Fire" but extended to recognize that fire can be generated in various ways and has three manifestations: *flame, burning charcoal* (in sense of glowing combustion) and *light.* It is interesting to remember one of *Theoprastu's* statements that if had the term moisture replaced by flammable volatiles would be an acceptable description for flaming combustion even today. For those times, *Theoprastus* gave a rather accurate account of the slaking of lime when he noted that quicklime evolved more heat when wetted that when left alone, and for the heat stored in the lime is analogous to the fuel required by lamps (note that if "heat" is replaced by "chemical energy" and "fuel" by a "liquid reactant" this statement still stands). Moreover he remarked that old quick-lime does not release as much heat as would new because time had reduced its energy through the contact with air, and that finely-grounded material evolved little heat because of its small particles (noting larger surface area in contact with the atmosphere). These remarkably acute observations not only show the quality of teaching at that time but also demonstrate that the foundation of todays's thermochemistry was laid as early as in the fourth Century BC by an implicit distinction of fire, flame and heat. The most significant contribution of *Theoprastus* was in his book "On Stones", in which he not only gave the first classification of minerals (into metals, stones and earth) but revealed a rough form of thermal analysis used as an aid in identifying stones and earth, the latter can be identified with clay minerals, relying only on the five senses, not an easy task even for today's technology. It was written *'some stones can be melted while others cannot, some are combustible while others are not and, in the very process of combustion or, rather, of exposure to fire, stones exhibit many differences ... for earth indeed may undergo melting and if, as some maintain, glass is made from vitreous earth, so too it is firing that causes this earth to become glass'*. *Theoprastus* described the burning characteristics of certain stones, possibly lignite or bituminous shale, melting of asphalt, conversion of pitchstone to perlite, and the firing stone of Siphonos (possibly steatite) that is *'soft enough to be turned on the lathe and carved, but when it is dipped in oil and fired it becomes extremely dark and hard'*.

Later *Vitruvius* produced his practical treatise "De Architectura" curiously noting the phenomenon of a thermal gradient across the material *'when the sun is keen and over-bakes the top skin it makes it seem dry, while the interior of the brick is not dried ... bricks will be more fit for the use if they are made two years before ... to dry throughout'*. From the fact that larger stones floated on mercury whereas small droplets of gold sink, *Vitruvius* made a clear enunciation of the principle of specific gravity. Clearly the practical knowledge of the effects of fire/heat had progressed more in the time between *Theoprastos* and *Vitruvius* than it had in several centuries previously, as would indeed be expected in a highly civilized Rome that paid much attention to aesthetic aspects of life and to personal comfort, which required the use, creation or manipulation of fire and/or heat.

3.3 Concept of four elements

As already mentioned, Greek philosophers played perhaps the most important role in the concept of fire trying to explain the attraction of different forms of matter introducing rather smart concept of opposing forces. *Empedocles* taught that originally all was 'The One'. All elements were held together in indistinguishable confusion by Love, while the force of Hate manifested itself as a separation of these elements. The four elements were kept in a 'sphere' of Love, while Hate surrounded the outside of the sphere. When Hate began to enter the sphere, Love was driven towards its center and the four elements were gradually separated from one another. The elements alone are everlasting, but the particular things we know are just unstable and temporary compounds of these elements. They are mortal because they have no substance of their own, their birth is a mixture and their death is their separation. He held fire as the rarest and most powerful compound of elements, which consumed the souls of all intellectuals, and which he thought was issued from a central fire, or the soul of the world.

Anaxagoras postulated a plurality of independent basic elements, which he called the seed (*'spermata'*) citing *"all things ('chremata') were together, infinite both in quantity and smallness; for the small too was unlimited. The science of all things was together, nothing was clear by reason of the smallness. For air and ether contained everything, both being unlimited. For these are the greatest items present in all things, both in quantity and in magnitude".* He thought that it was a mind, intelligence or pure reason *('nous')* that was the source of all motions as well as of the knowledge inherent within us. At the very beginning these seeds mixed without order but under the effect of a cosmic starter *'nous'* the non-arranged matter set itself into motion and began an orderly world *'cosmos'* that was created out of the initial chaos. In more recent times this idea recurs as *'deism'*, which is the belief in a God-architect, who, after creating the Universe, assumed no control over the creation. Such a deity is often used as the explanation of operations of a supernatural cause, which became popular even among some believing scientists. *Plato* seemed to distinguish between fire and heat as well as *Aristotle* apparently differentiated temperature from the quantity similar to heat even though the same word (*'thermon'*) was used for both. *Aristotle* and later philosophers paid attention to the notions of *"spirit, breath"* which by some were identified with ether and by others with fire that was always considered as a basic composition element. In Aristotle's view, any substance is a composite of form (*'morphe'*) and matter (*'hyle'*) originally meaning wood. Consequently this philosophical view was often called *'hylemorphism'* and it stated that matter without form cannot independently exist and so form cannot exist separately, which somehow rejects *Plato's* explanation of the universal Forms/Ideas as existing separately from individual things. He believed that things on Earth move because they tend to reach their natural places and argued that, although heavenly bodies have eternal motion, there cannot be an infinite series of movers and, therefore, there must be one, the big Mover – the Architect in the series, who is unmoved.

Demokritos and his teacher *Leukippos* imagined immense worlds that resulted from the endless multiplicity of moving atoms. The soul consisted of the smallest and roundest atoms, which were more subtle and globular in shape and could be identified with atoms of fire. Sensation was due to atoms from outside knocking up against the soul-atoms. The first formulation of the principle of causality can be seen in the words *"No thing comes about in vain without cause but everything for a reason 'logos' and by necessity 'anake'."* *Democritos* introduced the hypothesis of images or idols *'eidola'* as a kind of emanation from external objects that made an impression on our senses. He

Fig. 14. – Choice of symbolic epitomes related to Alchemy shows (left) the allegoric pictures associated with the early edition of books related to Hermetic philosophy and alchemy. The middle figure illustrates the secrete position of alchemy as a lady-figure sitting on the thrown with her head touching the sky (blessed wisdom), holding in the left-hand a scepter (as a symbol of power) and carrying in the right hand one open and one closed book representing the unlocked and secret sides of learning. The ladder symbolizes patience necessary to pull off all the steps to achieve self-acknowledged epiphany (attainment of the state of "Big Masterpiece"). In the right picture, the blindfolded figure illustrates the one who does not follow nature while the other, more appropriate one, is allowed to lionize the "Earth Interne – vitriol" following thus the rabbit, symbolizing 'prima materia'. The seven inherent steps are shown to go up, gradually climbing the mount, symbolizing the seven metals' tasks and aiming to the planetary Gods. After ascending the seven steps both principles (sulfur and mercury) are purified and, thus, are revealed as naked bodies, which symbolize the desired state of 'Heavenly Mercurial' (Cabala 1616).

believed in the shape and related arrangements of elementary particles, and similarly to *Pythagoreans,* he distinguished notions for *'matter'* and *'form'* linked through a process of development. In contrast, *Plato* and *Aristotle* believed that form had no separate existence but was immanent in matter; their philosophy and scientific ideas dominated Western thoughts for two thousand years until a radical change was brought about by the new discoveries in physics, astronomy and mathematics (*Copernicus, Bruno, Galilee, Descartes, Bacon or Newton)* that viewed the world as a perfect machine governed by exact mathematics.

In the Greek interpretation (Fig. 12. and Fig. 13.) all material things are a different combination of elementary *fire, air, water and earth* held together by integrative and structural essence *ether* that was a heavenly and imperishable matter (which was thought to make up the universe of fixed stars and firmament). The four elements were not only plain mixtures (quantities) but arranged as a balance of four qualities: *hotness, coldness, humidity and dryness* that defined each element using the pairs of opposites (dry/hot, dry/wet,

wet/cold and dry/cold). Hotness and coldness were active and the remaining two were submissive (secondary passive) qualities. Properties associated with dominant (active) qualities had a tendency to grow if the object is surrounded by either a hot or cold environment. It was, in fact, the first sense of a *thermal process*. Due to the enormous vastness of these relationships graphical representations became very popular (cf. Figs. 12, 13 and 14.) and later it was even believed that the formal manipulation with graphical symbols can help solve particular problems (cf. however, modern theory of graphs).

The hypothetical structure of matter based on such a scheme brings about an important consequence – the potential or intrinsic "thermal" property of all existing substances. Thus, e.g. alcohol, gunpowder and pepper are intrinsically hot substances, continuously active in this sense also with respect to other bodies, while opium and snow are examples of intrinsically cold materials. Moreover, the antagonistic nature (so called *'contraria'*) of different Elements and Qualities ensures eternal changes and movements of all things in the universe. These changes are, however, not completely free, but are submitted to the remarkable principle of *'antiperistasis'* that controls the relationship between the two active Qualities – coldness and hotness. It can be formulated as follows: *The properties of any body which is bound up with coldness (hotness) tend to increase in the case where the body is surrounded by the hot (cold) environment.*

This principle akin to the modern *Le Chatelier – Braun* principle provided, in many cases, correct qualitative predictions of the direction of thermal processes. Quoting *Oinipides of Chios* for a typical example consistent with the principle of antiperistasis: *"Water in a deep well shows in winter the smallest degree of coldness, while in very hot days is extraordinarily cold."* Interestingly, this statement is actually valid and is not only a consequence of our subjective feelings, but was confirmed by careful hydrological studies.

Besides numerous successful applications of the principle of antiperistasis, there were also cases where it completely failed. One example is the dissolution of black gun-powder containing saltpetre, which contrary to expectation does not warm up but instead cools down. Such exceptions were either neglected or, at best, provoked discussion about other weak points of the doctrine. The most important problem, crucial for the theory, was the so-called problem of *'primum frigidum'*. While there was no doubt in which element warmth dwells – of course in fire – the primary domain of coldness remained uncertain and, thus, made the conclusions of the theory not very plausible.

The later gathering of the vast practical experience with glass bowl thermometers resulted in the following of the *'peripatetical'* (i.e. Aristotelian) explanation of its function together with the above theory as a whole became rather diffident. Accordingly, the coldness in external air activates the hotness inside the bulb which then is likely discharged into the wall of the bulb. This process changes the ratio between 'Qualities' of the enclosed air; in other words, it changes its 'Form'. This depleted form of air has obviously a smaller volume and the resulting empty space has to be immediately filled by water due to the '*horror vacui*' – nature's abhorrence of a vacuum.

In spite of this fact, the concept of temperature was superfluous for the general description of natural processes within the frame of Aristotle's theory; the term '*temperatura*' was frequently used by ancient physicians well before *Avicenna*. Their idea of temperature was in closely connected to the individual's temperament and was determined by the relative levels of four Qualities that were necessary to maintain the form of the tissues of the human body in a proper healthy state – homeostasis. But, in fact, these old physicians did not appear to care about how to determine this evidently crucial parameter.

Certainly, matter ('*materia*' – potentia pura) was not distinguished from energy ('*energie*' – actus) such that it was proposed that *when heating a metal one was simply adding more "fire" to it. Theophrastos* proposed three stages of fire: *glow, flame* and *lightening* while *Galenos* brought in the idea of four degrees for warming and cooling with a "neutral point": equal parts of ice and boiling water. These four degrees were still accepted by medieval alchemists and *Mylius* [107] proposed a classification according to the Sun passing through *Aries* (signifying calcination), *Cancer* (solution), *Libra* (sublimation) and *Capricornus* (fermentation). The highest degree of fire was *burning as vehement as fusion* and *each twice as great as the preceding degree.*

Comenius, a well-known Bohemian educational reformer and philosopher of Czech origin, progressed to distinguish three degrees of heat (*calor, fervor and ardor*) and cold (*frigus, algor* and one unnamed) with a reference to an ambient (normal) temperature (*tepor*). The highest thermal stage, called "*ardor*", represented an internal degradation, i.e., "*a combustible substance that collapses inwardly and is dispersed into atoms*". The unnamed coldest stage was noted as "*a freezing at which a substance breaks up by constriction in the same way as the heat of fire decomposes it by burning*". In this manner *Comenius* actually, although unwittingly and unknowingly, hinted at the present-day concept of absolute zero. He also stated [11] an almost modern definition of *thermal analysis,* (genially interpreted [44] by *Mackenzie*) as "*…to observe clearly the effect of heat and cold, let us take a visible subject and let us observe the changes that occur while heated or cooled, so that the effect of heat and cold are apparent to our senses...*". *Comenius* was also the first to observe the "non-equilibrium character" of such thermal treatment and analysis, noting [3] "*...by a well burning fire we can melt ice to water and heat it quickly to very hot water, but there is no means of converting hot water to ice quickly enough even when exposed to very intense frost...*". It was an intuitive observation of the phenomenon now called latent heat, and possibly laid the foundation to the discipline of calorimetry.

3.4. Impact of alchemy

The history of fire cannot be complete without mentioning the subject of *'alchemy'* [108-111], a form of science that preceded, and arguably began, the more rigorous and modern discipline of chemistry [112]. The origin of the word alchemy is disputed and could be from the Arabic word *'al-kimijá'* – meaning treated by fire, the Hebrew word *'Ki mijah'* – meaning given by God, or the Greek word *'chemeia'* – meaning comprehension of wetness. The latter was also found in the writing of *Diocletian* as the art of making metal ingots and the term *'cheo'* means pouring or casting a liquid and *'chymeia'* means the art of extracting juices, herbal tinctures or generally saps in the sense of vitality. Also the current word 'chemistry' may be related to the Egyptian word *'kemet'* , which means 'black earth' or to something unknown but precious in relation to the Persian word *'khimia'*. Fire is even mentioned in the Christian teological lineage, e.g., the abbreviation INRI (*'Jesus Nazaretus Rex Judaeorum'*) was once interpreted as *'Igne Natura Renovatur Integra'*, i.e., through fire Nature is restored to its wholeness.

It is not generally known that alchemy was later subdivided into *spagyrii,* the art of producing medicaments, dyes, ceramics, etc., (often trying to transform matter into algebraic combinations), *archemii,* which focused on the development of metallurgy and transmutations of metals, and *Hermetic philosophy* (often synonymous to alchemy itself, see Fig. 14.**),** which is a sanctuary of learning *('prisca theologia')* build upon performance (*'traductio'*), explanation (*'exegesis'*) and interpretation (*'hermeneusis'* – nowadays giving substructure for a modern interpretative hermeneutic description).

Behind the legendary seven Hermetic principles there lie aspects of early cosmology and a verbally bequeathed book known as *'Kybalion'*. Some of the inherent principles of alchemy are, indeed, reflected in modern terms and ideas of current physics, which is often criticized because alchemists certainly thought within completely different frames and measures. Nevertheless let us abbreviate some of them:

i) Principle of spirituality – The Universe is a spirit or belief, and was created simultaneously with many other universes – their origin and destruction occurring in a twinkling of an eye.
ii) Principle of similarity – what is above is also below, birth of universes and fundamental particles, creation of works of art.
iii) Principle of vibration – nothing is motionless, everything is in permanent move, vibration of matter is slow while vibration of mind is too fast to evoke stationary state, rhythm switch leads to qualitative changes.
iv) Principle of polarity associates with vibrations, the higher the vibration, and the more positive the pole; everything has two faces, and it shows the antagonisms of order and disorder, war and peace, day and night. From this ethical edict there follows the transformation of hate to love.

v) Principle of rhythm – following the polarity as everything is oscillatory, coming up and away, breathing in and exhaling, arriving and leaving, inflowing and out flowing; circumambulatory come-and-go.

vi) Principle of causality – every action has its consequence, and coincidences do not exist but are the effect of an unknown law, people are subjected to a lawfully ordered universe.

vii) Principle of gender – sexuality is involved in everything, counterwork of masculinity (positive pole) and femininity (negative pole); process of generation, sexual energy in spiritual alchemy, God mate (often identified with the previous point iv).

Alchemy, however, prescribed to the idea of existence of a "first mover or God as creator" (Almighty, Deity) and believed in an evident order installed in the world. This argument, probably formulated by the Iranian prophet *Zoroaster*, was expressed in the form of the question: *"who could have created the heavens and stars, which could have made the four elements, except God?"*. In the Christian Europe the alchemy was similarly presented as *'donum Dei'* [113].

Hermetic learning might well have been ascribed to astrology by the *Hermes Trismegistos* but, later, it was extended to medicine and alchemy, where the obscure Byzantic *'Tabula Smaragdina'* [114] became a favorite source for even medieval alchemists. It contained seven famous principles, and from these scholars obtained knowledge of the laws of sympathy and antipathy by which the different parts of The Universe were related. Hermetism was extensively cultivated by the Arabs and through them it later reached, and consequently greatly influenced, the Western culture although often misinterpreted. Hermes' stick *'rhabdos'* was ornamented by two twisted snakes showing "waving water and blazing fire" as a unity of contradictions and thus becoming a symbol of life. In the present time it may be speculated that it resembles the double helix structure of DNA structure.

Although it is difficult to be certain of its origins, alchemy most likely emanated from China (as early as in the 8th Century BC and better inscribed by 144 BC in connection with an enterprise older than metallurgy – medicine). Chinese practitioners of alchemy generally considered its ultimate objective to be the development of medical practice and not the generation of gold from base metal. Alchemy in China was all but destroyed when *Kublaj-chan* ordered the burning of the *Taoist* writings, thus destroying most alchemistic records (which included that of the earlest recipe for gunpowder). For a long period of time, it was believed that physical immortality could be achieved through the taking of alchemical mixtures (drugs), an idea that probably vanished when the Chinese

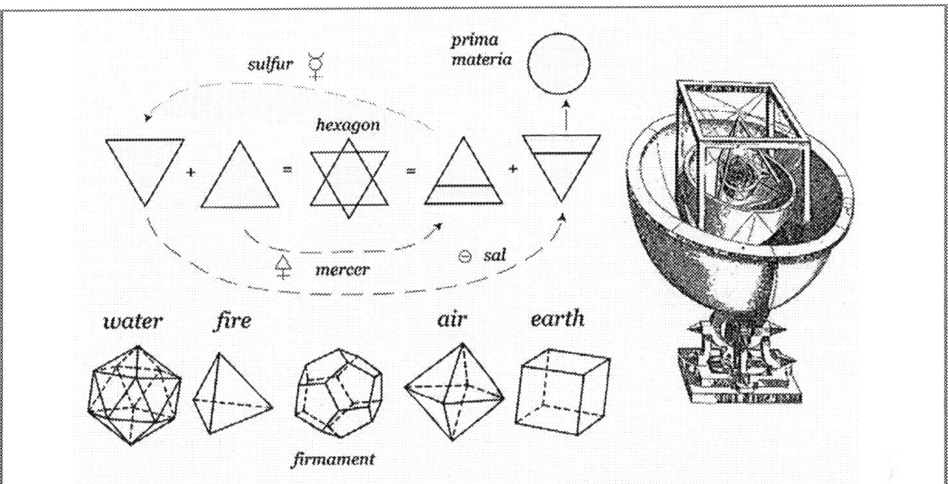

Fig. 15. – Symbols traditionally used for the depiction of the four elements, upper line shows the composition of the plane triangles while the lower line is made up of the explicit geometrical bodies. On the right, the geometrical model of an aggregated universe is revealed with the outermost sphere belonging to Saturn, and with Mercury and the Sun in the center. (The illustration is the imitation drawn after the Keppler's picture in his 'Mysterium Cosmographicum', printed in about 17th Century).

adopted *Buddhism*, which offered another, less dangerous avenue to immortality via meditation. This shift in belief left the literary manifestation of early Chinese alchemy embedded in the residual *Taoist* canons. One of the important but less well-known treatises was the book on therapy by nourishment [115] published in about 670 AD. It described the process of distillation and the quality determination of distillates that could be accomplished by the process of freezing out.

However, the oldest known Chinese alchemical treatise is supposed to be the "Commentary on the *I Ching*", which had an apocryphal interpretation of the 'classic of changes', and was especially esteemed by the *Confucians* who related alchemy to the mystical mathematics of hexagrams (six-line figures used for divination). Ancient Chinese natural philosophy was thus based on a flat, square Earth with the centered Chinese empire surrounded by the eight trigrams as the symbol of the Universe [116]. There were five elements ('*wu hsing*'), Fig. 13, that were related by two complementary principles '*yin*' and '*yang*'. Mutual transformation of the five elements could, e.g., give the birth of metals from Earth, or change metal to water, because by proper treatment metals can turn to the liquid state, etc. The elements were mutually antagonistic because fire is extinguished by water and water is stopped by earthen dams, etc. Yin represents the female principle (darkness, passivity) while Yang, the male principle, is connected with light and energy. Even numbers were considered as Yin and odd numbers as Yang. Each number was associated with one of the elements – the central number 'five' being crucial. It symbolized the Earth and, simultaneously, represented the most prominent position between the five geographical directions – the center (originally referring to the mainland of China).

This approach has also qualified in the so called '*magic squares*' – a famous object of philosophical, alchemical or mystic speculations [116]. It is known as a square-shaped array of numbers (or even letters) exhibiting certain properties, i.e., the sum of the numbers in each row, column or diagonal being equal. Because of its conspicuous worth it was kept surreptitious until the Arabic alchemist *Jabir ibn Hayyan* refined it into a very detailed but complicated system in his 'Book of the balances'. A numerical value for basic properties was assigned to each letter of the Arabic alphabet. When a word for a substance was subsequently analyzed letter by letter, the numerical values of the letters determined its composition. For example lead was composed of the 11 outer qualities (3 parts of coldness and 8 parts of dryness) and 6 inner qualities (1 part of warmth and 5 parts of humidity), altogether 17 parts. If, in a similar manner, the qualities were summed up for gold then a somewhat insignificant difference between lead and gold was found, i.e., for gold also 17 parts of 11 outers and 6 inners again, indicating that a proper alchemical treatment (transmutation) could change the nature of such 'interrelated' metals. It was certainly applicable to the given alphabet and thus could have been freely interpreted.

Later in *Agrippa's* and *Paracelsus'* medieval books, this magical application was rediscovered but using the squares that were constructed by the Islamic method (although no direct link was traced with Islamic sources). It appeared as a promising way to achieve novel discoveries of new relations between things and their properties. The myths of dignified numbers persisted in China until recently [117], which idea might be penetrating the denomination of the periodic table of the elements where the 'prime numbers' and 'numbers of symmetrical principle' are though to be the restrictive islands of the element stability.

Also in India there is evidence for alchemy in Buddhist texts. It came to be associated with the rise of *Tantric* religious mysticism and was recorded in the writings of "*Rasaratnakara*" (in about the 8th Century AD) and 'Treatise on Metallic Preparations' (~ 1100), which recognized *vitalism* ('animated atoms') and *dualism* ('love and hate' or 'action and reaction'). The earliest records from 500-300 BC mention that the theory of nature was parallel to the conception of rudiments ('*tejas*' – fire, '*vádžu*'- wind, air, '*ap*' – water, '*prthvî*' – earth and '*akáša*' – wood) but without a more definite influence comparing the role of elements in other alchemical images. In *Theravada's* view there was a plurality of universe surrounded by water and mountains having three planes of *materia form* (physical body), of *desire* (mental body) and of *immateriality* and/or formlessness (body of law). In practice, the Indians had begun to exploit metal reactions to the extent that they knew as many as seven metals (and already subdivided five sorts of gold). They supposed that metals could be "killed" (corroded) often to provide medicinal bases but not "resurrected" as was the custom of later European alchemy.

However, the birthplace of known alchemy is arguably Egypt and the Greek god '*Hermes*' (identifiable with the Egyptian god '*Thoth*'), who is possibly represented as the alchemy father in the largely indecipherable *Emerald Tablet* of 150 BC (as a part of a larger book of the secrets of creation), which existed in both the Latin and Arabic manuscripts. The history of Western alchemy may go back to the beginning of the *Hellenistic* period and is represented by *Zosimos of Panopolis* (3rd Century AD) who focused on the idea of a substance called "tincture" that was capable to bring about an instantaneous and magical transformation. The earliest notable author of Western alchemy, designated by scholars with the name *Bolos of Mende*, was a Hellenized Egyptian who is often represented by his indefinable treatise called 'Natural and Mystical Things' ('Physica kai mystica') that contains obscurely written recipes for dyeing and coloring as well as murky grounding of gold and silver.

The *testing* of materials was understood in a double sense: experimental and moral. Gold was considered noble because it resisted fire, humidity and being buried underground. Camphor, like sulfur, arsenic, and mercury, belonged to the 'spirits' because it was volatile. Glass was assumed to be a metal because it could be melted, a property associated with the seven known metals. The 13th Century AD pseudo-epigraphic book on chemistry, known as 'Synopsis of faultlessness' ('Summa perfectionis'), contained the terms alcohol (older meaning of powder), alkali, borax or elixir and also suggested that metals were compound bodies made up of a mixture of mercury and sulfur. It recognized '*prima materia*' (first matter) as being a *fixative* (visible and solid – earth, represented by sulfur), *quintessence* (personification – salt) and *evanescentive* (implicit, hidden – air, represented by mercury). It was characterized by triangles that pointed up (escaping) or down (falling), cf. Fig. 15.

It was close to the Platonian geometrization that represents fire as a tetrahedron, air as an octahedron, water as icosahedra[2]* and earth as a

---

[2]* It is clear that life on the Earth depends on the unusual structure and anomalous nature of liquid water. Its small embryos of four molecules tend to come together to form water bicyclo-octamers which may cluster further to cyclic pentamers expanding to a curious network of ordered icosaheda (as foretold by Plato), which are in a certain dynamical equilibrium with its collapsed, but more frequent form, a more symmetrical dodecahedron. The solid, denser form of water is the hexagonal ice and has a structure that is uncanonical with respect to the 'pentagonal like' symmetry of a solidifying liquid that contains a high number of nuclei of icosahedra and dodecahedra. It may help to explain why warmer water freezes more rapidly than cold water – the cold water at the temperature near freezing is densely packed with these, for the ice incommensurable nuclei, which are thus capable of easier and thus deeper undercooling. The water at higher temperatures gradually disintegrate them possessing thus a greater number of smaller fragments, which can survive rapid cooling being more easily compatible with the highly symmetrical configuration of ice.

hexahedron (cf. Fig. 15). By analogy it was also ascribed to show spheres of Mars, Earth, Venus and Mercury with the Sun in the center. Dodecahedron was assumed to play a role of the fundamental structure of firmament. Identification of elements with certain geometrical bodies led, however, to mechanization and mere abstraction of four elements was not thought to be clear enough for the need of alchemist's teaching so that it was supplemented by additional three (imaginary) principles: *sulphur* (as representing combustibility), *mercury* (as fusibility and ductility) and *salt* (as durability).

In the Middle Ages, European alchemy was chiefly practiced by Spanish and English monks (namely *Bacon* and conceivably also by the Christian mystic *Lulla*), who were seeking to discover a substance more perfect than gold (philosopher's stone) as well as a potable gold (elixir of life). Worth noting is Swiss/German *Paracelsus* who remarked that it is the human body that is the best of all 'alchemists' as it is capable of transmuting food into a variety of vital compounds. He also tried to understand alchemy as a kind of art. He highlighted the mystifying 'alkahest' as a universal medicament and he stated four pillars of medicine: *philosophy* (knowledge about nature), *astronomy* (knowledge of the macro-cosms), *alchemy* (in the terms of *Spagirii* as production of medicaments) and *virtue* (honesty of physicians). He held that the elements of compound bodies were salt, mercury and sulfur representing earth, water and air, respectively. He is also thought to be responsible for reintroducing the term alcohol from Arabic '*al-kuhl*' (originally meaning fine powder, Greek '*xérion*') as an early medicament otherwise known as '*tincture*', which he thought was a cure for everything. The Bohemian alchemist *Rodovsky* depicted alcohol as '*aquam vitae*', a medicinal elixir that is best obtained by a procedure involving 14 repeated distillations of wine and the addition of 15 various herbs. It provided more real grounds to the mystical world '*elixir*' (from Arabic '*al-iksír*' meaning gemstone).

Fire was regarded as imponderable or nonmaterial and alchemists used heat lavishly and most of their illustrations include some indication of fire, furnace or a symbol of sulfur. Despite the crudity of the above-mentioned degrees of heat, alchemists laid more emphasis on an accurate temperature control of furnaces (*Norton* [118]) necessary in early metallurgy (*Agricola* [119], cf. Fig. 16.). Adjustment of temperature was, however, purely manual, oil lamps with adjustable wicks, water and sand bathes for lower temperatures and variation of fuels for higher temperatures. In various processes, such as firing ceramics or melting glass, it was vital to keep certain temperatures constant. This required the introduction of early scales for experimental practice. Moreover, fuels giving moist or dry heats were distinguished with an awareness of how to produce high temperatures (also used in burning glasses and mirrors). They also accumulated a vast store of knowledge of the effects of heat on various substances, even if they were unable to interpret the results and satisfactorily measure the temperature.

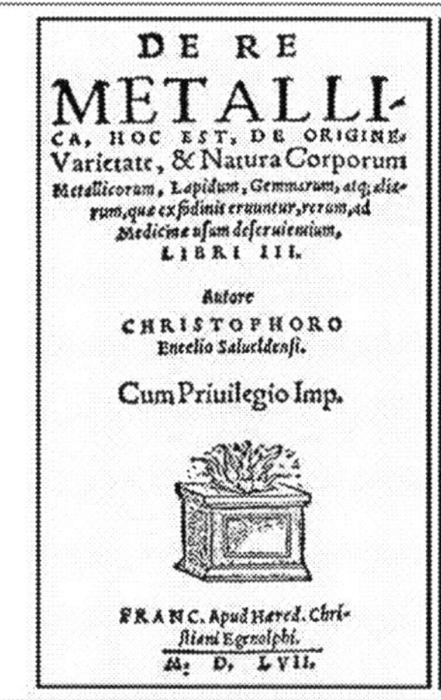

Fig. 16. – Reproduction of the title page (front prolegomenon) from the very influential book "De re Metallice". Additional two selected (inside) pages (right) describe the process of processing glass and illustrate the period furnace with glass pipe-blowing custom. Note there a smart way of drying the fuel wood by pushing it around the furnace wall that enabled the minimization of humidity and thus achieving the wood higher caloric value

The most celebrated process was calcinations, i.e., firing substances to turn into powder known as '*kalk*' or '*calx*' sometimes called '*alcool*' (of gold). The calcinations furnaces were customary named '*athanor*' (from Arabic '*at-tannur*') when sand in a bath was replaced by ashes.

While Arabs had only weak acids, European alchemists of the 16th Century learned to prepare and condense strong acids like `*aqua fortis*` (nitric acid) and spirits of salt or `*vitriol*` (hydrochloric and sulfuric acids), their mixtures capable of dissolving even gold. They became a powerful instrument that made it possible to produce and also characterize ever more varied salts and their spiritual parts separated by distillation enabling to create a more exacting relationship between the identity and way of its testing. An unidentified element, often termed as `*alcahes't*` (i.e., '*alkali-est*' in the sense of a universal solvent resembling today's action of catalysts) was believed to exist as grounds of the four basic possibly to act as an all-purpose medicine.

At the turn of the seventeenths Century, alchemy flourished remarkably during the region of the Bohemian emperor *Rudolph II* and Prague became a home of many famous alchemists, among others there were *Hajek* or *Rodovsky* of Czech origin as well as noteworthy *Stolcius* and *Marcus Marci* (cf. Fig. 17.). They wrote – for that time – very advanced books [120,121], which possibly foreshadowed some laws (such as the refraction laws of light and intuitively moved thoughts toward coming within reach of the conservation laws). *Marci*, however, was strongly convinced that white light was the simplest element (*'quinta essentia'*), which, interestingly, was close to the subsequent concept of 'elementary waves' propounded about fifty years later by *Huygens* in the wave theory of the basis of light. There, however, is inconsistent information about *Marci`s* educational activity, he was possibly rector of the famous Charles University, which was founded 1348 in Prague as the first university in the middle Europe. There, perhaps, a world first specialization called *"chimiatrie"* was unveiled, which was conceivably taught as an unusual subject with regards the traditional university disciplines: major *'artes liberales'* and minor *'artes mechanicae'* (i.e., learning common crafts such as warfare, see-voyage, business, agriculture, hunting, medicine or veterinary) but not in *'artes incertae'* (which was a part of the habitually rejected *'equivocal arts'* associated with occultism, which traditionally involved alchemy).

However, medieval learning is difficult to recapitulate in a condensed form but even in the contemporary world, full of progressive technologies, this elderly philosophy has retained its incomputable role. It remained engaged in the challenge of trying to maintain a sustainable world, on both levels of matter and mind, for the next generations. *Popper* recently recalled *'Tria Principia'* of cosmic evolution grades pointing out three internal appearances of the Universe: (i) a world of physical contradictions (challenge, personal precariousness – *'sal'*), (ii) a world of significance (implication, subjective experience – *'sulfur'*) and (iii) a world of energy (vivacity, creation of the human mind and ingenuity – *'mercer'*). It is somehow related to the interdicted and almost forgotten Hermetic philosophy, with prophecy (God) having the highest, and matter (Earth) the lowest state of eternal vibrations, everything there undergoing processes of dissolution (*'solve'*), and integration (*'coagule'*) within three levels: (i) exploitation of raw materials or digesting of food on a physical level (life), (ii) breathing, energetically based on a spiritual level (love), and (iii) meditation, thought based on a heavenly level (wisdom).

3.5. Emerging new ideas

The modern scientific world is far beyond using any mystic concepts, and many areas of research are now public property [42,122-126]. Early mystical philosophy did not really need exact science. It did not look for measurable quantities and therefore scholarly knowledge was deliberately kept a secret, often for moral reasons. The human mind, however, needs a bit of both. Hermetic philosophy admitted that the universe is calculable, separating quality and quantity at the same time, i.e., harmony is best when sensually perceived but when expressed by numbers. Measurement was thought to be associated with the consciousness of the person who actually makes it, but nowadays it comes close to the ideas of quantum mechanics. *Bohr* said *"there does not exist a quantitative world, there exists an abstract description of quantum physics. The task of physics is not a search how nature is, but what we can say about nature."*

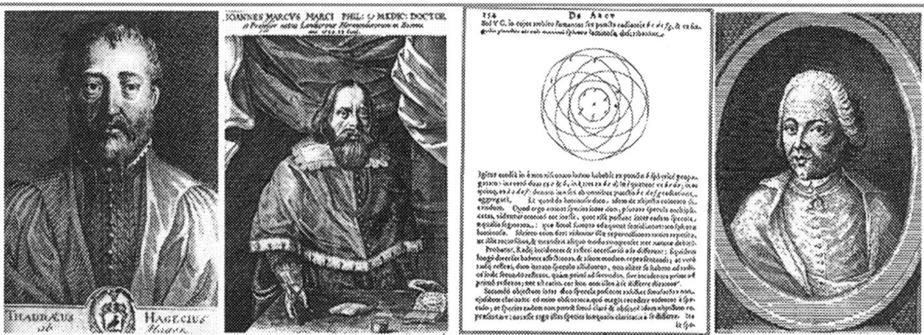

Fig. 17. – Some famous scholars from the Middle Age Bohemia: (left) *Tadeáš Hájek z Hájků* (1525-1600) known as Thaddaeus Hagecius ab Hagek , author of many books (geodesy, botanic, medicine) particularly acknowledged for the first concise book on the beer-making ´De cerevisia´ (brewery 1585), who became famous as a doctor of the *Rudolph II* (1552-1612, Rome Emperor and Bohemian King) during the flourishing period of alchemy in Prague. (middle left:) *Jan Marek Marků* (1595-1667) known as Ioannes Marcus Marci who probably helped to reveal the fundamental properties of the spectral colors that emerge when light passes through glass prism already aware of their monochromatic properties, i.e., any succeeding refraction or reflection did not change colors. He also studied the color change in rays when spectral colors are mixed. He passed light through a glass prism twisted into the form of a bracelet, which he called the 'trigonum armillare'. He pondered on the diffusion of light in circular spheres, and the way in which the intensity of the light diminished with increasing distance from the source, but only to a certain, terminal distance. He, however, was strongly convinced that white light was the simplest element, the 'quinta essentia'. Interestingly, his ideas were close to the subsequent concept of 'elementary waves' propounded about fifty years later by Huygens in the wave theory of the basis of light. (middle right:) Facsimile of the page 154 of his book 'Thaumantias liber de arcu coelesti' (right) explicates the Marci's two-dimensional model of the propagation of light spheres from a spherical source: on a circle with center (a) he chooses six regularly distributed point designated with the letters (b – g). From these points (as it were centers) he draws circles of the same diameter as a picture of the spheres of propagating light. The points of intersection the circles designated by (h – n) indicate the increasing intensity of the light in the direction of its source, and the set of light spheres is closed in an 'envelope'. (right:) *Prokop Diviš* (1698-1765) known as Wenceslau Procopius Diviss Bohemus from Pram a famous inventor of lightening rod (1754) and investigator of electrostatic effects.

As pointed out in the preceding chapters, fire was always kept in a central position of human awareness, in its view as a primary element. It was well-known that the orderly employment of fire provides warmth and pleasant conditions in which to think about, e.g., how to order things or how to gain easy energy. Wild fire was feared to destroy everything, creating chaos by means of the destruction of material possessions held by society, as well as in terms of the destruction of the human mind. Fire leaves a fingerprint in all types of evolution! Let us again, briefly, mention alchemy, as an example of an old endeavor for fire. Within the modern world of science, alchemy is considered to be rather archaic and without a real scientific base, often subjected to ironical comments. We, however, should recall that alchemical philosophy was close to the science of Causation's; it tried to perfect matter whilst being aware of nature as a model. That is to say, it respected the order of nature, somehow resembling present day thoughts on living nature (ecology). Alchemy was undoubtedly related with the process of self-recognition, and success in the laboratory led to individualization, and, *vice versa,* individuality guided laboratory mastery. Alchemy was a universal art of vital chemistry, which by fermenting the human spirit purified and finally dissolved itself into a kind of philosophy. On the other hand chemistry, as a consequently derived true science of facts, is primarily oriented to the utilization of nature, freely processing and exploiting basic/raw materials and trying to dominate nature – it shamelessly enforces order on nature and neglects its consequences.

Paradoxically, perhaps, it was a mystic (*Agrippa, Meyer, Stolcius, Paracelsus*) and practicing (*Agricola, Valentinus, Rodovsky, Sendziwoj, Libavius*) alchemists who really believed in the physical transmutation of metals and not the theoreticians who contributed most to the scientific progress during the medieval alchemical area [126]. The allegorical complexity of alchemical notations and the impossibility of knowing whether the author understood what he was writing, or merely had copied a text that was obscure to him, made alchemists the butt of criticism and mockery. Thereby, its rationalization of chemistry and its marginalization of acculturation aspects are inseparable from the invention of easily available and distributable forms of written messages. The role of printing in alchemy is commonly neglected although crucial. As and could thus be confronted and challenged by more modern authors still full soon as ancient texts were published they become accessible to the wider public of various allegoric depictions. In 1597, *Libavius* published a book about fire (*'Pyronomia,*) where he already enhanced the associated role of human skillfulness and proficiency. However, it was *Helmont* (epigone of *Paracelsus)* who, at the turn of seventeenth Century, rejected the persisting theory of four elements, as well as that of the three primary bodies. Instead he believed water to be the unique primordial substance, and looked for two sources of natural development of minerals and metals (*'seminal spirit'*), which were responsible for creating all objects in various shapes. He distinguished a kind of 'universal' gas from liquefied vapors, even identifying *'spiritus silvestris'* (CO_2) that he found emerged from certain natural substances when consumed by fire. In the book "*Sceptical Chymist*" (1661), *Boyle* extended the attack on the theory of the four elements and planted the modern roots for natural sciences and the concept of chemical elements. The associated salt theory, which originally subverted the idea of a salt as an alchemic principle, helped in the understanding of the phenomena of solubility as a process. It became, correlatively, a way of further separating salts that finally helped *Lavoisier* to arrive at the modern definition of *affinities.*

It is worth noting that *Plato* already possessed a similar view of an element based on *'chóra'* (analogous to Indian *'amah'* that is understood to become fire or water containing *'proté hylé')*, i.e., continuous dynamic transformation of elements within themselves (resembling a quantum vacuum in the framework of bootstrap and/or particle democracy. The four elements were then identified with the macroscopic phases of gas (from the Dutch contortion of the Greek world *'chaos'*) and Latin derived liquid and solid.   Fire   became comprehended as heat and recently even better related to plasma, which accentuated and furthered thinking in the direction of yet other forms of energy that lie outside the aims of this book (e.g. nuclear).

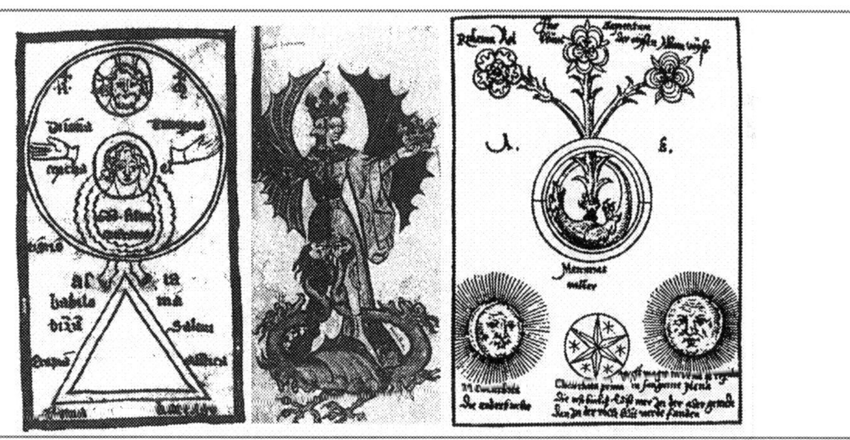

Fig. 18. – Symbolist use [9], for example, triangle (left) symbolizing both the Earth as well as fire (with its corner pointing up). It may be read in such a way that water and earth would be freed by fire out from 'prima materia' and transformed to 'kvitesency' symbolized by a superposed circle (about 14th cent). Middle: the demonic 'Hermafroid' stands on a dragon (symbolizing 'prima materia'). Masculine part holds sword (justice) while feminine holds crown (boast). It is noteworthy that this replicated picture is from about the 15th Century and is supposed to be a part of a gift from the Franciscans to emperor *Zikmund* at the occasion of the Constancy council that condemned the famous Bohemian scholar and reformatory priest, *Jan Hus* to the stake. At the right, notice inside the Hermetic egg, the laying 'Uruboros' speared to three flowers – golden, silver and central blue (which was assumed to be the Magian's flower – 'flos sapientum'), which was later assigned as the reassignment of Romaniticsm (Pandora 1588). Courtasy of *Helmut Gebelein*, München, Germany.

Boyle was also known to strongly criticize the traditional, so-called "dry way separation". He pointed out that when compounds were put into a fire that was 'too hot', the substances obtained were not necessarily components of them but were "creatures of fire" that revealed nothing about the original compound (products what experimenters call "artifacts" today). He preferred slow agitation and gentle heat that could have time to transform textures and therefore produce elements different than those produced from other methods of forcible decomposition. *Boyle* was also the first to use flame testing, probably the first *thermo-chemical analysis,* making it possible to recognize a substance by the color of its flame when it is burned as well as the property of evolved air called "elasticity" (its volume being inversely proportional to the pressure).

Although the 17th Century was a time of notable scientific progress, the scientists of the day were themselves far less respected and far less listened to than today's scientists. Some of them, such as *Newton* and *Leibniz* who are responsible for the introduction of many modern theories, were also devoted alchemists. The idea of *"fire fluid"* (globular particles that attach easily only to combustible objects) persisted for another two hundred years and assumed that when a substance is burnt, the non-substantial essence (*'terra pinguis'*) escapes. The definition of the laws of conservation needed more precision on account of the action of traditional *'vital* and *mortal'* forces. Conservation was assumed to be something general between and within the system as probably first noted by the non-cited Czech educator *Marcus Marci* [121].

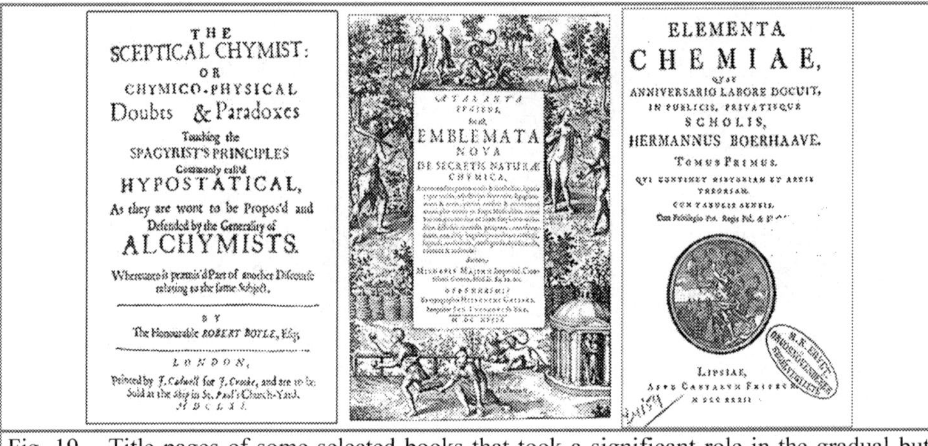

Fig. 19 – Title pages of some selected books that took a significant role in the gradual but lifelong maturity of an educational attempt to improve the learning of chemistry: (upper raw) from the yet alchemy-effected book by *Boyle* (1661) and *Majero* (1618) to the early chemical textbook by *Boerhaave* (1732).

Descartes [127] played an important role, even though he first believed that the Universe is filled with matter in three forms: fire (Sun), transparent matter (Heaven) and dark matter (the Earth). All observable effects in nature were assumed to happen due to the eternal movements of matter, which form gigantic or quite small whirls – *tourbillons*. Such a theory can, of course, explain everything, but unfortunately cannot predict anything. A little better was the theory proposed by *Gassendi,* a devoted follower of *Democritos,* who identified heat and coldness with microscopic material particles, adjoining atoms. Accordingly, the substance heat consists of spherical and very fast atoms, while the atoms of coldness are lazy tetrahedrons with sharp edges causing pain and destroying solid materials. The compatibility of this "substantial" theory with mathematical treatments probably helped it to survive with minor changes until the 19th Century. Interestingly, the premise of *Bacon* and *Descartes* that heat is a kind of motion, contradicting the opinion of *Gassendi* that heat consists of particles (`quasi-particles`) was unified, in a sense, by the modern kinetic theory of matter.

Some associations of the above mentioned ideas can be observed latterly in the symbolic lessons of Amerindians within their innate principles, for example, the magic number 'four' representing world cardinal points, four traditional elements, four animal legs, etc. Indians used to evoke magical circularity of shapes like cone-shaped tent (tepee or wigwam) or rounded stone because the nature provides globular forms (sun and moon), living creatures sustain the rounded appearance without corners showing thus the harmony of nature and being. The circle is also kept as a symbol of human belonging (while sitting around a fire when the Indian chibouk moves from hand to hand in support of peace). On the other hand, the implanted world of whites was assumed to be represented by squares like buildings, rooms, banknotes, televisions or cars, which made Indians feel sad and sicken of modern civilization. Indians complain that the actions of whites were not in the vital conformity with long-established wildlife, producing too much unnecessary waste, harming thus nature and producing disharmony for future generations. On the other hand, Indians tried to live within the rules traditionally given by the natural world as, for example, they hunted bison not only for food but they used almost everything to the last piece while the whites came to shoot bison just for choice meat and the pleasure of killing.

4. CONCEPT OF HEAT IN THE RENAISSANCE AND NEW AGE

4.1. Phlogiston – deciphering combustion

The various phenomena of combustion [52,128-134] were known for many centuries before any attempt was made to explain them or, indeed, before any attempt was made to investigate them. One of *Boyle's* contemporaries, *Mayow,* assumed that atmospheric air contained a substance which he termed *'spiritus igno aereus'* which combined with metals to form *'calces'* and which was associated with the process of respiration. This assumption contained the germ of discovery of the new approach to explain combustion later enriched by *Becher* who assumed that all inorganic substances consist of three 'earths': the mercurial, the vitreous and the combustible. The last term was named as *'terra pinquis'* and when any substance was burned, this essence escaped.

Stahl gave to us a special characteristic of chemistry called the "blend union" or the *'mixt'*, which was distinguishable from aggregations (mechanical unions), and their analysis became the entire task of the chemist alone. It was later proposed that a mechanical agent or "instrument", i.e., fire (or heat), water (solvent) or air, bridged the gap between the *'mixt'* and aggregates. Fire put the `phlogiston` (renamed `terra pinquis`) in motion, air blew off the most volatile parts and water put the parts into solvated motion. Fire was therefore the instrument, and the phlogiston was the element entering into the composition of the `mixts`. It explained combustion as well as the transformation of *'calx'* into metal and, *vice versa,* (i.e., metal = calx + phlog.). Phlogiston was, thus, a revolutionary element since it suggested that both combustion and corrosion aide in the same operation of *oxidation,* which is the inverse operation of the process now called *reduction.*

During the so called *'phlogistic'* period (lasting for about one hundred and twenty years) the science of thermochemistry was enriched by the labors of many eminent investigators, e.g., *Cavendish, Pristley, Scheele.* Although it is easy for modern scientists to ridicule the absurdities of such a phlogistic theory, it must be borne in mind that this idea was very much contributory to the better understanding of early views of energy conservation and it served to stimulate a vast amount of experimental research. The downfall of this theory was caused by the observed fact that products of combustion retained a greater weight (mass) than the combustible substances from which they were derived. The ingenious attempt to explain this phenomenon by assuming that *phlogiston* possessed a negative weight did not, however, survive later rational protests. The final overthrow of this idea may be thought as the marking of the beginning of a new era that began modern chemistry but the phlogistic period should also be associated with the discovery of the more important constituents of the atmosphere (dephlogisticated air – oxygen, inflammable air – nitrogen, fixed air – CO_2, phlogisticated air – hydrogen and compound nature of water).

The true attacks on phlogiston would become significant only in a larger context when the theory of gases arrived, which was essentially developed by the work of *Lavoisier*. The key factor in his theory was the new substance of heat or matter of fire, called *caloric,* which crept in among the constituent parts of a substance and gave it expansibility. If the physical state of body were explained by the quantity of caloric, then consequently air would lose its essential function as a principle. Although caloric differed from phlogiston because it could be measured with an apparatus called a *calorimeter* (designed by *Wilcke* and later used by *Laplace, Lavoisier* and others), it was nevertheless an imponderable element with its own properties.

The belief that the elastic properties of gases could be accounted for by supposing that gas particles were stationary and subject of mutually repulsive forces was wholly Newtonian approach and it was readily acknowledged by most of the 18th and 19th Century writers. It was related to the *Principia* where *Newton* had shown how Boyle's Law could be predicted on such a basis if it was assumed that the repulsive force between any two adjacent particles of gas expanding or contracting isothermally was inversely proportional to the distance between them.

Lavoisier expounded the view that gases and vapors resulted from a combination of 'matter of fire' and with a 'base' which could be either a liquid or other volatile solid. The mechanism of this combination he resembled that of a normal chemical union. Heating occurred in the process of combustion, for example, simply because the base of 'air vital' (oxygen) had a greater affinity for the inflammable substances than for the matter of fire and so combined with it, allowing the fire to escape and became free. When fire was combined with ordinary matter it was undetectable, and it was only when it was free that it affected the thermometer and produced the sensation of heat. He defined the true measure of hotness simply as the quantity of free (uncombined) fire in a body.

Some year's later the Irish chemist *Higgins* described fire as 'elastic fluid' whose elasticity was the result of repulsion, which was attributed to the 'charges of the repellant (fiery) matter' which formed 'distinct atmospheres' round the 'grosser parts' and so caused them 'to recede from each other contrary to their inherent and incessant attractive power'. *Higgins* went even so far that the density of the atmospheres varied 'reciprocally as the distances from the central particles, in a duplicate or higher ratio'.

Boerhaaves's description of fire in his *Elements of Chemistry* [54] showed that it was composed of particles, which were the smallest and most solid of all bodies yet known, and were also weightless, a fundamental property

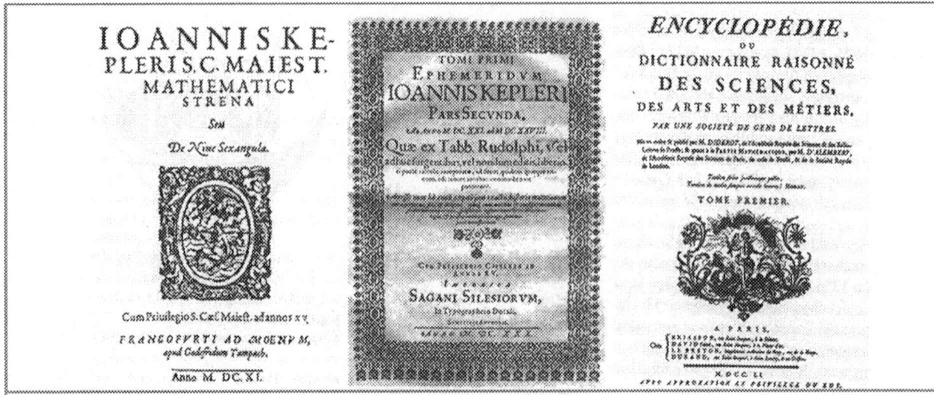

Fig. 20. – Less known book by *Kepler* "Strena seu de nive sexanguka" (1619) known as 'Christmas gift or about hexagonal snow' and (right) the front page of the famous 'French encyclopedia' (1751), which first dealt concisely with natural sciences.

for the later formulation of caloric theory. In his opinion it was the motion of particles of ordinary matter (conceived as a vibration) which was responsible for the phenomena of heat and it was the function of fire, by its own movement, simply to cause and sustain this motion. *Boerhaave*'s fire, both in its function and structure, resembled *Descartes'* subtle fluid rather more closely than it did *Lavoisier's* caloric. Indeed, it seems that belief in the materiality of fire was a part of standard doctrine, especially among chemists, long before *Lavoisier* received his sanction for the caloric theory.

4.2. Rise of the theory of caloric

By the time caloric theory emerged [134] there were already a number of well developed theories of electricity, magnetism and light that were based on the existence of subtle elastic and often imponderable fluids with properties remarkably similar to those of caloric. For example *Franklin's* description of electricity in the late 1740s conceived it as being composed of small weightless particles which, although being mutually repulsive, were attracted by the 'particles of common matter'. All bodies were though to contain a certain quantity of such electric fluid, even when they were electrically neutral, becoming charged only when an excess or deficiency of fluid was created. By the 1770s the fluid theories of electricity and magnetism were well known and generally accepted.

Unfortunately, the great pioneers, *Irvine* and *Black*, published almost nothing in their own lifetimes [133] and their attitudes were mostly reconstructed from contemporary comments and essays published after their death. *Irvine* supposed that heat was absorbed by a body during melting or vaporization, simply because at the melting- or boiling-points sudden changes

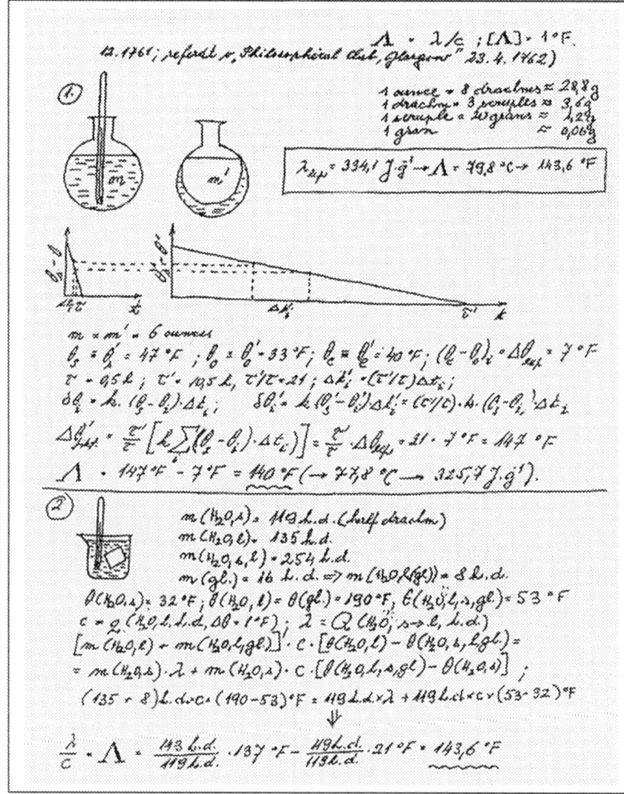

Fig. 21. – Recreation of Black's method [9] for the measurements of latent heat of ice melting, λ, (upper 1), was a special set up that was luckily made according to advice of a practical whisky distiller, citing *"when his furnace was in good order, he could tell to a pint, the quantity of liquor that he would get in an hour"*. So that *Black* made his experiment in such a way *"boiling off small quantities of water and found that it was accomplished in times very nearly proportional to the quantities, even although the fire was sensibly irregular"*. It described an almost linear change of temperature (θ) with time, which is probably the first record of thermal analysis. An allied calculation, according to Black's laboratory notes, is the incorporated and correct value of λ as revealed in the frame (in the relation to temperature increase of $\Lambda = \lambda/c$). Below (2) is shown the principle of Black's method for determining of latent heat of water melting, already using a kind of mixing calorimetry. He concluded *"I imagined that during the boiling, heat is absorbed by the water, and enters into the composition of the vapor produced from it, in the same manner as it is absorbed by ice in melting, and enters into the composition of the produced water. And as the ostensible effect of the heat, in this last case, consists not in warming the surrounding bodies, but in rendering the ice fluid, so in the case of boiling, the heat absorbed does not warm surrounding bodies, but converts the water into vapor. In both cases, considered as the cause of warmth, we do not perceive its presence, it is concealed, or latent, and I gave it the name of 'latent heat'.* (Courtesy of *Ivo Proks*, Bratislava, Slovakia).

took place in the ability of the body to contain heat. *Irvine's* account that the relative quantities of heat contained in equal weights of different substances at any given temperature (i.e., their 'absolute heats') were proportional to their 'capacities' at that temperature. It is worth noting that the term 'capacity' was used by both *Irvine* and *Black* to indicate specific heats, cf. Fig. 21. They introduced the term 'latent heat' which meant the absorption of heat as the consequence of the change of state.

In the calculation of the latent heat of a given mass of ice, *Irvine* took the quantity of heat that would raise the temperature of the same mass of water by 140 °F and the values for the specific heats of ice and water were thus 0.9 and 1.0 respectively. Where the melting point was x°F above the assumed 'absolute zero' it followed from Irvine's assumption that the total quantities of heat in the ice before melting and the water after the change of the state, both quantities being measured at 32 °F, were proportional to 0.9x and 1.0x respectively and hence, on the assumption that heat had been conserved, that x = 0.9x + 140 = 1400 so that the zero point in this calculation would therefore be -1368 °F, the value being very sensitive to the assumed figure of specific heats of ice, later corrected to 0.85. *Lavoisier* and *Laplace* were careful to emphasize that small errors in the specific heats could seriously affect the calculation having also little sympathy for such basic `Irvinist` assumptions as the proportionality between capacities of various bodies and the quantities of heat that they contained.

Similarly *Dalton* [133,135] chose to consider the expansion of air at constant pressure between 55 and 212 °F (the range of his own experiments) assuming a proportionality between the total heat content that would have increased from Q_{55} to Q_{212} and the change of inter-particle force from F_{55} to F_{212}, i.e., $Q_{55}/Q_{212} = F_{55}/F_{212}$. By further relating F to the distance, d, between the particle centers, where d^3 were, of course, proportional to the volumes that the air would have occupied at given temperatures it followed that $d_{55}/d_{212} = T_{55}/T_{212}$ providing an estimate of absolute zero at -1515 °F. In 1801 *Dalton* mentioned "...*this remarkable fact that all elastic fluids expand the same quantity in the same circumstances, plainly shrews that the expansion depends solely upon heat; whereas the expansion in solid and liquid bodies seems to depend upon an adjustment of the two opposite forces of heat and chemical affinity, the one a constant force in the same temperature, the other a variable one, according to the nature of the body; hence the unequal expansion of such bodies...*"

An important approach was introduced by *Cleghorn*, who in 1779 related the material theory of heat in his book '*De igne*' as follows: "...*since the quantity of fire distributed among bodies increases with the attraction for fire that the bodies exert and decreases with the repulsion between the fire particles themselves, it follows that if in any body the former quantity is diminished or the latter increased, then the fire will flow from that body until equilibrium is again restored. Heat is then said to be generated. On the other hand, if the attraction of any body were to be increased or if the repulsion between the fire particles were diminished, more fire would flow into the body and in this case cold is said to be generated...*"

Like *Dalton, Avogadro* believed that gases were composed of particles of ponderable matter each surrounded by a sphere of caloric that was retained by an attractive force between it and the particle. It was this force, he maintained, which determined the quantity of caloric in any given molecules and he thus differed from *Dalton*, for whom the quantity of caloric was the same in all molecules under similar conditions of temperature and pressure, irrespective of the nature of the gas and the magnitude of the force. The size of a gas molecule was determined by its chemical structure, and here gain *Avogadro* differed. So *Avogadro's* view of gas structure provided him a new and virtually unique tool for the development of caloric theory. *Avogadro* was evidently assuming a close analogy between the mechanisms that retained caloric round a molecule and that which, according to *Berthellet* [136], governed the chemical combination between substances. He even proposed a purely empirical relation involving refractive power (rp) and the affinity for caloric (A), such as: rp = 0.419 A + 0.581 (A)1/2 and he devised a scale of the affinities of all compounds to vary from 1.3 to 2.49. Between these limits he laid the affinities of certain acids and alkalis, true neutrality being at 1.7

The greatest benefit of caloric theory was that it supplied an obvious solution to the problem of thermal expansion and contraction. Heating a body had the effect of adding fluid caloric to it and, consequently, the body expanded. On cooling the opposite occurred, involving the removal of fluid caloric from the body. Many of the properties of heat were explained by considering each particle to be surrounded by an atmosphere of caloric, whose density is related to the intensity of the gravitational attraction between it and the center of any particle. The gravitational attraction was considered to be inversely proportional to the square of the distance between the centers of the particles involved, while the caloric atmosphere, which caused the repulsion, was assumed to obey a logarithmic law in the analogy to the Earth's atmosphere. In liquids, the caloric content was sufficiently high so that the atoms were not held in a rigid position by mutual gravitational attraction and in gas this attraction was considered negligible. Thus, it was predicted that the expansion of a gas would be much greater than that of a liquid and than that of a solid. It also explained that the expansion coefficient increased with temperature more rapidly for liquids than for solids. In certain views on the physical behavior of gases, particles were even assumed to be stationary, the pressure keeping them such and being derived from the tension of caloric.

The theory of caloric played probably a similar role in making the observation that gravitational force does not pull all things together to form one very dense body as a current dispute about the invisible dark matter (and/or energy), which hypothesis is used to explain its yet undefined antigravitational forces that keeps the Universe expanding instead of only contracting under the omnipresent gravity.

A careful distinction, however, was drawn between the intensity of heat and the quantity of heat [133,134]. All atoms did not have identical caloric atmospheres, and although they all had a logarithmic dependence of caloric density on distance, the rate at which the atmospheric density reduced, varied from substance to substance. The quantity of heat required to produce a given change of temperature for a given amount of material, was called the specific heat of a materials, by analogy to the term of 'specific gravity'. Heat could take two different forms, sensible and latent. Caloric was considered to combine with atoms in a fashion similar to how atoms bind together, and with such combinations the caloric changed from its sensible form and became latent. Such a chemical combination of an atom with caloric produced a 'new' compound in which neither the original atom nor the caloric retained its identity. No heat was considered to be lost in the process since it was reversible – cooling a body down returned the caloric back to its sensible form.

Let us consider two erstwhile examples; when a piece of iron is compressed by hammering, the sensible caloric is squeezed out and the surface of the iron becomes hot, or if a gas is compressed it emits caloric and becomes hotter. It was thought that sensible caloric could be squeezed from a body by artificially pushing the constituent atoms into closer proximity to one another than what the mutual repulsion of their caloric atmospheres would allow. Therefore, if pressure was put on a substance near its boiling point, some of the sensible caloric would be lost from the substance and a higher temperature would have to be applied before sufficient caloric was available to the atom for a vaporization to occur. The less caloric a body had, the greater was the attraction between the atoms of that body and its surrounding caloric fluid. In adding the caloric to one end of an iron bar, the atoms at the heated end required more caloric than their neighbors and by having more, their attraction for this caloric were less. Thus, the neighboring atoms attracted the caloric away and continued to do so until all the atoms of the substance had achieved the same caloric at atmospheres. The facility with which caloric could be passed from one atom to another depended upon the structure and composition of the iron (substance). It is worth noting that a more careful study of what is currently being taught and used as the world quantity of heat concerning its behavior and phenomena shows a striking similarity with the above-discussed caloric theory of the past

There was another important account by assuming the *Berthollet's* analogy between the behavior of caloric and that of acids, which allowed certain additional simplification. Thus it was no longer necessary to ascribe special properties to each of the forms of caloric. The latent heat and temperature changes was thus viewed as the result of opposition to the naturally expansive force of caloric in ponderable matter and it was the force of cohesion between the molecules which resisted this expansion force and so caused heating when caloric was added. *Berthellet* said in 1803 *"...If one hesitates to regard this similarity between the properties of caloric and those of a substance entering into chemical combination as a conclusive proof of caloric's materiality, one cannot but agree that the hypothesis that it exists presents no difficulties and has the advantage of involving only general and consistent principles in the explanation of phenomena...".*

4.3. Decline of the caloric concept

The weakening of caloric theory begin with the work of *Fourier* on heat conduction and with *Joule's* work of the early 1850s, which gave a trivial corollary of the equivalence of heat and work that, however, yet needed to determine the type of motion that was thought to constitute heat. Such uncertainty [136] was felt concerning the nature of the motion of heat and the process by which the matter was resolved was slower and far more complex than is suggested in most histories of that period. There was an interesting approach by *Rankine* (1850) to interpret the new thermodynamic principles in terms of an elaborate vortex model of the atom, the motion in this case being associated with an `elastic atmosphere` that revolved or oscillated about the `nucleus or central point` of any atom of matter. The quantity of heat was thus nothing more than the *vis viva* of the revolutions or oscillations perfumed by these atmospheres. Some years later (1867) another vortex theory was proposed by *Thompson,* who suggested that the atoms might be nothing more than centers of vortex motion in all-pervading fluid ether.

Already in the year 1812 *Davy* said [135] *"...The immediate cause of the phenomena of heat is motion, and the laws of the communication of motion... temperature may be conceived to depend upon the velocities of the vibration; increase of capacity of motion being performed in greater space; and the diminution of temperature during the conversion of solids into fluids or gases, may be explained on the idea of the loss of vibration motion, in consequence of the revolution of particles around their axes..."*

Rumford wanted to prove that heat has no mass and that it can be produced, without limitation, by friction. He also wanted to show that thermal motion of particles occurs also in liquids. He tried to explain heat transfer in a vacuum by the vibration of material particles that cause a 'wave motion of the ether', capable of propagating in a vacuum. *Rumford's* ideas contradicted the accepted understanding of such a heat transfer, which was thought to be the consequence of repulsion of the caloric particles in its 'non-ideal' solution, their high tension tending to 'redistilled' caloric through a vacuum from a warmer to a colder body.

Probably the turning point was *Clausiu*`s explanation (1857) in which he gave a lucid and most lucid and convincing account of various motions – rotational, vibration and translational – of which he believed the molecules of solids and liquids to be capable. He argued that the observed specific heats of gases could only be predicted theoretically by taking account not only by the translation *vis viva* of the gas molecules but also their rotational *vis viva*. He also introduced the concept of the mean free path of gas molecules and he used to explain the slowness with which the gases were known to diffuse through one another. It was left for *Maxwell* in 1859 to add the ingredients of our modern kinetic theory that was most obviously lacking in *Clausiu's* treatment, i.e., the statistical distribution of velocities among the particles of gas [52].

4.4. Heat and energy

In the year 1647, *Descartes* [127] became the first to propose the conservation law of the quantity of motion, presently called *linear momentum*, and showed it to be proportional to *mv*. Subsequently this idea was extrapolated by *Leibnitz,* who introduced the terms *'vis viva'* for the quantity mv^2 and *'vis mortua'* further related to *Newton's* gravitational forces. *Leibnitz's* idea of a 'vital force' was extended to an isolated system by *Lagrange* [136], when he assumed invariance of the sum of this quantity with the function of the coordinates of a particle, i.e., with their potential energy. The terms *'energy'* (ability for virtual work) and *'virtual displacement'* were used for the first time by *Bernoulli,* were weaving their path to a wider appliance very slowly and yet at the turn of 20th Century the term *'vis viva'* occurred still quite commonly. *Waterston* reported in 1845 [52] that the *"quality of perfect elasticity is common to most particles so that the original amount of 'vis viva', or acting force of the whole multitude, must for ever remain the same. If undisturbed by external action it cannot, of itself, diminish or increase.... striking against and rebounding from each other they undertake any possible mode of occurrence such that the particles are move in every possible direction and encounter each other in every possible manner during so small an elapsed interval of time that it may be viewed as infinitesimal in respect to any sensible period"*. It was *Coriolis* who started to use the name 'lives force' for a half of its original value, mv^2, and thus simplified the relation with work. However, it was *Joule* who entitled the principle of work as mechanical power. In 1853 *Rankine* introduced the term 'potential energy' and thanks to *Tomson* the outlasting *'vis viva'* was finally renamed as 'kinetic energy'.

Not knowing what fire actually was, *Black* avoided speaking of it, but he studied the specific relations between the two measurable quantities of heat and temperature, which were not new in themselves. *Amontons* first used the effect of heat when he made an air thermometer. In fact, the use of this new instrument brought about the very question of how to define the quantity of heat separately of temperature, since its measurement is a function of both quantities. *Black* was interested in the way in which heat was fixed in bodies and he called this heat as *"latent heat"* – the heat absorbed or released by a body during a change of state without a change in temperature. Contrary to the standard understanding of heat absorption as the penetration of a fluid (caloric) through a porous body, the way of absorption of latent heat is not comparable and must be understood as a

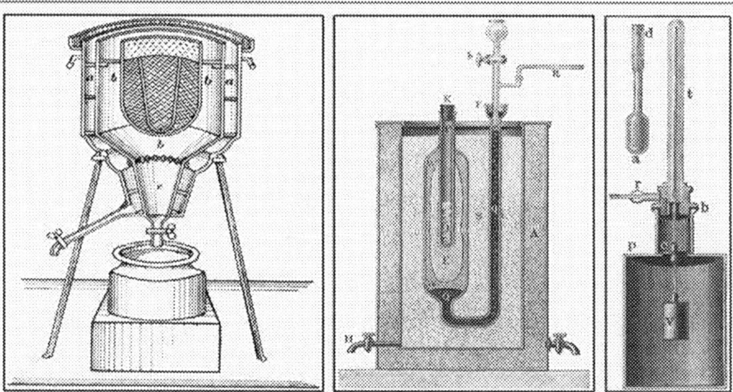

Fig. 22. – Time-honored ice calorimeter, which was first intuitively used by *Black* and in the year 1780 improved by *Lavoisier* and *Laplace*. The heated body is cooled down while placed in ice and the heat subtracted is proportional to the amount of melted water. In the year 1852, *Bunsen* proposed its more precise variant while determining volume instead of weight changes (middle). The cooling calorimeter was devised 1796 by *Mayer, Dulong* and *Petit*, but became known through the experiments by *Regnault*. Thermochamical measurements were furnished by *Favre* and *Silbermann* in 1852 using the idea of Bunsen ice calorimeter but replacing ice by mercury the volume measurement of which was more sensitive.

different, rather combined process that entails both the ideas of melting or boiling. *Black's* elegant explanation of latent heat to the young *Watts* became the source of the invention of the businesslike steam engine as well as the inspiration for the first research in theoretical thermochemistry, which searched for general laws that linked heat, with changes of state. Heat and its measurement were, however, to belong to mechanic-physics, were they were integrated into the economy of chemical transformations.

With little doubts, until the work of *Black* and *Irvine*, the notions of *heat* and *temperature* (from *temper* or *temperament* first used by *Avicena* in the 11th Century) had not been distinguished between. *Black's* work, together with that done by *Magellan*, revealed the quantity that caused a change in temperature but which was itself not temperature – the modern concepts of *latent heat* and *heat capacity*. They explained how heat is absorbed without changing temperature and what amount heat is needed to increase a body's temperature by one unit. Worth noting is the original proposal of a name for one unit of heat to be a *therm* (sufficient heat to warm 1g of water by 1° C) or the proposal by *Groffith* to name it after the lesser known physicist *Rowland*. The description of latent heat and heat capacity answered, to some extent, the warning given by *Boerhaave* [54] in the beginning of 18th Century ...*"if we make a mistake in the interpretation of what is fire, this deficiency can afflict all disciplines of physics and chemistry, because in all natural creations fire is involved, in gold as well as in emptiness.."*.

Rumford presented qualitative arguments for a fluid theory of heat with which he succeeded to evaluate the mechanical equivalent of heat. This theory, however, was not accepted until later approved by *Mayer* and, in particular, by *Joule*, who also applied *Rumford's* theory to the transformation of electrical work. The use of customary units called 'calories' was introduced by *Favren* and *Silbermann* in 1853. The characterization of one kilocalorie as 427 kilogram-meters was first launched by *Mayer* [136] in the year 1845. Regarding some impediments associated with the wider application of (only now traditional) units, such as calories and recently joules [J], the establishment of some practical links with traditional work [W=J/s] became necessarily accentuated. Therefore, an innovative unit *'horsepower'* (HP) was necessary to bring in as the performance measure thanks to *Watt,* who was unable to sell his steam engine to mines without telling the engineers how many horses each engines would replace, because the owners traditionally used horses to drive pumps that removed water. Jointly with another norm work power by then, produced by a healthy brewery horse (introduced and premeditated by *Joule* working as a Brewster), a horsepower unit was defined as 550 ft-lbs (~ 75kgm) of work every second over an average working day. It is clear that from that time the strong miner and brewery horse got off the use due to modern carriers and current 'average' horsepower would have to turn into somehow 'weaker' status of present 'racing horses'. Nevertheless it is steadily kept in traditional rating engines of cars and trucks. The other horse 'dimension' was otherwise important in mankind's history as was the separation of two horse backs, which became a factor for establishing the distance of wheels in Greek wagons, and later, it correlated with the separation of railways and profile of tunnels. Consequently it even gave the size of the external rockets in the US space program as their size was limited by the separation of railways and associated size of tunnels through which the rockets were transported to the launching sites

Nonetheless it took two centuries to replace the fluid theory of heat (*caloricum* or sometimes *thermogen)* by the vibration view (of the state of inherent particles) that was substantiated by the advanced theory of combustion by *Lavoisier. Sadi Carnot* provided the theory of the four cycle device [137]. However, the idealized theory of a heat engine was proposed on the confused basis of heat transport taking place in the form of a fluid (caloricum) discharging from the state of higher to lower tension ('conservation of materialistic caloricum') and in that time was also supported by *Clapeyron.* Notably *Carnot* excluded the existence of perpetual thermal motion, formulating an important efficiency theorem on the *"moving force of fire".* Such a force does not depend on the medium used for its outgrowth but just on the temperatures of bodies between which the transfer of heat is conveyed.

Following the 1826 textbook by *Pencelet* [138,139], *Rankine* introduced in the year 1853 the term *energy* – actuality (*'ergon'* – actus and *'energeia'* –

activity in the opposite sense to the possibility *'dynamis'* – potentia). Simultaneously to the development of the separated field of electricity and magnetism, another important field of thermal sciences was introduced and named *thermodynamics ('thermos'* – heat and *'dynamis'* – force), first made known by *William Thompson* and preceded by *Maxwell's* concept of thermal equilibrium [50]. Yet towards the end of the 19th Century [129,130], *Helmholtz* and *Laplace* described both theories of heat to be equally suitable to comply with a theory of temperature measurements because it was only determined by the state of the system under investigation. A similar understanding was anticipated for the internal motion of particles because heat was also a measure of the transfer of motion from one system to another – kinetic and potential energies being qualitatively different forms of motion (scaling variance in the degree of self-organization). Nowadays the transfer of heat is sometimes described in terms of non-integral dimensions of fractals.

The creation of the notion of entropy in thermal physics was a response to a similar problem in thermochemistry: the principle of the conservation of energy, always verified by physical-chemical transformations, could not be used to simply determine which transformations were possible and which were not. The state of equilibrium was defined by the fact that any spontaneous transformation that would affect it by diminishing entropy was impossible. Whilst a real steam engine functioned by real heating and cooling, the function of the ideal steam engine necessitated the fiction that two bodies of different temperatures would never be put into direct thermal contact. Thus the system is supposed never to leave the state of equilibrium whilst undergoing a displacement from one state of equilibrium to another, which are, moreover, infinitely close to each other and which is determined by an infinitely progressive variation of the controlled parameters. The problem of such a change description bears the same character as the time change during chemical reactions. *Duhem* called such changes 'fictitious transformations', which are reversible and entirely directed from the exterior when passing from one state of chemical equilibrium to another. *Van't Hoff* and *Le Chatelier* associated this idea with the law of displacement of equilibrium and chemical thermodynamics received its central principle that dealt no longer with mere 'energetic' but introduced thermodynamic potential as a minimum at chemical equilibrium. Physical chemistry, as an autonomous science in relation to mechanical physics, which has neither reactional events nor the second (entropy) principle, has brought thermochemistry to a level much richer than physics alone, see early calorimeters, Fig. 22.

Far from equilibrium [140,141], physical chemistry puts the accent on the global behavior of a population with local interactions. Such an idea may be also applied to the behavior of various societies, but the exact field of thermochemistry is more honored for two reasons. Firstly, it may encircle a great variety of cases capable of nonlinear coupling, and secondly, any creation of molecular structures can even take place independently of the processes itself (like hydrodynamic whirlwinds). Characteristic of such systems is the nonlinear coupling between inherent processes (that produce entropy), which often shows the capability of self-organization [9,13,57,141]. The spontaneous production of spatial differentiations and temporal rhythms, are often called dissipative structures. Structure is here understood in a more general context and is most likely to exhibit coherent spatial-temporal activity, and is dissipative in nature because it occurs only under the condition when the dissipative process is forcefully maintained.

Links between the production of entropy and the production of coherence leads to the association of two sciences: thermo-dynamics and thermo-chemistry. Coherence, however, touches not individual molecules (that can be in the framework of quantum principles) but effects the population of whole molecular aggregates. Coupling of the rates of simultaneous chemical reactions, may even bring the possibility to arrive at a world characterized by the quantum-mechanic-constants [142-144] (*Planck quantum length, thermal length*). Here belongs *Buffon's* original idea of 'animated molecules' (now called *Brownian motion),* which was, after all, refused by *Brown* saying '*the motion of particles in a fluid cannot be due, as others had suggested, to that intestine motion which may be supposed to accompany its evaporation*` and such a cooperative motion can be seen (according to *Perrin* and *Nernst* [144]) as equivalent to the action of fluctuations as a natural consequence of randomness of molecular motion.

In the other words, chemistry can produce stable structures that store a kind of memory of their formation conditions. Standard crystallographic structures can function as a kind of relay between histories of different types of their formation and the make up of individual materials. The history can be re-constructed on the basis of a physically measurable property characteristic of these structures. The singular link between chemistry and plurality of interwoven times for a structure's build up was, therefore, the center of alchemical preoccupations.

When the validity of the conservation law of mechanical energy was generally recognized, the French Academy of Sciences did not accept any new proposal for the construction of a mechanical '*perpetum mobile*'. At the same time, the widespread caloric hypothesis achieved important supporting results: the derivation of sound velocity with the use of Poisson constant (*Laplace),* mathematical description of heat conduction (*Fourier* 1822) and the proof of the same maximum efficiency for all thermal machines (exhibiting the impossibility to construct any perpetual mobile of the second generation). Its author, *Carnot* also derived the direct proportionality, $f(T)$, between the works performed in the 'fall' of the heat unit during an infinitesimal cyclic change of given working substance and within the corresponding temperature difference between the hotter and cooler reservoirs. It was written as the relation, $f(T) = C/(C_1T+C_2)$, where C's were constants and T is the Celsius temperature. The expression for $f(T)$ (corresponding to the related Clapeyron's term, $1/C$) was later named as Carnot's function, μ, equal to $(1 - \delta/\rho)\ dp/dT\ x1/\Delta H$, where δ, ρ and ΔH are densities of vapor and liquid, and the heat evaporation respectively. In fact, it was the first quantitative description of the equilibrium coexistence between the gaseous and liquid phases, currently known as the Clapeyron equation.

In 1850, *Clausius* published his first treatise on the new discipline on thermal science (not using yet the term 'thermodynamics', which was already introduced by *Thompson* in 1849) in which he reformulated and specified more exactly the two laws of thermodynamics and better specified the Carnot-Clapeyron relationship. In his famous publication "*On the modified form of the 2nd law of mechanical heat theory*", *Clapeyron* incorporated the possibility of compensation of the enforced negative processes by spontaneous positive processes. To the first class, the impossibility of transfer of heat from lower to higher temperatures was classified spontaneously (and similarly to the conversion of heat into work). The second class was aimed at the transfer of heat from higher to lower temperatures and the generation of heat by work. By mere intuitive and logical reasoning (which might remind us of a current field known today as synergetic) *Clausius* was able to prove that the implementation of processes can be quantified by an algebraic summation of the proportions of the heat content accumulated in the course of investigated process to the corresponding value of the absolute temperature.

For such a classical field of thermodynamics (understood in the sense of being a not too popular, but adequate term for 'thermostatics') the closed system defines and describes the energy conserving approach to equilibrium, which is, factually, thermal death (certainly, if neglecting ever-presented fluctuations). It bears similarity with the mechano-dynamics of an isolated system, i.e., fixed state of planets (or molecules) kept together by stationary orbiting. The next step is the true thermodynamics with thermal bath, capable to draw or reject energy as to balance the coexistence between the two phases divided by a phase transition. The counterpart in mechanical-dynamics is the existence of an external field acting, for example, on a stone dropped from a tower. The external fields are the source or sink of energy – momentum without being modified. Any disequilibrium must be created by, e.g., artificially created experiments (by 'experimentationists'). However, for the case of increasing number of thermal baths we have no such equivalent in mechanic-dynamics but we can account for the possibility of thermal cycles and get closer to the description of a situation met in nature. This is the curiosity of thermodynamics, which thus diverges from that familiar in scholarly mechanic-dynamics.

True reversible processes, however, served as an abstraction as they do not actually exist for macroscopic bodies under our control. By making interfering frictional effects small, we can diminish the inherent entropy change almost infinitesimally – as much as we please. To do this we usually require slower and slower process, i.e., longer and longer time for its completion. Therefore, the reversible process has interest only as a standard of comparison, not as a practical or desirable end in itself. We may cite *Sadi Carnot "we should not expect ever to utilize in practice all the motive power of combustibles. The economy of the combustible is only one of the conditions to be fulfilled in heat engines and within many other aspects it may become only secondary"*.

4.5. Atomists and matter

In 1675, *Lemery* published [145] "The Course of Chemistry", which had an enormous impact on its progress because it finally relinquished most references to occult qualities. For *Boyle* in his time-honored "Sceptical Chemist" (1661), the consequence of mechanist atomism was that all chemical bodies, whether we could resolve them or not, were produced by 'different textures' of a 'catholic or universal' matter, i.e. an arrangement of particles 'without qualities' would be responsible for what we call properties (characteristic qualities). Boyle's definition that holds as *'I now mean by elements certain primitive and simplest bodies, or perfectly unmingled bodies, which not being made of any other bodies, or of one another, are the ingredients of which all those called perfectly mixt bodies are immediately compounded, and into which they are ultimately resolved'* and was later known as a 'negative empirical concept' enabling modern definition of the element.

According to *Newton*, atoms were solid, heavy, hard, impenetrable and mobile particles that God made at the Beginning, but he did not accept the idea that atoms were characterized only by the force of inertia – a passive principle in virtue of which bodies remain in movement or at rest meaning that there would be neither destruction nor generation (as with life). Such a corpuscular chemistry as a site of conceptual experimentation on the sequence of atomism, assumed that the constituent elements did not continue to exist 'potentially' in a compound but composed it actually. Chemical transformation had to be thought of in terms of the separation and combination of particles, which were thus assumed invariant and incorruptible, existing prior to combinations and remaining themselves in *'mixts'*.

From the atomist point of view, *Aristotle's* concept of passage from potentiality to actuality no longer made sense. Correspondingly, genesis (*genesis*), destruction (*phthora*), and alternation (*alloiosis*) no longer referred to qualitatively different processes but to a kind of quantitative change, that *Aristotle* called locomotion (*phora*). Atomism, was seen as a *'methaphor'* for the alphabet, and the 'tiny bricks', which can provide a solid construction, were thought to be atoms, which became a principle of both reality and knowledge. There, however, remained a question: what caused the homogeneity? The concept of mixture reemerged, recalling what *Aristotle* stated as `stoicheia`, which are the constituent elements of a body and can be transformed during the process of decomposition. *Aristotle* also asserted that the '*dynamis*' of the elements remained in the '*mixt*', in other words, the properties of a mixt reflected the properties of its constituent elements. From the point of view of chemical operations, corpuscular chemistry tended to favor those procedures that were reversible. With the distinction between primary properties (extension, form, mass) and the secondary ones (heat, sound, color) the mechanist version of atomism denied all qualitative differences to atoms and only granted such differences to geometrical attributes.

As a consequence of the opposition between 'simple' and 'compound', there is the famous *Lavoisier's* definition of an element "*if we give to the name of elements or principles of bodies the idea of the last step that analysis can reach, all substances that we have not been able to decompose by any way whatsoever are elements for us, they act to as simple bodies and we must not suppose them to be decomposed until the time when experiment and observation will have furnished the proof*".

We certainly should not forget *Lomonosov* who was the first physical chemist to view chemistry from the standpoint of physics and mathematics. He considered matter to consist of minute imperceptible particles "*the oscillation and rotation of these particles increases on heating and when rotary motion reaches a certain stage the material melts or volatilizes. The particles in gases move more rapidly than in liquids and collide with each other*". His ideas were ahead of those of *Lavoisier* on the nature of heat and on its resultant effect on materials.

The Swedish chemist *Berzelius* [146] explained in the 1810s the action of the newly invented electrical battery and associated process of electro-deposition. He defined each simple substance and each compound body by a positive or negative polarity whose intensity varied according to the nature and the number of the positive and negative charges carried by the atoms of each element. The voltaic battery was not seen as a simple instrument but as a 'principle of intelligibility' where electricity was understood as the primary cause of all chemical action. The "electrical fulfillment" was actually the original Newtonian dream of mechanical actions between atoms. On the basis of opposite forces it was possible to design a simple method for predicting the degree of affinity: the scale of mutually reacting elements from the most electropositive to the most electronegative.

Dalton [135] made use of *Proust's* law as the basis for a new atomic hypothesis and suggested that chemical combinations take place in discrete units, atom by atom, and that the atoms of each element are identical. These atoms differed from *Newton's* corpuscles because they presupposed neither the void nor attraction and made no attempt to explain properties of simple bodies in terms of a complex architecture whose ultimate constituents would be atoms. After the guide of *Dalton's* new system of chemical philosophy, *Gay-Lussac* announced that the volumes of gases, which combine with each other, were in direct proportion – the volumetric proportions thus confirmed by the gravimetric ratios. The first law, to fashion a network of experimental facts from different disciplines, was formulated in 1811 by *Avogadro,* stating that equal volumes of different gases contain the same number of molecules.

Dulong, with his young colleague *Petit,* undertook a study of the quantity of heat needed to raise the temperature of one gram of a substance by one degree Celsius. They determined heat capacities of each atom to be nearly the same (convenient notion is gram-atom which was earlier specified as 6 cal/^{o}C^{-1} gram-atom^{-1}, today's expedient as about 25 J mol^{-1} K^{-1}). They concluded, *"the atoms of all simple bodies have the same heat capacities"*. It is now called *Dulong* and *Petit's* law of specific heat and was based on a new field – the above-mentioned calorimetry. This law could not lead directly to the atomic-weight values but it presupposed them. It was followed by the theory of chemical proportions derived upon isomorphism where only the number of atoms determined the crystalline form, and little angular variations attributed to the nature of individual atoms.

In a jungle of obscure and exotic terminology, it was often difficult to read chemical texts. In 1860, during probably the first international Chemical congress, held in Karlsruhe, participants tried to put an end to these deep differences on words and symbols that harmed communication and discussion. It raised a fundamental theoretical issue for the agreement on figures and formulas to be subordinated as well as the definitions of basic concepts: atoms, molecules and equivalents. Moreover it materialized the existence of an international chemical community and defined the rules for its functioning

In 1865, the English chemist *Odling* published the table of elements, which was almost identical as that published five years later by *Mendeleev* [147], although they did not entirely believe in the existence of atoms as particular species. The individuality of the chemical elements was for *Mendeleev* an objective characteristic of nature, as fundamental as Newtonian gravitation; it was both a condition and a product of his periodical classification having a real, and not just a theoretical meaning. His periodic classification was far from being prophetic of the electronic structure that is today so well known and, ironically, those defenders, who were in favor of the existence of the previously popular unique 'primordial' element, criticized it. *Mendeleev* protested against this misuse of his discovery and when he learned about radioactivity and about the electron, he even advanced his explanation in terms of vortices of the ether, which take place around the heaviest atoms. Thus he initiated an unfortunate hypothesis about the ether as a definite element and placed ether above the column of rare gases. Periodic classifications, see Fig. 23,

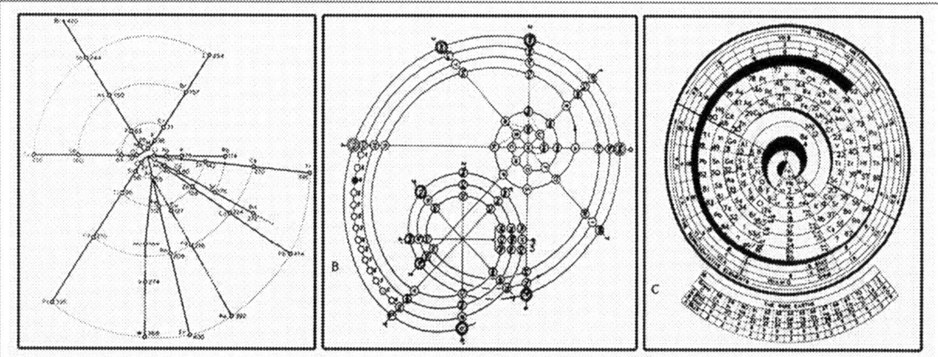

Fig. 23. – Various systematization of the elements periodicity [9] preceding and following the accepted periodic system introduced by *Mendeleejev* 1870. Left, simplified system pioneered by *Hinrich* in the year 1867, the twisting modification challenged by the German chemists *Nodder* (1920) up the recent spiral alteration (right) anticipated by the US chemist *Emerson* (1944). It is worth noting that these modifications show some signs of the ancient attempts of counting numerical symbolism as well as a natural tendency to take shape of spirals (cf. Fig. 64). Courtesy of *Vladimir Karpenko*, Praha, Czechia.

drew a lesson from the chemistry of substitutions and, correlatively, substitution lost its subversive status – it remained no more than one mode of combination among others.

In 1750, the German physician *Eller* published the first tables presenting solubility data of various salts and their mixtures in water at different temperatures (and indication of atmospheric pressure). The first database, however, was most likely done by the Swedish chemist *Bergman* (1783) who put in to order several thousand chemical reactions (containing 49 columns with 27 acids, 8 bases, 14 metals and others, and discriminated between reactions by a wet method in solution and a dry method by fire). Besides foregoing recent databases he also helped to circumscribe nomenclature. Reactions were no longer 'means' but became 'phenomena', which countered anomalies and had to extend between the chemical affinity and the other physical factors that emerge obstacles in its completion. Only the clairvoyance of the Russian chemist *Beilstein* laid the true grounds for modern databases when he originally published a survey of 15 000 organic compounds as a series of books published in the period 1880–1883, the continuation of which exists as a well respected database until today.

The important process of synthesis was not first distinguished according to the nature of the product (natural or artificial). Synthesis was thought to be merely total or partial. Substances were made from elements (commonly C, H, O) or from other, simpler compounds. *Wohler's* laboratory synthesis of urea in 1828, a substance made previously only by the action of a living organism, was celebrated as an event of unprecedented importance, demonstrating the nonexistence of the 'vital force' that had previously been necessary to create organic molecules. In 1847 *Frankland* discovered a new class of organometallic compounds, a concrete argument in favor of a reunion of organic and inorganic chemistry, which also became a starting point for the modern theory of valence (introduced as 'atomicity'). In fact it was a fruitful time for new scientific images such as that of *Kekule* who said that he owed his career to the vision of the ring structure of benzene as a snake chewing on its own tail (cf. Fig. 12.).

After the invention of bakelite, prepared from phenol and formalin in 1907, *Baekeland* introduced the generalized term 'plastics' in 1909 to design a larger class of products conceived as replacement products for natural substances, which were either rare or costly. They were a triumph of substitution chemistry and in the 1950s seemed to take on aesthetics of their own. Polymers were technological items before they were objects of knowledge. Laboratory chemists were annoyed when syrups did not crystallize or with solids that were impossible to melt and they mentioned them in notes as substances of unknown structure. Thermo-hardening properties were used to fashion a variety of objects, their structure yet unknown. Although there was a hypothesis that molecules joined together in chains by ordinary interatomic bonds (advanced for isoprene in 1879), most chemists from the beginning of the nineteenth Century thought that a pure body must be composed of identical molecules of small size. Macromolecules were explained as an aggregation of small molecules, which was later verified by crystallographic research.

Without being self-confident that crystals are internally arranged in a periodical manner *Bravais* mathematically described, in the year 1850, fourteen geometrical figures that can be periodically arranged in the space and characterized them as a combination of one or more rotations and inversions in a lattice, which is understood as a regular array of points (each point must have the same number of neighbors as every other point and the neighbors must always be found at the same distances and directions. All points are in the same environment. His idea was later approved by X-ray diffraction (*Röntgen* 1912) and this approach has been advantageously applied until now so that the so called 'Bravais Lattice' is understood as a three dimensional network, which tiles space without any gaps or holes (there are 14 ways in which this can be accomplished). Microscopic investigations made by *Reinitzer* when studying cholesterol in 1988 extended the developing basis of crystallography to the sphere of liquid crystals (later widespread to generality by *Lehmann*).

In 1855, the young pathologist *Fick,* wrote a work entitled "*Uber Diffusion*" published in Zurich "*Annalen der Physik*". Surprisingly, *Fick* was an experimental physiologist, but his work on diffusion became theoretical and his approach would today be called a phenomenological liner-response theory applied to mass transport. He started observing "*diffusion in water confined by membranes is not only one of the basic factors of organic life, but is also an extremely interesting physical process and, as such, should attract much more attention than it has so far"*. As a matter of fact the carrier particles involved in their current have to flow against a concentration gradient, which is analogous to Ohm's Law of electric current and Fourier's Law of heat flow. Assumingly, Fick's phenomenology missed the probalistic point of view that is central to statistical mechanisms and it was Einstein who 50 years later derived the diffusion equation from the postulates of molecular theory, in which particles move independently under the influence of thermal agitation.

The two most traditional actors in chemistry, the chemical reaction and heat, were joined to conceive 'thermochemistry'. Just as the fall of a body is characterized by the work of mechanical forces, the decrease in potential energy and the creation of kinetic energy, a chemical reaction must be defined by the work of chemical forces and the decrease in potential of these forces. Work and decreases in potential were measured by the amount of heat released by reaction. The state of equilibrium thus became the state in which the potential of chemical forces had reached its minimum value. It was a transposition of the old doctrine of effective affinities and corresponds to the discrimination among chemical reactions. The natural chemical reaction was the one spontaneously giving off heat while the endothermic reactions were considered constrained by an external action, by the chemist who adds the heat (preference to higher temperatures).

In 1867 *Guldberg* and *Waage* proposed a law that abolished any distinction between exothermic and endothermic reactions and created a new analogy with physics. They put forward the idea of 'active mass' in analogy with Newtonian force, i.e., the chemical force of a reaction was defined as the product of the active masses, and equilibrium was reached when the forces of opposite reaction became equal. However, the relation between forces and masses involved a specific coefficient of activity. It was a great success but it left open the question of its interpretation. In this hypothesis equilibrium was no longer defined as a state in which a reaction stopped but the state in which reaction rates were such that there effects compensates for each other. Equilibrium was not the state where forces, and the rates they determined, both vanished and therefore it was nothing special – just as the analogy with reactive collisions between molecules, which determined the given reaction, were on the whole as numerous as the collisions that determined the inverse reaction. The central concept of the current field called 'kinetics' is thus the probabilistic notion of frequency dependent on temperature.

Finally we come to deal with structures, ideal and real, associated with variety of defects, impurities, vacancies, interscialities, interfaces, etc. – these terms imply the usual hierarchy, the ideal being the prototype from which we can know about real things and their inevitable deviations. But henceforth it is the *defect* that is interesting, for the specific properties it gives the single crystal

– in the larger scale it is its matter-separating surface specifying its outer form and interfacial properties. The individual crystalline body is no longer a mere imperfect version of the ideal prototype but the reflection of the singular history of its formation, growth and orderliness. The possible multiciplicity as well as both the scientific and industrial interests for particular properties linked to the digression (accidental deviations) from the established laws (predictable state) were disposed to remain between the boundary status of 'defects' (a non-hierarchical group of cases, each associated with the circumstances that favor it) and 'normal cases' (as demonstration of rules).

By penetrating the micro-, supra- or nano-molecular world [88,89] and playing with the interactions among molecules in the standard three- but also in two- and even in the one-dimensional range (quantum wells or dots), the physical chemists has become a new kind of architect of matter, facetiously saying 'tricky designers for tricky matter'. Among others we may mention optoelectronics, superconductors, magnetics or, particularly, alloys with shape memory, self-adjusting photochromic glasses, bioactive ceramics and cements or quantum-well semiconductor microelectronics. Such sophisticated materials are often called *intelligent materials* [87], in the sense, as if the chemists had breathed life into matter and accomplished and old dream of alchemists.

We are aware of the great environmental cycles [70,77,84] of nitrogen, oxygen, carbon, phosphorus or sulfur and identified the succession of transformations, each one consuming what was produced previously and producing what would be consumed afterward. It is a beneficial self-recyclation (cf. Fig. 7), like a desired industrial assembly line in perpetual motion, which is looping back on its beginning [83]. But what is the a thermo-chemical balance if it does not integrate the many time horizons of those different processes that together create the overall global transformation [9,77,81,82] – we still have to learn a great deal!

4.6. Underpinning of thermodynamics

Towards the end of the 17th century *Papin* succeeded in proving that water can exist in a liquid state even at temperatures exceeding its boiling point if heated under increased pressure. It preceded the discovery of the *critical state* of substances and the determination of their critical values of pressure, volume and temperature (often named as 'disliguefying' or 'absolute boiling' points after *Andrews, Faraday* or *Mendelejev*). Experimental evidence [135,146,147,148] initiated shaping of appropriate state equations of a real gas intending to replace unsatisfactory technical analysis using the state equations for an ideal gas (already introduced by *Regnault* in the year 1847). The '*virial*' of external forces (~3pV), suggested by *Clausius* twenty years later, descibed the difference between the behavior of real and ideal gases. It was centered on the mean free path of a molecule and the potential of intermolecular forces was taken as being proportional to $1/V^2$, thus replacing the complicated function of *Laplace* that was based on the model of stationary molecules repelling each other due to their caloric envelopes (though satisfactorily explaining the surface tension, capillarity and even cohesion). The use of reduced values by *van der Waals* (1880) enabled him to postulate the *law of corresponding states* based on the state parameters expressed in the units represented by their critical values.

The most important personality in thermodynamic history was credibly *Gibbs* who discriminated that a system of r coexistent phases, each of which having the same independently variable components, n, is capable of $(n + 2 - r)$ variations of phase, known until now as the famous 'phase rule', that factually unveiled that the whole is simpler than its parts. It followed that for temperature, pressure and chemical equivalents ('potentials' later specified as 'chemical potentials' by *Ostwald*) the actual components bear the same values in the different phases and the variation of these quantities are subject to as many conditions as there are different phases (introduction of partial derivatives). This important work on the theory of phase equilibria was published in the period 1873 to 1878 in an almost unknown journal "Transaction of Connecticut Academy" and its insufficient publicity was fortunately compensated by the proper recognition of renowned scientists [135,146-151], such as *Maxwell, Duhem, Ostwald* or *Le Chatelier,* also mentioning the Dutch school of thermodynamics, that must be particularly credited with the broader application attempts aimed at the problems of general chemistry and technology.

One of crucial focuses was the clear-cut meaning of entropy often specified to considering the case of mixing two gases by diffusion when the energy of whole remains constant and the entropy receives a certain increase. We may cite *Gibbs* quoting [151] "*...in the mixture of two identical and ideal gases by diffusion an increase of entropy would take place, although the process of mixture dynamically considered, might be absolutely identical in its minutes details (even with respect to the precise path of each atom) with process which might take place without any increase of entropy. In such respects, <u>entropy stands strongly contrasted with energy</u>. Again, when such gases have been mixed, there is no more impossibility of the separation of the two kinds of molecules in virtue of their ordinary motions in the gaseous mass without any special external influence, than there is of the separation of a homogeneous gas into the same two parts into which it has once been divided, after these have once been mixed. In other words, the impossibility of an uncompensated decrease of entropy seems to be reduced to improbability...*".

Another important area of Gibbs's interests was the depiction of a fundamental dependence of the intrinsic energy of a one-component system versus volume and entropy, often called the '*thermodynamic surface*'. It helped

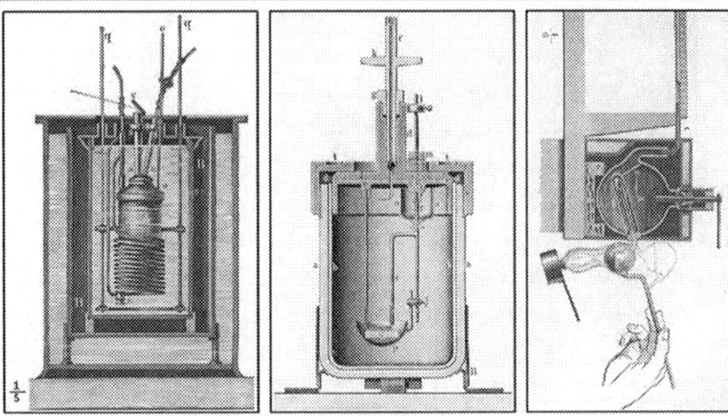

Fig. 24. – *Favre* and *Silbermann* are not widely known for their early construction of a combustion calorimeter, which was adjusted for higher pressures by *Berthelot* (known as today's calorimetric bomb).

in the distinguishing of individual areas of stability in single- and multi-component systems that are in an equilibrium coexistence and which yield a whole what is now called *"surface of dissipated energy"*. This method for investigating the equilibria of many-component systems, with the use of equality of the potentials of all the involved components in all the existing phases, became widely accepted after the introduction of the quantities of the *escaping tendency* (1900) or 'fugacity'(1901). Finally the term *'activity'* was lastly introduced by *Lewis* in 1907 as relative [150] and established in the current meaning in 1913. Curiously *Lewis* was involved in various thermodynamic applications reaching even to the realm of economics [152].

In the last quarter of the 19th century the theory of *metastable phase equilibria* marched under a more serious consideration although the formation of non-equilibrium states of pure substances and their mixtures had been experimentally proved long before (e.g. *Fahrenheit* published in 1724 experimental results on undercooled water). Gibbs was the first to use the term *unstable* equilibrium while the authorship of *metastable* is ascribed to *Ostwald* who in parallel with the term *labile* presented an exact definition of its meaning as early as in 1897. *Van der Waals* modified the original Gibbs's terminology of *limit of absolute stability* to 'binodal' (points of contact of the common tangential planes within the Gibbs's thermodynamic surface were named as 'nodes' according to mathematician *Cayley)* and the *limit of essential instability* were called 'spinodal' (i.e., curves dividing the concave-convex surface into areas convex in all directions and those remaining completely concave). It is also worth noting that these purely theoretical considerations led to the discovery of two rather extraordinary and less known phenomena, such as 'retrograde' condensation and the 'barotropic' effect.

At the turn of 20th century a more modern nomenclature also emerged, such as the notion of a *eutectic* mixture and a eutectic point introduced by *Guthrie* (derived from the Greek expression used already by *Aristotle* in a similar sense of being easily melted – 'eutektor'), followed by 'peritectic' reaction (*Lehmann)* or 'eutectoid' (*Howe*). The progressive nature of both the theoretical and experimental treatments of this period is manifested in the fact, that the described phenomena were not yet fully understood yet. It was helped by the new approach called *thermal analysis* (*Tammann* 1905) that enabled the determination of composition of the matter without any mechanical separation of crystals just on basis of monitoring its thermal state by means of its cooling curves – the only method capable of the examination of hard-to-melt crystal conglomerates. It brought along a lot of misinterpretations, the legendary case of the high-alumina regions of the quartz-alumina binary system continuously investigated for almost hundred years. It, step by step, revealed that the mullite phase irregularly exhibited both the incongruent and congruent melting points in dependence to the sample course of equilibration. It showed that mere thermal analysis is not fully suitable for the study of phase equilibria, which settle too slowly. In 1909 there was elaborated another reliable procedure of preserving the high-temperature state of samples down to laboratory temperature, factually *freezing-in* the high-temperature equilibrium as a suitably 'quenched' state for further investigation. It helped in the consistent construction of phase diagrams when used in combination with other complementary analytical procedures, such as X-ray diffraction or metallographic observations.

Among the generalized applicability of the fundamental laws of thermodynamics, the description of the equilibrium coexistence of a mixture's phases became important. *Kirchhoff's* relation, which enabled the calculation of *'the amount of heat, which is liberated when a certain mass of water dissolves the minimum amount of salt as it is capable to dissolve'*, represented the first of these (cf. Fig. 24.). Another case was Raoult's law for the decrease of the melting point of a solution as well as the decrease of pressure of the solvent's saturated vapors over the solution. In 1884, the relationship, earlier derived by *Gulberg* for the depression of freezing point and the solute-solvent ratio, was taken over by *Van't Hoff* in his extensive treatise when calculating, for the first time, the osmotic pressure of a solute. Here also belonged the first derivation of the 'liquidus' in the phase diagram of the condensed binary system for the region of the low contents of solute (say for the phase 2) familiarly known in the form of $dT/dx_2 = RT^2/\Delta H_{2\,(melt)}$ and later derived also by *Planck*. However, the author of a more general relationship was *Le Chatelier* in 1885 who proposed a logarithmic function of composition of a saturated solution in the form $dx/x = (k/\sigma)Q \times dt/T^2$ where x, k/σ and Q are the ratios of amounts of substances of solute and solvent, the proportionality constant related to gas constant and the molar heat of dissolution, respectively. A few years later it was improved by

Shreder in his description of a solvent by the equation, $log\ x = -\Delta H_{(melt)}\ [T_{(melt)} - T]/(R\ T_{(melt)}T)$ since then known as the *LeChatelier-Shreder* equation. It was based on the assumption that the molar heat of dissolution over the whole range of temperatures and compositions is constant and equals to the molar heat of fusion of the pure component, i.e., showing the relationship $\Delta H_{(sol)}/T \approx \Delta H_{(melt)}/T_{(melt)}$.

Important studies were performed by *Lewis* who defined the 'perfect' (later termed as 'ideal') solution already in 1901, thus following van't Hoff's idea, which became extraordinary fruitful in the further development of thermodynamics of mixed systems since it enabled rational thermodynamics to sort out the solutions based on the differences between the behavior of real and ideal mixtures. Rault's law for extremely diluted solutions involved the use of an equality of differentials of logarithms of the absolute activity, a, and that of the mole fraction of a solvent, x, yielding $d\ log\ a = d\ log\ x$. *Guggenheim* classified solutions according to the experimentally determinable parameters, employing the dependence of the excess of Gibbs energy of mixing, $(\Delta G^{ex})_{mix}$ proposed as a combination of symmetrical and asymmetrical functions. *Lewi's* model of an ideal solution was applied to molten salts in 1945 by *Těmpkin* who took the limit of solution of this kind for a mixture of two independent ideal solutions of ions with a coinciding sign of charge.

Adjusted field of innovative statistical thermodynamics reached important authority thanks to *Boltzmann*, who kept persuading (and succeeded to convince his followers only after his 1906 suicide) the validity of logarithmic relation between phenomenological entropy and microscopic complexion, which equation was, factually, not his invention as it has already been applied to games of chance a century earlier by the French mathematician *DeMoivre*. Nevertheless *Boltzmann* found its new utility in thermodynamics and later *Shannon* yet another efficacy in communication, both applications dealing with a new specification called *uncertainty*. It has become an everlasting discussion what is the nature of uncertainty and its interconnectivity between individual topics. It was accepted that for thermodynamics uncertainty reflects something fundamental about the system as one cannot know in which microstate such a thermodynamic system happens to reside at any instance because it fluctuates stochastically among an ensemble of such microscopic possibilities. On the other hand, in information, the uncertainty was not seen thus fundamental as it rather reflects the fact that a given message or sequence is but one of many that might have been generated from a given symbol set so that once the sequence is specified the uncertainty seems to fade away.

Apparently, it was found apprehensive to know a thermodynamic system in general for which we routinely need some representation about certain in-depth make-up (microscale processes), which, however, may be dependent on the model applied. We may recall *Gibbs* citing "*...in spite of certain* *incomplementarities of thermodynamic and statistical approaches they are inseparablyconnected with each other, i.e., thermodynamic values are averaged on the whole system means of physical values, which are considered in detail by statistical physics. In this sense the statistical approach would justify the thermodynamics from the mechanical point of view. On the other hand, the same thermodynamic values, which are observed in physical reality, can be considered as some kind of guidance through the labyrinth of statistical theory, although such a kind of guiding can be rather blind...".*

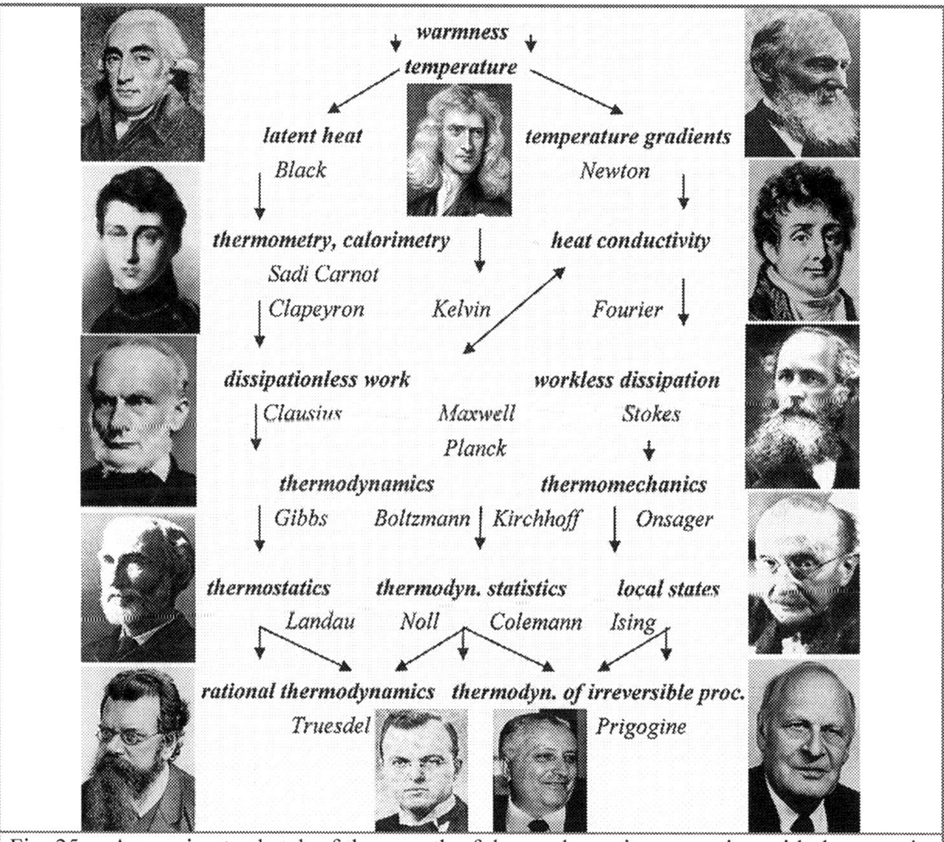

Fig. 25. – Approximate sketch of the growth of thermodynamic conception with the portraits of some famous pioneers, left column from above: *Joseph Black* (1728-1799), *Sadi Nicholas Carnot* (1796-1832), *Rudolf Julius Clausius* (1822-1888), *Josiah Willand Gibbs* (1839-1903), *Ludwig Eduard Boltzmann* (1844-1906), right: *Kelvin*, Baron of Larges, Lord Williams Thompson (1824-1907), *Jean Baptiste Fourier* (1768-1830), *James Clark Maxwell* (1831-1879), *Max Carl Planck* (1858-1947), *Lars Onsager* (1903-1976), middle: *Sir Issak Newton* (1642-1726), *Clifford Ambrose Truesdell* (1921-) and *Ilya Prigogine* (1917-2003).

In the middle of the 1920s, there arose an important period of the *Onsager* revolution initiated by his article on "Reciprocal Relations in Irreversible Processes" [153], which actually followed a kind of quantitative thermal analysis. It was based on the accomplishments made by *Ising* when describing the thermal behavior of a linear body consisting of elementary magnets under the assumption of interaction of their nearest neighbors. It was preceded by the definition of order-disorder transitions and the formulation of a general model of the second order phase transition according to which the system reveals the so called λ-point of transition, which is easy to characterize by an order parameter with zero and non-zero values above or below this transition temperature. In this treatise *Landau* also proposed an expansion of the *Gibbs* energies into a power series and entered, thus, the world of broadened phase transitions. It was just the starting point of modern theories of irreversible processes (*Coleman, Prigogine, Truesdel,* cf. Fig. 25).

Quantum mechanics brought about a real revolution [154]. It may be said that 'Hamiltonian physics' was kept alive in the halls of Academies while the 'theme of entropy' subsisted in the workshops of the smiths where it was actually born. The predictive power of quantum mechanics was upheld in connection with the concept of periodic states of motion; it returns and repeats itself, in this way carrying recognition. In the modern inflowing concept of lively disequilibria, the concept of entropy became important, showing that dissipative systems are profoundly different from Hamiltonian systems, which can be formalized from the unifying principle of minimum action. Hamiltonian systems come, in general, from a well defined variation principle; it provides an important apparatus for conservation laws and symmetries and produces great simplifications as well as bringing one close to a solution. Initially distasted systems of yet mysterious dissipation do not have an equally general formulation, they are open, non-isolated, interact with the outside and exhibit generalized properties, possibly the reason why they are sometimes considered scientifically 'less aristocratic' and, thus, worthy of increased attention. In contrast to the classical perception of equilibration, for a dissipative system the state of motion literarily dies out unless energy is constantly provided.

4.7. Thermal radiation and the modern concept of vacuum

In the year 1874, the Italian experimental physicists *Bartoli* [155] put forward an idea of bringing electrodynamics and thermodynamics into the treatment of heat radiation, which is often considered a major source of inspiration for the later concept of *Stefan-Boltzmann* Law of blackbody radiation [156]. Light pressure, however, was not a new subject and played important role with respect to the competing theories of light: the corpuscular emission was taken to imply the existence of light pressure whereas the wave theory was usually regarded as incompatible with such a pressure. In 1870, *Crookes* constructed his radiometer ('light mill') and interpreted its motion by result of mechanical energy of radiant heat. In 1873, *Maxwell* noted in his new electromagnetic theory that *"in a medium in which waves are propagated there is a pressure in the direction normal to the waves ... and that falling rays might perhaps produce an observable mechanical effect on delicately suspended things in vacuum...and he calculated that for the maximum sunlight to be about 8.82 x 10⁻⁸lb per square foot"*.

At the same time, *Maxwell* also enlightened *Kepler's* observation that the comet stream is tailed away from the sun upon the pressure of the rays falling from the sun. In 1924, the Russian scientist and rocket theoretician *Ciolkovskii* first realized that the pressure of sunrays could be used for propelling a cosmic vehicle furnished by suitable radiation-reflecting sail in the moment when it is lanced out to the cosmic space. As early as in the year 1970 the effects of sunrays were utilized to successfully orient the Mercurial probe 'Mariner'.

The recent calculations provided by NASA revealed the estimate that a cosmic sail-ship could be accelerated to the speed five times faster than that achievable by conventional rocket drive. If such a sailor would be launched in the year 2010, it would meet early cosmic probe 'Voyager' in the year 2018, i.e., within eight years it would exceed the distance that 'Voyager' needed forty years to pass. Currently there are several successful (and also unsuccessful) missions, which employed such a sunrays drive for its acceleration, for example, the UNESCO spacccraft 'Star of Tolerance' with the projected sail area of 1600 m^2 or the sailing vessel 'Cosmos' proposed by the American Planetary Society. Such assignments, however, are still under the scientific disputation whether an absolute reflection of solar photons on the finished mirror would follow the laws of thermodynamics (requesting upon the gain of mechanical energy a change of temperature through the change of the photon wavelength).

However, the examination of light balances (rather than light mills) was actually due to the need to distinguish between the direct action of radiation and the indirect action caused by convection currents in the residual molecules in vacuum, resulting from heating. *Bartoli* adopted the *Clausius* version of the second law of thermodynamics restricting heat transfer from a colder to a warmer body without a compensating amount of mechanical work being performed. *Bartoli* imagined a perfectly evacuated system consisting of four concentric shells. The outermost and the center shells were firm black (fully absorbent) bodies while the intermediary two shells were contractible (and upon the need also removable) with a perfectly reflecting surfaces on both sides. The spaces between two outer and inner shells, were thermally isolated while the outermost and central shells, when the other two were in thermal equilibrium. Its function, however, required some artificial operations: at a given moment the destruction of the second outer shall produces radiation of heat in the entire space in between outmost shell reached thermal equilibrium, the second shell is re-created and the third shell is destroyed. Thereupon the second shell is contracted as far as its radius becomes equal to that of third. On repetition by cyclic operations, a definite quantity of heat is taken from the outer and transferred into an inner shell. Such transfer allows an unconstrained assumption that the temperature of the third shell can be higher than that of outer one.

Assuming that the inner shell is a finite system with an intrinsic quantum property to expand, the process will be "spontaneous". We have to do with the case where "the heat passes spontaneously from a colder body to the hotter one", which is in a clear contradiction with the second law of thermodynamics in the original *Clausius'* wording. Such a straightforward argument is rather naïve (however, not too rare in the literature) and does not stand the confrontation even with a more advanced *Clausius'* formulation of the second law of thermodynamics claiming:

"a passage of heat from a colder to a hotter body cannot take place without compensation".

The "compensation" here means a change within the system, which eventually has in consequence in the "natural" heat transport from a higher to a lower temperature. At first glance, it is very tempting to judge that the "compensation" is realized here by the expansion of the inner shell followed by the heat flow from the hotter soot to the colder calorimeter. However, the inner shell itself is considered to behave like a usual mechanical-thermodynamic system conservatively closed, which was provided at the moment of its creation (with some initial potential energy and then being a part of the system). Even in this case, one can conclude that there is nothing strange and that no violation of the second law takes place. It has had important consequences in various experimental and thought proposals to construct such a devises [157] and even has led to the possibility to violate the second law of thermodynamics, the discussion of which has been lasting until now and even growing deeper [158].

Such an original, qualitatively thought experiment seemed merely to have served as a heuristic guide for *Bartoli* to consider the existence of light pressure. Later he approved this highly artificial construction by a more realistic variant in which the spherical shells were replaced with a classical cylinder (with permeable and reflecting walls) operculated with crest and tail lids (fixed as blackbody) and operating with two moving pistons as heat reflectors [156]. The net result is again a transfer of heat between the cylinder's closures assuming the amount of heat, q, to be proportional to $K R S_A / v$ where S_A is the surface of lids with the emissive power K, piston diameter R and radiant heat velocity, v $(\equiv c)$. Consequently, *Bartoli* derived the relation for radiation pressure, $p = 2 Q E / v$, where Q is the power received by the unit area and E is the mechanical equivalent of heat, enabling him to calculate the solar pressure to have the value

8.4 x 10^{-4} g/m^2 (which, however, was about 2000 times less than required to power a radiometer). Few years later, *Bartoli* gave up his original idea deciding that there is no light pressure after all.

The absolute validity of the second law of thermodynamics was for long a hot topic of discussion and particularly *Maxwell, Boltzmann or Loschmidt* concentrated on the principle necessary for a mechanical explanation, in general, not referring to particular processes such as radiant heat, which might violate the law. In 1882, the US astronomer *Eddy* argued in a thought experiment of a system of apertures distributed equidistantly around three concentric cylinders (*'syren'*) that radiant heat (having presumably a finite velocity) could apparently be an exception. He evidently viewed his result as providing the support for the *Maxwell-Boltzmann* hypothesis, which in *Eddy's* interpretation stated that *'the second law was merely the mean result emerging from the laws of probability'* and which was an escape from the pessimistic prediction of the prognoses of heat death of the Universe.

Boltzmann later raised a critique against the details of Bartoli's process arguing by necessity to have to modify it to become reversible and adopted the assumption of radiation pressure as well as the validity of the second law of thermodynamics (implicitly rejecting *Eddy's* claim). He conveyed the relation $p(T) = (\pi/c)\ T \int \phi(T)\ dT/\ T^2$, where ϕ is the energy flux and T is the absolute temperature, and confirmed experimentally Stefan's Law of radiation energy in the form $p(T) = (\pi/3c)\ \phi(T)$ and $\phi(T) \approx T^4$. It literally followed *Maxwell's* old result written as $p(T) = (1/3)\ \psi(T)$, where ψ is the energy density. In 1893, *Wien* derived his displacement law by extending Boltzmann's reasoning [159] to cover the separate wavelengths of the blackbody radiation (scaling) so that $u_\omega = \omega^3\ f\ (T/\omega)$, where f is a universal function and ω is the frequency. Both these relations were temperature dependent so that at the absolute zero they would vanish, which would bring somehow 'disastrous' consequences on the stability of atoms factually leading to body's collapse [160].

Let us mention that any system is defined as being at absolute zero when no heat flow, Q, can occur out of the system during any reversible isothermal process performed on the system. Consequently, for the classical electromagnetic zero-point radiation process, only the nonzero spectrum is suitable for establishing an equilibrium state with the electric dipole oscillators at a temperature of absolute zero. This requirement ($Q=0$ and $T=0$) must also satisfy the third law of thermodynamics, i.e., the ratio of Q/T should also approach zero in the limit of $T{\to}0$, which places a further restriction on the spectrum of incident radiation.

Therefore, if a statistical equilibrium configuration is at all possible for a system of classical charged particles, then at a temperature of absolute zero must exist a zero-point classical electromagnetic radiation as well as a zero-point oscillating motion for the charges. Of course, zero-point field and motion are normally associated with quantum-mechanical systems and are alien to the traditional ideas of classical thermal physics. However, a qualitative way of understanding why zero-point fields and motion exist, should be a natural part of thermodynamic behavior of classically charged systems of particles, i.e., they *cannot* exist in a static, stable equilibrium.

Hence, if an equilibrium situation for charged particles is at all possible, then the charges must be following a fluctuating, oscillatory path in space. The oscillating charges produce fluctuating electromagnetic fields, which in turn act upon these charges. Thus, any possible equilibrium situation must involve the presence of electromagnetic radiation, as well as an oscillatory motion for the charges, even at a temperature of absolute zero. All motion of charges would then possess a stochastic character. These qualitative ideas correspond moreover to what we observe in nature when at the zero absolute temperature molecular activity does not cease but has a *zero-point* motion.

This has brought into serious consideration the concept of zero-point energy of electromagnetic field background, which was originally introduced into physics for the sake of consistency of experimentally observed spectral composition of the blackbody radiation with the assumption of discontinuous light emission [161]. In modern quantum electrodynamics, the zero-point energy arises rather from the non-commutativity of operators corresponding to the wave amplitudes of electromagnetic field [162]. The spectral distribution of the blackbody radiation, which conforms to this requirement, is represented by a complete Planck's formula: $u_{\omega T} = (\omega^2/\pi^2 c^3)\{\hbar\omega/[exp\,(\hbar\omega/kT) - 1] + \hbar\omega/2\}$. This formula consists of two additive terms, the first describing the purely thermal (i.e. conventionally temperature dependent) part of the blackbody radiation. The other one corresponds to the radiation surviving even at absolute zero temperature, down to the so-called zero-point radiation. Using homogeneity arguments, there is only one possible form of the spectral energy density, u_ω, which is Lorentz invariant; it reads as $u_\omega = (\hbar\omega^3/2\pi^2 c^3)$. This relation, which describes isotropic zero-point radiation, has two significant properties: namely, it is Lorentz invariant and, in contrast to the temperature dependent term in the previous equation, its integral taken over all admissible frequencies is divergent.

A serious disadvantage of this formula is obviously due to its divergence with respect of the integration over the infinite frequency range. In order to obtain a physically more meaningful figure a rather laborious work with infinities and/or the introduction rather arbitrary cut-off frequency is required. Moreover, any macroscopic model of partitions involved in thermodynamic thought experiments with electromagnetic radiation should not a priori ignore their microscopic atomic structure without serious danger of introducing a bath by zero point radiation. Physically, the zero point radiation can be interpreted as a random, highly pervasive background field existing due to the incoherent vibrations of distant charges dispersed somewhere in and over the Universe. It is exclusively responsible for the quantum behavior of minute particles (often called 'Zitterbewegung'). Hence, the zero-point radiation ensures the stability of ordinary matter by precluding the devastating effect of the Coulomb interaction.

In this respect it worth noting that there exist the force between uncharged conducting surfaces, which is called '*Casimir force*' and which was described as one of the least intuitive consequences of quantum electrodynamics [163-168]. Casimir force per unit area equals to $\pi^2\hbar\, c/(240\ r^4)$, where r is the sub-micro-separation of two parallel plates and was derived as early as in the year 1948 [169] by considering the electromagnetic mode structure between the two parallel conducting plates of infinite extent in the comparison with mode structure when plates are infinitely apart by assigning a zero-point energy of $\hbar$ $\omega/2$ to each electromagnetic mode (photon). The only inherent fundamental constants are $\hbar$ and c, while the electron charge, e, is absent, which implies that the electromagnetic field is not coupled to matter. The role of the speed of light, c, is to convert the electromagnetic mode wavelength, as determined by r, to frequency, while the *Planck* constant, $\hbar$, converts the frequency to energy.

This Casimir effect results, thus, from changes in the ground-state fluctuations of a quantified field that occurs due to the boundary conditions. This was not predicted by the *London's* in his well-established calculation [170] made back in the year 1930, according to which the *van der Waals* unretarded forces arise directly from the *Coulomb* interactions between the molecules undergoing quantum fluctuations, yet later analyzed by *Boyer* [171] who reconstructed Casimir results for the energy change by proposing that the force between the plates arises from the zero point field subject to boundary conditions. It is worth noting that it has further followed the classical work by *Lifshitz* [172] on dispersion forces between dielectric bodies under field fluctuations.

It occurs for all quantum fields and it can arise from the choice of topology and is somewhat a current explanation of `vacuum` similarly to the Aristotelian `plenum`. It can be further extended to see the *Casimir`s* unsuccessful attempt to derive the fine-structure constant of the Universe by constructing an electron-based model upon the assumption that an electron is a sphere of uniform charge density, with its total charge equal to the electron charge and whose radius is determined by the balance between the attractive Casimir force (holding the electron together) and the Coulomb repulsion (tending the electron to expand). The effect of motion received attention in view of moving boundaries. An observer, who is in a uniformly accelerating frame could conclude that the frame (within the acceleration a) is reflected on the thermal bath of temperature $T = \hbar a/(2\pi ck)$ It calls attention to the acceleration, which promotes zero-point fluctuations to thermal fluctuations, similarly to the case if the plates of the Casimir experiment are accelerated away from each other enabling thus the photons generation in the gap [173].

5. UNDERSTANDING HEAT, TEMPERATURE AND GRADIENTS

5.1. Development of the concept of temperature

Temperature in view of *'temperament'* was first associated with the state of a human body [22,35,51,52,174]. Probably the first attempt to define the state of the human body by objective physical measurements came from a group of Italian scientists at the beginning of the 17th Century. *Sanctorius* experimentally studied the forces exerted by muscles, the content of various fluids in the body and the frequency of pulses using a *pulsologium* – an apparatus devised by *Galileo*. He tried, also, to measure the instantaneous characteristic of temperament, i.e., temperature, by means of a modified version of a very old device called a `*thermoscope*`. In fact, *Philon of Byzantium* (about 3rd Century BC.) constructed an instrument that demonstrated the expansion of air on heating, thus laying the foundation of thermometry without actually measuring temperature. Some years later *Diocles* developed the theory of the parabolic burning mirror and *Heron of Alexandria* (about 1st Century AD.) first used a thermoscope for a more practical use. A medieval form of this instructive instrument consisted of a large bulb hermetically attached to a thin glass tube, the end of which was immersed into water in another vessel. To prepare the apparatus for experiments, a few bubbles were driven out of the tube by a slight heating of the bulb. After that, the instrument worked as a gas dilatometer, sensitive to changes of temperature (but also to changes of the external pressure). The addition of a regular scale, made of glass pearls, to the pipe of the thermoscope enabled *Sanctorius* to judge the degree of the patient's temperature and then to choose the proper medical treatment. This conversion from a curious toy into a powerful measuring device, which provided data that could be applied, gave the thermoscope all the features of an effective discovery.

By the 17th Century knowledge of the *'weatherglass'* (the common name for thermoscope) was widely spread among educated people, either due to the new edition of Heron's papers or upon the accessibility of excerpts of Arabian alchymistic manuscripts. Assigning only a single "true" inventor of thermometer, from among persons such as *Galileo, Segredo, Fludd, (F.) Bacon, van Helmont, Boyle* and others, is practically impossible. Among these inventors was also *Goethe*, who more than a century later (1732) had patented a *'virginomorphic'* glass bowl filled with wine and containing a very strange pipe – the device was more worthy of deep psychoanalytical study than for a "reliable forecast of weather". However, during the second half of the 17th Century there were in use some advanced forms of thermometers for medical and meteorological purposes, namely those constructed by *Guericke* and by the members of the *Accademia del Cimento* in Florence who also invented the closed fluid-thermometer.

The second half of the 17th Century may be characterized as an era of the differentiation between pure theoreticians and experimentalists. Typical of the theoreticians, represented e.g. by *Bacon, Descartes and Gassendi,* was a very prudent and critical approach to new experimental facts, a deep interest in new methodology that was more reliable than that used by medieval scholastics, and the will to construct universal theories. Regardless of the progress made by the theoreticians, it would have all been futile it were not for the extensive work by other experimental scientists. The words of *Fludd* that *"...the thermometer became a mighty weapon in the Herculean fight between 'Truth and Falsehood'"* were prophetic. The most distinguished person who was trying to use this 'weapon' for quantitative measurements was *Boyle.* Unfortunately, the main problem concerning his experiments was the absence of sufficiently reproducible fixed points that characterized the given thermal states, and consequently he was able to perform only relative measurements.

However, *Römer* and *Fahrenheit* satisfactorily solved the serious problem of suitable scales much later, at the beginning of the 18th Century, by their use of the first sealed precision mercury-in-glass thermometer. They introduced fixed points such as the freezing point of an aqueous solution of salmiac (0), the freezing point of water (32), the normal temperature of the human body (96) and the boiling point of water (212). The intervals between the fixed points marked on the scale of such a fluid thermometer were divided regularly into degrees. This calibration, which was for some time *Fahrenheit's* personal secret that ensured a very good reproducibility of results for a number of different instruments. In 1746 the Swedish astronomer *Celsius* introduced a new temperature scale with zero set at the boiling point of water and 100 for its freezing point. After his death *Linne* changed it to the standard scale form of <0–100> that we still use to this day.

The difficulty with setting up the precise 'freezing' points was linked with the pervasive effect of water undercooling in its dependence on the environment and type of measurement, which variously effected the precise temperature determination of water freeze up. Delayed solidification was caused by the properties of solid ice, which is a more packed but lighter form of water, having the hexagonal structure, which shows evidence of distinctive incommensurability with the pentagonal-like symmetry of solidifying liquid water containing high number of fluid-like centers of ice-incommensurable icosahedrons and dodecahedrons, therefore capable of undercooling below the standard freezing point (now $0^{\circ}C$). Therefore other complementary systems were searched that were, as expected, based on every day practice such as that with salty water (experience with ice formation over sea) or mixtures of water with alcohol (freezing of wine).

One of the earliest attempts to put high temperatures on a mathematical scale was done already by *Newton* (1701), who described a thermometer on the basis of oil and calibrated it by taking *"the heat of the air in winter when water begins to freeze"* as 0 and *"blood heat"* as 12 so that on this scale water boils at 34. The melting points of alloys were determined by an extrapolation and his logarithmic temperature scale was proposed for high experimental temperatures on the basis of the equation, $\theta = 12\{2^{(x-1)}\}$, where θ was the temperature in units of the above scale and x represented the logarithmic temperature. Actual values on the arithmetic and logarithmic scales were determined by interpreting the time that an iron-bar took to cool down to "blood heat" in terms of Newton's law of cooling.

An analogous method for the construction of a thermometric scale was devised independently by *Amontons* [22,35,175] who made experiments with a constant volume gas thermometer. By extrapolating the regularly-divided (into 100 degrees) scale between the boiling and freezing points of water below the freezing point, Amontons noticed that there should be a point corresponding to the zero pressure of a gas in the thermometer. He called this point (lying approximately 270 degrees below the freezing point of water) the absolute zero or point of 'ultimate coldness' (*'l'extrême froid'*) and suggested its use as a natural fixed point. Mankind is indebted to Amontons for another fruitful idea; the invention of the gas thermometer. Whilst it was not a very convenient instrument, it, nevertheless, appeared as a reliable and very important tool for the calibration of more practical liquid thermometers.

Fahrenheit's and *Amontons'* scales have a lot of common features with modern thermometric scales, which enabled the fundamental problems in scientific thermometry to be solved; namely: to assign a number θ, called the *empirical temperature*, to any given thermal state, to decide whether two bodies have the same temperature or not, and to determine which body has the higher temperature. Later *Maxwell* recognized that for thermometry to be a logically closed system it is necessary to add a concept of thermal equilibrium and another theorem, sometimes called the zero law of thermodynamics, according to which: *"two bodies which are in thermal equilibrium with a third one are also in thermal equilibrium with each other."* By establishing this theorem, which encompassed the form of Euclid's first axiom, the development of the concept of empirical temperature was practically completed.

The enduring process of enhanced awareness was clearly affected by the caloric theory of gases, which emerged in about 1770s as a result of fusion of the well-established geometrical view of Newtonians to the gas structure with the emerging theories of electric and magnetic fluids. One inherent problem, in which *Laplace* took a particular interest, was the search for a rational scale of temperature. The chief difficulty was that there was no guarantee that the mercury in a conventional thermometer expanded uniformly or, as *Gay-Lussac* put in terms of current caloric theory that "equal divisions on its scale represents equal increment in the tension of caloric". He had later shown that the expansion of the gas was proportional to the readings on the mercury thermometer (assumingly between 0 and 100 °C) and such measurements were strongly advocated by *Dulong* and *Petit* as a standard way for a proper comparison of temperatures. In this claim the following argument was given by *Laplace* "*... if the temperature of the air is supposed to increase while its volume is kept constant, it is quite neutral to suppose that its elastic force, which is caused by heat, will increase in the same portion as the temperature. If the pressure of the air is now altered to the value that it had initially, its volume also will increase in the same proportion as the temperature. Hence it seems to me that the air thermometer gives a precise indication of variation in quantity of heat...*".

While *Laplace* was trying to improve *Newton*'s expression for the velocity of sound he arrived to an important relation relating adiabatic process (*a*) with the pressure increment that would have been obtained under isothermal conditions (*i*) $(\Delta P/\Delta V)a = (c_p/c_v)(\Delta P/\Delta V)i$ where (c_p/c_v) are the ratio of specific heats at constant pressure (*p*) and constant volume (*v*). As a consequence the velocity of sound was given as $\{(c_p/c_v)(P/ro)\}^{1/2}$. Another important studies were also associated with the effect of adiabatic heating and cooling, besides of *Watt, Wedgwood* or *Priestley*, its possibly best account was given by *Darwin* who curiously applied this knowledge even to meteorological phenomena ascribing the cold at high attitudes to the expansion of air as it rose and experienced a diminishing pressure.

The intuitive caloric theory of heat [133] was misapprehended by scientists and historicist as it was discredited at that time by acceptance of the principles of energy conservation. It showed, however, that those who studied and made use of the caloric theory including *Lavoisier, Dalton, Laplace, Poisson, Carnot,* or *Avogadro* did so because caloric gave them a satisfactory basis for the study of heat, whether in physics or in chemistry. We should keep in mind that many foundation of thermodynamics were laid while this theory held sway, particularly between the years 1750 to 1850.

In 1824, whilst investigating the theoretical optimization of steam engines, the devoted colorist *Carnot* devised an idealized heat engine capable of performing virtual, fully computable, cyclic processes [22,176,177]. His concept consisted of a thermally-insulated cylinder containing a gas which is provided with a movable piston. The bottom of the cylinder can be either insulated or, in turn, put into contact with two bathes (A or B), which have different empirical temperatures ($\theta_A > \theta_B$). These operations may be performed in such a way that only isothermal and/or adiabatic changes are involved, so that it can be proved mathematically that the work done during one cycle is maximal. Using these conclusions, and the conjecture about the impossibility of a perpetual motion (*'perpetuum mobile'*) being generalized for thermal phenomena, *Carnot* formulated the following important theorem: "*The moving force of the fire (i.e. useful work) does not depend on the agent used for its development and its magnitude; it relies only on the temperatures of the bodies between which the transfer of heat takes place*".

It is worth noting even today almost all the mechanical energy produced is created by conversion of thermal energy in some sort of a heat engine [3,9,131]. Such man-made engines are literally *energy transducers* designed to transform chemical energy, held within the electron configuration and stored in the form of fuel and oxygen, into mechanical work – including the realm of thermodynamics, which is the necessary transformation via the intermediate step of heat. As said above, the engine's thermal operation can usually be approximated by an ideal thermodynamic power cycle of some kind to be make up of a series of processes that must return the working substance (often fluid) to its initial state. During almost all of such processes involved, one property (temperature, T, pressure, P, volume, V, heat flow, Q and its potential S, etc.) is commonly held constant. All reversible adiabatic processes are naturally isentropic ones. It is common practice to plot the processes composing the cycle on a graph of property coordinates, such as T-S or P-V or so, and for such an illustration the cyclic integral of work is equal to the cyclic integral of heat.

The figure of merit is defined as the ratio of the desired energy gain to the energy that is input, i.e., the thermal efficiency η equals the ratio of the net work output to the heat added at high temperature. Repeating in short that $\eta = 1 - T_{out}/T_{in}$, which means that it is impossible to build a device whose sole effect is to produce work while exchanging heat with a single thermal reservoir (in) only, without having a low-temperature sink (out). In practice, there are four phenomena that render any real thermodynamic process: friction, unrestrained expansion, mixing of different substances and heat transfer across a finite temperature difference, the last being most important. Only three ideal power cycles can be assumed reversible, often called *externally reversible power cycles*, to include the *Carnot* (power optimum), *Stirling* and *Ericsson* cycles. However, the Carnot cycle is not a practical power cycle because it has so little specific work that any friction would almost eat up the network output and *Stirling* engine is a rather complicated to run (although its renaissance is on progress reconsidering its worth for economical exploitation of low value waste-heats, sun-heated air and T-difference between light and shadow rocket surfaces). Beside two isothermal processes, *Ericsson* cycle replaces two originally reversible isentropic (adiabatic) processes by two reversible isobaric processes and the heat transfer is than taking place in all four processes, which requests an internal component called a *regenerator*. The engine is practically composed of a compressor, turbine and counter-flow heat exchanger serving as the mentioned regenerator [9,76].

It was *Kelvin's* excellent idea that every thermometer may be treated as a special kind of thermal engine working between a bath kept at the temperature, which has to be measured, and another one at the reference temperature. According to Carnot's theorem, the efficiency (i.e., normalized 'useful' work) of any reversible cycle is dependent only on these two temperatures, regardless of the working (thermometric) substance. By taking this efficiency as a measure of the temperature, the absolute scale (i.e. independent of device and/or of material constants) can be constructed. In order to find additional conditions to which such a construction must be submitted, *Kelvin* constructed a mental experiment that had three reversible engines simultaneously working with three baths at different empirical temperatures $\theta_1 > \theta_2 > \theta_3$. Application of *Carnot's* theorem to all combinations of these cycles provided a functional equation with solution, $Q_1/Q_3 = \varphi(\theta_1)/\varphi(\theta_3)$, where $\varphi(\theta)$ is a monotonic, positive, definite and real function of the empirical temperature, θ. The simplest choice, which later became the basis of the international temperature scale, is that it corresponds to the relationship $\varphi(t) = \alpha T$ where α is a universal constant and αT is that for heat, Q, exchanged between the bath and thermometer. This convention is fully consistent with the empirical scales induced by isochoric (so called *Amontons'* scale) and/or isobaric (*Avogadro's* scale) equations of state.

How important the choice of $\varphi(t)$ was for the further interpretation of concepts in thermal physics will be apparent from the following example. *Dalton*, in analyzing not very reliable measurements [178] of the thermal expansion of fluids, found a quadratic dependence between supplied heat, identified by himself as temperature, θ, and the increase in volume, V, of the fluid with respect to that at its freezing point. Using this conjecture as a basis for the construction of a temperature scale, he was able to fit the isobaric equation of state of any permanent gas by the formula $V/V_0 = exp\ \{\beta(\theta-\theta_0)\}$ where β is also a universal constant. The initial state, characterized by the volume V_0 and by both temperatures θ_0 and T_0, will be changed to a new equilibrium state that corresponds to the volume V. It is easy to show that the difference $(T-T_0) = const\ (V-V_0)$ is directly proportional to the work done by the gas against the external pressure, P. On *Dalton's* scale, the temperature difference factually measures the increase in entropy (in its usual sense) of a gas in a thermoscope, because $ln\ (V/V_0) = \beta\ (\theta - \theta_0)$.

It is remarkable that both the arbitrarily chosen quantities of temperature, related to just the same experimental situation (volume change induced by heating), can have such a different interpretation. There is an enormous freedom in how to choose the function φ but it is a very difficult task, and a matter of intuition, to anticipate whether such a choice will be of practical value in the future. Thus the intuitive opinion of *Dalton* that the temperature should reflect something closely related to the content of heat in a given body actually corresponds to his idea. On the other hand, the simple interpretation of heat conduction and the evaluation of efficiency (of, e.g., steam engines) require temperature to behave like the potential of a heated fluid. In this case, *a linear scale*, equivalent to the contemporary *Kelvin's* international scale, is the most convenient one.

At about the same time (1848) *Thomson* used the same basis of Carnot's thermal efficiency and tried to introduce a logarithmic temperature scale, *Th,* in the form $\eta = const\ dTh = dT/T$. After integration it follows that $Th = const_1\ lnT + const_2$, where both constants can be determined by using the traditionally fixed points of the melting and boiling of water. This scale, however, would dramatically change the customary concept of thermodynamics. The meaning and interpretation of the third law of thermodynamics would ease if the zero temperature would be replaced by infinity ($T = -\infty$) but the traditional degree of freedom, $\sim 1/2\ kT$, would revolutionize to embarrassing proportionality, $T \sim exp\{(Th - const_2)/const_1\}$, etc. It shows the significant role of a suitable convention, which we can illustrate using a hypothetical case of Newton's definition of force, *F.* Instead of the contemporary expression $\{d(mv)/dt\}$ a more complex relation could have been adjusted, such as $\{d(mv)/dt\}^\pi$ or even $ln\{d(mv)/dt\}$, which would essentially have changed our tradition and thought .

As indicated, temperature reveals itself to be a potential of heat, illustrating its similarity to mechanical work and its well-known complete conversion into heat [22,50,179]. There, however, is no a single process enabling a complete reverse conversion of a given amount of heat back into mechanical work without the accompaniment of other changes. The very absence of this opposite transformation logically excludes the possibility that *heat is equivalent to any other kind of energy,* which is characterized by the unlimited fluctuations amongst its particular forms. This serious inconsistency in classical thermodynamics is compensated for by the traditional introduction of a new quantity (called entropy) and of a new axiom (known as the second law of thermodynamics), which states that this entropy is indestructible ("never decreases") and is always increasing (factually being "created" during every real, irreversible route). It is, therefore, no wonder that a meaningful physical interpretation of this quantity lacks easy demonstration. It has just the same properties as heat (in its common sense), the name of which has already been quite improperly used to label a certain special kind of energy. Indeed, if we *identify* heat with entropy, the mysterious second law of thermodynamics becomes quite intuitive and very easy to understand, stating, *"heat cannot be annihilated in any real physical process".*

For instance, in an experiment where heat (= entropy) is generated by means of friction between two blocks of any material, it is clear at first glance that the heat (= entropy) will not disappear by moving the blocks in opposite direction but will instead only increase heat further on. Indeed, just to be able slow down such a process, one would have to effectively decrease the amount of friction being generated, i.e. by suppressing the generation of heat (= entropy) during the movement, however, there is no process involving the mentioned blocks that could reverse the system and destroy the heat (= entropy) that had already developed. Moreover, the substitution of the word "entropy" by the word "heat", which is no longer regarded as a kind of energy, would enable a more intelligible interpretation of temperature [9,22] as the desired potential in a closer analogy with other potentials (electric, gravitational) used in other branches of physics.

5.2 Heat transfer

In the 1850s, the Scottish physicist *Maxwell* [50] initiated convincing investigations into the mechanical theory of heat. He argued that the velocities of point particles in a gas were distributed over a range of possibilities that were increased with temperature, which led him to predict, and then verify experimentally, that the viscosity of a gas is independent of its pressure. In the following decade *Boltzmann* began his investigation into the dynamical theory of gases, which ultimately placed the entire theory on firm mathematical ground. Both men had become convinced that the novel notion of entropy reflected molecular randomness. *Maxwell* expressed it this way: "*The second law of thermodynamics has the same degree of truth as the statement that, if you throw a tumblerful of water into the sea, you cannot get the same tumblerful of water out of it back again.*" These were the seminal notions that in the following century led to the field of non-equilibrium thermodynamics.

Although *Euler* had already formulated a mathematical theory for the convection in fluids in 1764, the scientists of the 19th Century started to treat the subject of heat transfer with the simple notations of Newton's laws of cooling [180] (even if we now consider it insignificant in comparison to his other works on mechanics and optics). We may repeat "*the greater the outer conductivity and surface of a cooled solid body the faster is its cooling. On the other hand, the body's thermal inertia, given by the product of its specific heat and mass, slows down its cooling*". It, however, did not yet distinguish between different modes of heat transfer. Perhaps, that was how 'irreversible heat flow' acquired its unpleasant moral connotation where the term 'dissipation' was synonymous with 'intemperate or vicious mode of living'. Nonetheless, the natural behavior of heat entails a distinction between past and future or even illustrates extrapolation to an ultimate heat death of the Universe.

In 1818 *Dulong* introduced an improved law of cooling, and *Petit* suggested that a body absorbs and emits heat at the same time. They argued that the rate of cooling, $\underline{v}$, of a body at temperature, $T + \theta$, in a vacuum surrounded by a container at temperature, θ, must be determined by the general formula $v = f(T + \theta) - f(\theta) = m \, a^{x} \, (a^{y} - 1)$ where f is a power function and the parameters $\underline{m}$,

148

_a_, _x_ and _y_ are 2.037, 1.0077, θ and T, respectively. This formula was later simplified to describe the thermal effect measured by a thermoelectric instrument, _z_, in the form $z = a\,T^2\,(T - \theta) - b\,(T - \theta)$, thus combining the terms responsible for radiation and Newton's previous cooling that accounts for the influence of the surroundings. One consequence of this exponential-like form was that the heat radiated by a body would never become zero for the finite values of temperature.

In 1860 it was _Maxwell_, who borrowed from _Clausius_ the concept of the mean free path traveled by a molecule from one collision to the next, and who introduced the modern view that heat conduction is a special case of a general transport process of which diffusion and viscosity are parallel examples. The T^4 term was later found to fit the corrected _Dulong-Petit_ data leading to the gradual establishment of the _Stefan-Boltzmnann_ law of radiation. Worth of note is _Leslie's_ work on the rate of cooling that furnished the relationship, $dT/dt = - a\,T$. If the initial temperature of a hot body is T_o (at t = 0), then $T = T_o\,e^{-at}$. The constant _a_ represents the range of cooling and can be estimated by measuring the time taken to cool down an object to the value of $T_o/2$. The rate of cooling was found by dividing 0.693 by the time required for T to decline to $T_o/2$. This, in fact, provided the early basis for theoretical calorimetry.

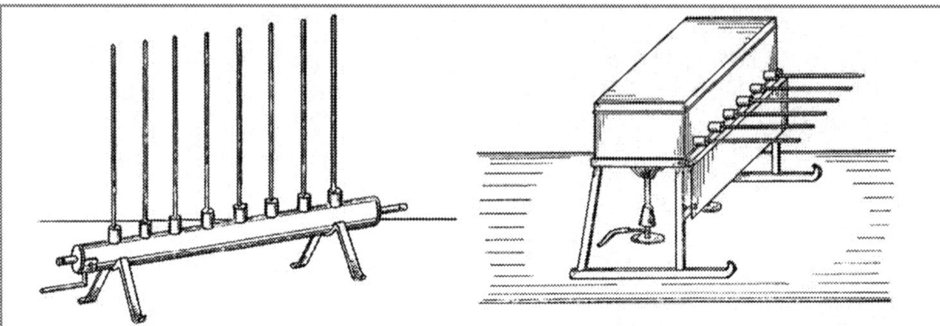

Fig. 26. – Two instrumental versions of the _Ingenhaousz's_ apparatus formerly used for the heat conductivity measurements at the beginning of 19th Century. The rods made form different metals (Ag, Cu, Zn, Pb, etc.) were immersed into the vessel filled with boiling water (or flowed steam). In order to follow the gradual temperature changes of individual rods their surface were covered by either the layer of wax (to observe the melting) or mixture of $Cu_2J_2+HgJ_2$ (to observe the change of red color to yellow). It was stressed out that '_rods must be sufficiently lengthy; vessel heating ought to be prolonged and intensive so that the temperature differences are well visible to the eye. However, potential experimenters are apprehended of possible investigation artifacts that even though the silver is a better heat conductor than copper, the silver rods from bazaar show often heat retardation because of their contamination – unseen mix up with copper_'. The experiment could be reversed so that the advancement of coldness was possible to observe if the heated water was exchanged for the mixture of ether with snow particles of gaseous CO_2.

The study of temperature distribution was initiated by *Biot,* who found that if a copper rod and an iron rod are joined together end to end, and the coupled ends are placed in heat sources [12], then the heat will conduct through the copper end more quickly than through the iron end, because copper has a higher proportionality constant, k equal to 92, whereas, iron has a value of only 11. He introduced a method to determine the relative thermal conductivities λ (*de facto* thermal diffusivity, the term used by *Kelvin* as *a*) of two well-connected metallic bars of similar shape that are put under a steady heating of their join. A measurement of the distance from the join where the temperature is the same on both bars, x, provides the heat conductivity value in the form $\lambda_2/\lambda_1 = (x_1/x_2)^2$. An even more intelligent insight was the estimation of λ on the basis of average temperature measurement at three arbitrary but equidistant points, T_1, T_2 and T_3. If bars are equally separated they provide a useful average – the dimensionless value of *n* equal to $1/2\ (T_1 + T_3)/T_2$. It aid to determine λ_2/λ_1 by distinguishing, x, by help of the logarithmic form, $ln[n + (n^2 – 1)^{1/2}]$ providing the first insight into the thermal exponential distribution. For both our amusement and conceding the difficulty inherent in the path to better understanding, consider the early method of thermal measurements in Fig. 26.

Perhaps the more important contribution came from *Fourier,* who in the year 1822 judged the above mentioned thermal properties in detail and summarized them in the renowned *Fourier's Law* of heat conduction in the form of $Q = – \lambda s\ \Delta T/\Delta x\ t$. It states that the amount of heat, q, which passes through a surface (as a cross section), *s*, in time unit (sec) is proportional to the temperature difference, ΔT, per given length, Δx (i.e., to the unite temperature gradient of 1 degree per 1 meter). The proportionality constant λ stands for the standard *coefficient of thermal conductivity* while a $= \lambda/(c_p\ \rho)$ expresses *thermal diffusivity* where c_p is the specific heat capacity and ρ is the specific mass. It complies well with the generalized physical observations of one-dimensional heat flow that results from the temperature non-uniformities, i.e., heat flow and temperature gradients are exactly opposite one another in direction, and the magnitude of the heat flux is directly proportional to this temperature gradient. It is worth noting, however, that this early account of the ratio of heat over temperature (in cal/1°) was presupposed to bring the same meaning as 'mass'.

It is clear that we should understand generalized heat flux, *q*, as a quantity that has a specified direction as well as a specified magnitude [35,175,180-183]. Fourier's Law summarizes this physical experience succinctly *as* $Q = – \lambda_i\ \nabla T = \lambda\ \Delta T/\Delta x$, where the temperature gradient ∇T is represented by three (i,j,k) – dimensional derivatives, i.e., $\nabla T – i\partial T/\partial x + j\ \partial T/\partial y + k\ \partial T/\partial z$, and where the standard thermal conductivity can now be found to depend on both the position and temperature, $\lambda_i = \lambda\{(i,j,k)(T)\}$. Fortunately, most materials are nearly homogeneous so that this equation simplifies to $\lambda = \lambda\ (T)$ and for a one-dimensional flow is further simplified to the difference $\Delta T/\Delta x$. It is evident that λ may vary from one substance to another, which is accounted for in part by varying electron and photon conductance. One feature that substances have in common with one another, according to empirical evidence, is that λ is always a positive constant. This, of course, makes intuitive sense, at least if the molecular concept of temperature is invoked; the heat (kinetic energy at the microscopic scale in another sense) tends to flow from regions of high internal energy to regions of low internal energy, which is consistent with the above statement that heat flow is opposite the gradient direction of temperature, being consistent with the laws of thermodynamics. To show this we must emphasize that heat flow is independent of velocity gradients or gradients of concentration, such as would exist in a fluid mixture, so that it is required that $\lambda/T^2(\nabla T)^2 \geq 0$ approving the previous request of $\lambda \geq 0$ and therefore consistent with experimental results.

Fourier's Law has several extremely important consequences in other kind of physical behavior, of which the electrical analogy is most striking. Let us first consider *Ohm's law* describing the flux of electric charge, I/A, by the electric current density, $J_e = -\sigma \nabla V$, where σ is the electric conductivity, V is the voltage, I are the amperes of electric current and A is an area normal to the current direction (vector). For a one-dimensional current flow we can write Ohm's Law as, $J_e = -\sigma \Delta V/\Delta x$. Because ΔV is actually the applied voltage, E, and R_e is the electric resistance of the wire equal to $\Delta x/(\sigma A)$, then, since $I = J_e A$, the equation becomes the form of Ohm's Law, $V = R_e I$. The previously written formula, $\Delta x/(\sigma A)$, thus assumes the similar role of thermal resistance, to which we give the analogous symbol, R_t, and which shows how we can represent heat flow through the slab with a diagram that is perfectly identical to an electric circuit.

Another analogous relationship is that of mass transfer, represented by *Fick's law* of diffusion for mass flux, J, of a dilute component, 1, into a second fluid, 2, which is proportional to the gradient of its mass concentration, m_1. Thus we have, $J = -\rho D_{12} \nabla m_1$, where the constant D_{12} is the binary diffusion coefficient and ρ is density. By using similar solutions we can find generalized descriptions of diffusion of electrons, homogeneous illumination, laminar flow of a liquid along a spherical body (assuming a low-viscosity, non-compressible and turbulent-free fluid) or even viscous flow applied to the surface tension of a plane membrane.

In view of general thermophysical measurements, there are some other important subjects that, whilst outside the scope of this chapter, are worth mentioning. For example, the coefficient of thermoelectric tension (*Seebeck cofficient*, α) describes the proportionality between the heat generated at a material's contact when a certain amount of electric current passes across it (*Peltier coefficient*, P) and the composite heat, Q, due to the electric current, I, at time, t, generated within the material held under a temperature gradient, ∇T (*Thompson coefficient, $\mu = Q/(J_e \nabla T) \approx Q/\{\Delta T\ T\ t\}$*). All these coefficients combine to form certain complex parameters such as the proportion of thermal and electric conductivities, $\lambda/\rho \approx (k/e)^2\ T$ (where k is the *Boltzmann* constant and, e, is the charge of electron), or the quality factor of thermo-electric material, $Z = \alpha^2 \rho/\lambda$, or some other mutual interdependency like $P = - \Delta\alpha\ T$ or $\mu = - T\ (\partial\alpha/\partial T)_p$.

When developing methods that combine thermal determinations [16], it is necessary to take into account a number of factors, many of which either do not occur or leave only a small influence on the entire measurement of other, for instance electric, quantities. There are no perfect conductors nor do absolute thermal insulators and most thermal conductivities vary in a relatively narrow range (within approximately five orders of magnitude). Furthermore, at higher temperatures these values converge and it is, therefore, difficult to realize a well-defined thermal flow, particularly when assuming the practical lengthiness of a thermal process, which, in comparison to electric flow, takes a rather long time to achieve equilibrium.

It is more important to note that within phenomenological theories any flux can be generalized as the kind of the dissipation function, Φ. Thus it can be written as a sum of all the thermodynamic fluxes, J_i, and their conjugate forces, X_i, i.e., $\Phi = \sum_i J_i X_i$. The fluxes are unknown quantities in contrast to the forces, which are known functions of the state variables and/or their gradients. It has been found experimentally that fluxes and forces are interwoven, i.e., a given flux not only depends on its own conjugate force but may depend on a whole set of forces acting upon the system as well as on all thermodynamic variables (T,P,..). Various fluxes can be considered the rate of change of an extensive variable, X_i ,i.e., $J_i = \partial X_i/\partial t$. Conjugate forces are identical in form to the phenomenological flow equations shown above and can be expressed, close enough to equilibrium, in the linear form [153] of the *Onsager relations*, or $\partial X_i/\partial t = J_i = \sum_k L_{ik} X_k$, where the constant coefficients, L_{ik} , are called the 'phenomenological coupling coefficients'. Written in this fashion, thermodynamic forces are differences between the instantaneous and equilibrium values of an intensive variable (or their gradients).

For example the difference *$(1/T - 1/T_{eq})$* is actually the thermodynamic force conjugate to the internal energy, which is the familiar Fourier Law, i.e., $q = - \lambda_i\ \nabla T$ or *Stokes Law*, i.e., $p^v = - 2\ \eta\ \nabla v$. The traditional Fick's Law is then obtained by introducing the condition that one works at a constant temperature and pressure, while Ohm's Law requests supplementary constraints provided that the magnetic induction and all couplings are ignored. Similarly we can predict a linear relationship between the rate of advancement of a chemical reaction and the affinities, which is correct only within a very narrow domain around equilibrium.

5.3. Non-stationary heat diffusion and the Schrödinger equation

In general, any of these phenomena involves the movement of various entities, such as mass, momentum, or energy, through a medium, fluid or solid, by virtue of non-uniform conditions existing within the medium [3,16]. Variations of concentration in a medium, for example, lead to the relative motion of the various chemical species present, and this mass transport is generally referred to as diffusion. Variations of velocity within a fluid result in the transport of momentum, which is normally referred to as viscous flow. Variations in temperature result in the transport of energy, a process usually called heat conduction. Besides the noteworthy similarities in the mathematical descriptions of these three phenomena; all three often occur together physically. It is in the case of combustion, where a flowing, viscous, fluid mixture is undergoing chemical reactions that produce heat, which is conducted away, and that produce various chemical species that inter-diffuse with one another.

Description of a more frequent, non-steady progress requires, however, the introduction of second derivatives. These are provided by a three-dimensional control of a finite region of a conducting body with its segmented surface area, denoted as s. The heat conducted out of the infinitesimal surface area ds is $(-\lambda \nabla T)(k\,ds)$, where $\underline{k}$ is the unit normal vector for the heat flux, $q = -\lambda \nabla T$. The heat thus generated (or absorbed) within the underneath region, v, must thus be added to the total heat flow into the surface, s, to get the overall rate of heat addition. Therefore we need integration. Moreover, the rate of the increase of internal energy, U, for the given region, v, is granted by the relationship, $dU/dt = \int_v \{\rho\, c_p\, (\partial T/\partial t)\}\, dv$, where the derivative of T is in the partial form because T is a function of both, v and t. Applying Gauss's theorem, which converts a surface integral into a volume integral, we have to solve the reduced formula of the form, $\int_v \{\nabla . \lambda \nabla T - \rho\, c_p\, (\partial T/\partial t)\}\, dv = 0$

Since the region, v, is arbitrary small, the integrand must vanish identically so that the heat diffusion equation in three dimensions reads, $\nabla_x \lambda \nabla T + dq/dt = \rho\, c_p\, (\partial T/\partial t)$. If the variation of λ is small, λ can be factored out leaving a standard but more complete version of the heat conduction equation (i.e., the second law of Fourier) $\nabla^2 T = (1/a)\, (\partial T/\partial t)$, where a is the thermal diffusivity, and the term $\nabla^2 T \equiv \nabla_x \nabla T$ is called the *Laplacian* and is arising from a *Cartesian* coordinate system $(i\partial/\partial x + j\, \partial/\partial y + k\, \partial/\partial z)\,(i\partial T/\partial x + j\, \partial T/\partial y + k\, \partial T/\partial z) = i\partial^2 T/\partial x^2 + j\, \partial T^2/\partial y^2 + k\, \partial T^2/\partial z^2$. This is the standard second order form of the heat equation showing that the change of temperature over time is proportional to how much the temperature gradient deviates from its linearity. In the other words, the bigger the protuberance in the temperature distribution, the faster is its compensation. However, its mathematical solution is complicated and is not the aim of this text. Just as an example, consider a massive object of temperature T_o (at the initial conditions of $T\{x,0\}$) that would affect the temperature of its environment, T_c. The solution in time, t, and space distribution x gives a solution in the form, $T(x,t) = T_c + (T_o - T_c) \Phi [x/2(kt)^{1/2}]$, where Φ stands for the *Bessel* functions, available from tables but not easy to work with.

The fact that the second Fourier Law bears a general form of a *distribution law* is thus worthy of special note. The significant reciprocity can be found for the famous *Schrödinger* equation when taking into account a diffusion process as something fundamental [154,184-188] and intimately related to light and matter (close to the physical intuition of *Feynman*). It is usually written in its simpler form as $d\psi/dt = i\,c'\,\Delta\psi$, where ψ is the wave function, i stays for the standard complex number and c' is an 'imaginary' diffusion constant. The different paths along which diffusion with an imaginary diffusion constant occur do not necessarily lead to the toting-up of the observed effects but can, instead, result in a destructive interference. This is asymptotic for a new phase in physics, intimately related to the mathematics of complex functions. It describes diffusion as probability amplitude from one point to the next along the line. That is, if an electron has certain amplitude to be at one point, it will, a little later, have some amplitude to be at neighboring points. The imaginary coefficient makes the behavior completely different from the ordinary diffusion equation such as the equation that describes gas spreading out along a thin tube – instead the real exponential solutions the results are shown in complex waves.

Physics, however, has already gone one stage beyond the complex numbers in order to describe weak and strong interaction in elementary particle physics (non-abelian gauge theory that uses "hyper-complex" numbers – multivectors, by working, for example, the SU(?) theory of particle quantum flavor dynamics). Such hyper-complex mathematics of diffusion may be related to the realm of life yet too complex to solve but recently applied, at least, to the kinematics of the human knee. The complex diffusion problem can also be recalled in view of the substrate-bound Maxwellian understanding of light, deeply rooted in the world view of many physicists even today, as well as to Steiner's view of electricity as light in a 'submaterial' state [184].

5.4. Practical aspects of heat flow – contact resistance and periodical outcome

As a matter of interest we can begin this section by mentioning the incident described by *Schwart* in 1805 when he experienced an unusual audio sound while an as-cast hot ingot of silver was placed on the cool iron anvil [12]. Later, it was closely studied by *Trevelyan* and *Faraday* who showed that any hot metal standing on two separated tips resonates with the cooling substrate (preferably with an amply different coefficient of heat expansion). Practically speaking, the metal is pushed up by one of its tips and gets back to another one, and so on, which thus endows oscillation continuity.

No two solids will ever form a *perfect thermal contact* when they are pressed together and the usefulness of the electric resistance analogy is immediately apparent when looking into the interface of two conducting media. Since some roughness is always present, a typical plane contact will always include thermal impedance such as tiny air gaps, structural irregularities, etc. Heat transfer thus follows at least two parallel paths across such an interface. We can treat the contact surface by placing an interfacial conductance, $\boldsymbol{h_c}$, (in $[W/(m^2 K)]$) in series with the conducting material on the either side. It accounts for both the contact material and the surface's mutual finish executed by a joining force (deformation), near-surface imperfections and interface temperature profile, $T = T(r)$ providing $R_{total} = (1/\boldsymbol{h_1} + 1/\boldsymbol{h_c} + 1/\boldsymbol{h_2})/A$, where $h_{1 \text{ and } 2}$ are the respective conductivity of materials on the both sides (i.e., $h_1 \approx \lambda_1/d_1$, d being the thickness).

A frequent task is to find the temperature distribution and heat flux for a sample holder granted as the long enough hollow cylinder with a fixed wall temperature – inside temperature T_i for radius r_i and outside T_o for r_o . For $T = T(r)$ we have $(1/r)\, \partial/\partial r\, (r \partial T/\partial r) = (1/a)\, (\partial T/\partial t)$ to provide general integration in the form $T = c_1\, ln\, r + c_2$. The solution gives $(T - T_i)/(T_o - T_i) = ln\,(r/r_i)/ln\,(r_o/r_i)$. It would be instructive to see what happens when the wall of the cylinder becomes very thin, or when (r_i/r_o) is close to unity. In this case $ln\,(r_i/r_o) \approx (r - r_i)/r_i$ and the gradient has a simple linear profile $(T - T_i)/(T_o - T_i) \approx (r - r_i)/(r_o - r_i)$, the same solution that we would get for a planar wall. The heat flux falls off inversely with the radius.

We often want to transfer heat from the outside through composite resistances where we, again, can recognize Ohm's Law similarity and write the resulting thermal resistance in the form called the overall heat transfer coefficient, R_{therm}. It represents the series of actual thermal resistances in the same way as is traditionally made for an electrical circuit with multiple electrical resistances. Recalling the preceding example of a convective boundary condition of a cylinder we can write for heat Q a simplified equation, $Q = R_{therm}\, s\, \Delta T\, t$, where $R_{therm} = 1/\{ 1/h + r_o ln(r_o/r_1)/k\}$ is given in $[W/(m^2 K)]$ units and $\boldsymbol{h_c}$ is again the customary interfacial conductance.

The opposite approach would be a more detailed inspection of individual thermal processes taking place on the entire interface. Let us consider only two simple parallel thermal passes: one, most rapid solid conduction, carried out through the direct contacts and the other, so called interstitial heat transfer, which is much slower as intermediated through gas-filled interstices by convention, or even by radiation across evacuated gaps, as illustrated in Fig. 27. Surprisingly it may establish enormous temperature gradients along the interface – just assuming that the solid conduction takes place on one micron interface with one degree temperature difference then the resulting value for the term $\Delta T/\Delta x$ ascends to 10^4 K/m, which is comparable to the gradient produced on the wall of a high temperature furnace. It is clear that a series of such contacts arbitrary distributed along the interface give rise to a community of heat sources having a periodical resolution. The resulting heat wave is not planar and horizontally parallel to the conductive interface but has a sinuate profile. Further it is associated with the local volume changes due to the temperature dependency of expansion coefficient so that it is also the source for the above mentioned sound effect when a hot ingot possessing jagged surface with tips is placed on a cold metallic plate.

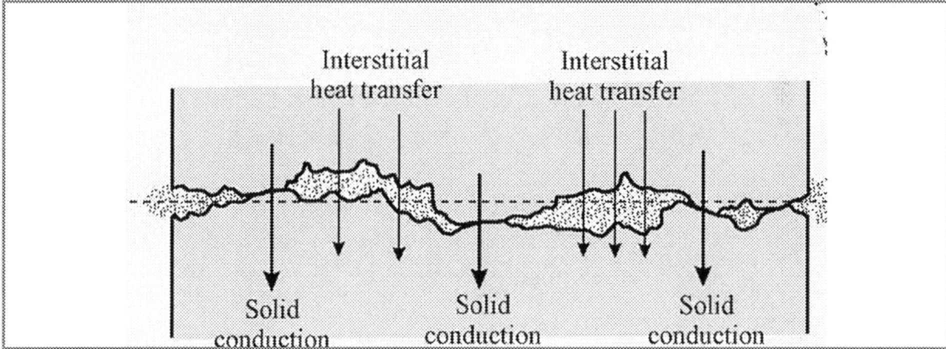

Fig. 27. – Illustrative diagram showing the overall heat transfer from side to side of the two (macroscopically) well-linked solid bodies. The separation plane represents a thermal contact between two real (often-uneven) solid surfaces eligible, however, to create spot-high temperature gradients, located at the individually disassociated thermal junctions. Miniature contact places exhibit either the good (conduction – thick arrows) or poor (mixed connectivity – thin arrows) heat transfer. Possible effect of such uneven distribution of intimate thermal contacts is often overlooked. In view of always existing 'rough-touch-character' of most thermal contacts the possibility of an internal heat pulsation cannot be excluded even for ordinary thermally-controlled experiments, which use the miniaturized sample-cells that possess relatively large surfaces of their associated mechanical contacts. Even the opposite (retarding) performance may be anticipated when the poor-connective-sites execute the damping role for the externally introduced modes of thermal oscillation.

A thermally fluctuating interface is a very common feature of many reacting interfaces. It may furnish a traditional solution, where the sinuous profile is extinguished by gradual retardation of thermal waves if the bulk (e.g. material layer or sample holder-cell) is large enough, or a less common solution, where such thermal fluctuations are enhanced by some processes, which multiply thermal effects (e.g. by befitting self-organization). This may occur even for special cases of thermophysical measurements when working in micro-dimensions (miniature samples or very thin walls of sample holders and material layers), which are not capable of compensating for such sinuous thermal fluctuation. Such a case is gradually manifested through routine observations by the detection of thermal noise. It may be the result of externally ordained temperature, often periodic alternations (temperature superimposed modulation) that may positively interfere thus enhancing the localized tendency to thermal vacillation. It can occur in the case of temperature-controlling methods applied by some TA instruments working under a specific furnace program of two-fold modulation (constant temperature increase with saw-tooth or sinusoidal modulation) often presented as renovation but yet without thorough analysis of the interference effects involved.

A periodical thermal source is, therefore, very important to analyze. It can act on a coupled body in two basic ways:

(i) when provided by an external (stationary) and large enough (unbounded) body of temperature, T, that is repeatedly changing in a particular manner, usually expressed according to the convenient goniometric function, $T(0,t) = A \cos \omega t$, where A stands for the (maximum) amplitude and ω is the angular frequency given by relation $\omega = 2\pi/T = 2\pi f$, where T is the period and f is the frequency,

(ii) when represented by an internal heat generator the heat production of which is again denoted by a goniometric function, usually understood as the time derivative of the previous cosines (due to dT/dt), then $T(0,t) = B \sin \omega t$, where B stands for the product ωA . For example, the solution for temperature steady state assuming a simple infinite plate in the fixed environment of such periodical changes of temperature, $Tp = Tp_o + Tp_{max} A \cos \omega t$, resolves in the analogous function, $T(x,t) = A(x) \cos (\omega t - \varphi_x)$, where φ_x represents the phase shift and $A(x) = f (\omega/2k)^{1/2}$ stands for a complex function, f, of angular frequency, ω, and heat conductivity, k. There follows some important points:

- The resulting amplitude of oscillation, $A(x)$, decreases exponentially with the penetration depth, x, i.e., $A(x) \approx A \exp -(\omega x/2k^2)^{1/2}$.

- The transferred temperature oscillations are time delayed that is again proportional the penetration depth $(\omega x/2k^2)^{1/2}$.

- The penetration depth depends on the oscillation periodicity, i.e., the smaller period the lesser penetration. The ratio $A(x)/A$ equals to $exp -(\omega x/2k^2)^{1/2}$ which also makes possible to estimate heat conductivity using $\lambda^2 = \omega x/\{2ln^2 A(x)/A\}$.

5.5. Turbulent boundary layers

We must remember that besides the flow of heat, there also exist other fluxes such as that of fluid, solutions of salt, the viscous stresses or even the flux of electrons or light. These laws are commonly based on the conservation of momentum and additional hypotheses, the justification of which lies in the satisfactory comparison of deductions from these laws with experimental observations. Such comparisons between theory and observation are limited, particularly in respect to the more complex situations that may arise when several different fluxes become alive simultaneously. It is conceivable and even probable, that such classical laws of heat conduction, diffusion, and momentum transfer must be altered to account for phenomena, which cannot be otherwise explained. In short, the state of our knowledge of macro- and micromolecular transfer of properties is not complete.

For example, turbulence in a fluid medium can be viewed as a spectrum of coexisting vortices in which kinetic energy from the larger ones is dissipated to successively smaller ones until the very smallest of these vorticular 'whirls' are damped out by viscous shear stresses. It is best illustrated by cloud patterns where huge vortices of continental proportion feed smaller 'weather-making' vortices of the order of hundreds of kms in diameter. They further dissipate into vortices of cyclone and tornado proportions that disintegrate further to smaller whirls as they interact with the ground and its various protrusions. Interestingly the process continues right on down to mm- and micro-scales. Practically we want to create a certain measure of the size of turbulent vortices ('length-scale'). This may be done experimentally by placing two velocity-measuring devices very close to one another in a turbulent flow field or using so called mixing length introduced by *Prandtl* as an average distance that a parcel of fluid moves between interactions. It bears a physical significance similar to that of the molecular mean free path. The associated mixing length, l, can help to define the so called eddy diffusivity (of the momentum ε_m). For notations of the instantaneous vertical speed of a fluid parcel and the velocity fluctuation that increases in the y-direction when a fluid parcel moves downwards (+) into slower moving fluid, we obtain measurable definition of $\varepsilon_m = l^2 |\partial u/\partial y|$. This value can be useful to characterize the turbulent fluctuations in the boundary layer (friction velocity, viscous sublayer or buffer layer dependent to the *Reynold's* number (see also chapter 12).

As with the existence of a turbulent momentum boundary layer there is also a turbulent thermal boundary layer, characterized by inner and outer regions. In its inner part, turbulent mixing is increasingly weak and heat transport is controlled by conduction in the sublayer. Farther from the wall, a logarithmic temperature profile is found, and in the outermost parts of the boundary layer, turbulent mixing becomes the dominant mode of transport. The boundary layer ends where turbulence dies out and the uniform free-stream conditions prevail, with the result that the thermal and momentum boundary layer thickness is the same. Fourier's Law might likewise be modified for turbulent flows by introducing the local average of the turbulent temperature, T', and the eddy diffusion of heat, ε_h , which suggests yet another definition, the turbulent Prandtl number, $Pr_t \equiv \varepsilon_m/\varepsilon_h$.

Knowing that the complex heat transfer coefficient, h_x, is a complex function (of k, x, c_p, u_{max}) and that the mean film temperature, $T_f = (T_w + T_\infty)/2$,

158

the derivation can be based on Fourier's Law again, $h_x = q/(T_w - T_\infty) = k/(T_w - T_\infty)$ $(\partial T/\partial y)_{y=0}$, certainly under certain simplifications such as the fact that a fluid is incompressible, pressure deviation does not affect thermodynamic properties, temperature variations in the flow do not change the value of k and the viscous stresses do not dissipate enough energy to warm the fluid significantly. By assuming the control volume, R_v , in a heat-flow and fluid-flow field we can write that the rate of internal energy increase in R_v is equal to the sum of the rate of internal energy and flow work out, net heat conduction rate out and the rate of heat generation, all related to R_v . It provides the so called 'material derivative', abbreviated as DT/Dt, which is treated in details in every fluid mechanics course and which represents the rate of change of the temperature of a fluid particle as it moves in a flow field. For the field velocity, u, and particle velocity, $\vec{v}$, it takes the energy differential form, $DT/Dt = \partial T/\partial t + \vec{v} \, \nabla T$, or in particular, $\partial T/\partial t + v_x \, \partial T/\partial x + v_y \, \partial T/\partial y = \alpha(\partial^2 T/\partial x^2 + \partial^2 T/\partial y^2)$, which is even more common in its dimensionless form, when T is substituted by θ, which is equal to the ratio of instantaneous versus maximum gradients, $(T - T_w)/(T_\infty - T_w)$, where subscripts w and ∞ mean the wall boundary and infinity.

This is a subject sample when the specific field of heat transfer involves complex motions which can be seen, however, as a remote discipline of classical thermal analysis but which, nevertheless, takes part in any complex process of self-organization.

5.6 Special aspects of non-equilibrium heat transfer

There is a vast amount of literature on heat conduction when counting with relaxation terms [189], which is trying to correct its paradoxical theoretical nature, i.e., the so-called absurdity of the propagation of a thermal signal with infinite speed. It follows from the solution of the heat and/or diffusion equation, which gives a definite construal of propagation for any time. This certainly contradicts the macroscopic observations and was one of the first incentives for the development of extended irreversible thermodynamics. Let us examine an isolated system, which is initially composed of two subsystems at different temperatures. At a given instant, they are put in thermal contact. Equilibrium thermodynamics theory predicts that the final equilibrium state will be that of equal temperatures. However, it does not give any information about the evolution from the initial to the final states, which could be either a pure relaxation or even oscillatory. Though in the oscillatory case the heat could paradoxically flow from cold to hot regions in several time intervals, it is not in contradiction with the second law of thermodynamics because of its global nature. The problem becomes acute when one tries to give a local formulation of this second law in particular time and in local space, which is crucial in second sound experiments. It leads to the definition of the localized production of entropy, s, and the formulation of an extended equation for the evolution of heat,

$q = \nabla \theta^1$, to be compatible with the required positive definiteness of s' and positive values of the parameter μ, i.e., $\nabla \theta^1 - \alpha \theta^1 q' = \mu q$, where $\alpha = \tau/\lambda T$ and $\mu = (\lambda T^2)^{-1}$. For small values of heat flux, the contribution of multiple order of $(q\ q)$ to the absolute temperature, θ, can be neglected, so that θ is taken equivalent to the local-equilibrium temperature, T. Then it provides the form of *Cattaneo* equation, $\tau q' = - (q + \lambda \nabla T)$.

It is important to realize that there is the generalized absolute temperature, θ, and not merely the local-equilibrium temperature, T, which is the relevant quantity appearing in the entropy flux and the evolution equation for the heat flux. Considering infinitesimally small thermal disturbances around an equilibrium reference state, the vales of θ and T become identical with each other up to the second-order corrections and the evolution equation of T and q are as follows, $\rho c \tau\ \partial^2 T/\partial t^2 + \rho c\ \partial T/\partial t = - \nabla (q + \lambda \nabla T)$, where, for simplicity, relaxation time, τ, and heat conductivity, λ, can be assumed constant. Joining the preceding equations together it results in a new hyperbolic equation of the telegrapher type, namely $\tau\ (\partial^2 T/\partial t^2) + \partial T/\partial t = a\ \Delta T$. Under high frequency perturbations (or pulses) the first-order term derivative is small compared with the two other terms and the equation becomes a wave equation whose solution is known in the literature as second sound (for, e.g., dielectric solids in low temperatures). There is several other contexts known, such as in the analysis of waves in thermoelastic media, fast explosions, etc.

In the case of suddenly heated materials (dielectric discharge, laser pulses, fast exotherms or even picosecond-irradiation-induced nuclear fusion) the associated energy transfer cannot be adequately described using the classical Fourier's Law. In order to study such special responses of the system subject to a general perturbation due to an energy supply term, g(r,t), the above energy balance equation have to be extended and the right-hand side is replaced by $[\ g(r,t)\ +\ \tau\ \partial g(r,t)/\partial t\ ]/\rho c.$

It is important to note that the time derivative of the energy supply term has no counterpart in the classical theory of heat. The difference between the classical and extended formalism follows. In the classical description, the response to heat perturbation is instantaneously felt in all space, because the term $\Delta T(x,\ t)$ is non-zero at any position in space for $t>0$. In contrast, $\Delta T(x,\ t)$, derived upon the mathematical manipulation with the energy added equation, vanishes for $x > \sqrt{(\rho c/\lambda \tau)}/m^2$, so that a perturbation front propagates with the speed $U \approx \sqrt{(\rho c/\lambda \tau)}$, while the positions beyond the front are unaltered by the action of pulse.

The description based on a generalized hyperbolic equation predicts the propagation of a front peak, which is absent in the classical parabolic model. It means that the latter underestimates the high temperature peak, which may be present in the wave front, which means that the distribution of thermal energy is altered. In the classical parabolic case it is scattered within the whole spatial region, whereas there is a sharp concentration of energy in the actual case of the hyperbolic solution. It has serious consequences in understanding the absolute non-equilibrium temperature, θ, which is continuous across the ideal wall between the system and thermometer, while their respective local-equilibrium temperatures, T's, may be different. A more intuitive understanding may be achieved in viewing the situation where two hypothetical thermodynamic systems are thermally connected through an ideally connecting plate. One system is at equilibrium temperature, T, while the other is at a non-equilibrium steady state under a high heat flux generated by a temperature difference across its system body. According to classical theory, no heat should flow from one system to another if equilibrated. However, a contrary flow is to emerge, which is proportional to the gradient of the non-equilibrium temperature, $\nabla\theta$, because one equilibrium side has $\theta_1 = T_1$ while the other $\theta = T (1 - \gamma q^2)$ where γ is proportional to $k^3 T^3$. It yields the disequality $\theta < T$. This example corroborates the previous statements, according to which a thermometer indicates the generalized temperature: heat transfer between the two systems will take place until they have reached the same temperature, rather than the same local-like equilibrium. This idea provided an incidental basis to the definition of a "coldness" function assumed to depend on the empirical temperature and its time derivative.

5.7. Human response as the warm-cool feeling

As was accentuated in Chapter 1, the most decisive process of our instinctive responsiveness is the spontaneous sensation of warmness. So it is not unusual to also pay attention to the transient thermal properties that play an important role in our positive sensation of our clothes wrapping. In particular, it applies for flat textile fabrics, as not only the classical parameters like thermal capacity and thermal diffusivity are involved, but also transient thermal parameters, which describe the *warm-cool feeling* [190] or *thermal-contact feeling of fabrics*. It occurs during the short or periodical thermal contact of human skin (thermo-sensors) with any object of a different temperature. In the thermo-physical instruments market are two commercially available measuring apparatuses, the *'THERMO-LABO'* instrument and the *'ALAMBETA'* instrument, which can express the level of thermal-contact feeling in terms of the so called 'thermal absorptivity'. The generally accepted measuring methodology or reference values are, however, missing and the thermal-handle of fabrics, whose level strongly affects customers when buying clothes, is still evaluated subjectively.

The principle of an automated instrumentation relies on the application of a direct ultra thin (0.15 mm) heat flow sensor, which is attached to a metal block of constant temperature which differs from the sample temperature. When the measurement starts, the measuring head, containing the mentioned heat flow sensor, drops down and touches the surface plane of the measured sample, which is located on the instrument base under the measuring head. At this moment, the surface temperature of the sample suddenly changes (i.e., the boundary condition of first order is worked out), and the instrument computer registers the heat flow course. Simultaneously, a photoelectric sensor measures the sample thickness.

The computer display then shows the steady-state values for characteristics such as the thermal conductivity λ [W/(mK)], thermal resistance R [m^{2}K/W] and thickness of the sample s [mm], but also the transient (non-stationary) parameters like thermal diffusivity and so called thermal absorptivity b [Ws$^{1/2}$/(m^{2}K)], Thus it characterizes the warm-cool feeling of textile fabrics during the first short contact of human skin with a fabric. It is defined by the equation $b = (\lambda\rho c)^{1/2}$, however, this parameter is depicted under some simplifying conditions of the level of heat flow q [W/m^2] which passes between the human skin of infinite thermal capacity and temperature T_1 . The textile fabric contact is idealized to a semi-infinite body of the finite thermal capacity and initial temperature, T_o, using the equation, $q_{dyn} = b (T_1 - T_0)/(\pi \tau)^{1/2}$, which is valid just for the short initial time, τ, of thermal contact between the skin and fabric. For longer times, exceeding a few seconds (where the minimum time is given by the square of thickness s divided by 12,96 and also divided by thermal diffusivity, a), the heat flow, q, loses its dynamic (transient) character and its level sinks to the steady-state level given by the relation $q_{st} = (T_1 - T_0)/ R = \lambda(T_1 - T_0)/ h$. The higher the level of thermal absorptivity, the cooler the feeling it represents. Practical values of this parameter for dry textile fabrics range from 30 to 300 [Ws$^{1/2}$/(m^{2}K)]. It was approved that this parameter (which was formerly also used in civil and mechanical engineering) characterizes with good precision the transient thermal feeling which we get in the moment, when we put on the undergarment, shirts, gloves or other textile products. It was found that 'grey' fabric always exhibits the highest thickness, highest thermal resistance (and consequently lowest thermal conductivity) and lowest thermal absorptivity (warmest contact feeling). During the subsequent stages of any chemical treatment, practically all the above mentioned thermophysiological properties become worse. Singeing, for example, makes the outstanding superficial fibres shorter, which reduces the thermal-insulation air layer entrapped amongst the fibres, during any resin treatment the air is substituted by a polymer of higher thermal conductivity and capacity, etc.

Thermal contact sensation by means of some reliable measuring instrument opens new ways for the optimization of this relatively new fabrics-surface characteristic, which often presents a factor that influences the choice of cloth or garment by a customer, and whose method of objective assessment deserves an international standardization. Since the touch sensation strongly affects our choice, the objective assessment of this sensation became very important in the last decade and became a part of generalized thermal analysis.

5.8. Non-equilibrium fluid systems and our cosmological 'engine'

On the other hand, we may look at 'thermal feeling' of our planetary ecosystems, which, in general, is influenced by positioning in interstellar space [9,191]. No planet is a closed system neither is it in contact with two or more thermal baths. Each planet is in contact with a hot radiator (sun $\sim$ 5800 K) and cold radiation sinks (outer space $\sim$ 2.75 K). Each planet therefore realizes a kind of a 'specific cosmological engine'. Being a sphere is crucial because of its rotation and revolution modes with an inclined axis, which are responsible for the richness and complexity in the behavior induced by the solar influx. The atmospheric fluid motion adds the spice of chaos so that the thermodynamic events, which take place within a planet, are sustained by non-equilibrium flows, which must obey the fundamental laws of non-equilibrium thermodynamics.

If our atmosphere were stiff, its thermodynamics would not contain any chaotic phenomena. If the Earth were a rigid homogeneous sphere with no atmosphere and consequently neither albedo nor downward diffusion, the direct solar influx could be calculated without any special experimental measurements. To each point of the surface the solar influx would be a trigonometric factor times the solar constant strictly changing along the exposed face. The neighboring planet Mars is active in this way, so that if we were to colonize it we would need first to create its albedo by manufacturing the retroactively reflecting atmosphere upon introducing CO_2 from mineral sources, or simply by delivering 'Earth-type pollution' (as in our atmosphere) necessary to increase the surface temperature and to enable melting and evaporation of soil-buried water. This process could be relatively brief in comparison to the next step, which should be the creation of a livable environment by an appropriate process (planting) in order to introduce oxygen. There, however, are many other obstacles to tackle, such as too high levels of dangerous radiation (which is screened out from the Earth by its magnetic field shield). The surface of the planet Venus is, on the contrary, extremely hot because its atmosphere is dense and opaque due to high concentration of back-reflecting components and particles (containing even droplets of H_2SO_4) exhibiting, therefore, extremely violent atmospheric motion. Because of a very special nature of the external cosmological state, the boundary conditions for the planets determine the possibility of being an ecosystem, namely a dissipative system with vitality steadily maintained. We can expect a very complex circulation of availability over a planet. We can define a global physical ecosystem to be a system, which satisfies the following properties: it is closed with respect to mass exchange (occupying a finite domain) but is open regarding energy (outgoing flux is matching that which is received) and it is in a globally steady state for a longer period (energy balanced).

So far, the cosmic ecosystem offers only one example, Earth, and it may be its _only one_ realization. The Earth has developed as a heterogeneous thermodynamic system with many components in all three phases of gas, liquid and solid, and with their interfacial boundaries of a complex character. Its mathematical solution, even categorically reduced, is infinitely complex, either the simplest one using just the single component and one phase in homogeneous view or another, with two components that do not mix, assuming a heterogeneous concept, or many others including chemical reactions and sophisticated evaluations by finite elements, lattices, etc.. Additional variety of biochemical cycles became as complex as the definition of life itself. The problems of predictability with increasingly complex non-linear models always tremendously inflate.

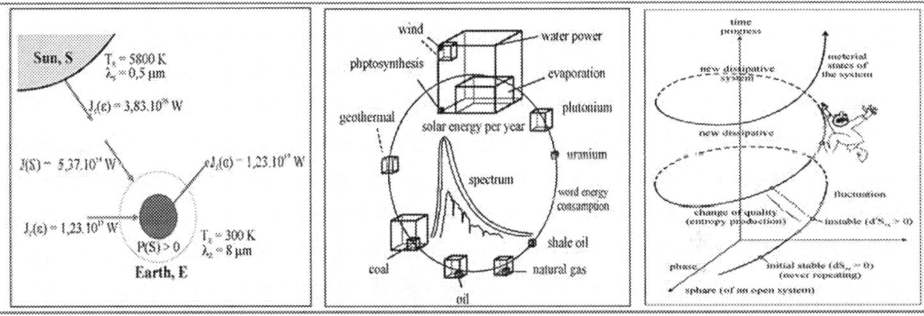

Fig. 28. – Left, the entropy relation between the Sun and the Earth is depicted while the right sketch offers the entropy issues, which cause that our world keeps in a gradual but repetitively coiled progress. Middle, the global view of the income energy delivered from the sun in comparison with the one-time capital resources of energy available on the Earth as schematically shown by the mutually amount-related cubes (the inset illustrates the original spectrum launched by the sun and its stepwise changes while reaching the outside atmosphere and finally collected at the sea level).

For example, let us see the temperature field and its reduction, and follow the temperature at a given spatial element. After a period of one year, or averaged in ten years or so, we expect this temperature to repeat itself with certain accuracy. Violations of its expectancy can be attributed to local chaotic interactions (fast irregularities), global chaotic interactions (slow irregularities) and finally to man-made modifications to the energy balance such as occurs by the injection of excess CO_2 on the terrestrial surface. Considering now the density field of other components (SO_2, aerosols) and its finite element reduction, we again expect only small changes; otherwise the Earth would become thermally or chemically poisoned. Certainly, it can become a result of an abrupt volcano eruption or a sudden impact of meteorite but recently it is *steadily* provided by man, who is thus a strange occupant of the Ecosystem and, at the same time, most flexible and very vulnerable. He cannot stand changes of pressure, chemistry of environment and particularly of temperature. The healthy body temperature variation is about one degree and thus man needs a complex body-control system plus a shelter – house and clothing. There is an evident difference between the living creature (a self-developed control system) and the entire ecosystem of Earth (determined by solar radiation).

The power per square meter of the Earth's surface is not a large number, but this modest gift is what the Nature has given us. Any particularly stable society could get along with; the generations of dinosaurs lived in equilibrium with this number for a hundred million years, humankind also managed for several millennia until starting to expand its population and advanced its technology. Finally, in the last thirsty century, a blink with respect to history, a small percentage of humankind has developed a trend for fossil energy consumption, which has become out of tune with the natural orders of magnitude of power that likely the cosmic evolution has decided to give us. Sometimes, it is hard to believe that we are not clever enough, acting without scientific conscience, just simply driven by allurement of increasing growth.

In the first approximation let us consider the Sun as a stable source of Planck's distributed photons at the temperature, $T_S = 5800$ K, at its outer radius R_S. The photons that leave the Sun travel in all directions and without interactions sink at infinity or, better, go into equilibrium with the fossil bath at about 2.7 K. We can define the temperature T_d, at the distance d, which is $T_d = T_S \sqrt{(d/R_S)} > 1$. For the Sun's entropy emission similar relation holds: $S_d./S_S. = \sqrt{(d/R_S)} > 1$. If the solid Earth were a black rigid sphere with radius R_E, and with transparent atmosphere, the inward and outward fluxes would be given by simple boundary conditions of T's and R's. The relative motion of the Earth and the Sun produces a periodic function f, with the period p, which embodies the mass continuity, *Navier-Stokes* equation and energy balance relations for the atmosphere still very far from authenticity. The stationary distribution of the sustained equilibria corresponds to a capacity of mechanical power that may be expressed by a distribution of an infinite number of Carnot's engines whose global efficiency can be formulated in general terms by mathematics. It is approximated by the relation, $\eta_{global} = \iint \lambda \, (T_1-T_2)^2/max \, (T_1.T_2)]dS/R^2 \, \delta T^4$, where λ and δ are respectively the overall Newton thermal conduction (fixed as 8 w/m^{2}K) and *Stefan-Boltzmann* radiation constants (providing $\delta T^4 \approx 1384$ W/m^2). The calculated values of η_{global} are in the neighborhood of one hundredth.

The biological network interacts with the thermodynamic ecosystem (is factually 'thermodynamically living') so that it is not a passive stage but a livelihood architecture. The global treatment induces better understanding of the concept of global steadiness governing a distribution of local disequilibria. If the Earth had an irregular orbit, the problem would not admit a globally steady-like solution and if, in addition to the daily and annual periodicity, there were additional irregular movements, there would be neither global steadiness nor the less restrictive property of global periodicity. The ample complication being the variation in either the solar emission or mutual orbit geometry, which imply changes in the input conditions for the Earth ecosystem and which is known to happen in certain often-repeating short (30, 70 or 150 years) or lengthy (thousand and million years) intervals depending chiefly on the Sun, but having also many chaotic components (eruptions).

Let us mention that the Earth's orbit is fortunately close to a circle but not completely. Its ellipticity varies over the course of the year but its effect is too weak (3 %) to cause the instant seasons but strong enough over its severity, regressing by about one full day every 58 years. The Earth reaches perihelion (closest point towards the Sun) in early January but this date does not remain fixed but slowly regresses. Tropical year is measured between two vernal equinoxes (being the base of our Gregorian calendar) while between two perihelia lies the anomalistic year (about 25 minutes longer), which moves completely through the tropical year in about 21,000 years. Most of this difference is due to the slight change in the direction of the Earth's rotation axis in space, called precession, which is tilted from the normal to the orbit plane at an angle of 23 degrees and precesses once every 23,000 years. The eccentricity of the Earth orbit varies periodically with a time scale of about 100,000 years, so that climatologically it can become imperative if positively modulated together with the 21,000 cycle of perihelion. Another 41,000-years variation is obliquity, which is different from precession. These astronomical phases are often called *Milankovitch* cycles (after the theory of the Serbian civil engineer) but their true impact on climate changes is not yet clear.

However, within a cosmological scale, the concept of an isolated system becomes no longer practicable because the definition even of our neighborhood boundaries is tricky (as the Sun is enveloped by far reaching dispersion layer full of eddies and horrible jets and our terrestrial atmosphere is diffuse and full of motion, containing rarefied ions with outwards super-fast streams) and the equations of fluid motion fail to hold. These alterations in the input conditions bring about yet new dynamical regime of atmospheric and oceanic motions (apart from changes in composition, structure, etc.). As these modification are non-linear again, nobody can predict, or know, what could be the resulting effect with respect to what we have experienced or can resolve from history.

Let us move to an example of a renewable engine, and to this purpose consider now thermodynamics of the Earth. This is a system in approximate global steady disequilibrium state, which means that, more or less, inward and outward radiation fluxes balance. We can recall the standard thermodynamic approach to consider a spatial domain, Ω, with elementary volume $d^3r^{\rightarrow}$ with coordinates, $r^{\rightarrow}(x,y,z)$, boundary, Ω_B, and time, t, assuming the density, $\rho(r^{\rightarrow},t)$, fluid velocity, $v(r^{\rightarrow},t)$, and temperature, $T(r^{\rightarrow},t)$, which are not all locally stationary. Energy dissipation, w, (irreversibility or entropy production) comes from thermal conduction (∇T) and viscous shear (tensor, σ_{ij}), and contains a complicated maintained flow of disequilibria to which corresponds a dissipated availability for a *natural* system, $w^d_n = \sigma_{ij}\,\partial v_i/\partial x_{(k)} + \lambda\,(\nabla T)^2\,/\,T$ (given in [W/m^3]). The power density is constantly created and dissipated and this is the power embedded in the natural motion. If we know the solution of the natural dynamics, we could construct the quantity w, calculate it numerically and this would be the evaluation of the dissipated power of the Earth. Its volume integral would be a fundamental global number qualifying the planet's stationary global conditions, so to say, its 'health'. This is actually an extension of the story of a classical engine that produces mechanical work and this work is immediately dissipated into heat. Gasoline is solar radiation and ambient coincides with engine, because there is no "outside" to the global ambient. The right-hand-side of above-mentioned equation can be separated into two parts, the first being the shear availability and the second the Carnot availability. For an *active* system (familiarly called 'windmill availability') the second term combines with the term of mechanical power yielding $w^a = \lambda\,(\nabla T)^2\,/\,T + v^{\rightarrow}\,\nabla P$, which is technically the power source of opportune dams, windmills, etc., constructed by man, while the remaining dissipation, $\sigma_{ij}\,\partial v_i/\partial x_{(k)}$, apart from a negligible delay, changes nothing from the natural circulation. We can thus see the natural motions as a loop powered by the availability that creates motion, which in turn creates availability, which is in turn dissipated and creates motion again, and so on. The natural and the active systems are indeed different.

Their joint consideration is equivalent to the consideration of burning a certain amount of gasoline in nature, and see what happens if we burn it inside the engine. These motions can be periodic (or better chaotic) depending to the temperature gradients. In those places where $(\lambda(\nabla T)^2\,/T)$ and $(v^{\rightarrow}\,\nabla P)$ are particularly large, ejection of mechanical power may occur, typically in the form of the growth bubbles in a pot of hot volcano magma. It teaches us many things. We learned that in a stretched-out period, the solar radiation is not constant and its emission pattern is partly periodic and partly chaotic. The final result is that the Earth's ecosystem has an extremely complex thermodynamic configuration [191,192] with the presence of chaos overlapping a predominant periodicity. It shows the physical fact that the fundamental numbers arose from the pure use of the global formulation; in summary they are predominantly the temperatures of the Sun, the separating distance, and the Earth's spheroidicity. The given availability is actually a skin, or better an interface phenomenon particularly when introducing into the model the study of chemical reactions and detailed fluid motion, which come necessarily together because the overwhelming majority of processes take place in the fluid state. It is not clear what number of chemical reactions should be taken into the description, at least the biophysical cycle of CO_2 (growth $\rightarrow$ $6CO_2 + 6H_2O = C_6H_{12}O_6 + 6O_2$ $\leftarrow$ combustion) being most important. Nature itself has factually invented super-sophisticated chemical cycles and man is now endeavoring in his wealthy empire to develop somehow ignorant and counterpart violations.

It is clear that good ideas for the theory of ecological human shelter may come from the study of dissipative learning systems, i.e., such adaptive systems that are fit for survival. The Earth has the necessary vitality, complexity and equilibrium-like stability, which is prerequisite for its survival. The hierarchy of interactions may help to formulate the concept that man is not a deterministic system, but rather a finalistic system, or better a control system. So the species ecosystem has a simpler external interaction and the ecosystem, as a giant engine very complex inside, has its rather simplistic and almost ngligible interaction with the Universe itself, enabling to see it as a yet deterministic world. Therefore, it is the profound mystery to see the existence of islands of order that nurture the possibility of intelligent life. The study and wider application of truly non-equilibrium thermodynamics and the theory of chaos makes sense, but there is a great, unexplored domain beyond that particularly regarding non-equilibrium thermodynamics of the Earth environment itself.

6. HEAT, ENTROPY AND INFORMATION

6.1. Caloric as entropy and the information concept

It is clear that besides analytical mechanics and the theory of electromagnetic fields, thermodynamics is considered to be the entrenched and logically closed theory. Though it seems curious in the framework of so well-established fields of science there are relentless attempts and ongoing couloirs discussions concerning the validity of the Second Law of thermodynamics when the quantum nature of the system must be taken into consideration [158,194]. There are various axiomatic forms of the thermodynamics which seem to guarantee absolute clarity of the concepts involved. Nevertheless we may encounter some difficulty in finding a book where the subject is treated in a way that is genuinely comprehensible by an ordinary student. The origin of such a difficult understanding of thermodynamics is connected with an inconvenient choice of its conceptual basis established more than 150 years ago. Traditionally the most obscure is an artificial concept of entropy and rather exceptional form of the "Second Law" of thermodynamics. Whereas the universal laws have mostly the form of conservation laws, the logical structure of this second law is quite different. Ultimately formulated, it is a law of irreparable waste of "something" in every real physical process. This imperative negativistic and pessimistic character of the Second Law is a permanent source of dissatisfaction not only for philosophers but also for many active researchers in the sciences. Criticism of the Second Law has a history as long as the law itself and, moreover, in recent decades an unprecedented number of challenges have been raised against the Second Law from the position of quantum mechanics [158,194]. These arguments, however, are as a rule, enormously complicated with numerous approximations and neglects and consequently rather disputable.

It is a very old empirical fact that the thermal processes in nature are submitted to certain restrictions, which strongly limit the class of feasible processes. The exact and sufficiently general formulation of these restrictions is extremely difficult and sometimes even incorrect, e.g., the principle of *Antiperistasis* [195], Braun-le Chatelier's principle [196] as well as the second law itself but, in spite of it, are found very useful. That is why we believe [197] that the Second Law, as well as other laws which put analogous limitations on thermal processes, reflects experimental facts with an appreciable accuracy and thus it should be aptly incorporated into the formalism of thermodynamics. On the other side, being aware of the fact that the contemporary structure of thermodynamics with its somehow archaic conceptual basis may have intrinsic flaws, we venture to claim that the absolute status of the Second Law should not be criticized or denied from the point of view of another physical theory (e.g. quantum mechanics) before the correction of these imperfections has been made explicable enough.

A serious flaw in the conceptual basis of classical thermodynamics concerns even the so-called First Law of thermodynamics. The first step toward this law was made by *Rumford* by the generalization of his observations made at an arsenal in Munich (1789) [198]. Accordingly, practically unlimited quantity of heat was possible to produce only by mechanical action i.e. by boring of cannon barrels by a blunt tool and this experimental fact was analyzed by *Rumford* as follows: *"It is hardly necessary to add, that any thing which any insulated body, or system of bodies, can continue to furnish without limitations, cannot possibly be a material substance: and it appears to me extremely difficult, if not quite impossible, to form any distinct idea of anything, capable of being excited and communicated, in the manner the heat was excited and communicated in these experiments, except it be* <u>*motion*</u>*"*. The same idea that heat absorbed by a body, which is particularly responsible for, e.g., the increase of its temperature, is identical with the kinetic energy of its invisible components was further supported by arguments due to *Joule* [199]. Results of his ingenious and marvelously accurate experiments have been summarized into two points: The quantity of heat produced by the friction of bodies, whether solid or liquid is always proportional to the quantity of force expended. The quantity of heat capable of increasing the temperature of a pound of water by 1° F requires for its evolution expenditure of a mechanical force represented by the fall of 772 lbs through the space of one foot. (and note, that the term "force" has here evident meaning of energy). In spite of clearness of these correct statements, *Joule* did not stressed out explicitly the fact that in his experiment we have to witness the execution of only *one-way* transformation of work into the heat. Throughout his published thesis he, instead, tacitly treated heat as it were a physical entity fully equivalent or even identical with mechanical energy. It was probably due either the influence of *Rumford* or the reasoning that in the experiment heat appears just when mechanical work disappears and so these two entities ought to be identical. Such an extremely suggestive, but seemingly incorrect idea was later canonized by *Clausius* [46] who proclaimed the object of thermodynamics to be *"die Art der Bewegung, die wir Wärme nennen "*, i.e., the kind of motion we call heat [52].

In the history of thermodynamics objections have emerged in opposition to such an 'energetic' interpretation of the heat, unfortunately, these objections were only rare and without any adequate response, one of them being offered by *Mach* [51] or generalized by *Bridgman* [200] *"It does not seem obvious that not all formulations of the second law can be exactly equivalent"*. Accordingly, it is quite easy to realize a device of Joule's type where a given amount of energy W is completely dissipated and the heat in amount of Q=JW is simultaneously evolved, where J is universal Joule's proportionality factor. On the other side, as far as it is known, there is no single real case where the same amount of heat Q is transformed back into mechanical work ($W=Q/J$) by reversion the original process. Taking into account this circumstance together with the very generic property of the energy, which can be principally converted into another form of energy without any limitation, we must exclude the logical possibility that the heat is energy at all. Of course, traditionally postulating the apparent equivalence of energy and heat, a somewhat meaningful mathematical theory of thermal processes can be (and actually was) established. The price paid for the equivalence principle is, however, rather (intolerably) high. In order to make the theory consistent it was necessary to create somewhat synthetic and highly abstract quantities like entropy, enthalpy, free energy, and various thermodynamic potentials the meaning of which is more formal than physical. The mathematical manipulations with their over 720 derivatives and differentials (which are sometimes total and sometimes not) [197,200] actually provide the tool which interpretation is, however, rather matter of art than of science.

Astonishingly, there is an elegant way out of these problems, which was first suggested by *Callendar* [201] and later in a more sophisticated form by *Job* in his impressive book [202]. The main idea is that the heat in common sense (e.g. as a cause of elevation of temperature of bodies exposed to the heating) should not be identified with a kind of energy but with the entropy as is known from classical thermodynamics. This case the heat–entropy (ς) concept attains the content almost identical with that of Carnot's "caloric" ς [203] whereas the empirical temperature θ (i.e. 'hotness' or 'warmness', [51,52,133] automatically begins to undertake the role of its potential (we intentionally use the final Greek letter ς as it involves both, the usual S for entropy and C for caloric). For the increase of potential energy dϵ of the amount of caloric ς due to the increase of temperature by dθ we may write $d\epsilon =_\varsigma C(\theta)d\theta$, where $C(\theta)$ is the so called Carnot's function. It is a well established experimental fact that this function can be reduced to a universal constant equal unity if using instead of an arbitrary empirical temperature scale the ideal gas temperature scale T (i.e., the known absolute Kelvin scale) [22]. In this case, for the potential energy ϵ corresponding to the amount of caloric ς kept at the temperature T holds that $\epsilon = \varsigma T$.

The perfect analogy with other potentials known from physics, such as gravitational and electrostatic potentials, becomes then evident. After the terminological substitution of heat-energy by heat-entropy it is only a technical problem to reformulate two fundamental laws in a manner which is common in classical axiomatic thermodynamics [204], namely:

i) Energy is conserved in any real thermal process.

ii) Caloric (heat) cannot be annihilated in any real thermal process.

We should note that formulated in such a way the first and second law are conceptually disjunctive because caloric has nothing to do with energy. The possible link between these laws and quantities provides, however, the above formula $\epsilon = \varsigma T$.

It is not our intention to discuss here the application of theorem (ii) as applied to particular cases known from empirical observations of real processes (see, e.g. [202]) but, instead, we proceed further making use of a well established connection between entropy and information [205]. Accordingly, the information has a character of negative entropy (i.e., we can write $I = -\varsigma$) and therefore, in the old-to-new provisional terminology, we can identify the production of caloric with the destruction of information and the flux of caloric with the information flux in an opposite direction. Theorem (ii) can thus be reformulated in terms of information as

Information (I) is destroyed in any real thermal process.

The veracity of this theorem seems to be very obvious at first glance as almost everybody has experience that by burning of newspapers in a stove or combusting petrol in a car engine these materials are lost for ever, together with the information involved. On the other hand, it is little convincing that such a "tiny thing" as the information is and can really be able to control natural thermal processes. So that we assume that the validity of above postulate is quite general and apt for substitution of the classically assumed second law of thermodynamics.

Assuming that the above theorem is decisive for all thermal processes, we can even paraphrase Rumford's original analysis of his observations made during boring cannons with a blunt borer in the following way: *"It is hardly necessary to add, ……. , except it be perished information."*

In order to involve the information into the physical reasoning it is first necessary to convert information coded, as usual, in binary units I_2 (bits) into the information I_p expressed in physical units. This relation obviously reads $I_p = (k\ ln2)I_2$ where k is the Boltzmann's constant ($= 1.38 \times 10^{-23}$ J/K). It should be stressed, however, that by choosing this particular constant as a conversion factor the absolute Kelvin temperature scale was simultaneously chosen for temperature measurements.

We can assume now that there is no information *"an sich"* or in other words information needs in all cases a material carrier. From the point of view of macroscopic thermal physics there is, however, fundamental difference between e.g. genetic information inscribed in the DNA and information provided by a gravestone inscribed with personal data. Whereas in the former case for coding of information structural units on molecular level are used, which should be described by microscopic many-body formalism, to the later case rather a macroscopic description in terms of boundary–value problem is adequate. To distinguish without ambiguity between these two extreme cases we need, however, a criterion which, having a sign of universality specifies what the "molecular level is". As far as we know, a good candidate for such a criterion is modified *Sommerfeld's* condition distinguishing between classical and quantum effects [193,206,207]. It read as $\Omega \le 2\pi\hbar$ where Ω is the phase space occupied by a structural unit '*qubit*' where minimally 1 bit information is stored and $\hbar$ is the Planck's universal constant (= 1.05×10^{-34} Js). Direct computation of the action Ω corresponding to one atom built in an ordinary crystal, liquid or gas confirms the validity of such conditions. It provides evidence for the fact that every atom together with its nearest neighborhood should be treated as a quantum-structural-unit responsible for information storage on a "molecular level". Generalizing this result, we can conclude that the very nature of Carnot's caloric is the destructed information originally coded in occupied quantum states of structural units of which the macroscopic system under investigation consists.

The mechanism of information transfer through the macroscopic system is assumed to be due to erasing information in one particular structural unit, which is influenced by the neighboring one in the field of long-range forces defined by macroscopic system as a whole. There are, however, limitations of such a process. First, as the information storage in both neighboring structural units is submitted to the same condition it is impossible to exchange more information from one unit to another than about 1 bit per $2\pi\hbar$ of the occupied phase space. Second, the exchange of information must be in agreement with boundary conditions put on the macroscopic system as a whole, which are locally realized, e.g., by long-range forces. It may thus happen that the transfer of some information from one unit to the neighboring unit is incompatible with these external conditions and information is lost. The loss of information physically means that some characteristic pattern of structural unit has disappeared and a wider class of quantum states becomes accessible. In the frame of the presented model any loss of information should be accompanied with the development of energy and we need to explain where the energy comes from?

We are inclined to interpret the stability of quantum objects as a result of existence of zero-point electromagnetic vacuum fluctuations exactly compensating energy loses due to the recoil radiation from this object. Such an approach well known from stochastic and quantum electrodynamics [208], confines our considerations to the systems controlled only by electromagnetic interactions, namely, low temperature plasma, gases, condensed matter and chemical reactions in these systems. Accordingly, the cohesion energy of any such a system is nothing but the energy of electromagnetic modes of the background zero-point radiation accommodated in such a way that they fit the geometry, i.e., 'non-symmetricality' of the system.

Characterizing the dimensions of the quantum ($\hbar$) electromagnetic (c) system by a single length parameter a, we can apply dimensional analysis [24] and immediately obtain for the cohesion energy the Casimir's type formula $\epsilon \approx \gamma\,(\hbar\,c/a)$, where the dimensionless parameter γ should be determined from a particular geometry of the system (usually ranging from 0.1 to 0.001 [208]. The change of dimension, a, or the complete thermal destruction of the structural unit with given energy has a consequence that just this amount of energy is developed at the given place. As this energy is, in fact, a modified energy of the all-pervasive universal zero-point background, we have to operate with an energy supply from a practically inexhaustible non-local source of energy. Therefore, within the frame of stochastic electrodynamics every thermodynamic quantum system should be interpreted as an open system even in the case where it is finite.

In order to make the presented system of quantum thermodynamics more intelligible, we give three examples in an attempt to illustrate how some common observations within the frame of the above system should be interpreted.

1) *How does the heat engine work?* A heat engine, in the sense of original Carnot's theory, is simply a kind of mill driven by caloric ς falling from a higher potential T_1 (boiler) to a lower potential T_2 (cooler). Information thus flows from the cooler with condensed water (better ordered than steam) to the cylinder of engine where the information is destroyed (by weakening of correlations among molecules during the expansion) giving rise to useful work originating in zero-point background. Then the residual information continues to flow to the heater where it is dissolved during ordering of configuration of the steam. Note that the flow of information and the flow of water are opposite in this case and that the question how the boiler is heated is put aside. In a typical combustion engine at low temperature the fuel with high information content flows into the cylinder. During the combustion of fuel the information which is coded in its structure is destroyed and the useful work from the zero-point quantum electromagnetic energy is produced there. The information, however, flows inside the combustion space for this type of engine also through the exhaust-pipe so that a special attention must be paid even to this part.

2) *How does the Bunsen's ice calorimeter work?* As this apparatus functions at a well defined temperature T_M (i.e. at the melting temperature of ice equal to 273 K) it, in fact, measures directly inputted caloric and may thus serve as a kind of "entropymeter". Indeed, the information destroyed and the latent energy of melting is connected here in an especially obvious way. An estimate of the latent energy ϵ_M per one mol of ice can be obtained as follows. We make use of the important fact that the H_2O molecules retain their integrity in both water and ice, simultaneously neglecting the effect of clustering of molecules at temperatures well above the melting point which is responsible for non-trivial

macroscopic behavior of water (as discussed already a long time ago [209]). Within the frame of such a simplified model it would be necessary for the melting of ice to break down four well–oriented bonds per every water molecule and substitute them by a quasi-continuum of states (clustering for $T>T_M$ is neglected). Such a transformation corresponds approximately to the destruction of $I_2 = 4$ bits of information per molecule [205] and we can thus write $\epsilon_M = (Nk\ ln2)T_M\ I_2$ where N is Avogadro's constant ($= 6.02 \times 10^{23}$ mol^{-1}). The estimate of ϵ_M then reads about 6288 J/mol, which is in an almost excellent agreement with the experimental value of 6007 J/mol.

3) *Can we ever make any generalized quantification?* There are different microscopic parameters characterizing the configuration of a structural unit where the information is stored which can in principle be constructed from quantum numbers describing this system. The relation connecting these microscopic parameters and macroscopic boundary conditions is evidently very complicated. If we, however, as above, confine ourselves only to a single parameter a – a characteristic dimension of the structural unit, this relation can be found in an explicit form and compared directly with experimental data. Combining the above formulae, the temperature change of the potential energy of a structural unit is given by the relation $d\epsilon\ /dT \approx - (k\ ln2)\ I_2$ which is caused by the erasing of information I_2. Substituting for ϵ the Casimir's quantum of cohesion energy we immediately obtain an estimate for the corresponding relative expansion of the unit, which is $d\ ln\ a/\ dT \approx a\ (k\ ln2\ /\gamma\ \hbar\ c)\ I_2$. Assuming that the thermal process is homogeneous and isotropic, this coefficient must be within the order of magnitude identical with the macroscopically observed coefficient. For typical condensed matter where bond length $a \approx 4 \times 10^{-10}$ m and $I_2 = 1$ we obtain for the coefficient of relative thermal expansion the value of $1.2 \times 10^{-7}/\gamma$, which is near to the values experimentally observed (typically $\approx 10^{-5}$), provided that $\gamma \approx 0.01$.

Even if it is not common yet in the current treatises on classical thermodynamics we can become aware that some alternative exposition of innovative structures of quantum thermodynamics [193,197] are plausible as build on the standard Carnot's theory where the fluxes of caloric are identified with the negative fluxes linked with information. For that reason, the thermal energy evolved by thermal processes becomes identical with the electromagnetic zero-point background energy evolved by the destruction of information inscribed in a structural unit ('qubit').

We can mention a noticeable concurrence with some early ideas such as those presented by *Tribus* in his exceptional book [210] that we can consider that the entropy only measures the extent of our ignorance about the detailed behavior of a system, or, as *Gibbs* put it [151], shows that our 'mixed-uppedness' increases. If, for example, at a particular instant of time we have contrived to bring two bodies of different temperatures together, then at that time we can make some fairly definite statements about how much energy the two bodies have and how much each one has. The process of heat transfer serves to literarily 'wipe out' some of this specific information. At time progress, the energy distributes itself between the two systems and we therefore know less about where it is than before. Of course, our knowledge has not gone to zero as we can still say on the average how much energy each system has, but the number of microstates that the two bodies can assume is enormously larger than before so that our knowledge of these microstates is less sure.

In the late 1990s the Dutch Nobel prize laureate *'t Hooft*, who has merit for bringing the hypothesis of 'hidden variables' back to life, argued that the salient difference between quantum and classical mechanics is the information loss. A classical system contains more information that a quantum system, because classical variables can take any value, whereas quantum ones are discrete. So far a classical system, to give rise to a quantum one, must lose information and that can happen naturally because of friction or other dissipative forces. As an example we consider stones, which are thrown out of a high tower at different speeds. Air friction causes the stones to approach the same terminal velocity, which an observer would measure at the sidewalk. i.e., information of initial speed became a 'hidden variable'. A wide range of starting conditions leads to the same long-term behavior, known as an *attractor*, and these attractors are discrete just like quantum states. And the laws they obey derive, but differ from, the classical Newtonian laws.

6.2. Idealized heat engine working with zero-point electromagnetic radiation

Thought (so called 'Gedanken') experiments with idealized Carnot's engine belong traditionally to the most powerful tools of classical thermodynamics. Since the time of *Bartoli* [155,156] they are also used with appreciable success for the theoretical investigations of interaction between electromagnetic radiation and matter. A crucial role in these studies is played by the concept of adiabatic wall (partition) which is, as a rule, realised by means of an absolutely reflecting mirror made of a "perfectly conducting material". Application of such an abstraction to theoretical treatment of the properties of black-body radiation enclosed in a cavity enabled one to introduce the concepts of temperature and entropy of radiation into classical thermodynamics and, eventually, led to the derivation of the correct form for the dependence of the integral radiation density on temperature (*Stefan-Boltzmann* Law). Among other results of these pioneering studies, an interesting theorem related to the subject of the present work should be mentioned [211], namely: *"Expansion or compression of a cavity with adiabatic walls do not change the entropy of the radiation enclosed"*. It is a typical feature of this approach that the electromagnetic radiation was considered to be a self-contained entity, which might be only slightly influenced by the shape of the cavity and which was essentially independent of the quality of its walls. The very importance of the physical nature of the walls of the cavity on the processes involved was first realized only later by *Planck* [212], who simulated the physical properties of the walls by a finite set of abstract "oscillators". It was his extensive research devoted to the black-body radiation which started the development of quantum mechanics and eventually led to the discovery of the so called zero-point (ZP) energy of his "oscillators". Curiously, as recognised appreciably later [208], the existence of the said ZP fluctuations, taken as 'an Ansatz' added to the classical electrodynamics, brings about a theory which provides a satisfactory description of effects traditionally treated by quantum electrodynamics (QED). As the present paper is just a contribution to this theory, known in the literature as random or stochastic electrodynamics (SED) [208,209], we systematically ignore the concepts and ideas related to the QED description of physical vacuum in this paper. Thus, we assume that there is a real fluctuating electromagnetic field, existing quite independently of the source and thermal electromagnetic fields and persisting even in the absolute zero temperature limit where classically all motions cease. Furthermore, this 'all-pervasive' electromagnetic radiation of *a priori* unknown origin is homogeneous, isotropic and its spectrum is invariant with respect to the group of conformal (*Lorentz*) transformations. It is interesting enough to notice that the latter property is decisive even for the analytical shape of the ZP electromagnetic spectrum. As was proposed by *Boyer* [213] on the basis of homogeneity arguments, there is only one possible form of spectral energy density $\rho(\omega)$ which is the Lorentz invariant, namely $\rho(\omega) = \hbar \omega^3 / 2\pi^2 c^3$, where ω is the frequency and c is the speed of light.

The difficulties encountered by the application of this formula to the solution of particular problems are mainly due to its divergence with respect to the integration over the infinite frequency range. To obtain from it a physically meaningful figure, a rather laborious work with infinities [213] or an introduction of a more or less arbitrary cut-off frequency [208] are required. It is true that e.g. remarkably good prediction of ponderomotoric forces existing between two infinite perfectly conducting planes was obtained in the original exposition of this so called Casimir's effect [215], just by the computing of a finite difference of two infinite radiation forces acting on different sides of the said planes. In spite of excellent agreement of these results with experiment [216], such reasoning is unphysical. The corresponding mathematical procedure is based on an exact cancellation of semi-convergent infinite series containing the terms rather sensitive to the boundary conditions on the infinite planes, which cannot be, however, well-defined in principle [217]. Moreover, any macroscopic model of partitions involved in thermodynamic thought experiments with electromagnetic radiation should not ignore their microscopic atomic structure without threatening serious danger of introducing a fatal error.

It is evident that the formal manipulations with infinite number of terms describing high frequency electromagnetic modes, resulting just from the interaction of radiation with the partitions, is, in the case where the wavelengths are smaller than the inter-atomic distances, physically questionable and may provide the correct results only by chance or on the basis of unknown reasons. That is why we are convinced that the incorporation of intrinsic material properties of the reflecting or absorbing partitions into the thought experiments with radiation, at least to a certain degree of approximation, is an inevitable part of such considerations.

In our recent contribution [157,219] we tried to analyse a simple thermodynamic thought experiment performed with a heat engine containing ZP radiation, in which the reflecting walls are made of a material having an intrinsic upper cut-off frequency more realistic than a "perfect conductor" is. The spectrum is divergent (i.e., $\hbar\omega\, u_\omega \to \infty$) and thus must have a cut-off point, the suitable limit of which is the *Compton* frequency of electron, $\omega_c = mc^2/\hbar$. The necessity of the existence of the upper cut-off frequency for the interaction of ZP electromagnetic radiation with ordinary matter can be explained as follows [208,209,219]. Setting aside the problem whether the electrodynamics can be extended to arbitrarily high frequencies or not (admitting e.g. formula (1) without limits), we discuss only common materials in which the response to the external electromagnetic radiation is mainly due to the electrons. It is obvious in this case that at very high frequencies $\omega \geq c/b$, where $\underline{b}$ represents the extent of the structure and c the velocity of light, the electrons are not able, because their speed is limited by c, to follow the electromagnetic vibrations of appreciable amplitude and the strongly oscillating field has to uncouple from them. Reasonable estimate for the upper frequency limit of such a decoupling is Compton's frequency $\omega_C = mc^2/\hbar$. Accordingly, it is assumed that all the parts of heat engine considered are made of a material which is, up to a certain cut-off frequency $\omega_K \leq \omega_C$, a perfect conductor (mirror) and is simultaneously fully transparent for frequencies $>\omega_K$. Moreover, this constitutive quantity ω_K, should be, in order not to violate the principle of relativity, obviously Lorentz invariant.

The engine itself consists of a cylindrical cavity provided with a piston which can move without friction. The position of the piston is measured by the distance x between its inner side and the bottom of the cavity. The thought experiment is performed where $T = 0$ with an engine bathed in ZP radiation. Let us now expand the cavity by moving the piston with a constant velocity $V \ll c$ from its starting position at $x \approx c/\omega_K$. Because in this case ω_K is the gravest frequency, there is at the beginning no other electromagnetic radiation except the isotropic ZP radiation of frequencies $> \omega_K$ freely penetrating through the walls of the cavity. The beams with frequencies belonging to a certain narrow band just above ω_K, however, when meeting the inner side of the moving piston, will reflect from it, because their frequencies observed from the moving coordinate system of the piston will be, due to Doppler's effect, smaller than ω_K. Similarly, also the beams with frequency just below ω_K falling on the outer side of the moving piston will penetrate into the cavity. All such reflected and penetrated beams will remain trapped there, because their frequencies are smaller than the cut-off frequency ω_K, moving within the cavity to-and-from. To be more specific, the frequency shift of a beam reflected from the inner side of moving mirror is given by the formula $\omega_1 = K\,\omega_0$, where ω_0 is the original frequency, ω_1 the frequency after the first reflection, K Doppler's factor which depends on velocity V (or $\beta = V/c$) and incidence angle θ. (For the piston moving outward from the cavity K is evidently < 1.) Because reflection belongs to the group of conformal transformations, i.e. just to the group which preserves the spectral composition of ZP radiation, relation (2) should map any band of ZP spectrum (1) onto another band of the same curve. Thus any narrow band of ZP radiation lying just below ω_K can be transformed by a number of multiple reflections from the moving piston to the ZP radiation extended down to the gravest mode $\omega_G \approx c/x$ of the cavity. The process follows the formula $\omega_N = K^N \omega_0$, where ω_N is the frequency of an original beam after N reflections from the moving piston. To assess the limiting behaviour of the process just described, some approximations are necessary. For example, for quasi-stationary displacement of the piston (i.e. $V > 0$) the following estimate is valid $K \approx 1-2\beta \cos\theta$.

At the distance x between the piston and the reflecting bottom of the cavity the beam will suffer, during the time corresponding to the displacement of dx, N reflections from the piston where $N \approx \cos\theta\, dx\,/\,(2\beta\,x)$. Consequently, with increasing N, $\omega_N = \omega_0\,(1-2\,\beta\,\cos\theta\,)^N \to \omega_0\,\exp\,(-\cos^2\theta\,dx/x)$. From this formula it is obvious that for small β the explicit dependence of the limiting frequency on velocity disappears and that for sufficiently large expansion, dx, the cavity is filled practically down to zero frequency by ZP radiation. The process is, due to the time reversibility of beams, reversible also in a thermodynamic sense. Particularly, for already filled large cavity ($x \gg c/\omega_K$, $\omega_G > 0$) any change $dx \ll x$ is fully analogous to the classical reversible adiabatic process. Indeed, because in this case the spectral composition corresponding to the distribution of ZP fluctuations given by (1) is preserved during expansion or compression, the entropy change $dS = 0$ as well as $dT = 0$, which is fully in accordance with the definition of zero absolute temperature in classical thermodynamics.

The behaviour of small cavities (formally $x \approx c/\omega_K$) is, however, qualitatively different. The lower cut–off frequency corresponding to the ground mode of the cavity ($\omega_G \approx c/x$) reaches high values comparable with ω_K in this case. The number of admissible modes within the interval (ω_G, ω_K) thus changes during the expansion or compression of the cavity appreciably and the process can be no more treated as adiabatic. From this point of view it is apparent that partitions with upper cut-off frequency are equivalent to the classical adiabatic partitions only if the cavity is large enough, or, in other words, if $\omega_K \gg \omega_G$. For $\omega \geq \omega_K$ the spectrum of ZP radiation, regardless of whether it is inside or outside of the engine, is not influenced by its presence. Alternatively, the difference between behaviour of large and small cavities may be also evidently ascribed to the very existence of Casimir's forces in small cavities. Then, '*mutantis mutandis*', for the computation of Casimir's forces arising during the working cycle of our idealized engine, the only relevant must be the electromagnetic modes from the finite frequency band (ω_G, ω_K), where ω_G depends on the detailed geometry and absolute dimensions of the engine.

In a brief summarizing, we have introduced a new model [219] of first-step approximation for the reflecting partition with a sharp upper cut-off frequency ω_K, which has been used for the construction of an idealized heat engine working with ZP electromagnetic radiation. Analysing semi-quantitatively a simple process performed with this engine at $T = 0$, we have shown that there exists a mechanism of filling real cavities with the ZP radiation which is based on Doppler's effect. We have proved that for large cavities ($x \gg c/\omega_K$) the behaviour of partitions with the cut-off is essentially the same as that of adiabatic partitions known from classical thermodynamics [200,204,210]. In contrast, for small cavities where $x \approx c/\omega_K$, the corresponding Carnot's process in an idealized heat engine ought to be strongly non-adiabatic.

We have further claimed that for considerations of ZP radiation in small real cavities the high frequency tail ($\omega > \omega_K$) is irrelevant. This statement is e.g. in apparent conflict with a standard SED approach to the computation of the Casimir's forces in small systems with perfectly reflected (adiabatic) boundaries.

6.3. Heat engine realized on a molecular level

Certainly, there are other microscopic models the enumeration of which [9,158,194] is not the aim of this chapter but we cannot avoid mentioning here the famous article by *Szilard's* showing the classically-assumed model [220] of one-particle gas, executing a work cycle with the aid of an "intelligent demon", which has had, perhaps, the greatest impact in the development of the idea of *Maxwell's* demon already known since 1866 [55]. Let us imagine that the particle is in a box of length, L, whose walls are maintained at temperature, T. A partition is inserted at the halfway point, dividing the container into two chambers of length $L/2$. The demon determines which chamber holds the particle, and the partition is then replaced by a movable, frictionless piston fitted with appropriate pulleys to enable the particle to do work, W, on the piston, thereby lifting a weight. The piston is then removed and the particle is once again in the container of the length, L. In the process, the energy, $E = W$, goes from a constant-temperature reservoir to the particle, and the net result (ignoring

180

the demon for the moment) is a decrease of the reservoir's entropy, $\Delta S = k\, ln\, 2$. However, the realization that the demon's state has changed by the information added to its memory makes it clear that the memory must be cleared. Landaer's principle *"erasure of one bit of information sends entropy, k ln 2, to the environment"* [120] implies that the deletion of the memory content sends, at least, the elementary entropy, ΔS, to the environment, and this is sufficient to save the second law showing that the strong role of entropy is obvious. Yet an outside observer, who cannot see the particle, does not know which chamber houses the particle. From quantum mechanical perspective, the particle can be considered to be in both chambers, with the probability ½ for each, until a measurement is made. Once the measurement is completed, the entropy does not seem to drop, but only for the demon, who alone knows the result of the measurement. Looking for the answer, where is actually produced the entropy that is sent to the environment during erasure, we have taken into account that the phase space consists of the state of the particle (left, right, or both) and that of the demon's memory (left, right, or a standard state) [222].

In a sense, the memory is a redundant copy of the system's state that does not carry extra entropy. In contrast, after the work process, the entropy values of the memory and the systems are independent of one another and just are added together. We can see that any analysis that focuses on information gathering is often done incomplete. (As matter of curiosity, in the extreme quantum state, when the particle and the reservoir are in an entangled-quantum-state we have to assume a special case where the total entropy cannot be written as a sum of the system and reservoir entropies, see next) *Bennett* [223] argued that the measurement itself carries no threshold cost. Instead, it is necessary to consider a complete thermodynamic cycle, in which information is gathered, stored, and eventually erased to restore the initial configuration. *Bennett* also suggested the use of algorithmic information theory making possible to define entropy without using an ensemble. Algorithmic entropy, I_s , is thus defined as the length, in bits, of the shortest computer program that runs on a universal computer and fully specifies a system's state. For a memory state, s, simply a string of <0> and <1> so that the algorithmic memory is maximal when a string, s, is random, in which case, I_s, is the length of the string in bits. Perhaps of most importance, algorithmic entropy provides a definition of entropy for a microstate, just as an energy 'eigenvalue' is defined for each microstate.

Another example can be constructed on basis of the statistical traffic of individual molecules, which move more or less at random and which may create certain organization through the convection cells. In such a case the order-lining source is rooted in the viscosity of the fluid – negligible viscosity makes convection impossible. Cohesion within the system is essential and in the so called *Bénard* cells this cohesion originates in viscosity (for more details see our previous book [9]). It seems clear that a force must be applied throughout the system. There are two potentials present: downward aiming gravity and upward pointed gradient of temperature due to external heating (man-made heat or sunshine warmth). Although the latter has its origin outside of the system, the gradients exists inside the system in a much stronger sense than the gravitational potential. Unlike the gravitational potential, which is conservative, the temperature gradient induces a non-conservative entropy gradient within the system and convection is an attempt to reduce this gradient. The temperature differential must be large enough to allow the formation of fluctuations sufficiently large necessary to overcome viscosity in a faster rate than such instability is dissipated. This creates unsteadiness resulting in a fluid that is not moving on the microscopic level only. Buoyancy then drives the convection cells to create a new regime, which is stable but moving [9]. This process is reflected in the delicate balance between the variables involved (Rayleigh number) and can even be simplified as a series of hypothetical, coupled, reversible and extremely miniature Carnot's engines.

From the book by *Feistel* and *Ebeling* [224] we can perceive that such a dissipative structure can be considered from a thermodynamic point of view as energy transforming system. In comparison with the Bénard's convection involved a classical heat engine has to include certain presumptions. First of all we have to recognize that in both, such as Carnot's as well as Rayleigh-Bénard's machine-like cases, we can find a fraction of the incoming heat transformed into an upgraded form of energy. That is mechanical work as the ordered motion in the former and the structural and orderly form of rolling in the latter case. Notwithstanding this similitude, an essential difference can be pointed out between these two kinds of this higher quality of energy. Carnot's engine produces mechanical work affecting its surroundings through the outcome of the directionally lifting of a weight while the Rayleigh-Bénard's motion manifests itself in the generation and subsequent conservation of the ordered arrangement (represented by geometrical structures). This structure is continuously appearing and disintegrating, in other words, a fraction of heat is repetitiously upgraded into structure and degraded back to heat (that ends up in the colder reservoir). This operational characteristic of Bénard's convection rolls corresponds with that of those systems, which are termed '*autopoietic*', i.e., with such procedures that are not concerned with the production of any output, but only with its own self-renewal (in the same process structure). Of a special concern is thus a supra-Carnot efficiency of these systems, which cannot be traced to the efficiencies of individual coupled heat engines but to the way in which their operations are coupled to each other (self-organisation).

In this intent we can mention the previous objective of Clausius' second law analysis, whose result was by him expressed in terms of what he called "*the principle of the equivalence of transformation*". According to *Iñigues* all the possible operations of a simple cyclical process can be subsumed in the relation describing ΔS_u, which represents an entropy change of zero for the universe and is the central equation of the negentropic reformulation of the second law of thermodynamics [225]. ΔS_u consists of three terms, $-Q_h/T_h + Q'/T_c - w/T_h$. The last one is seen to give the limiting reversible execution as a negentropy operation with an associated entropy change equal to $(-w/T_h)$. The second term corresponds to the limiting irreversible operation in which all of the heat coming out of the hot reservoir (**h**) finds its irreversible way into the cold reservoir (c). In this situation no work is produced ($w=0$) and we can equal the heat given off by the hot reservoir Q_h with that reversibly transferred to cold reservoir, Q'. The irreversible transfer of an amount of heat from the hot to the cold reservoir, corresponds to an entropy change, $Q_h/(T_h - T_c/T_hT_c)$, which is an entropic operation. It follows that the universe's entropy change for a simple cyclic process transit from a positive to a negative value as we move from the entropic (zero efficient operation represented by the irreversible limit) to the efficient, negentropic reversible limit. Between these limits there exists an operation with an efficiency η', i.e., $0 < \eta' < \eta_{rev}$, which can be identified with solving the efficiency of Q/Q_h that is found as $\eta' (= Q/Q_h) = (T_h - T_c)/(T_h + T_c)$. Thus, all the operations with efficiencies smaller than η' will be entropic, while those with efficiencies greater than η' will be negentropic. The particular operation for which $\eta = \eta'$, occurs with $\Delta S_u = 0$. The fact that the entropic and negentropic efficiency regions respectively arises due to the preponderance of unordered over ordered energy (and *vice versa*), allows us to speculate. The emergence, the coming into being of self-organizing phenomena, might obey a similar mechanics, i.e., the emergence can be seen concomitant to the universe's transition from entropic to negentropic, with the onset of organization taking place at that 'umbral' point at which $\Delta S_u = 0$.

6.4. Entropy, order and information

Entropy is a unique concept for science for at least two reasons. First, it is non conserved property, something produced in natural processes. Second, in its statistical treatment, it reflects uncertainty concerning the microstructure. Let us recall a routine system containing N particles, which do not depend on each other and can be combined arbitrarily in different combinations. The number of implicated combinations, which we can nominate as possible states is equal to $W_c = N^N$, so that $S = kN\ lnN$. If the N_i is the number of particles in the i-th box, (i.e., the number of particles possessing such values of the properties which define i-th box), the total number of this kind of distribution is equal to $W_c = N!/\prod_{i=1}N_i!$. For large enough sets it can be approximated by *Stirling's* formula $N! = (N/e)^N$ so that $S = kN\ \Sigma^n_{i=1}\ p_i\ lnp_i$. The frequencies $p_i = N_i/N$ can be considered as *probabilities* of a given particle to have the i-th set of given properties. At fixed N it follows the S is maximal when the probabilities are equal to each other, any particle can be detected at any point of the space of properties, i.e., the system is not ordered. When all p_i are equal to zero, the system is ordered as all particles are concentrated in the *k-th* box.

The paradox here is that if entropy is a state property of a system it cannot depend on what we happen to know about the system. Quantum mechanics has a similar-sounding, but quite different epistemological problem, which, in principle, placed limits on the precision by which certain pairs of properties are measured. Since measurement involves experimental design and choice of parameters of interest, in the quantum framework the observer is required to complete the phenomenon. In statistical thermodynamics, however, entropy is microscopic uncertainty and if we interpret entropy as lack of microscopic information about the macroscopic thermodynamic state we seem to get involved in the identity of that state alone, which would be a conflicting standpoint. Therefore, let us discuss all such viewpoints and inherent differences often arising from not fully congruous ideas, which try to enlighten the true interdisciplinary of the notion of entropy.

It appears that any better organization of a system drives the system into a less disordered state, less chaos, which certainly requires some definite effort (structured *input*), such as some kind of work. Remember that 'work' is a transient phenomenon, actually a process, which cannot be put in storage as such, only its product may be stored as a change of energy content of the system acted upon (as we literally assume heat on a similarly virtual basis of work).

Therefore, the application of energy to a system may result in four possible changes: (i) Energy merely absorbed as non-specific heat, thereby increasing the entropy; (ii) Input of energy causes the system itself to become more highly organized (causing, e.g., the atom to achieve a richer 'more improbable' thermodynamic state); (iii) System performs a physical (mechanical, electrical) work (mostly on bases of human innovative ideas – transducers) or (iv) Information work is implemented, creating thus a more organized organism (such as an organic living cell).

Accordingly, we need to analyze more closely the phenomenological meaning of the term 'entropy', which was the earlier accepted measure of an infinitely small increase of heat related to given temperature. The randomness, probability, organization and/or information, which have been later attributed to greater or lesser entropy of a given substances [226-233], are somehow a perceptual manifestation of the same basic phenomenon, which is the incorporation (absorption) of (thermal) energy. Entropy, however, is a mathematical function, which does not possess a direct 'physical reality' characteristic of material bodies, but when multiplied by the temperature, for which the entropy has been calculated, the product becomes the quantity of thermal energy that must be absorbed for a substance to exist at the temperature above absolute zero.

Though entropy holds the same units as energy it stands, however, in its contrast and the logarithmic relation between phenomenological entropy and microscopic complexion was actually a stimulation impact of the earlier game theory of chance. Thus entropy has been related to information following the well-known paper by *Shannon* on a mathematical theory of information [235] without a detailed definition of *what information really is* and *what basic properties it bears*. Consequently, we assume that the change in entropy may be brought about not only macroscopically by changing the heat content of the system, but also by altering the micro-organization of a system's structure, in other words, actually adding order or better information. We may disorganize a system by directly applying heat or alternatively by otherwise disordering its structure upon hypothetical 'withdrawing' its ordering information.

By identifying entropy with 'missing information' we can even get a deeper way of approaching the central question *"why does the second law hold"* in the sense *"why does missing information increase with time"*? Rather then puzzling over why heat flows as it does we are now curious how nature defines questions that can be asked, the number of answers that can be given to those questions, the extent to which the answers are known in some way, and how these things change with time. So, analyzing the situation that may challenge the second law of thermodynamics, we have a new way to look at them, in terms of whether there is some aspect of the system that becomes better defined without a compensating loss of information somewhere else. This opens up new avenues for understanding these systems and for seeing expressed through the second law.

It was *Maxwell* [50] who laid the groundwork for connecting information with thermodynamics by hypothesizing a mechanism for violating the second law by a demon who could control and separate differently moving molecules. As noticed above the resolution of this conundrum was begun by *Szilard* [220], who argued that the demon acted on the information provided by gas molecules and converted that information to entropy. *Brillouin* [227] further showed that for that demon to acquire the information $kln2$ necessary for the binary decision of stop-go operation it would have to expend an amount of energy no less than $kTln2$, thereby increasing the system's entropy by at least $kln2$. Information to operate the control required compensating the expenditures of negentropy resulting in concept penetration to thermodynamics to show the inviolability of the second law from possible microscopic intruders connecting somehow the microscopic to the macroscopic.

Back in 1929, *Szilard* [220] presented several famous 'gedanken' the fundamental restrictions experiments from which one can assume that additional information about a system can lead to decrease its entropy, even for a single molecule imprisoned by a piston-and-cylinder boundary. It may well provide the concept that there is a thermodynamic cost for *any* measurement. Eventually, it can be enumerated and seen to be regularly paid when a measuring instrument is reset, turning off its standard interaction [229]. In the above-mentioned micro-scale, however, the crucial step is to realize that fluctuations of miniature piston are as important as those of the molecule filler.

Let us take an example from everyday physics using the simple case of water [209]: on heating the solid structure of ice it first melts, then boils and on further heating (far above the melting point of platinum at 1700 C) it continues to transform to plasma of ions and electrons [236]. At even higher energy levels, the integrity of the atoms themselves becomes compromised and at sufficiently high temperature matter can be observed to 'boil' again to yield a plasma of quarks and gluons (so-called 'quagma'). Entropy is eventually coming to tie in the extreme state of complete randomization. On the other hand, if we cool ice down to absolute zero, where entropy is supposed to eventually become zero, we reach an ideally ordered state of perfect crystal with supposingly no thermal vibrations (which would be inaccessible due to inevitable scrunch of charged particles so that any collapse can be avoided only by some residual motion – athermal orbiting innate to vacuum, cf., zero-point electromagnetic radiation). The thermal energy cannot be apparently decreased by further withdrawing heat. However, it was experimentally shown that for certain inorganic molecules or states (glasses) there may exist exemptions for a particular crystalline modification, such as N_2O, which can have either $S = 0$ or $S = 5.76$ [J/K], both at $T = 0$. It is comprehensible that there may remain information about its ultimately perfect structure and its definite occupation of space, so that there is no theoretical reason why one may not additionally cause a further 'decrease' of entropy by certain hypothetical 'incorporation of spatial information' [235,236].

It seems to be impossible, but we can propose a force field which could 'freeze' the constituents to total immobility at higher temperatures, as well. In fact, we can actually identify such a phenomenon as we can approximate the supposed force field to hold atoms in a relatively immobile state at elevated temperatures recalling certain organic molecules in which the resonating π – ("pi") clouds of electrons act as an inter-atomic force stabilizing the positions of atoms. The best example, however, are met in all biological systems that, for example, use the absorbed heat in order to maintain a stable temperature so as to minimize externally induced entropy changes. Whether looking at the DNA molecules and related genetic or metabolic systems, cellular organization, the evolution of organisms or other ecosystems, the process is always the same: there is an entropy limitation by inserting certain 'information' as the simple systems become more complex, more differentiated and integrated, both within the internal organization and with the environment outside the system, evolving itself to become *thermodynamically increasingly improbable* [235].

On the other hand, we have to realize that the absorption of thermal energy has a constant effect on entropy that is a function only of the temperature and therefore seemingly no influence on what we perceive as randomness or organization. We cannot regard qualities to be independent of heat particularly when considering, for example, the triple point of water where the same state functions equal for the three phases that can coexist at equilibrium temperature. The only way that ice can be converted into water is for the environment to have a temperature greater than that of melting point, whereupon heat will pass from the environment into the system. The question is whether the heat exchange can be considered active or passive. It is not entropy that is exchanged, as entropy is a quantity that is acquired or lost, which does not perform a function or does work.

Such an inquisitive concept of adding information into the zero temperature state of an ideal crystal would certainly violate a traditional understanding and would also appear intuitively false as there is no way how to achieve this feat anyhow. We, however, may admit that it would be somewhat possible to add more information to a system that is already perfectly ordered by making it *more complex*. Therefore we cannot fully refuse an idea of the existence of a kind of very hypothetical conversion-like process of energy to *structural information* and hence the possibility to accept the concept of negative entropy (so-called '*negentropy*') [227,236].

Let us continue to analyze the statistical meaning of entropy, $S = k \, logW_c$, starting with traditional *Boltzmann's* investigations of W_c, assumed as '*complexion*', i.e., the number of possible microstates of a system[1]. It was

---

[1] Totalitarian regimes were always trying to suppress free dissemination of unsafe ideas by traditional burning of books. Let us take a habitual case of burning a bible with its mass about 0.5 kg (having paper heat contents about 10^7 J/kg). The heat produced is roughly five thousand of kJ. Concerning the bible information content possibly evolved in the form of accountable 'heat' we can assume that its text accommodates about million combinations (n) coded via 64 (o) letters that can be used to calculate $W = n!/[(n/o)!]^o \cong n \, log \, o = 10^{-10}$ kJ. There is evidently a great incomparability of such obtained values, i.e., $\Delta Q_{book} \sim 10^6$ J >>> $\Delta Q_{text} \sim 10^{-13}$ J, the difference approaching as many as 19 orders of magnitude is far below any detectable fluctuations and would become accountable only along with simultaneous burning of unimaginable 10^{12} books. There we ought to ask what about the case of two or more identical books or whether a bible has more valuable information content than a scientific book or a novel, if the same subject matter is written in different languages, etc., etc.. We can similarly proceed to analyze energy conservation curiosity looking upon the source of everlasting fire in Hell. Would it be possible to account for the heat associated with information brought in by souls and estimate how many souls would thus be necessary? First we have to postulate that, if souls really exist, they must have some 'mass' and can possess information content. Then we can account on different religions that often state that 'if you are not a member of that entire religion you would go to Hell'. Since there are more than hundred religions and one does not usually belong to more than one religion we can project that all souls must go to Hell. With birth and death rates we cannot only expect that the number of souls in Hell is to increase almost exponentially but we can also estimate about 1% year contribution from the population of 10^9 , i.e., 10^7 souls. Assuming average brain to bring

modified by *Schrödinger* who suggested in his book that a living organism is "fed upon a negative entropy", arguing if W_c is a measure of the *system disorder*, its reciprocal, $1/W_c$, can be relatively considered as a *measure of order, O_r*. According to *Stonier* [235] organization or structure is factually a reflection of order. Organization and associated information, I, can thus be seen as naturally inter-linked, though information is a more general and also more abstract quantity bearing an conceptual necessity to be freely altered from one form to another (structure of a written text can be different due to various alphabets but the information contained is the same). In the first approximation, we can assume a linear, d, dependence so that $I = d\ O_r$ or $W = 1/O_r = d/I$. By rearrangement of traditional entropy $S = k\ log\ (d/I)$ and $I = d\ exp\ (-S/k)$ which can define somewhat a more fundamental relationship between information I and entropy S, see Fig. 29.

In consequence we can assume that the inherent parameter $\underline{d}$ represents certain information constant of the system at zero entropy. Recalling the ideal crystal of ice at 0 K, $d = I_o$, which is not so imaginary assuming that I_o is a constant for all values of I and S within the system but may not be a constant *across* the systems. This becomes intuitively apparent when one compares such a single inorganic crystal with an organic crystal of DNA. Surely, at any comparable temperature below melting or dissociation (including 0 K), the molecule of DNA would unquestionably contain more information than that of ice. Returning to the fundamental equation rewritten as $S = k\ log\ (I/I_o)$, assume it is another quantitative expression of the system disorder. The ratio between this so-called 'information content' of the system when its entropy is zero, and the actual information content of the system at any given entropic value, S may remain as a generalized probability function to correspond the *Boltzmann's* original W, but now known from information theory, as the *number of possible coding*. We, however, are familiar with the original derivation made for gases in which I would never exceed I_o and therefore S would never become negative.

Let us turn again to the example of ice, now calculating the change of entropy between a perfect crystal of ice at 0 K (S=0), and its vapor state (S_{vap}) at the boiling point (373 K). It yields a value of about 200 J/K per one mole of

---

in memory in Terabites so that the total soul contribution is 10^{19}, which would still be so diminutive donation not exceeding a unit joule production per year, yet significantly deficient for keeping a noteworthy fire. We can proceed still one hypothetical step more to see whether the Hell is exothermic or not. Adhering to the Boyle's Law and assuming that the temperature and pressure in Hell is to stay the same, the ratio of mass and volume of souls needs to maintain constant. If the Hell is expanding at a slower rate than the rate at which souls enter the Hell, it will break loose and, on contrary, if the expansion is faster, the temperature and pressure will drop until Hell freezes over. So that there is not only lack of energy for the everlasting fire but also the soul rate must be well controlled, which does not comply with religious teaching and beat out any effort to make a more scientific analysis.

water. Now we can make use of two possible representations of the proportionality constant, k, in the relation of S = k logW. In physics it has the meaning of the *Boltzmann* constant, $k = 1.38x10^{-23}$ J/K, while in informatics it can either retain unity (yet a nonstandard unit called '*nats*') or reach the value of $1/ln2$ known as previously introduced '*bites*'. It may provide a 'linkage' between information (coding) and entropy (complexion) so that and we can formally write $I = I_o\ exp\ (-1.45x10)^{25}$ or, on the binary base, $I = I_o\ 2\ \{-(2.1x10)^{25}\}$, where the exponent is to the base 2 so that it may be stated in bits. It is worth noting that the increase in information/entropy manifests itself as a negative sign in the exponent indicating a loss of information. Consequently, the information needed to organize a mole of water from the chaotic state of steam into a state of perfect crystal of well-structured ice, which would require an input of about 35 bits per single molecule. One may even calculate that an entropy change of approximately 6 [J/K] is required to bring about the required loss, on average, of one bit per molecule, so that we can agree to the general relation:

$$J/K = 10^{23}\ bits.$$

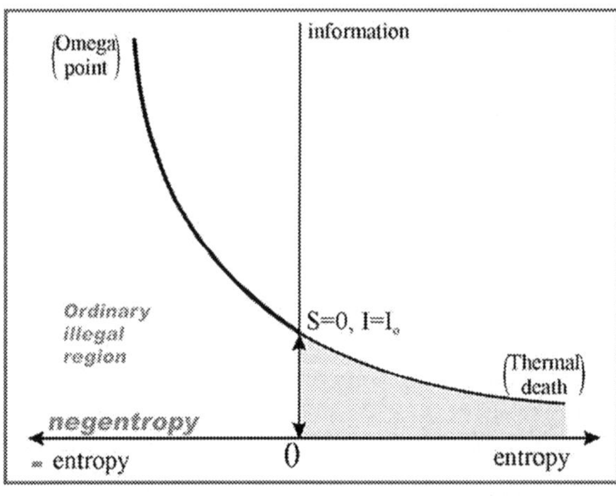

Fig. 29. – Underlying relationship between the entropy, *S*, and information, *I*, linked by the formal equation $S \cong ln\ (d/I)$ where the constant, *d*, represents the apparent information content I_o at the zero value of entropy.

Certainly, this covers a complicated process of trapping gaseous molecules existing in a cloud state by means of complex arrangement of electric and magnetic fields. The randomly moving molecules become fixed by the force fields into a regular array. This is a general characteristic for the formation of a crystalline state, in general, and there is a spectrum of crystalline states ranging from a powder up to a perfect crystal usually classified by customary X-ray diffractions patterns. There are certainly greater angles of the spread of information content. Moreover, a gas consisting of molecules also contains information residue because the organization intrinsic to molecules affects their behavior as a gas, like ions in plasma. Similarly, we can consider the formation of a single crystal, letting, for example, silicon to grow from the appropriate solution remembering the quest of chip industry, which is based on the insight that it is possible to obtain silicon of define purity and structure. Such a crystal is acting as a template growing according to thermodynamic rules or may be hypothetically viewed in the role of a progression where the growth information is provided by reading them from its environment (temperature, solution, tension and inherent gradients).

As shown above, in biology information and thermodynamics meet most closely and the problem of Shannon's entropy becomes especially acute [233]. If we consider a nucleotide sequence, with the sample-description space $X = f\,(A,\ T,\ C,\ G)$ indicating the available four-fold alphabet $A,\ T,\ C$ and G. If the chance of any base appearing at a locus is ¼ , the relation $\log 4 = 2$ measures its before-the-fact uncertainty. This symbol-uncertainty constitutes the basis for assigning the nucleotide sequences entropy. Thus a DNA sequence 100 units long has, assuming symbol 'equiprobability', Shannon's entropy of 200 bits. If we happen to know what that sequence is, then that entropy becomes information and such reasoning compromises the concept of entropy at any front.

The development of an organic nucleus is metabolic and interstitial and thus far more complex. For example, the DNA molecule produced as a string of simpler molecules of nucleotides, amino acids (whose sequences represent a series of messages), may be isolated in a test tube as a crystal to contain all the necessary information for reproducing a virus or baby. Here we can envisage that such a structure, thermodynamically as improbable as a perfect crystal at absolute zero, can be created to stay alive at room temperature in the form of a perfectly organized organic crystal. Restricting our protein as composed of the 21 essential amino acids only, we arrive at 21^{200} possible primary structures, in binary terms equivalent to approximately 878 bits per molecule. (In contrast, if language would consist of only ten-letter words, the total vocabulary available would amount to 26^{10} to require mere 47 bits per word). The entropy change can also be measured for denaturation of a molecule of enzyme from biologically active to inactive states, which involves about 900 J/K. The exponent of the information ratio can be interpreted again to represent an information change of 155 bits per molecule. If these information assumptions prove to be correct, that the bits per molecule lost when a perfect ice crystal is vaporized (35) is much lower than that for the inactivation of an organic trypsin molecule (155) completed, however, within a much narrower temperature interval.

Blumenfeld [237] tried to estimate mechanically the amount of information accommodated in a human organism by assuming 10^{25} amino acids ordered in proteins as contained in 7kg of human body nitrogen. This roughly corresponds to 10^{27} bits of information assuming that the other contributions are appreciably lower. For example, 150g of DNA contains only 10^{23} bits and for about 10^{13} individual cells (10^{14} bits) having 10^{8} ordered polymetric molecules

(10^9 bits) we obtain after multiplication mere 10^{22} bits. It, certainly, lies at the level of zero approximation because the information in living organisms is not only the estimates of the number of amino acids but also their consequential order.

In the axiological sense [233] a measure of information, I, for a system, R_s, is some measure of changes caused by I on R, or better for a more strict sense, an alteration taking place in the *infological* system $If(R_s)$. A unique measure of information exists only for oversimplified systems because any complex system R_s with a developed infological subsystems $If(R_s)$ possess many parameters that may be changed so that such a system demand many different measures of information in order to reflect the full variety of the system's properties as well as conditions in which the systems is functioning. It follows that the problem of finding one universal measure for information is somehow difficult and almost unrealistic.

Shannon deliberately excluded from his investigation the question of the meaning of a message, i.e., the reference of the message to the thing of the real world citing *"frequently the message have meaning; that is they refer to or are correlated according to some system with physical or conceptual entities. These semantic aspects of communication are irrelevant to the engineering problem. The significance aspects are that the actual message is one selected from a set of possible messages. The system must be designed to operate for each possible selection, not just the one which will actually be chosen since this unknown at the time of design"*. Certainly in a societal information design an item of information is an interpretation of a configuration of signs for which member of some societal group are accountable while the utility of information give the measure of information which can be called the quality of information. If I is some portion of information, then the quality of this information is equal to that caused by I change of the probability $p(R_s,g)$ of achievement of a particular goal, g, by the system R_s. The corresponding (infological) system, $If(R_s)$, is the state space of the world in which the system R_s is capable to properly function.

Let us be more pragmatic in our forethought [238] and see, for instance, one gram of dried *Saccharamyces cerevisiae* cells to hold entropy of about 1.3 [J/K] (at room temperature) while the same mass of crystalline a-d-glucose, which is a common organic substrate used to grow this yeast, has entropy of just 1.16 [J/(K g)] [239]. Does it mean that cellular fabric is more random, more probable, and less organized than the substrate from which it is formed? One should be very careful about mechanical comparisons based on mere entropic values, which do not seem to make logic, particularly dealing with variability in solids. The difference between biological purposeful and purposeless information is yet intriguing [239] and the idea whether we may relate biological order and complexity to thermodynamic entropy has not gone uncriticized [240]. Citing *Battley: "it is hard to believe that the absorption or loss of entropic thermal energy is anything other that a purely passive phenomenon, and that the driving force behind a spontaneous reaction is not the entropy change, but the chemical and physical event that cause the change in free energy...it may not necessarily contribute to a greater randomness or a lesser organization, depending on how we as observers conceive these terms to mean"* [238].

In this respect there is a strong call [231] then again, to have a completely appropriate *alternative* for entropy in information theory, which can be *complexity* so that what the Shannon formula actually measures, is the complexity of structural relationship. Factually in such an algorithmic approach to information theory the entropy of the symbol ensemble is replaced by the program of information required to specify sequence. The entropy of a thermodynamic system is a measure of the extent to which it is not tied down to a particular matter-energy distribution. A complex system, in contrast, requires a structured relationship among elements and such a system must have before-the-fact alternatives, such that one could not predict those structural relationships just from the properties of the elements concerned. Yet once these elements and relationships have been specified, the indeterminacy vanishes.

6.5. Information and organization

The information, which is processed by the solidifying silicon solution when growing a semiconductor single-crystal or by a human cell interacting with a strand of DNA, is the organizational pattern of the carrier of information independent of human minds. Information existed for billion years prior to advent of the human species and for billion years that information has been processed. DNA itself would be useless unless a living cell processes the information. If the information is an independent reality, which means that it exists regardless to the capability of decoding the inherent text. Information is considered to be distinct from the system which interprets (there is dichotomy between the information intrinsically contained by a system and the information which may be conveyed by the system to some acceptor). Information may even organize information, which is a process occurring in our brains or may take place in our computers in not too distant future. It, however, became next to impossible for a single mind to fully command more than a small, specialized portion of information.

In general, we can imagine that *information has similar behavior as energy*; both exist independently and do not need to be perceived to exist neither understood. Energy is defined as the capacity to perform work and it is the *nature* of energy supplied or withdrawn, which determines the nature of the changes occurring in the system acted upon. Information is defined as the capacity to organize a system or maintain it in an organized state. Similarly to the existence of different forms of energy (mechanical, chemical, electrical, heat) there exist different forms of information. Energy is capable of being transferred from one system to another likewise information may be transformed from one form to another. The information contained by a system is a function of the linkage binding simpler units into more complex units. Thus, all organized structures contain information and, on the contrary, no organized structure can exist without containing some form of information. Information is an *implicit component* of virtually every single equation. Therefore, there must be existent an information flux being proportional to the question and to the gradient between information fields. From the thermodynamic and causality considerations a general upper bond on the rate at which information can be transferred in terms of the message energy was theoretically inferred (i.e., $<10^{15}$ operations per second as to avoid thermal self-destruction) [229].

Information is governing the laws of physics, i.e., in order to measure distance and time properly, an organized frame of references must be available and thus the measurements of space and time establish information about the distribution and organization of matter and energy. An organized system occupies space and time, therefore, the information content of a system is proportional to the space it occupies. *The universe may be seen organized into a hierarchy of information levels* [235,236]. Energy and information may put on view certain signs of invented inter-convertibility. In contrast to heat, however, all other forms of energy involve organized patterns and may be said to contain information. The application of heat, by itself, constitutes no contribution in terms of information. Structure represents the product of information interacting with matter. All forms of energy other than heat exhibit, or are dependent on, some sort of organization or pattern with respect to space and time (e.g. an immature steam engine versus refined electric alternative motor).

We need a particular service of the so-called *energy transducers,* which define two necessary conditions for the production of useful work: they create a non-equilibrium situation and they provide a mechanism of countervailing forces necessary for the production of useful work. For a real physical system, such as a steam engine, the energy contained in heated gas is not converted into work until it is contained and modulated by an engine that has accumulated, within it a history of invention, certain information content in the form of know-how of produced machinery. A very advanced information system, such as living cells enables to transform an input energy into useful work under circumstances not possible in other physical system. For example, the cell can provide the dissociation of water and/or enables stripping electrons off hydrogen atoms at room temperature, which for another unorganized system would require heating above 1000 °C to destroy water molecules by violent collisions.

The structural information content of any energy transducers is the same at the end of a process as it was at the beginning neglecting its wear out. The information kinetics created by the steam engine when it has produced an asymmetric distribution of high and low energy molecules of water in the two chambers separated by a piston is associated with potential energy. The apparent loss of 'kinetic' information that is the part of the information, which has been degraded to rest heat, can be associated with the dissipation of entropy. For energy science it is important to measure the process efficiency (η) given by the ratios of the work output to the heat input, $\eta = W/Q$. The maximum possible efficiency is limited by the temperatures of input (T_{in}) and output (surroundings T_{uot}), so that $\eta_{max} = 1 - T_{out}/T_{in}$ and we may be interested to find the relation of the actual (η_{act}) and maximum efficiency to follow that $\eta_{act}/\eta_{max} = f(I_{in})$ or $I_{in} = -log(1 - \eta_{act}/\eta_{max})$. Using the above conversion factor (1 [J/K] = 10^{23} bit) and by measuring the ratio of the differences it would become possible to ascribe a real value to the above introduced value [235,236] of *information inputs (I_{in})*.

Regarding ecosystems, we should point at the information problem of species diversity of a community with regards to its stability, which is undertaken by ecologists as an axiom for evaluation. It shows that communities, which are more complex in structure and richer in comprising species, are more stable because deferent species adapt differently to environmental variations. Therefore, a variety of species may respond with more success than a community composed of a small number of species. Perhaps this motivates the fact that *Shannon's* information entropy was suggested as a measure of species diversity, where p_i is related to the number of specics in the community and population size of the i-th species. In this respect let us mention a familiar relationship [241] $dK/dt + (K + \xi)(d\xi/dt) \geq 0$, where K is the so called *Kullback* measure of the increment of information $(\approx \Sigma p_i ln(p_i/p_i^o)$ and ξ is related to thne content of biomass akin to $ln(N/N_o)$. If the positivness of $d\xi/dt$ means an increase in the total biomass in the course of evolution then the positiveness of *dK/dt* can be interpreted as an increase in the specific information content (per biomass unit). It is then obvious that if both the total biomass and its biomass is constant *($d\xi/dt=0$)* then the system can evolve only if the information content of its biomass is growing, which can be interpreted as the *growth of diversity*. On the other hand, the information content can decrees (dK/dt<0), but if the total biomass growth sufficiently fast *($d\xi/dt>>1$)* then useful energy ('exergy' [77], see paragraph 8.5) is growing, and the system is also evolving. At last, there can happen paradoxical situation when useful energy is increasing while total biomass and its information content are decreasing when guarantee $|\xi| >> |K|$ and *dK/dt* $<< 1$.

Accepting that heat is a certain 'antithesis of organization' and by implication certain duality behavior of non-identical energy and entropy as information we can venture to imagine a speculative possibility that energy and information may interact to provide an 'assortment', which might be viewed as 'energized information' or alternatively 'structured energy'. Consequently, information and energy may not be viewed as the opposites of a bipolar system, rather they can be hypothetically considered as the partners (like matter and energy), on assuming two sides of a triangle, with matter comprising the third. Such a conceptual model might be conceivably used to define hypothetical 'boundaries' of our newly supposed physical but highly forethought universe [235].

Almost sci-fi consequences of such an extreme portrait could be read as follows. If the upper corner would consist of pure energy, E, in an infinite entropy state, and the right bottom corner of unadulterated matter, m, the adjacent right side of the triangle is justifiable within the Einstein equation, $m/E \cong \mu_o \varepsilon_o$ (i.e., characterized by the values of μ_o and ε_o as the permeability and permittivity of vacuum, which may predicate certain structure of our Universe).

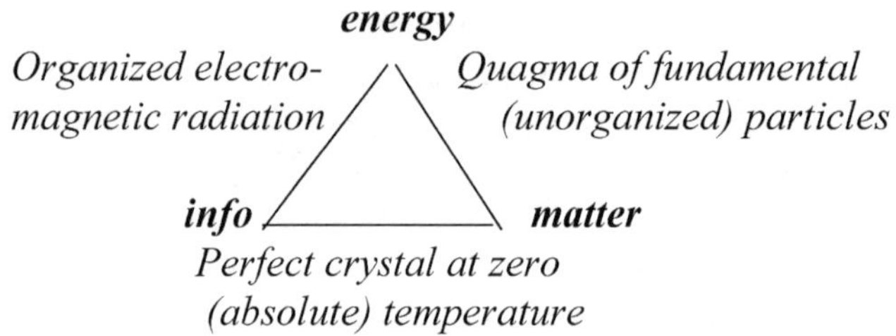

Associated condition of the absence of information would display a state of randomly distributed, unlinked 'foundations' (neither imaginably inherent particles nor any form of mass would exhibit no organization whatever) moving aimlessly, with no landmarks, no events or space. It would become impossible to measure either time or space, as such form of information would not exist (it might be somehow assumed to take place in the center of a black hole). Left triangle side would consist of absolute vacuum traversed by force fields and pulses of energy, possibly capable to measure time, but distances may have no meaning being without landmarks. Bottom triangle side might enable to measure the distances but perhaps no time as in the absence of any motion, or events, time might be frozen. In some way, curious left corner, containing neither matter nor energy, might consist of unspecified information particles. Let us preliminarily label them 'infons' to be forcibly envisaged as photons, which appear to have stopped oscillations propagating through the empty space down to absolute zero Kelvin. This is an idea as fantastic as was the concept of 'tachyons' with their imaginary masses. The field of relativistic theory left the problem how to explain motion solely in terms of matter. Dirac's theory implied negative energy states and his mathematical equations anticipated the existence of anti-matter. *Feynman* mathematically demonstrated that a positron might be regarded as an electron moving backward in time and some yet not well defined 'super-particles' of an inter-galaxies size are foreseen in astrophysics – possibly to be even some relict 'strings' of early matter grown in time.

On the other hand, the above triangle can also represent some philosophical aspects, i.e., mental relations when respectively replacing energy, matter and information by power, property and spirit, which endows with new interrelations linked with the each triangle side being the strife, creativeness and love, reminding some Greek philosophical beliefs discussed in Chapter 3.

We can wind up the recent impact of computers, aware that it has changed our perception of information as something purely static because inside computers information may once appear to gradually achieve a dynamics of its own. Already now, we are not sure where and how we pay compensation for the information often lost accounting for waste heat when imprinting or erasing memory contents. Information written to some memory register by a computer program must be discarded to the environment in order to reset the register and complete the cyclic process. We can even suggest that each energy transfer between two separate physical objects is a measurement. We can thus believe that measurement is the transfer of information between the sample – an examined physical object and the apparatus, and that any energy transfer will carry information about the shift in energy state of the objects under observation.

It is almost sure that the domain of information will soon dramatically change similarly to the comprehension of traditional physics, which learned in the past century how to assimilate surprising ideas, such that energy may be converted into matter and *vice versa*. It brought an important consequence into nuclear physics and possibility to gain new forms of energy. We should assume that even the empty space, when enclosed in any organized structure, may bear a significant piece of information (like a space dividing words in most languages) so that the absence of structure within a structure may carry information as real as the structure itself. The Barrow's made-believe verdict [242]" *if life will engulf the entire universe the so-called Omega Point would be approached and the infinitive amount of information hitherto stored would logically reach its completion*". However, fantasy was not the aim of this chapter although we cannot dodge completely out this type of query. We cannot keep away from this type of discussion in detailed analysis of advanced views on science of heat. Anyhow we cannot pass over the everlasting question *"Is there as purpose in the nature"* [43,341-244].

6.6. Quantum information processing

Besides the energy enquiries pointing down towards the yet fully unexploited world of quantum mechanics there is another associated neighborhood of the emerging field of quantum information processing and technology (often abbreviated as QIPT) [245,246]. Following extrapolation of the exponentially decaying number of electrons per an elementary device on a chip it ought to get to one electron per device somewhere around the year 2020.

It is clearly too naïve and oversimplistic but, at least, gives us a hint. Eventually, we will get to scales where quantum phenomena rule, whether we like it or not. If we are unable to control these effects, then data bits in memory or processors will suffer errors from quantum fluctuations and device may get malfunctioning. The quantum research has already shown that the potential exists to do much more and instead of playing a support role to make better conventional devices, quantum mechanics could take the center stage in new technology that stores, processes, and communicates information according to its own laws.

Most information manipulation is done digitally, so data is processed in the form of bits. The two states of a conventional data bit take many forms such as magnetic orientation, voltage, light pulses etc. At any time, a bit is always in one state: written in suggestive quantum notation, it shows $|0>$ and $|1>$, hence the name, although bits get flipped as data is processed or memory rewritten. However, the quantum analogue of a conventional bit, so-called *'qubit'*, has a rather more freedom. It can "sit" somewhere in a two-dimensional Hilbert space. We can picture it as the surface of a sphere with a general form $|\psi> = cos\ \alpha\ |0> + exp\ i\ \phi\ sin\ \alpha\ |1>$ parameterized by two angles. Factually, the conventional bit has only the choice of the two poles while a qubit can live anywhere on the surface of the sphere. States such as (1) are superposition states, they have amplitudes for and thus carry information about the states $|0>$ and $|1>$ at the same time. Similarly, the register (collector) of N qubits can have exponentially many (2^N) amplitudes, whereas the analogous conventional data register can hold one of these states at any given time. Clearly, if it is possible to operate, or compute, simultaneously with all the amplitudes of a quantum register, there is the possibility of massively parallel computation based on quantum superposition.

Our standard practice shows that we can read ordinary information (e.g., book) without noticeably changing it, which is not so simply for quantum information. If a qubit in the state (1) is measured to determine its bit value, it will always give the answer 0 or 1. This is a truly random and irreversible process with respective probabilities of $cos^2\alpha$ and $sin^2\alpha$, and afterwards the qubit is left in the corresponding bit state $|0>$ or $|1>$. It thus would be impossible to read, copy or clone, unknown quantum information without leaving evidence of the intrusion. There are many types of usable qubit exits: such as two adjacent energy 'eigenstates' of atoms or ions separated by microwave, vacuum or single photon state of a mode in small optical or superconducting microwave cavities, orthogonal linear or circular polarization of a traveling photon or a weak light pulse, energy eigenstates of a spin in magnetic field or of an electron (or 'exciton') in a quantum dot or two charge states of a tiny superconducting island – the practical realization being defined by the application. A resulting multi-qubit processor would enable massively parallel quantum computing, i.e., interference between all the amplitudes, and such a device could, yet theoretically, be arranged to provide solutions to certain tasks that we will never be able to perform with even the most sophisticated conventional supercomputers.

An intriguing possibility is *spintronic* chips, where we can flip electron spins back and forth coherently, which goes again beyond the traditional binary digits producing the new phase digits so called '*phits*'. The more precisely we can read the phase, the more dramatically we can increase the density of data storage. Thanks to the decades of work on magnetic resonance imaging, which detects the spin of atomic nuclei we became skillful enough how to read the involved angles precisely enough.

The irreversibility of quantum measurements enables, in addition, two correspondents, say Alice, A, and Bob, B, to communicate with a guaranteed security using photon qubits and public communication. Related quantum cryptography is secure against eavesdroppers, even if they have their own quantum technology. Two qubits, A and B, can exist in a state like $| \psi >_{AB} = 2^{-1/2} (| 0>_A | 1>_B + | 0>_B | 1>_A)$, which cannot be factored, so neither qubit has a state of its own, independent of its partner so that all the quantum states of a multi-qubit register contain entanglement. It follows that there is no reason for the two qubits to be in the same register or indeed physical location, and in such cases, distributed entanglement provides for a revolutionary form of communication, so-called quantum teleportation. It may bring remarkable consequences when compared with conventional communication: entangled states can be used as a resource for teleporting quantum states [208] destructing them in one place and rebuilding them in another location.

Teleportation is the name given by sci-fi writers to the feat of making an object disintegrate in one place while a perfect replica appears somewhere else, as first suggested by *Bennett* in 1985 [247]. Today's theory is based on the famous effect, which is often called '*Einstein- Podolsky- Rosen* correlation' or conveniently 'entanglement' (in the meaning of superposition, embroilment or mix up) known since 1930s [248]. In 1993 this intuition was confirmed by showing that teleportation is possible in principle by destructive observation (a photo-detector measures photon in a destroying manner by its absorption and conversion into electric signal) or by new difficult-to-execute technique known as quantum non-demolition measurements (where a photon in a cavity is probed without absorbing any net energy from it). It is worth noting that teleportation is a fundamentally different process with respect to the classical transfer where the object must be seen and understood and its image (photograph, order chart, digital scan, etc.) transmitted to the receiver who uses the instructions to build the object and even duplicate copies. During teleportation, however, the object can be simply perceived to be first 'destroyed' to fundamental particles and than 'reconstructed' from particles available somewhere else. Thus, in the quantum teleportation process, physicists take a photon and transfer its properties (such as polarization, i.e., the direction in which its electric field vibrates) to another photon, even if the two photons are at remote locations.

6.7. Quantum diffusion and self-organization of oscillatory reactions

Periodic chemical reactions, known from the second half of the 19s Century [249-253], perform a curious class of reactions generating macroscopic patterns periodic both in space and time [9] (e.g., known as *Liesegang's* rings [249] and *Runge's* dyeing figures [251]). They are mostly considered to be a spectacular demonstrations of self-organisation [9,47,57,253] due to the non-equilibrium nature of thermodynamic processes involved. As these reactions violate traditional view on chemical kinetics characterized by the natural tendency to reach equilibrium by the shortest way, they were (and are) interpreted as a precursor of life processes [254,255]. *Maupertuis'* principle of least action [256] can be applied [252,258], citing *'when some change takes place in nature, the quantity of action necessary for the change is the smallest possible being the product obtained by multiplying the mass of the bodies by their velocity and distance travelled'* (similar to *Fermat's* principle of least time quoting "*the nature acts via the easiest and the most accessible way reached within the shortest time*") and thus differently interpreted elsewhere [9,36,252,257-259].

Fig. 30. – Examples of common, natural co-centric patterns and believed to arise by the same generalized mechanism. From left, there are examples of processes taking place within long-, middle- and short-range time intervals. The cross-section of the natural semiprecious stone agate, shaped by hydrothermal reaction in geological formations. Middle left, shells and the strips initiated by dye centers, which are responsible for the continuous pigmentation process and the cross-section of the trunk of about a 100 year old pinetree, indicating the northern direction by denser rings due to the less favored growth conditions (lower temperature, etc., typical for the growth-ring separation in general). Right, *Liesegang's* rings of silver chromate crystallites formed during the diffusion controlled solid-liquid chemical reaction in a Petri plate when placing a crystal of silver nitrate at the center of a glass plate coated with gelatin containing a dilute solution of potassium dichromate. In all such cases the ring-band separation is related to the instant propagation velocity at which the diffusion infiltrates the space available, its geometrical arrangement and the size of propagating species, all constituting the proportionality, which is governed by the value of about 6.6×10^{-34} [J s] that exhibits a striking coincidence with the *Planck* constant (acknowledged to control the wave relations in the deep microcosms). For more details see ref. [36,197,252,258,265].

Besides the fact that the kinetics of periodic reactions is controlled by diffusion (so called *Nernst–Brunner* kinetics [260], another peculiar and somewhat enigmatic feature of these reactions was discovered in the 1930s [261], which may be concisely expressed as $Mv\lambda \approx \hbar$, where M is the molecular weight of precipitate, v the speed of spreading of reaction fronts, λ the length parameter of reaction patterns and $\hbar$ ($= 1.05 \times 10^{-34}$ Js) is the Planck's universal constant. For particular configuration of the system in which reaction take place should be its left-hand side completed by a geometric factor (e. g. two for three- and $1/\pi$ for two-dimensional cases) and by so called 'tortuosity' characterizing the topology of the system [252,257]. As a rule, the resulting factor is of order of unity.In addition to periodic reactions revealing the behavior depicted above (where the chemical nature of processes involved is well established, like Liesegang's rings [249,253] or Belousov–Zhabotinsky's [262] waves [9,254,255, 263,264]) there, moreover, exist cases where a straightforward interpretation in terms of periodic reaction is somewhat questionable, nevertheless, the equation $Mv\lambda \approx \hbar$ is justifiable with appreciable accuracy. As a good example it may serve a rather curious interpretation of annual growth rings of trees, which has been proposed in 1954 by *Schaabs* [265] (see Fig. 30.). He assumed for the case of tree rings that we have to do with a special kind of Liesegang's reaction with cylindrical (beam-like) symmetry for which he expected relation $M\lambda^2/\pi\tau \approx \hbar$, where M is again a molecular weight of precipitating cellulose, λ is the distance between neighbouring annual rings and τ is the growth period. Taking then into account that the glucose-based polymer cellulose having empirical formula $(C_6H_{10}O_5)_N$ is known to create in wood the chains of average polymerisation degree $N \approx 400$, its molecular weight may be determined immediately as $M \approx 400 \times 162 \times 1.67 \times 10^{-27} = 1.08 \times 10^{-22}$ kg. Admitting further for the distance between annual rings a value $\lambda \approx 5 \times 10^{-3}$ m and for the duration of the season an estimate $\tau \approx 8 \times 10^{6}$ s we obtain an "incredibly exact" figure for Planck's constant, namely $\hbar \approx 1.07 \times 10^{-34}$ Js.

Of course, similar periodic patterns having this "quantum-like" property [9,36,252,259] have been encountered in the nature astonishingly frequently (see Fig. 30.) and are undoubtedly of high scientific significance. We believe, however, that the identification the periodic chemical reactions without ambiguity would require much more critical analysis of chemical processes behind them [36,258] as it is done in the example given above.

Spectacular images of these periodic phenomena suggested that there ought to take place some wave process unseeingly involved [254] and, of course, in the case where $\hbar$ is at play this wave process should be a quantum one. That is why practically all early explanations in a somewhat simplistic way, made a direct use of the concept of *de Broglie* pilot wave [266]. Accordingly, the concept of de Broglie's wave, controlling the space probability distribution of a molecule, enhances the precipitation at points where the probability of sojourn of the molecules is maximal, i.e., at points departed by certain multiples of de Broglie's wavelength λ. In spite of the fact that this idea seems to be compatible with above relations there have persisted a strong feeling that we have to do with yet non-justified extrapolation of quantum concepts to the range of macroscopic phenomena.

On the basis of elementary analysis of diffusion equation (i.e. Fick's law), which is for the sake of simplicity written here for one dimension only as $\partial n/\partial t = D\ \partial^2 n/\partial x^2$, where n is the concentration of reactant and D the diffusion coefficient, one can conclude that the movement of "average" Brownian particle (i.e. single molecule of the reactant, *Smoluchowski-Einstein* theory [267]) is controlled by exactly the same equations as the movement of concentration maximum induced by spreading out from a limited point source [268], namely $x = \sqrt{(2Dt)}$, where x has meaning either of the mean position of particle or of the concentration maximum. In terms of diffusion speed defined as $v = \partial x/\partial t$ the equation reads as $xv = D$, which fulfils the role of a specific relation of Heisenberg's quantum uncertainty, $xv \geq D$. Note that v has an alternative meaning of the mean stochastic velocity of single molecule and thus the movement of diffusion reaction front provides in this sense a macroscopic picture of the movement of an "average" single molecule [267]. What remains then is just to clarify the conditions under which the diffusion constant D may attain the Fürth's value $D_Q = \hbar/2M$ (stochastic quantum mechanics).

A certain dissatisfaction with various attempts to treat quantum motion as some kind of a stochastic process touched some basic quantum concepts [186-188]. The missing key can provide a rather close analogy between the diffusion equation and the Schrödinger equation [143] observed by Fürth [269]. Accordingly, these equations may be mapped one onto another by substituting for the diffusion coefficient, i.e., the value $iD_Q = i\ \hbar/2M$, and/or identifying tentatively the universal noise source behind the assumed stochastic process with electro-magnetic zero-point fluctuations of vacuum [208,270]. However, these attempts at the interpretation of quantum mechanics as a stochastic theory bring about serious problems. Classical Einstein–Smoluchowski's description of diffusion (i.e. as special case of Brownian motion of a small particle) is essentially a description of Markovian stochastic process (configuration space i = 1). However, any stochastic process in phase space, which is assumed to underlie motion of a small quantum particle, cannot, supposedly, bear Markovian character, because, by definition its state at t = 0, determines the probability of states at any later time t. In other words, the particle possesses "memory" in the phase space.

On the other side, intermitted measurements performed in the configuration space reduces repeatedly the wave packet of a particle causing that the "memory" is being lost and the process can thus be treated as Markovian [258]. From the length of observed course $L_k = \sum l_k$ and its resolution $\Delta l_k = \sqrt{<l^2>_k}$

we may obtain relation for D using uncertainty $xv = D$ i.e., $D = \Delta l_k(k\Delta l_k)/\tau$. Consequently we can derive the so called Hausdorff's length $\Lambda = L_k(\Delta l_k)^{d-1} = D\tau/(\Delta l_k)(\Delta l_k)^{d-1}$ where d is the fractal dimension, which value is identical for both processes, i.e., for Brownian and quantum motions in the configuration space ($\Lambda = \hbar\tau/2M\Delta l_k(\Delta l_k)^{d-1}$) and which is equal 2 [36,271].

The fact that the Brownian and quantum diffusions are indistinguishable by intermitted measurements performed in the configuration space (and this is just the equivalent of experimental techniques by means of which the periodic chemical reactions are investigated) justifies the direct comparison of empirical diffusion coefficients with the Fürth's value $D_Q = \hbar/2M$. Assuming that the noise sources behind classical and quantum stochastic behavior are independent the obvious formula for diffusion coefficient may be written as $D = D_S D_Q /(D_S + D_Q)$. Physical meaning of such accessibility of the Fürth's limit, formally expressed by the condition [271] $D_S \gg D_Q$, is thus the sufficient partial decoupling of particles from the source of classical noise.

By comparison with empirical data it is now easy to show that there are numerous cases where $D \approx D_Q$ and where the realistic valuation of Einstein-Smoluchowski relation [36,267] ($kT/3\pi\eta a \gg \hbar/2M$) is satisfied as is the case of e.g. Na^+, K^+, Ca^{2+} and Ag^+ ions (in aqueous solutions at room temperature). It naturally applies for the proton (H^+) alone, which may be a new guide to all cyclic processes in the sphere of biology.

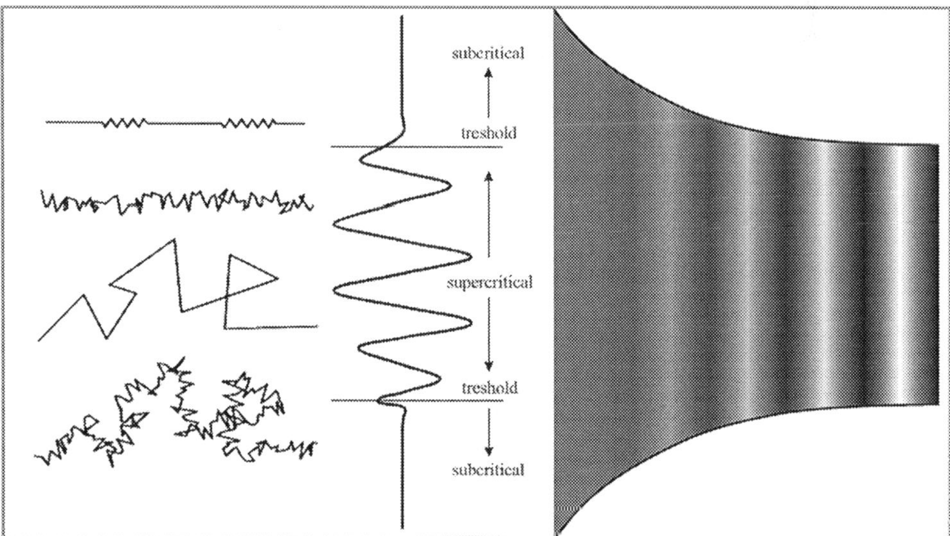

Fig. 31. – Schematic diagram showing the subcritical and critical oscillatory regimes of a computer scheme (customary called *Brusselator*) with a two-dimensional illustration of the wave's gradual development (right). Left upper the equivalent circuit of $D_Q + D_S$, the path below showes possible course of a particle following quantum and yet below the Brownian motion, the bottom pathway stays for their superposition [258].

Fig. 32 – Quenched-in morphology of the directionally solidified dielectric eutectic of $PbCl_2$-AgCl processed at a high melt undercooling (ΔT=50 K) and growth rate (v = 10 mm/hr) at both the microgravity ($G_o \rightarrow$ 0, left) and terrestrial conditions (G_o=1, middle). It reveals the diffusion-controlled oscillatory growth of $PbCl_2$ lamellae in the crystallographic direction [100] parallel to the reaction interface (left). The lamellae that solidified directionally at terrestrial conditions (middle) exhibit, however, characteristic growth defects due to the interference of mass flow with gravitational field that is absent for the space experimentation (left). Time-dependent effect of prolonged lamellae coarsening provides then a common pattern of rough morphology regularly achieved for traditionally lengthened equilibration of sinters (not in scale)

.

Below: Rotating spiral wave as a simple case of a reaction-diffusion pattern. It is a kind of *Belousov – Zhabotinsky* 2D-waves experimentally observed for the system H_2SO_4-$NaBrO_3$-$CH_2(COOH)_2$. This illustrative view shows co-rotating (retrograde) meandering wave shaped in spirals to appear at low malonic acid concentration. Left, the spirals undergo retrograde meandering which for higher concentration turns opposite. At low sulfuric acid concentration the spiral are stable while for higher concentrations the distance between the successive wavefronts starts to vary spatially until a large number of small spirals around the edge of the stable spiral emerge. This process can lead to the turbulent state (middle) beyond the convective instability of the spiral. Enlargement of the turbulent region is called as the *'Eckhous'* instability, which is a quite common phase instability of yet periodic pattern (right – not in scale).

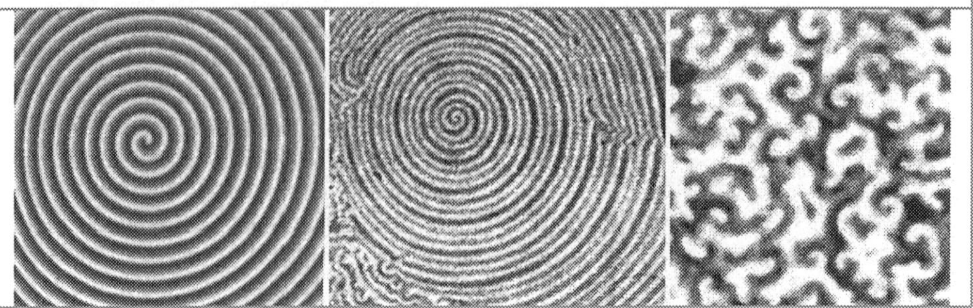

Based on the classical Einstein–Smoluchowski's description [267] of diffusion (as a particular case of Brownian motion) and accounting for the fact that the Brownian and quantum movements are indistinguishable by intermitted measurements in configuration space, there follows an important fact [36,258] that a certain class of self-organized periodic reactions [36] can be realistically characterized by the empirical dispersion relation, $Mv\lambda \approx \hbar$, which is factually controlled by the Fürth's quantum diffusion of reactants. In its fundamental sense we can challenge to say that oscillation processes in aqueous solutions are likely caused by quantum motion of protons (while the oscillations in solids are caused by electrons).

7. THERMODYNAMICS AND THERMOSTATICS

7.1. Principles of chemical thermodynamics

Unlike other branches of physics, thermodynamics in its standard postulation approach [272] does not provide direct numerical predictions. For example, it does not evaluate the specific heat or compressibility of a system, instead, it predicts that apparently unrelated quantities are equal, such as $\{(1/T)(dQ/dP)T\} = - (dV/dT)P$ or that two coupled irreversible processes satisfy the Onsager reciprocity theorem ($L_{12} = L_{21}$) under a linear optimization [153]. Recent development in both the many-body and field theories towards the interpretation of phase transitions and the general theory of symmetry can provide another plausible attitude applicable to a new conceptual basis of thermodynamics. In the middle of Seventies Cullen suggested that `thermodynamics is the study of those properties of macroscopic matter that follows from the symmetry properties of physical laws, mediated through the statistics of large systems' [273]. It is an expedient happenstance that a conventional 'simple systems', often exemplified in elementary thermodynamics, have one prototype of each of the three characteristic classes of thermodynamic coordinates, i.e., (i) coordinates conserved by the continuous space-time symmetries (internal energy, U), (ii) coordinates conserved by other symmetry principles (mole number, N) and (iii) non-conserved (so called `broken`) symmetry coordinates (volume, V).

Worth mentioning is also the associated and intriguing means of irreversibility, or better, of the entropy increase, which are *dynamical* and which often lie outside the scope of standard edification. Notwithstanding, the entropy as the maximum property of equilibrium states is hardly understandable unless linked with the dynamical considerations. The equal *a priori* probability of states is already in the form of a symmetry principle because entropy depends symmetrically on all permissible states. The particular function of entropy is determined completely then by symmetry over the set of states and by the requirement of extensivity. Consequently it can be even shown that a full thermodynamic (heat) theory can be formulated with the heat, Q, being totally absent. Nonetheless, the familiar central formulas, such as $dS = dQ/T$, remains lawful although dQ does not acquire to have the significance of energy. Nevertheless, for the standard thermophysical studies the classical treatises are still of the daily use so that their basic principles and the extent of applicability are worthy of brief recapitulation.

The conventional field of a mathematical description of the system, Ω, defined as a physical and macroscopic object (see Fig. 33.), in which certain

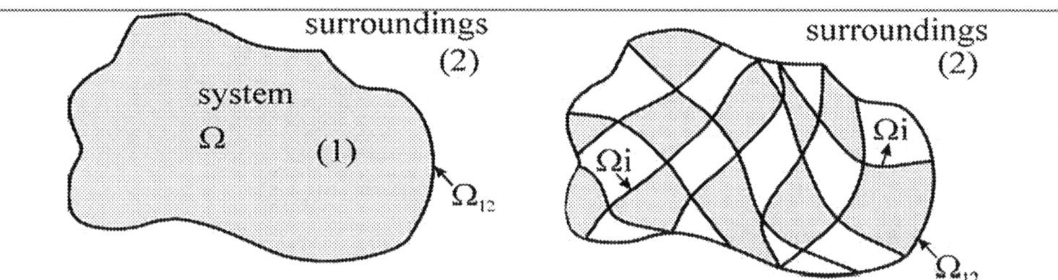

Fig. 33. – Circumscription of a basic thermodynamic system, Ω, with its central homogeneous interior (1) and the surface circumference, Ω_{12} , separating it from its outer periphery (called surroundings – 2). Right: a successively developed heterogeneous system (right), which subsists more complicated features, but is also more common in natural case as any variant of non-ideal homogeneous system. The interior is broken to smaller areas (grains or cells) separated by authentic interfaces, which properties are different from those of its yet homogeneous cores. Thoughtfully, we can also imagine such a (total) system, Ω, to consist of numerous (identical and additive) homogeneous subsystems, Ω_i's, all being in good mutual (thermal and other) contact, which thus make possible to simply communicate the concept of an active (non-equilibrium) system under flow.

(phenomenological) physical quantities (variables, *observables*) can be directly or indirectly measured, is traditionally acknowledged constituting the entire domain of thermodynamics [191,200,218,272,274-277]. Perhaps, it is one of the most general theories developed for the direct application to physical-chemical systems, where attention focuses on the investigation of thermal properties of materials based on the choice of certain parameters selected to represent the system. The basic concept involves the assumption of the existence of a stable, so-called *'equilibrium state'*, where each out-of-equilibrium situation of any system or sub-system evolves reversibly towards equilibrium, but not *vice versa*. The privilege of such *reversible* processes is that they can be described with the differential forms of linear equations while, on the contrary, irreversible transformations can be treated only by inequalities (and often as non-linearity).

The equation of energy conservation in thermodynamics is routinely a continuity equation, $dU = dQ - dW$. We understand it in such a way that within the domain, Ω, with its outer $1 \Leftrightarrow 2$ boundary, Ω_{12} , there is the *internal energy* variation, dU, due to the fluxes of energy in two forms: *heat across* the interface Ω_{12} ($dQ > 0$ is incoming) and *mechanical work* delivered by Ω itself ($dW < 0$, if work is received). It is worth noting that this already includes several aspects of irreversibility [9,191]: i.e., the thermal flow, $\Delta \Sigma^{flow}$, from T_1 to T_2 (with $T_1 > T_2$) and the transformation of dW to dQ (at fixed T and P) are irreversible while the transformation dQ to dW (at fixed T and P) is factually impossible. It is clear that *disequilibrium* makes things moving. For instance, thermal discontinuity, $T_1 > T_2$, is needed to push the heat flow through the $T_1 \rightarrow T_2$ interface, but certainly not for the opposite $T_2 \rightarrow T_1$ or not even for T_1 equal T_2. This argumentation coincides with the irreversibility of the reverse heat flux.

There are variables such as temperature, T, and pressure, P, which are identified with the *intensive, controllable field forces* (I) and related to the intensity of the effect. They are the same for any subsystems (Ω_1 and Ω_2) and for any choice of Ω. There are also the *extensive variables* (X) like volume, V, and entropy, S, which have the character of *measurable deformations* or displacements variable along with the extent of the system. Namely for any choice of Ω_1 and Ω_2 it holds that $X_1 + X_2 = X$ (e.g., $V_1 + V_2 = V$, assuming no contribution from the interface Ω_{12} (in-between Ω's). Extension to further choice of the other matching variables to encompass the system description in a more complete (as well as complex) way can be made, including, e.g., the magnetic field, H, or mole number, n, paired with the magnetization, M, and chemical potential, μ.

The above mentioned variables are considered to be the *state variables* and thus are defined by means of the internal energy, U, by its derivatives, e.g., $(\partial U/\partial S)_V = T = T(S,V)$ or $(\partial U/\partial V)_S = P = P(S,V)$. It yields the so-called *equations of the state*. In simple terms, for an ever present, multi-directional (*disordered*) motion (associated with heat flow), U *always* links with the product of T and S, while for an unidirectional (*ordered*) motion, U is moreover represented by other products, such as P and V, or H and M, or N and μ. In addition, some distinct forms of *thermodynamic potentials*, Φ, (as the state functions, again) are established, traditionally named after the famous thermodynamists, such as *Gibbs*, G, or *Helmholtz*, F. This yields, for example, the renowned relations for entropy, i.e., $S = S(T,P,\mu) = - (\partial F/\partial T)_{V,n} = - (\partial G/\partial T)_{P,n} = (\partial \Phi/\partial T)_{I,X}$

What we are able to handle theoretically and compare to experiments are the *response functions* that are listed in Table 6.I. Their behavior is independent of the immense multiplicity of the underlying components, that is, the variety of their different physical-chemical properties, which arises from experimental discovery, not as a theoretical axiom and factually confirms the conceptual power of thermodynamics.

In view of a non-stationary character of our often only `equilibrium-adjacent` thermophysical studies we need to distinguish three gradual stages of our thermodynamic estimate, i.e., the description particularly related to the experimentally adjusted status of temperature [3,277].

Classical Equilibrium	Near-equilibrium	Non-equilibrium
$I\ (dI \to 0)$	$I\ (\Delta I,\ dI/dt = \text{const})$	$I\ (dI/dt,\ d^2I/dt^2 \approx \text{const})$
$T\ (dT \to 0)$	$T\ (\Delta T,\ dT/dt = \text{const})$	$T\ (dT/dt,\ d^2T/dt^2 \approx \text{const})$

Classical thermodynamics of a closed system is thus the counterpart of dynamics of a specifically isolated system. The closed system defines and describes the energy-conserving approach to equilibrium, which is factually a 'thermal death'. The counterpart in dynamics is the reality of external fields which give and take away energy or momentum to the system without being modified. Still further, we can consider thermodynamics of the open system with two thermal baths providing the possibility of thermal cycles. We may go on assuming several baths getting closer and closer to the description of a real multiplicity situation in nature.

Temperature coefficients derived from the relationship between the generalized forces, I, and deformations, X.

Measurable Extensive, X,	Controlled Intensive, I,	dX/dI	dX/dT	dI/dT
Entropy, S,	Temperature, T,	Heat capacity $C_P = T(dS/dT)$		
Volume, V	Pressure, P,	Compressibility $\varphi = dV/dP/V$	T-expansion $\alpha_V = dV/dT/V$	P-coefficint $\beta_V = \alpha_V/\varphi$
Magnetization, M, Mag. field, $\mathcal{H}$,		Magnetic Susceptibility $\chi = dM/d\mathcal{H}$	Magnetocaloric Coefficient $\alpha_M = dM/dT$	Thermal mag. Succeptibility $d\mathcal{H}/dT = \alpha_M/\chi$
El.polarization, P, Elec. field, E,		Dielectric Susceptibility $\chi_P = dP/dE$	Pyroelectric Coefficient $p_P = dP/dT$	Thermal dielec. Susceptibility $dE/dT = p_P/\chi_P$
Mech.deformation, ε, Mech.strain, σ,		Elasticity Module $K_\varepsilon = d\varepsilon/d\sigma$	Therm. coef. of Deformation $\alpha_\varepsilon = d\varepsilon/dT$	Therm. coef. of Strain $K_{\sigma T} = d\sigma/dT$

Therefore, in classical thermodynamics (understood in the yet substandard notation of *thermostatics* [272,274,275,279]) we generally accept for processes the non-equality $dS \geq dQ/T$ accompanied by a statement to the effect that, although dS is a total differential, being completely determined by the states of system, dQ *is not*. This has the very important consequence that in an isolated system, $dQ = 0$, and entropy has to increase. In isolated systems, however, processes move towards equilibrium and the equilibrium state corresponds to maximum entropy. In true non-equilibrium thermodynamics, the local entropy follows the formalism of extended thermodynamics where gradients are

208

included and small corrections to the local entropy appear due to flows, making $dS/dt \geq dQ/dt\ (1/T)$. The local increase of entropy in continuous systems can be then defined using the local production of entropy density, $\sigma(r,t)$ [278,279]. For the total entropy change, dS, consisting of internal changes and contributions due to interaction with the surroundings (source, i) we can define the local production of entropy as $\sigma(r,t)\ \equiv d_iS/dt \geq 0$. Irreversible processes [279-281] obey the *Prigogine* evolution theorem on the minimum of the entropy production and $S = S^{dis} + \Sigma^{source}$, where $\Sigma^{source} > 0$.

We can produce disequilibrium operating from the outside at the expense of some external work, $\Delta W^{ext} > 0$ (using the Gibbs terminology) and once the system is taken away from its equilibrium we can consider ΔW^{ext} as $\Delta \Phi^{max}$, now understood as the maximum obtainable work. We can relate the ratio of ΔW^{ext} to $\Delta \Sigma^{source}$ $\{= \Delta Q(1 - T_o/T) - (P-P_o)\ \Delta V\}$ as the inequality greater than zero. For $\Delta W^{ext} \rightarrow 0$ we can assume the ratio limit to catch the equality $\Delta \Sigma^{source}/\Delta W^{ext} = 1/T = \partial S/\partial U$. It is important as it says that the arrow of thermodynamics goes in the direction of the increased entropy (or dissipated energy) that was embedded in the disequilibrium situation. This is another representation of the second law of thermodynamics as it leads us in a natural way to think in terms of vitality of disequilibrium. If there is no disequilibrium we must spend energy to create it. Nevertheless, if disequilibrium is already given, we may think of extracting energy from it. As a result, the ratio $\Delta \Sigma^{source}/\Delta W^{ext}$ can be understood in terms of heat flow in view of the efficiency of *Carnot's* ideal conversion ($\eta = 1 - T_2/T_1$) to become $\Delta \Sigma^{flow}/\Delta W^{ideal} = 1/T_2$. It comes up with a new content: before we talked of energy that was dissipated but here we point out the energy, which is actually extracted. Thermodynamics is thus a strange science because it teaches us at the same time both: *how both The Nature and our active artifact system behave.*

7.2 Effect of the steady temperature changes (rate of heating)

For the standard thermal analysis examinations, the *heat exchange Q'* (= dQ/dt) between the sample and its surroundings, must be familiarized as a fundamental characteristic, which specifies the experimental conditions of all thermal measurements [3,277]. As such, it must be reflected by the fundamental quantities when defining our extended system, i.e., the principal quantities must be expressed as functions of time, $T=\boldsymbol{T}(t)$, $P=\boldsymbol{P}(t)$ or generally $I=I(t)$. Therefore a sufficient description of the sample environment in dynamic thermal analysis requires specification, not mere values of T or P and other $\boldsymbol{I}$, but also particular inclusion of the time derivative of temperature, T' (= dT/dt), respecting thus the kind of dynamic environment (usually according to the kind of heating employed). Please, note that the *apostrophe* (') signifies derivatives and the *bold italic letters* ($\boldsymbol{I, T, \Phi}$) functions (as used further on).

Hence, the state of material can be characterized in terms of material (*constitutional*) functions [282] of the following type $V = V(T, T', P)$, $S = S(T, T', P)$, or generally, $\Phi = \Phi(T, T', P)$. Let us write the basic energy equation in forms of fluxes, $U' = Q' - PV'$ or as a general inequality relation, $0 \geq U' + TS' - PV'$, where primes represent the time derivatives. Substituting the general thermodynamic potential, Φ, the above inequality, eventually complemented with another pair of ($I \leftrightarrow X'$), is changed to, $0 \geq \Phi' + ST' - VP'$ (eventually added by $X \leftrightarrow I'$). Now we can substitute the partial derivatives from material relations into the inequality of the state function, $\Phi(T, T', P)$, or $0 \geq [\partial\Phi/\partial T + S] T' + [\partial\Phi/\partial P - V] P' + [\partial\Phi/\partial T'] T''$.

According to the permissibility rule [277], the above relation must hold for any allowed process and for all values T and P, and their derivatives, which can thus be chosen arbitrarily and independently. For $T=0$ and $P=0$, it is reduced to its last term, i.e., $0 \geq [\partial\Phi/\partial T] T$, which can be solved for all T only if $[\partial\Phi/\partial T] = 0$. Consequently, the state function $\Phi(T, T', P)$ cannot depend on T' and its form reduces to mere $\Phi(T, P)$. In the same way we can eliminate the second and any other term by accounting the pairs $T - P$, or generally $T - I$.

However, the analysis of the entropy term of the first term is more difficult [3,277] because it can be split into two parts, i.e., equilibrium related entropy, $S_{eq} = S(T, T'=0, P)$, and the complementary term, $S_i = S - S_{eq}$, or $0 \geq [\partial\Phi/\partial T + S_{eq}] T' + [S - S_{eq}] T'$. For the fixed values of T and P it takes the of an analytical relationship, $0 \geq aT' + b(T')T'$, for the variable T', with $b(T')$ approaching zero if $T' \rightarrow 0$. Such an inequality can be satisfied for arbitrary T only if $a=0$ and if $b(T')T' \geq 0$, i.e., if $\partial\Phi/\partial T = S_{eq}$ and $[S - S_{eq}] T' \leq 0$. The resultant relation represents the dissipation inequality. Provided the *term* $S = S(T, T', P)$ is negligible or, at least, sufficiently small. This is a portrayal of a *quasistatic process* for which the standard relationships of (equilibrium) thermodynamics are valid to an agreeable degree.

7.3. Thermal properties and measurable quantities

Changes in the thermal state of our system are, as a rule, accompanied by changes in nearly all macroscopic properties of the system and, vice versa, a change in a macroscopic property results in a change in thermal properties [1,3,9]. This fact underlines the fundamental importance of the logical structure of the mathematical description, which mutually connects all variables regardless if they are specified or not. In addition to the previously accounted basic pair of parameters, T-S and P-V, that are adjusted and controlled externally, we can enlarge the definition of experimental conditions by including other experimentally measurable quantities, X. We can choose them to represent the (instantaneous) state of individual samples under specific study, cf. Table. Such variables must be readily available and it is obvious that they depend on the size and structure of material (so that they must have extensive character). For practical purposes we often use the function called enthalpy, $H = H(S, P)$, and we can write it as a function of controllable intensive parameters only, at least the basic pair T and P, so that $H = H(S, P) = H(S(T, P)P)$. Applying Maxwell transformations, e.g., $(\partial V/\partial S)_P = (\partial T/\partial P)_S$, we get $dH = (\partial H/\partial S)_P [(\partial S/\partial T)_P dT + (\partial S/\partial P)_T dP] = T [(\partial S/\partial T)_P dT + (\partial V/\partial T)_P dP] + V dP$, which can be rearranged using coefficients shown in Table. In practical notions we introduce the coefficients of thermal capacity, C_p , and thermal expansion, α_v via $dH = C_p dT + V(1 - \alpha_v T) dP$.

If the system undergoes a certain change under the action of an external force field, F, then the work done, dW, by the system is given by the product of the generalized force and the component of the generalized deformation parallel to the direction of the acting force. It is obvious that the work resulting from the interactions of our system with its environment will be decisive for the thermodynamic description.

Let us replace volume by magnetization, M, and pressure by magnetic field, $\mathcal{H}$. Then $dU = TdS + \mathcal{H}dM = dQ + \mathcal{H}dM$, which is the most useful relationship describing the dependence between the thermal and magnetic changes in the material under study (such as the magnetocaloric effect). The corresponding Gibbs energy $G = G(T, \mathcal{H})$ is given by $dG = -SdT - Md\mathcal{H}$ and the previous equation acquires an additional term associated with magnetocaloric coefficient, α_M, i.e., $dH = C_p dT + V(1 - \alpha_v T) dP + (M + \alpha_M T) d\mathcal{H}$.

Similar approach can be applied to the systems, which exhibits electric field, E, paired with electric polarization, P, adding thus the term $(P + p_p T) dE)$ for the pyroelectric coefficient of dielectric polarization, p_p. Mechanical deformation, ε, paired with mechanical strain-tension, σ, contributes the coefficient for mechano-elastic properties, α_ε , as $(\varepsilon + \alpha_\varepsilon T) d\sigma$. Chemical potential, μ, paired with mole number, N, adds the term $(N + K_{nT} T) d\mu)$ for the coefficient of chemical activity, K_{nT}.

By using all pairs of variables involved, we can continue to obtain the generalized function, Φ, already termed as the general thermodynamic potential, which is suitable for description of any open system, even if all external force fields are concerned. Although a thermophysical experiment does not often involve all these variables, this function is advantageous to localize all possible interrelations derived above as well as to provide an illustrative description for generalized experimental conditions of a broadly controlled thermal experiment, for example,

$$d\Phi = -S dT + V dP + \sum_i \mu_i dN_i - M d\mathcal{H} + P dE + \varepsilon d\sigma + \text{any other } (X \leftrightarrow dI).$$

7.4. Chemical reactions

The course of a chemical reaction, which takes place in a homogeneous (continuous) system can be described by a pair of new quantities [3,274,275]: extensive chemical variable, ξ, denoted as the *thermodynamic extent of reaction* (and corresponding to the amount of reactant), and intensive chemical variable, A, called the *affinity* and having a significance similar to the chemical potential (i.e., $\xi A \Leftrightarrow N \mu$). The significance of ξ follows from the definition equation $d\xi = dN_i/v_i$, expressing an infinitesimal change in the amount, n_i , due to the chemical reaction, if v_i is the *stoichiometric coefficient* of the i-th component. If the temperature and pressure are constant, the expression for N_i $(= v_i \xi)$ yields the equation for the Gibbs energy in the form of a summation over all components, i, $\Delta G_r = \Sigma_i \mu_i v_i = (dG/d\xi)_{T,P}$. This quantity becomes zero at equilibrium and is negative for all spontaneous processes.

The difference between the chemical potentials of the i-th component in the given system, μ_i, and in a chosen initial state, μ_i^o, is proportional to the logarithm of the *activity*, a_i, of the i-th component, i.e., to a sort of the actual ratio of thermodynamic pressures. It thus follows that $\mu_i ^- \mu_i^o = RT\, ln\, a_i$ and on further substitution the *reaction isotherm* is obtained, $\Delta G_r = \Sigma_i v_i\, \mu_i^o + RT\, ln\, \prod_i a_i^{vi} = \Delta G_r^o + RT\, ln\, \prod_i a_i^{vi}$, where $\prod$ denotes the product and a_i is given by the ratio of so-called thermodynamic *fugacities* (ideally approximated as p_i/p_o). The value of the *standard change in the Gibbs energy, ΔG_r^o* , then describes a finite change in G connected with the conversion of initial substances into the products according to the given stoichiometry (when $\xi = 1$). It permits the calculation of the *equilibrium constant, $K_a = \prod_i a_i^{vi}$* and the associated equilibrium composition if $\Delta G_r^o = -RT\, lnK_a$.

Chemical reaction is accompanied by a change of enthalpy and, on a change in the extent of reaction, ξ, in the system; the values of both G and H of the system change too. If the relationship, $G = H - TS$, is applied in the form of $H = G - T(\partial G/\partial T)_{P,\xi}$ and when assuming that the quantity $(\partial G/\partial \xi)_{P,T}$ is zero at equilibrium, then $- (dH/d\xi)_{T,P} = T \partial/\partial T\, (\partial G/\partial \xi)_{P,v(eq)} = T\, \partial/\partial T\, (\Sigma_i v_i\, \mu_i\,)_{P,\, eq}$ term is obtained. The equilibrium value of $(dH/d\xi)_{T,P}$ is termed the *reaction enthalpy, ΔH_r* , which is positive for *endothermic* and negative for *exothermic* reaction. The *standard reaction enthalpy, ΔH_r^o* is the quantity provided by a process that proceeds quantitatively according to a stoichiometric equation from the initial standard state to the final standard state of the products.

It is convenient to tabulate the values of ΔH_r^o and ΔG_r^o for the reaction producing compounds from the elements in the most stable modification at the given reference temperature, T^o ($\cong$ 298 K), and standard pressure, P^o ($\cong$ 0.1 MPa). Tabulated values of G's and H's relate to one mole of the compound formed and are zero for all elements at any temperature.

The overall change in the enthalpy is given by the difference in the enthalpies of the products and of the initial substances (so-called *Hess'* law). The temperature dependence is given by the *Kirchhoff* law, i.e., upon integrating the relationship of $(\partial \Delta H^o / \partial T) = \Delta C_P$ in the form , $\Delta H_{T,r}^{\ o} = \Delta H_{\theta,r}^{\ o} + {}_\theta\!\int^T \Delta C_{Pr}^{\ o}\ dT,$ where the $\Delta H_{\theta,r}^{\ o}$ value is determined from the tabulated enthalpy of formation, mostly for the temperature $\theta = 298$ K. For $\Delta G_{T,r}^{\ o}$ it holds analogously that $[\partial (\Delta G_{T,r}^{\ o} /T)/\partial T]_P\ =\ (\Delta H_{T,r}^{\ o})/T^2$ and the integration and substitution provides $\Delta G_{T,r}^{\ o} = \Delta H_{\theta,r}^{\ o} - T\ {}_\theta\!\int^T (1/T^2)\ [\ {}_\theta\!\int^T \Delta C_P]\ dT\ +\ T\ \Delta G_{\theta,r}^{\ o}$, where the integration constant, $\Delta G_{\theta,r}^{\ o}$, is again obtained by substituting known values of $\Delta G_{298,r}^{\ o}$. On substituting the analytical form for C_P $(\cong\ a_o\ +\ a_1 T\ +\ a_2/T^2)$ the calculation attains practical importance . Similarly the equilibrium constant, K, can be evaluated and the effect of a change in T on K is determined by the sign of ΔH^o_r – if positive the degree of conversion increases with increasing temperature (endothermic reactions) and *vice versa*.

7.5. Heterogeneous systems and the effect of surface energy

If the test system contains two or more phases then these phases separates a phase *boundary*, which has properties different from those of the bulk phases it separates. It is evident that the structure and composition of this separating layer must be not only at a variance with that of both phases but must also implement certain transitional characteristics in order to match the differences of both phases it divide. Therefore, the boundary is a very important and crucial object, which is called the *interface,* and which is accompanied by its surface area, *A*. The area stands as a new extensive quantity, which is responsible for the separation phenomena, and which is indispensable for the definition of all *heterogeneous* systems. The internal energy, $U = \boldsymbol{U}\ (S,V,A,N)$, holds in analogy to the previous definitions and the corresponding intensive parameter is the surface Gibbs energy, γ, $(=\partial U/\partial A)$, which has the meaning of a *tension*. According to the depth of study, such a new pair of variables, $A \Leftrightarrow \gamma$, can further be supplemented by another pair, $\varpi \Leftrightarrow \varphi$, where ϖ is the curvature of the interface and φ is the surface curvature coefficient. The existence of both these new pairs of variables indicates the thermodynamic instability of phases (1) and (2) in the vicinity of their separating interface. It yields the natural *tendency to contract the interface area and decrease the surface curvature.*

Changes of interface characteristics enable a certain nominalization of the contact energy, which is in the centre of thermodynamic descriptions of any heterogeneous systems putting on display the phase separating interfaces. A direct consequence is the famous *Laplace* formulae, $P^{(1)} - P^{(2)} = (1/r^{(1)} - 1/r^{(1)})$, which states that the equilibrium tension, *P*, between the phases (1) and (2), is affected by the curvature of the interfaces, i.e., by the difference in radii $r^{(1)}$ and $r^{(2)}$ that is, more or less, mechanically maintained at the equilibrium. It can be modified for a spherical particle of the phase (1) with the radius, $r^{(1)}$, located in the phase (2), so that $P^{(1)} - P^{(2)} = 2\gamma/r^{(1)}$. By introducing chemical potentials, μ, the relation converts to conventional, $\mu^{(1)} - \mu^{(2)} = 2\gamma V/r^{(1)}$, where V i the partial molar volume and γ is the interfacial energy. With a multicomponent system, this equation can be written separately for each component, i (on introducing the subscripts, i), which may result in the formation of concentration gradients across the interface layers during the formation and growth of a new phase.

Physics

Thermostatics | **Thermodynamics**

sample is characterized by a set of extensive quantities, X_i

generalized thermodynamic potential, Φ, (and its definition by constitutive equations)

$$T\ (dT) \longrightarrow T\ (\Delta T)\ dT/dT\ (=T')$$

effect of heating rate

$$\Phi = \Phi\ (T, J_i) \qquad \Phi = \Phi\ (T, T', J_i) \cong \Phi\ (T, J_i)\ if\ T' = \phi$$

experimental conditions are given by the action of external forces, J_i

effect of the second phase nucleation

$$\text{surface energy} - d\gamma \ \text{——} \ A - \text{surface}$$

nucleus

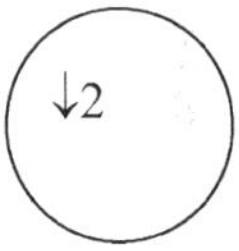

$\downarrow 2$ **r**-diameter

Heterogeneous system

$$\Phi = \Phi\ (T, \gamma, J_i) \qquad 1\uparrow$$

diffusion

$$\text{chemical potential } \mu \text{——} N - \text{mole number}$$

$$\downarrow 2$$

$$\text{——— interface}$$
$$1\uparrow \quad \mathbf{r} = \infty$$

$$\Phi = \Phi\ (T, \mu_i, J_i)$$
$$\Downarrow$$
$$\Phi = \Phi\ (T, N_i, J_i)$$

introducing the description of a reaction, phase $1 \Rightarrow 2$, intensive parameters be completed by extensive ones to represent the instantaneous state of the sample

$$d\Phi/dN_1 - d\Phi/dN_2 = d\alpha/dt\ \Delta\Phi$$
constitutive equations of kinetics
$$d\alpha/dt = f(\alpha, T)$$

$$dT/dt = f(\alpha, T) \quad = 0\ \text{(isothermal)}$$

Chemistry (kinetics) $\qquad = \phi$ (nonisothermal)

Another consequence is the well-established form of the *Kelvin* equation, $log(p/p_o) = 2\gamma V^{(2)}/(RTr^{(1)})$, which explains a lower pressure of vapors of the liquid with a concave meniscus, $r^{(1)}$, in small pores. This effect is responsible for early condensation phenomena (or local melting of porous materials) or for overheating of a liquid during boiling (as a result of the necessary formation of bubbles with a greater internal pressure than the equilibrium one above the liquid). An analogous equation holds for the growth of crystal faces during solidification, i.e., $log(p/p^{(ap)}) = 2\gamma V^{(s)}/(RT\boldsymbol{h})$, where $\boldsymbol{h}$ is the longitudinal growth of the given crystalline face. When an apparent pressure (ap) inside the crystal is considered instead of the original partial pressure (o), it discriminates smaller pressure for larger crystals and results in the spontaneous growth of greater and disappearance of smaller crystals. With common anisotropic behavior of crystals, it is worth mentioning that the surface energy depends on the crystal orientation which, according to the *Wulff* law, leads to the faster growth in directions characterized by greater, γ, according to the ratios, $\gamma_1/\boldsymbol{h}_1 = \gamma_2/\boldsymbol{h}_2 = \dots$, where $\boldsymbol{h}_i$ is the perpendicular distance from the *Wulff* coordinate system center of the crystalline phase.

The surface energy can also decrease because of the segregation of various components into the interface layer, whether by concentration of admixtures relocated from the main phase, or by adsorption from the surroundings. In accordance with the formulation of the general potential, Φ, the following relation can be written for the interface in a multicomponent system at equilibrium: $\Sigma_i\,\mu_i\,dn_i - SdT + A\,d\gamma = \Phi = 0$, or, $d\gamma = -\Sigma_i\,\Gamma_i\,d\mu_i + s^{(2)}\,dT$, where $\Gamma_i = n_i^{(2)}/A$ and $s^{(2)} = S/A$ are important experimentally measurable coefficients often employed in adsorption measurements, i.e., $\Gamma_i = \partial\gamma/\partial\mu_i$, as the specific surface adsorption and, $s^{(2)} = \partial\gamma/\partial T$, as the specific molar entropy. A practical example is the *Gibbs* adsorption isotherm, $\Gamma_i = -C_2\,d\gamma/(RTdC_2)$, which describes the concentration enrichment of the surface layer, where C_2 is the admixture concentration in the main bulk phase. This can also become important with polycrystalline solids, because the surface layer of individual microcrystalline grains can contain a rather different concentration of admixtures than that coexisting in the main parent phase, similarly to the gradients developed by temperature non-equilibration. Owing to the variation of experimental conditions, the resulting precipitates may thus exhibit different physical properties due to the differently inbuilt gradients, which can manifest themselves, for example, in the measurements of variability of the total electric resistance of such polycrystalline samples.

Heterogeneous systems, which are more common in nature and dealt with in the field of solid-state chemistry [3,283,284], exhibit supplementary intricacy of the resultant properties, as compared to homogeneous systems. In contradiction with the often idealized (but less frequent) homogeneous systems,

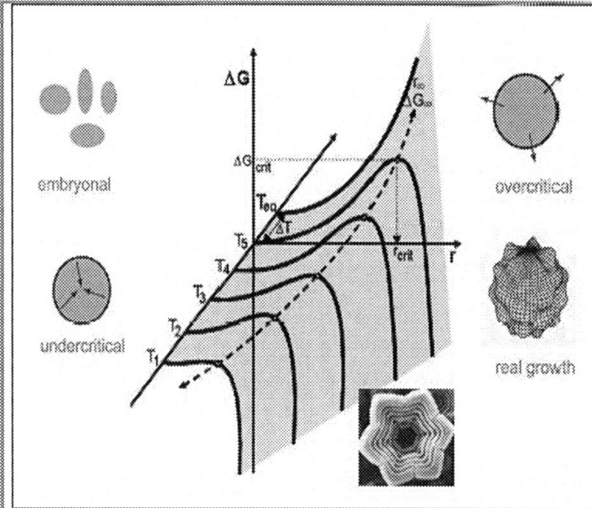

Fig. 34. – Dependence of ΔG for the new phase formation on the nucleus radius, r, and on the temperature, T. With increasing undercooling, ΔT $(= T - T_{eg})$, the critical values (i.e., the threshold size necessary to uphold the nuclei to exist) decrease (dashed line) as the change $\Delta G^{1 \to 2}$ is generally proportional to the ratio $\Delta T/T_{eq}$. Assuming $T_{eq} \cong T_{melt}$ and introducing reduced values $T_r = T/T_{melt}$ and $\Delta T_r = (T-T_{melt})/T_{melt}$ the following approximate solutions are convenient for practical applications:

Model	ΔH	ΔS	ΔG	surface energy, γ	critical ΔG_{crit}
Constant	ΔH_r	$\Delta H_r/T_r$	$\Delta H_r/\Delta T_r$	ΔH_r	$\Delta H_r/(\Delta T_r)^2$
Linear	$\Delta H_r\,T_r$	$\Delta H_r\,T_r/T_r$	$\Delta H_r/\Delta T_r\,T_r$	$\Delta H_r\,T$	$\Delta H_r\,T_r/(\Delta T_r)^2$ or $\Delta H_r/(\Delta T_r\,T_r)^2$

Schematic stages of new phase embryonic formation: left upper, clusters of the locally organized bunch of melts that may serve as prenuclei sites, it continues throughout the under-critical nucleus tending to disappear and to the super-critical nucleus ready to grow up spontaneously (upper right). The real process of growth is often irregular due to deforming forces arising from both the internal and external upshots (below). The real image of a crystal is often layered, as shown on the bottom photo.

heterogeneous systems must must include the action of inherent interfaces, which makes the description more convoluted, see previous table. A specific but generally challenged problem is the existence of boundaries, which spherically separate each particle of the newly formed second phases within the matrix. Another consequence is the well-established form of the *Kelvin* equation, $log(p/p_o) = 2\gamma V^{(2)}/(RTr^{(1)})$, which explains a lower pressure of vapors of the liquid with a concave meniscus, $r^{(1)}$, in small pores. This effect is responsible for early phase. Such an interlayer interface is factually performing the elementary role of a *disturbance* within the system continuality (i.e., true homogeneity defects). Within any homogeneous system, any interface must be first created, which process accompanies a particular course of the new phase formation called *nucleation*. The *surface energy* plays, again, a decisive role because it acts contrary to the spontaneous process of transition proceeding from the initial (homogeneous, mother) phase (1) into the newly produced, but regionally separated, phase (2). In the bulk, this is accompanied by a definite decrease of Gibbs energy, $\Delta G_v^{1 \to 2}$. Disregarding the energy of mixing, the Gibbs energy of a

216

spherical nuclei with volume, r^3, is negative, but is compensated for the positive surface energy, γ, which is dependent on, r^2, and which value predominates for all small nuclei at $r < r_{crit}$. The resultant dependence, see Fig. 32, thus exhibits maximum for a critical nuclei radius, r_{crit}, proportional to the simple ratio $\gamma/\Delta G_v^{1\to2}$, with the critical energy barrier, $\Delta G_{crit}^{1\to2}$, and amounting to the powered ratio 3:2 for the ration, $\gamma^3/(\Delta G_v^{1\to2})^2$. *For the equilibrium temperature, T_{eq},* (where $\Delta G_v^{1\to2} \to 0$ and $\Delta G_{crit}^{1\to2} \to \infty$) follows that the *nucleation of a new phase is thermodynamically impossible* (even if athermal nucleation can occur owing to implicated fluctuations). Nucleation can thus happen *only under distinct non-equilibrium conditions*, i.e., during a definite charge of undercooling (or reverse supersaturation). Further on, at given thermodynamic conditions, all nuclei with $r<r_{crit}$ are unstable and have a definite tendency to disintegrate and disappear. Only nuclei with $r > r_{crit}$ are predisposed to grow spontaneously, see Fig. 34. A more detailed portrayal is given in the following chapter 8 in connection with the description of kinetic phase diagrams, cf. Section 7.8.

The surface free energy can be approximated by a function of the latent heat and molar volume, v, so that $\gamma = 0.065\, \Delta H_f\, v^{-2/3}$, where ΔH_f is the enthalpy of fusion. The free energy difference between the liquid and solid phase is proportional to the melting temperature, T_m, i.e., $\Delta G = \Delta H_f\, (T_m - T)/T_m$, assuming that the associated ΔC_p approaches zero. Afterwards we can obtain an approximate equation for the excess Gibbs energy of the critical nucleus formation, $\Delta G_{crit}^{1\to2} = (16/3)\pi(0.065)^3\Delta H_f\, T_m^2/(T_m - T)^2$. The critical temperature, T_o, where the nucleation frequency of the metastable phase (') is equal to that of the stable phase, is linked to the ratios of enthalpies, R $= \Delta H_f'/\Delta H_f$, which must be less than unity. If this is satisfied, then $T_o = [T_{mr} - \sqrt{(RT_{mr})}]/[T_r - \sqrt{(RT_{mr})}]$ where T_{mr} and T_r are the reduced temperatures, T_m'/T_m, and, T/T_m, respectively. We may generally expect that the nucleation frequency of the metastable phase is larger than that for the stable phase if (and only if) the undercooling (or equally, but less likely, supersaturation) is large enough.

7.6. Equilibria and generalized Clapeyron equations

In the physical chemistry of solids [3,217,283-285], data on the chemical composition, i.e., the concentration expressed in mole fraction, $N_i = n_i/n$, are most often used. If some components are part of the surroundings (partially of an open system), the components **i** can be divided into conservative and volatile ones. In most real cases, the studied system consists of several (at least two) phases **j**, denoted by exponents *s1, s2,* ... (solid phase) or *l1, l2,* ... (liquid phase), etc.. Generalized function, Φ, termed the general thermodynamic potential, can be of practical use when the effect of external force fields is assumed. Although a thermophysical experiment involving all these variables is not encountered in practice, this function is advantageous for finding necessary interrelationships, is shown by the fundamental equation:

$$\Phi = U - TS + PV - \sum \mu_i N_i - \boldsymbol{I} \Leftrightarrow \boldsymbol{X}.$$

Worth a special noting may be an innovative thermodynamic concept for the description of partially open systems [286]. This original models take the advantage of the separation of components into two distinct categories frees (shared) and conservative (fixed) and then suggests an adequate way of their description and analysis. Quasimolar fractions and quasimolar quotients are used to describe macroscopic composition of the system and quasimolar thermodynamic quantities are established to substitute molar quantities (otherwise customarily employed in closed systems). New thermodynamic potential, called hyperfree energy is introduced through Lagendre transformation to specify the equilibrium condition of such partially open systems. To assemble this hyperfree energy function, shortened thermodynamic quantities and quasimolar phase fractions are used, which makes also possible to derive a new form of Clapeyron-like equation. Such modified thermodynamics is suitable for better analysis of the behavior of complex systems with a volatile component studied under the condition of dynamic atmosphere as is, for example, the decomposition of oxides in air.

In a single-component system ($\mathbf{i} = 1$) containing one component, n, only, three different phases can coexist in equilibrium, i.e., $\Phi^g = \Phi^l = \Phi^s$. If only one intensive variable, $\boldsymbol{I}$, is considered in addition to T, it can be solved for all three phases, g, l and s, at the isolated points (T, $\boldsymbol{I}$) called *triple points* . For two phases, the solution can be found in the form of functions $T = T\,(\boldsymbol{I})$, i.e., for liquids, $T - P$, hypothetical solutions, $T - \mu$, dielectrics, $T - \boldsymbol{E}$, mechanoelastics, $T - \sigma$, and, $T - \mathcal{H}$, ferromagnetics. Particular solutions yield the well-known Clapeyron equations, shown in the previous Table 6.I. [3].

If the existence of two components A and B with the mole fractions N_A and N_B (or also concentrations) is considered, then we have to restore to chemical potentials, μ, so that $\Phi^A\,(T, I, \mu_A, \mu_B) = \Phi^B\,(T, I, \mu_A, \mu_B)$. The number of phases can be at most four, which however, leaves no degree of freedom to the system behavior (i.e., coexistence at the isolated points only). Through analytical expression needed to separate μ's, we obtain $(dT/\mu_A)_I = (N_A^{s1} N_B^{s2} - N_A^{s2} N_B^{s1})/(S^{s2} N_B^{s1} - S^{s1} N_B^{s2})$. Analogous expression holds for the case of $(dI/\mu_A)_T$ i.e., when S is replaced by the corresponding extensive quantity, X. The denominator is always positive and because μ increases with the increasing concentration, the above ratio depends solely on the sign of the numerator, i.e., on the ratio of the mole fractions. If $N_A^{s2} > N_A^{s1}$, the concentration of component A is greater in phase s_2 than in phase s_1 and the temperature of this binary

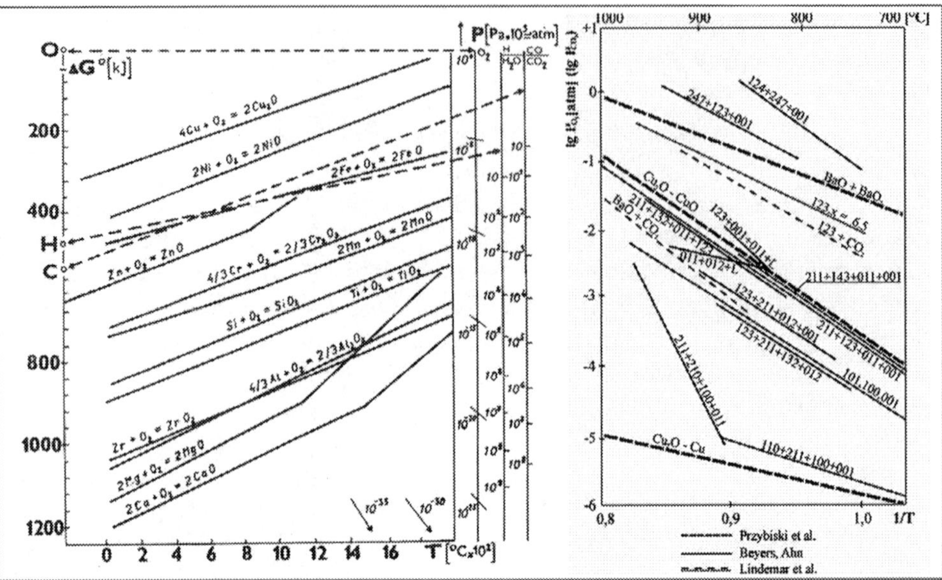

Fig. 35. – The popular *Richardson* and *Jefferson* plots [287] (left) of the standard free energy versus temperature is often acknowledged for the case of simple oxides formation at the given pressure of either pure oxygen (P_{o2}) or the its partial activity at given mixtures ($P_{H2/H2O}$ or $P_{CO/CO2}$). The set of straight lines originating in the point H and intersecting the corresponding scale on the right-hand side, from which the equilibrium composition of the gaseous phase can be found for the reduction of an oxide to the corresponding metal [3]. A special case (right) adopted here according to ref. [288] for the phases and reactions to occur in the system of high temperature superconductors (HTSC) exhibited by the coupled stoichiometric formula $Y_xBa_yCu_zO_\delta$ (abbreviated in the simplified cation ratios X:Y:Z). It illustrates the temperature dependence of oxygen pressure expressed for the various reactions drawn on basis of data surveyed and compiled by *Šesták* and *Koga* [288] including particular figures published elsewhere.

mixture increases with increasing concentration of *A*. The same reasoning can be applied for the behavior of component *B*.

Thermoanalytical methods often solve questions connected with the study of oxidation and dissociation processes. When working under certain oxidizing atmosphere, given either by p_{O2} or chiefly defined by the powers, *y*, of ratios of CO/CO_2 or H_2/H_2O in the form of the standard equation $\Delta G_o = -RTln(p_{CO}\,p_{CO2})^y$ we arrive for simple reactions at approximately linear functions. This attracted considerable attention in the fifties when applied to the description of equilibrium states for competing metal-oxide or oxide-oxide reactions, often known as *Jefferson* monograms [3,274,287]. It was of service even recently when refining reactions to take place in the copper sesquioxide systems necessary to care for the preparative methods in ceramic superconductors [288], see Fig. 35.

Such a classical approach towards the calculation of chemical equilibrium based on the construction of a stability condition for each independent chemical reaction is rather awkward for the entire computation of equilibrium of complex systems containing a greater number of species. The earliest non-stoichiometic algorithm base on the minimization of *Gibbs* energy on the set of points satisfying mass balance was initially proposed for homogeneous systems in [289,290] and survived in [291]. The first approximations to equilibrium compositions [292-294] played important role, which often assumes the convex behavior of *Gibbs* energy function and the total system immiscibility.

Among numerous various algorithms and commercial programs let me mention only one, the algorithm developed by Czech domestic teamwork [294,295], which applied simple premises: the *Gibbs* energy is a linear function of the number of moles, the mixing effects are removed by neglecting higher logarithmic terms of the *Gibbs* energy function, etc. As many others, it provides bases for a practical elaboration of the computer controlled evaluation and guides the establishment of databases, called "MSE-THERMO" [295], which thus become a serious partner to many different ready-to-use programs broadly available in contemporary thermodynamic literature. Such advanced studies were achieved on the basis of the good tradition of the Czech thermodynamic school, made available by outstanding domestic authorities [296].

7.7. Ideal and real solid solutions, phase diagrams

In the classical thermodynamic description of a system in chemical equilibrium, the given phase areas in the phase diagrams represent zones of minima of the Gibbs energy, whereas the phase coexistence is given by the *Gibbs* phase rule [3,297-299]. To construct a phase diagram requires mapping all available phases and locating their phase boundaries which, unfortunately, is not possible by mere measurement of temperature and/or concentration changes of the Gibbs potential. An equilibrium phase diagram, however, cannot say anything about the reaction, which transforms one phase to another, nor about the composition or the structure of phases occurring under conditions different from those that make equilibrium possible. In general, under the conditions of phase equilibria, when the chemical potential, μ, of the liquid phase (for each component) is equal to that of the solid phase, stable coexistence of both phases occurs, which is also the base of any theoretical description. A further step shown below is the indispensable awareness that there is not enough driving force concealed in their equilibrium difference ($\Delta\mu = 0$) necessary to carry on any change.

An important basis to understand phase diagrams is the analysis of thermal behavior of the formation of solid solutions (ss). This process is accompanied by a change in the *Gibbs* energy, ΔG^s_{mix}, called the partial molar *Gibbs* energy of mixing (assuming that $\Delta G \cong (1 - N_B)\mu_A + N_B\mu_B$, when $N_A +$

$N_B = 1$). Its value is considered as the difference in the chemical potentials of component A in the solid solution and in the standard state of solid pure component, i.e., $\Delta\mu^s_A = \Delta G^s_{mix} - N_B(\partial\Delta G^s_{mix}/\partial N_B)$. In the graphical representation, expressing the dependence of the Gibbs energy on the composition, it represents the common tangent for the solid phases, $s1$ and $s2$.

The derived character of the phase diagrams is primarily determined by the behavior of the ΔG^s_{mix} dependence, which is customarily shown as a subject of chemical potentials between the standard (o) and actual states, as $\Delta G^s_{mix} = N_A (\mu_A - \mu^o_A) + N_B (\mu_B - \mu^o_B)$, but it can be conveniently expressed by the simplified set of equations:

$$\Delta G^s_{mix} = \Delta H^s_{mix} - T\,\Delta S^s_{mix} = \Delta G^{ex}_{mix} + \Delta G^{id}_{mix}$$

$$\Delta G^{ex}_{mix} = \Delta H^{ex}_{mix} - T\,\Delta S^{ex}_{mix} = RT\,(N_A\,ln\,\gamma_A + N_B\,ln\,\gamma_B)$$

$$\Delta G^{id}_{mix} = -\,T\,\Delta S^{id}_{mix} = RT\,(N_A\,ln\,N_A + N_B\,ln\,N_B)$$

The term with superscript (*id*) expresses the ideal contribution from the molar Gibbs energy of mixing while that with (*ex*) represents its excess contribution. The extent of non-ideality of the system is thus conveyed showing the degree of interactions between the mixed components, A and B, which in the solution interact through forces other than those in the original state of pure components. It may result in certain absorption or liberation of heat on the real process of mixing. It is most advantageous to introduce the *activity coefficient, γ,* and the activity of component, i, is then determined by the product of mole fraction and the activity coefficient, $a_i = N_i\,\gamma_i$. Whereas the mole fraction can be directly measured, certain assumptions must be made to determine the activity coefficients. For an ideal solution, $\gamma = 1$, and the mixture behaves ideally, i.e., there is no heat effect of mixing ($\Delta H_{mix} = 0$). The literature contains many empirical and semiempirical equations [300,301] for expressing the real activity coefficients in dependence on the solution composition, which consistent description is a key problem for separation processes. It, however, fall beyond the capacity of this book. Most common is the single-parameter description based on the so-called *regular behavior* [302] of solutions (when $\Delta H^{ex}_{mix} >> T \Delta S^{ex}_{mix}$), i.e., $\Delta G^{ex}_{mix} = \Omega\,N_A\,N_B = \Omega\,N_A(1- N_A)$, where Ω is a temperature-independent (strictly regular) or temperature-dependent (modified) interaction parameter. It is evident that for a constant Ω value, the dependence is symmetrical at $N = 0.5$. However, in the study of solid solutions an asymmetrical dependencies are most frequently accoutered when introducing the simplest linear dependence in the form $\Omega = \Omega_o\,(1 + \Omega_l\,N)$, often called *quasiregular model* [3,300-302]. These descriptions are suitable for spherical-like molecular mixtures of metals.

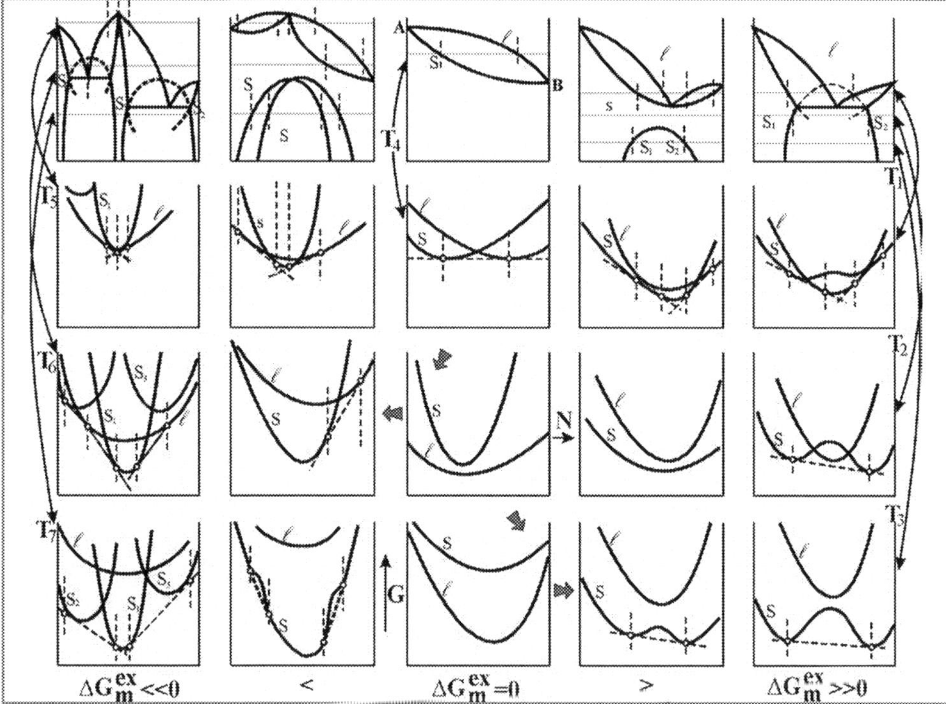

Fig. 36. – The illustrative sequence of formation of isostructural (the two right-hand columns, where $\gamma<1$, $\Omega<0$ and $\Delta H^s_{mix}<\Delta H^l_{mix}$) and non-isostructural (the two left-hand columns, where $\gamma>1$, $\Omega>0$ and $\Delta H^s_{mix}>\Delta H^l_{mix}$) phase diagrams for a model two-component system, A and B, forming solid solution. The change in the shape of phase diagram (particularly of the solidus, s and liquidus, l) results form a change in the magnitude of interaction parameter, Ω (starting from a model case of ideal solution, the middle column, where $\gamma=1$, $\Omega=0$ and $\Delta H_{mix}=0$). The boundary lines of the temperature dependence of the concentration are shown in the first row, whereas the following three rows give the dependence of the Gibbs energy on the composition at the three selected temperatures (denoted by the arrows at the side).

For more geometrically complicated molecules of polymers and silicates we have to take up another model suitable for *athermal* solutions, where T ΔS^{ex}_{mix} is assumed to have a greater effect than ΔH^{ex}_{mix} . Another specific case of the specific solid-state network are ionic (so called *Tyempkin)* solutions [303], which require the integration of energetically unequal sites allocation in the above mentioned concept of regular solutions. This is usually true for oxides where, for example, energetically disparate tetrahedral and octahedral sites are occupies on different levels by species (ions), so that the ordinary terms must be read as a summation of individual species allocation. Another specification may bring on thermal or concentration dependent, Ω, such as *quasi-subregular* models, where $\Omega = \Omega_o (1 - \Omega_T)$, or $= \Omega'_o (1 - \Omega_N)$ [3,6].

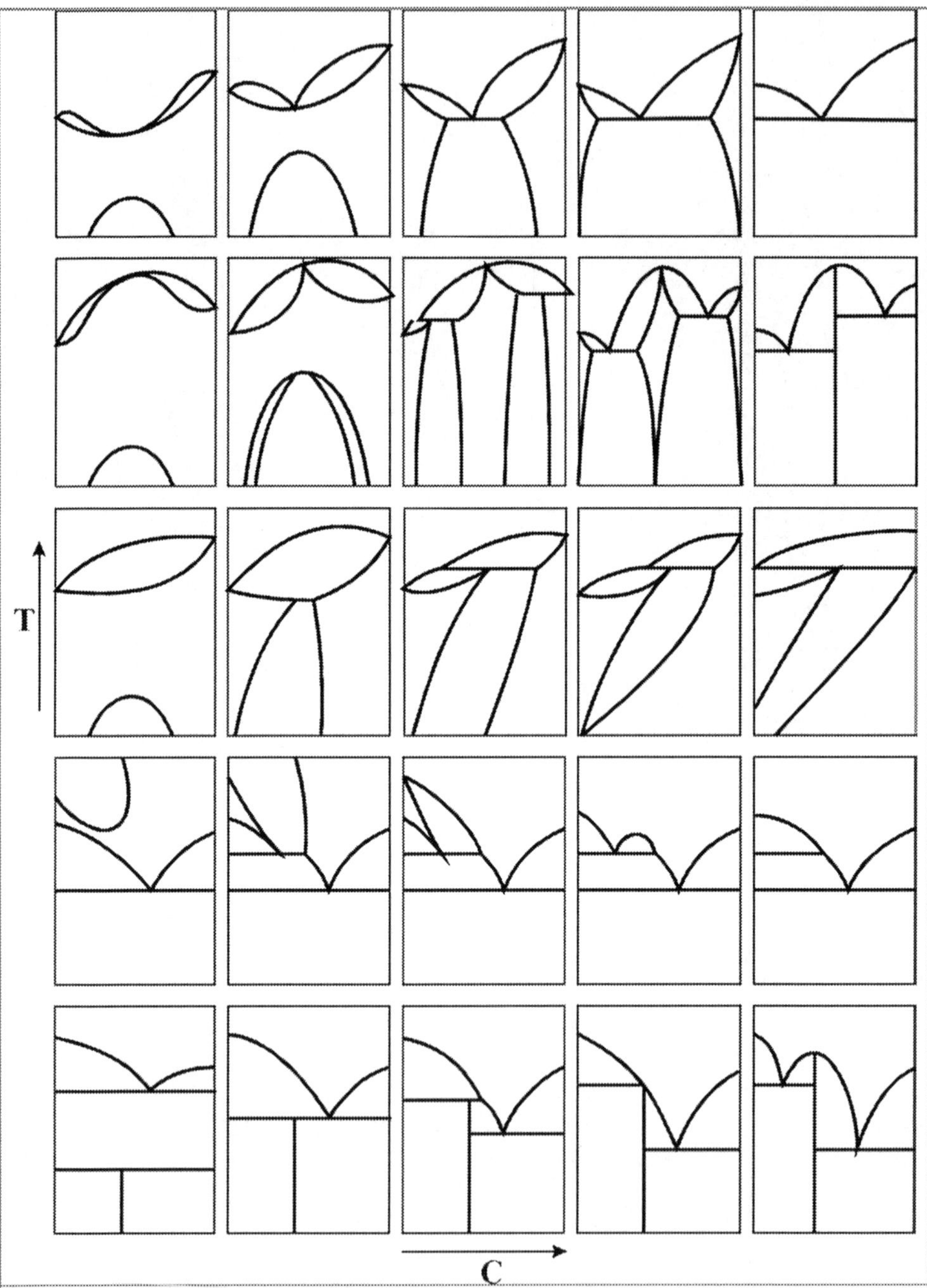

Fig. 37. – Topological features of a range of phase diagrams developed in the interactive sequences (left to right) based on the regular solution theory [6]

The most important minute is the impact of ΔG's on the shape of the phase diagram, see Fig. 37. which accounts the relative position of the Gibbs energy function curves for the liquid and solid phases, and the contribution from the individual ΔG's terms and, finally, the shape of the functional dependence of the molar Gibbs energy of mixing on the composition and thus also on the resultant shape of the $\Delta G = \Delta G'(N)$ function. It follows that in a hypothetical diagram of the G vs. N dependence the lowest curve represents the most stable phase at a given temperature. Therefore, for temperatures above the melting points of all the phases involved, the lowest curve corresponds to the liquid phase, whereas for temperatures well below the melting points, the lowest curve corresponds to the solid phase. If the two curves intersect, a two-phase region is involved and the most stable composition of two coexisting phases is given by abscissas of the points of contact of the curves with the common tangent. If ΔG^{ex}_{mix} is negative, than the resultant ΔG function is also negative, be cause the term $\Delta G^{id}_{mix} = - T\Delta S_{mix}$ is also negative. On the other hand, if ΔG^{ex}_{mix} is positive, the situation is more complicated because of the opposite effect from $T\Delta S_{mix}$ term, the value of which predominates, especially with increasing temperature. Therefore there are two basic cases: first the mixing of the components A and B is accompanied by a positive enthalpy change, ΔH_{mix}, which is actually contributing to an increase in ΔG_{mix}. It results from a stronger interactions in pure components A and B, i.e., A with A and B with B, in solution. The eutectic point is lower than the melting points of pure components. As the interactions in the solid state are much stronger than in the liquid, the solid solutions have a tendency to decrease the number of interactions by a separation process, i.e., by clustering phases richer in components A or in components B. On the contrary, if mixing of components accompanied by a negative enthalpy change, that further decreases the ΔG_{mix} value, the interactions between the unequal components predominates and the solution AB is more stable than the components A and B. Therefore the eutectic temperature is higher than the melting points of A and B. In solid solutions clustering again is advantageous but now through the enrichment of AB pairs at the expense of AA and BB. Important guide in better orientation in possible phase boundary shifts is the effect of pressure, see Fig. 38., which well illustrates various shapes of most frequent binary phase diagrams similarly as the gradual change of the interaction parameter.

For ternary diagrams the relationship is analogous but more complex when including three components A, B and C and the adequate sets of three equations [3,297-299]. Most numerical procedures results in the determination of *conodes*, i.e., lines connecting the equilibrium compositions, the terminal points of these conodes and the regions thus formed are then analogous to the previously described binary diagrams. A detailed explanation falls, however, beyond the simple illustrative intention of this chapter and is a common subject of numerous databases and computer evaluation procedures.

7.8. Nucleation phenomena and phase transitions

Any process of a phase transition (necessarily creating system's heterogeneity) does not take place simultaneously within the whole (homogeneous) system, but proceeds only at certain energetically favorable sites, called *nuclei*. We can envisage the newly created face (interface) to stick between the parent and second (different) phases as an *'imperfection'* of the system homogeneity, representing thus a definite kind of *defects*. Nucleation phenomena, i.e., formation (due to fluctuations) of distinct domains (embryos, clusters) of a new (more stable) phase within the parent (original, matrix) phase, are thus important for all types of the first-order phase transitions. It includes reorientation processes, compositional perturbations as well as the defect clustering (such as cavity-cracks nucleation under stress or irradiation of reactor walls or the development of diseases when a malady embryo overgrow the self-healing, etc.). Therefore the *nucleation* can be seen as a general appearance responsible for the formation processes throughout the universe, where the clusters of several atoms, molecules or other species are formed as the consequence of the fluctuation of thermodynamic quantities. As a rule, the small clusters are unstable due to their high surface-to-volume ratio but some clusters may bring themselves to a critical size, a threshold beyond which they are ready to step into a spontaneous growth.

The metastable state of the undercooled liquid (corresponding to $T < T_{eq}$) can differ from the state of stable liquid ($T > T_{eq}$) by its composition and structure. There can be found complexes or clusters of the prearranged symmetry (often met in the systems of organic polymers, common in some inorganics like chalcogenides and already noted for liquid water near freezing, which can either facilitate or aggravate the solidification course. The nucleation can occur in the volume (bulk) of the homogeneous melt (called *homogeneous nucleation)* or it can be activated by impurities, such as defects, faults, stress sites, etc., acting as foreign surfaces (yielding *heterogeneous nucleation* particularly connected with these inhomogenities). In principle, both processes are of the same character although heterogeneous nucleation starts at lower supersaturations in comparison with the homogeneous case because the formation work of the critical cluster is lowered.

Such a decrease of $\Delta G_v^{1 \rightarrow 2}$ is usually depicted by a correcting multiplication factor, proportional to *(2+cos θ) (1 − cos θ)²*, where θ is the contact (adhesion, wetting) angle of the nucleus visualized in the form of an apex on a flat foreign surface of given inhomogeneity. However, only the angles over 100° have a substantial effect on the process of nucleation and they can be expected when the interface energy between the impurity and the growing nucleus is small, chiefly occurring with isomorphous or otherwise similar crystallographic arrangements. Practically the critical cluster size is usually less than 10^{-8} m and that is why the nucleation on surfaces with radius of curvature larger than 10^{-7} is considered as a true nucleation on the plane surface. In fact, we are usually unable to establish correctly such associated physical parameters, like number of pre-crystallization centers, value of interfacial energy or allied formation work of the critical cluster) and that is why the homogeneous nucleation theory has been put on more detailed grounds than heterogeneous. Moreover, homogeneous nucleation becomes usually more intensive (thus overwhelming heterogeneous nucleation) at any higher undercooling, which represents most actual cases under workable studies.

In general, the fluctuation formation of the embryonic crowd-together is connected with the work that is required for the cluster creation, $\Delta G_n(T)$, and it is given for the most common liquid-solid phase transition by $\Delta G_n(T) = -n\Delta\mu(T) +A_n\ \gamma(T)+\Delta G_E$ where $\Delta\mu(T)$, n, A_n, $\gamma(T)$ and ΔG_E are, respectively, the temperature dependent supersaturation (i.e. the difference between the chemical potentials of the parent and crystal phases), number of molecules in cluster, the surface of a cluster (of n molecules, which equal to $\varsigma\ n^{2/3}$, where ς is a geometrical factor depending on the shape of the cluster), the temperature dependent interfacial energy per unit area and the Gibbs energy change due to external fields. If we know the nucleation work we can qualitatively determine the probability, P_n, for the formation of a given cluster according to $P_n \approx exp\{\Delta G_n(T)/kT$, where k is the Boltzmann constant.

From the conditions of the extreme (*) in $\Delta G_n(T)$ $\{d\Delta G_n(T)dn\} = 0$ it follows that $n^* = \{2\varsigma\ \gamma/3\ \Delta\mu(T)\}^3$ and $\Delta G^* = 4\ \varsigma^3\ \gamma^3/\ 27\ \{\Delta\mu(T)\}^2$ showing that both the critical cluster size, n^*, and the activation energy of nucleation, ΔG^*, decreases with decreasing supersaturation sensitive to temperature. The term ΔG_E can be affected, for example, by the polarization energy, **P**, connected with the formation of a nucleus with antiparallel orientation of its polarization with respect to the external electric field, **E**. It adds the multiple of **P**, **E** and σ, which leads to a decrease in the overall energy of the system. A more general contribution of elastic tension, σ, tends to deform a nucleus that is usually formed under ever-present matrix strain (e.g., due to the different volumes of parent and crystal phases). It become evident for any miss-shaped spheroid, with flattened radius, r_f, and thickness, 2δ, for which the deformation energy holds proportional to $r_f^2\ \sigma\ (\delta\sigma/r_f)$. The term $(\delta\sigma/r_f)$ has the significance of the elastic Gibbs energy. In correlation, the classical studies by *Nabarro* [304] are worth mentioning as he calculated the tension in a unit volume for cases when the entire strain is connected within the mother phase. He assumed various shapes of nuclei in the form of function, $f(\sigma/r_f)$, where for a spheroid it equals one, for

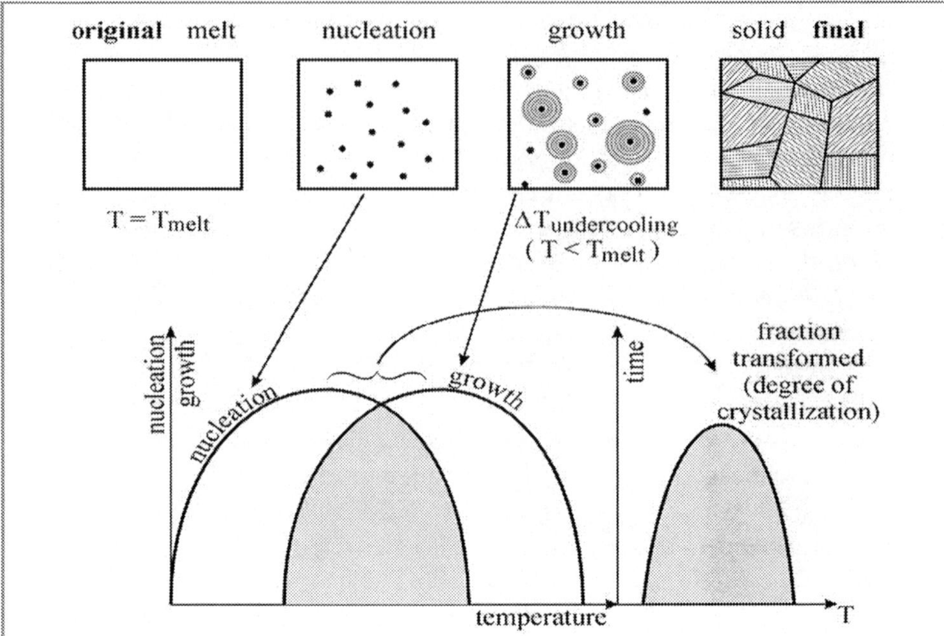

Fig. 35. – Inclusive diagram of the mutually overlapping patterns of nucleation and growth showing that if nucleation curve goes first before the growth curve (bottom left), the overall crystallization process (bottom right) is straight-forward (upper) towards the formation of solid crystalline product upon cooling. However, if their sequence is opposite (growth curve preceding nucleation one) the crystallization becomes difficult because the too early progression of growth is strongly hindered by the lack of nuclei necessary to seed the growth, which favorites greater undercooling. The cooperative position of the curves, and the degree of their mutual overlapping, then determines the feasibility of overall processing, e.g., the ease of crystallization and/or opposite process of vitrification, for the latter the nucleation curve should be delayed behind the growth curve as much as possible.

needles it is much greater than one and, on contrary, for discs it becomes much lower than one. Although the mechanoelastic energy is minimal for thin disks, their surface is relatively larger compared with their volume and thus the optimal nucleation shape is attained for flattened spheroids. There can be found further contribution to ΔG^*, such as the internal motion of the clusters which is sometimes included, but it becomes meaningful only for the nucleation of vapor. Equally important is the introduction of the concept of solid solutions, which brings complication associated with another parameter on which the nucleation becomes dependent. Namely it is the concentration, affecting simultaneous processes of redistribution of components, reconstruction of lattice and the redistribution of components between phases and particularly across the phase interface, which is balancing the composition inside each phase. The nucleation work become three-dimensional functions of n and C, exhibiting a saddle point [9], i.e., the point at which the curvature of the surface is negative in one direction and positive in the perpendicular direction, $(\partial \Delta G/\partial n = 0$ and $\partial \Delta G/\partial C = 0$ with associated ΔG^* and $C^*)$ [6,191,281,305].

The main objective of nucleation (experimental) acquaintance is the determination of the nucleation rate, which is the number of supercritical stable embryos formed in the unit volume per unit of time. Another objective is the transient time, τ, (otherwise called induction, incubation or delay time or even time lag), which is associated with the system crystallization ability and which non-stationarity arises from the time-dependent distribution functions and flows. It is extremely short and hardly detectable at the phase transition from the vapor to liquid. For the liquid-solid phase transitions it may differ by many orders of magnitude, for metals as low as 10^{-10} but for glasses up to $10^2 - 10^4$. Any nucleation theory is suffering from difficulty to define appropriately the clusters and from the lack of knowledge of their basic thermodynamic properties, namely those of interfaces. Therefore necessary approximations are introduced as follows:

- restricted degrees of freedom of the system relevant for the clustering process,
- neglecting memory effects (so called 'Markovian' assumption),
- ignoring interactions between clusters,
- disregarding segregation or coalescence of larger clusters (rate equations linear with respect to the cluster concentration) and
- final summations being substituted by integration.

The detailed description of various theories is shown elsewhere [5,6,305,306].

7.9. Effect of perturbations and features of a rational approach

Every system bears certain degree of fluctuations around its equilibrium state, i.e., is subjected to *perturbations*, δ. In the case of entropy, its instantaneous state, S, differs from the equilibrium state, S_{eq}. It can be conveniently depicted by using an extrapolation series $\delta S + \delta^2 S/2 + \dots$ etc.. The resulting difference can then be specified on the basis of internal energy, U, such as $(S - S_{eq}) = \Delta S = (1/T_1 - 1/T_2) \, dU + [\partial/\partial U_1(1/T_1) + \partial/\partial U_2(1/T_2)](\partial U)^2/2$. The first term on the right-hand side vanishes at the instant of equilibrium (i.e., $\delta S \to 0$ because $T_1 = T_2$) and the second term must be negative because $C_V(\delta T)^2/2T^2 = \delta^2 S > 0$. Such an average state is stable against thermal fluctuations because of the positive value of heat capacity, C_V. In a non-equilibrium stationary state, however, the thermodynamic forces (F) and the corresponding flows of energy, j_E, and matter, j_M, do not disappear but, on contrary, gain important status. Hence the first entropy variation is non-zero, $\delta S \neq 0$, and the second variation, $\delta^2 S$, of a given elementary volume (locally in equilibrium) has a definite value because the complex integrand has negative sign.

Thus the term $d/dt(\delta^2 S) = \Sigma_M \, dF_M \, dj_M > 0$ (simplified at the near-equilibrium conditions as $\Sigma_M \, F_M j_M > 0$) must be positive if the non-equilibrium stationary state is stable.

As shown above, the classical description has been undoubtedly useful; nevertheless, it has some drawbacks both from the fundamental and practical points of view. It is based on the macro- and/or local- equilibrium hypothesis, which may be too restrictive for a wider class of phenomena where other variables, not found in equilibrium, may influence the thermodynamic equations in the situations taking place when we get out of equilibrium. The concept is consistent with the limiting case of linear and instantaneous relations between fluxes and forces, which, however, becomes unsatisfactory under extreme conditions of high frequencies and fast non-linear changes (explosions). Then the need arises to introduce the second derivatives (''), such as $\Phi = \Phi \, (T, T', T'', P, \ldots)$ and the general thermodynamic potential, Φ, is can be no longer assumed via a simple linear relation but takes up a new, non-stationary form $\Phi = -ST' + VP' + (\partial\Phi/\partial T)T'_{T'', P} + (\partial\Phi/\partial T')T''_{T, P} + \ldots$ where all state variables become dependent on the temperature derivatives.

It is worth mentioning that an advanced formalism was gradually introduced by *Eckert, Maixner, Coleman, Noll* and *Truesdel* [309] in the form currently known as *rational thermodynamics* [140,204,274,307-310], which is particularly focused on providing objectives for derivation of more general constitutive equations. We can consider the absolute temperature and entropy as the primitive concepts without stressing their precise physical interpretation. We can assume that materials have a memory, i.e., the behavior of a system at a given instant of time is determined not only by the instantaneous values of characteristic parameters at the present time, but also by their past time (history). The local-equilibrium hypothesis is not always satisfactory as to specify unambiguously the behavior of the system because the stationary or pseudo-stationary constitutive equations are too stringent. The general expressions describing the balance of mass, momentum and energy are retained, but are extended by a specific rate of energy supply leading to the disparity known as the *Clausius-Duhem* fundamental inequality, $-\rho \, (G' + ST') - VT \, (q/T) \geq 0$, where ρ and q are production of entropy and of heat, ∇ stays for gradients and prime for time derivatives.

There are some points of its rationality worth mentioning:

i) The principle of equipresence, which asserts that all variables, which are present in one constitutive equation; must be a priori present in all other constitutive equations.

ii) The principle of memory, or heredity, which dictates that current effects depend not only on the present causes but also on the past grounds. The set of variables, x, is no longer formed by the variables of the present time but by their whole history, i.e., $x = f(t - t^o)$.

iii) The principle of local action, which establishes that the behavior of a given material point (subsystem), should only be influenced by its immediate neighborhood, i.e., higher-order space derivatives should be omitted.

iv) The principle of material frame-indifference, which is necessary to assure the notion of objectivity, i.e., two reference observers moving with respect to each other must possess the same results (usually satisfied by the Euclidean transformation).

8. THERMODYNAMICS, ECONOPHYSICS, ECOSYSTEMS AND SOCIETAL BEHAVIOR

8.1. Thermodynamics and its societal and economical applicability

Similarly to mechanodynamics, the field of thermodynamics was developed as a tool to help technical progress initiated by the invention of heat engine. It is clear that this progress has not been isolated as it interconnects several disciplines responsible for the evolution of civilization based on energy resources and later encompassing other relevant fields of ecology, sociology, biology, economy, etc.. Individual domains of application obviously brought various entry values, which are not independent but form an interdisciplinary information treasure, which is further exploitable for new claims.

Best examples are materials, or better goods, with various functional values such as utility worth, technological and scientific level, market price, moral values, etc. It interrelates not only nearest status (and production) of neighbors but also second and higher-order neighbors interaction, which result in a complex web encompassing production cost, manufacturing requirement, technology know-how, scientific newness, ecology impact, communication need, transport friction, competition challenge, price configuration, feasibility, innovation, sophistication or marketing and advertisement. This was reason why thermodynamic pattern has become a generalized method to describe wider phenomena than only those associated with the science of heat [41,311].

As an accepted example we can see financial markets, which are complex systems in which a large number of traders interact with one another and react to external information in order to determine the best price for a given item. Because the quantity of information is so large, it is difficult to extract a subset of economic information associated with some specific aspects. The difficulty in making prediction is thus related to an abundance of information in financial world and not on the habitually assumed lack of it, which are contrary manners in other branches of sciences. When a given piece of information affects the price in a market in a specific way, the market is not completely efficient so that from the time series of prices we may detect the presence and evaluate the conjecture of this information [62,312].

During the past 30 years, physicists have achieved important results in many fields, among others phase transitions, nonlinear dynamics or disordered systems. In these fields' proposition the power laws, scaling and unpredictable series are often involved essentially helping the interpretation of underlying physical phenomena. Almost hundred years ago the Italian social economist *Pareto* [313] investigated the statistical character of the wealth of individuals in a stable economy by modeling them using the distribution $y \approx x^{-a}$ where y is the number of people having income x or greater than x and a is an exponent that was estimated to be 1.5 applicable to various conditions and nations. Since that the power laws became dominant in various physical and societal sphere of in fluence. Empirical studies of the lower end of the wealth axis showed, however, that the distribution is rather exponential, while the high-wealth tail still maintains a power law [314,315]. It was interpreted as a result of conservative laws for total wealth leading to the robust Boltzmann-like exponential distribution, whatever the random wealth exchange remains in full analogy with the energy distribution in a classical gas of elastically scattering molecules. It guided formulation of a model of wealth production and exchange, where agents randomly interact pairwise governed by a kinetic equation for one-particle distribution function [316].

Economy, however, can be regarded as a complex many-body scheme so that the allied systems like stock exchange markets display scaling properties similar to the systems analyzed in statistical physics. Scaling behavior in currency exchange satisfies scaling with an exponent around 0.45 [317]. Self-similarity and associated fractal structure of financial signals have also been the target of analysis [318]. It should be remarked that the concept of a power-law distribution was first seen counterintuitive because it's apparent deficiency in a characteristic scale until the works by *Mandelbrot* [319] on scaling and fractals [53] and by *Stanley* [320] on scaling for thermodynamic functions. Another concept ubiquitous in the natural sciences was the *Einstein* conception of random work [321] later applied to the theory of speculation in financial market. Rather revolutionary development was Mandelbrot's hypothesis [322] that the price changes follow a Levy stable distribution [323,324] defined a basin of attraction in the functional space of probability density function, $p(x) \approx x^{-(1+a)}$.

A growing number of engaged physicists [325,326] have recently attempted to analyze and model financial markets and economic systems in general as initiated by the classical work of *Majorana* [327] and others [328,329]. In 1990s the new field of *econophysics* emerged [330] being complementary to the traditional approaches of mathematical finance [312].

Most current economic theories are deductive in its origin because they assumes that each participant knows what is best for him assuming that all participant are equally proficient in choosing their best action. It is somehow reminiscent to statistical physics where all particles adjust to certain preference (stability) conditions. In the real sociological world, however, the actual participants, or better players, do not have a perfect foresight and hindsight [331], most often their action are based on trial-and-error inductive thinking, rather than the deductive rationalism. In this respect the evolutionary games became important segment within the standard framework of game theory [332,333]. However, this approach realistically applied in economics is not as

232

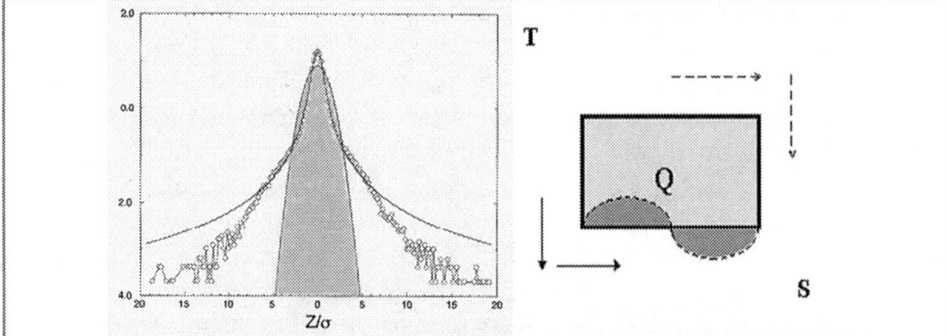

Fig. 39. – Left: Semi-logarithmic scaled plot of the probability distribution (vertical axis $P(x)$) showing the traditional Gaussian distribution (inner shaded area), which is adequate, for example, in describing the standard energy distribution among gaseous molecules but which in economics can be converged only at an asymptotic regime for large assess. Therefore, another type of distribution was found more suitable based on the summation of independent and identically distributed stochastic processes (x_i) characterized by the power-law tails ($P(x) \approx x^{-(1+\alpha)}$, see solid lane with $\alpha = 1.5$). Such *Lévy* stable symmetrical profile is thus obtained from scaling analysis and scale factor determination defining a basin of attraction in the functional space. It well describes the probability density function for incidence price changes (stock market indicated by open circles). Right: Economic application of the textbook *Carnot* thermodynamic cycle, where the solid arrows show traditionally the power cycle while the opposite cycle (dashed) shows the reverse heat pump (refrigeration process). Correspondingly, the assumed business cycle (when replacing conditions of minimum *Gibbs* energy by maximum economic prosperity) can be imagined by using a less conventional representation of T as the mean property of a society, while retaining the meaning of entropy S (societal order/disorder) the same. In other words, products are manufactured in a cheap (lower T) market and sold in a more affluent market (higher T).The shaded area represents useful work Q (either as heat or money) and the dotted line illustrates the more natural, non-equilibrium processing due to delays caused by, e.g., thermal conductivity (*Cruzon-Ahlborn* diagram) or business transport obstacles. It is worth noting that a similar non-linear backgrounds will be shown charmmetrical for thermophysical measurements where, for example, s-shaped zero line is acquired for a DTA peak (Fig. 81.).

convenient for generalization as we prefer to view a game involving a larger number of players, i.e., a more statistical approach capable to indicate and explore emerging collective phenomena. The study of game theories is a useful tool to simulate real situations, which helps to enlighten various phenomena of sociological and biological behavior. Let us see one instance in particular.

Chalet and *Zhang* [334] considered a population of N odd players each having finite number of strategies and having a choice to prefer one of two tenders (sides) based on his common knowledge of the past record. The payoff of such a binary and minority game is to declare that after everybody has chosen its side independently, those who are in the minority side win and all winners collect an award (except for playing in random). The game develops asymmetrically as the players adopt accordingly so that the apparent advantages disappears for those sitting at one side fixed. With increasing number of alternatives the players tend to get confused and may perform worse. Though the general tendency for best and worst players is rather consistent the gap between rich and the poor awardees appears to increase with time. This approach is open to all sorts of variations, such as situation-motivated payoff and game further structuring even forcing players to fight each other. Temporal record shows that there factually develops a kind of 'arm race' screening that the better strategy (kind of more brain) power leads to advantage. So in the scope of evolution of survival-of-the-fittest the players tend to develop bigger brains to cope with the evolution of survival-of-the-fittest and the richer players develop 'bigger brains' to get along with ever-aggressive fellow players. If extended to the effect of environment we may even speak about the emergence of intelligence so that touching the problem of biological evolution.

8.2. Efficiency, economic production and generalized Carnot cycle

Thermodynamic work is well known in stretching films, charging an electric system, compressing a gas or magnetizing a body, etc.. In every case it is possible to put the equation for the work W done in the form of the multiple of the generalized force, F, and generalized displacement, X, i.e., $dW = FdX$, which is a standard part of the continuity equation, $dW = dU + dQ$. It is worth repeating it involves three distinct concepts with separate notions of W (work), U (energy) and Q (heat), where the latter represents the transferal process only, i.e., heat as energy in transit [191,311], cf. previous Chapter.

Now there arises the question how we can bring it in the economic work or better production by proper replacing individual notions by fitting simulatives. Remember that economic production, like work in thermodynamics, is not an exact differential. We can introduce the output production function [312,335] as income $Y = f(K,L)$, which can be determined by capital (K) and labor (L). The difference of production (Y) and consumption (C) leads to savings (S). In economic sciences the balance of production can be generally calculated from a standard (exact) differential form if $dY = 0$. Usual solution is given by a power law $Y \cong aK^aL^b$ with the elasticity constant $(a+b) = 1$ determined by the production factors.

As in thermodynamics, many processes in engineering, economics, ecology, biology or agriculture rely on a closed production cycle so that $\oint \delta Y \neq 0$ (where non-exact differential form is marked by δ). If we call the first path Y and the returning path C, the output of cyclic production leads to non-zero profit (P) which may be invested or saved (S) [336]. The concept of non-exact differential forms leads to the proper relation of so called neoclassical theory [312], i.e., $Y-C=S$. According to the laws of calculus [337] this solution may be

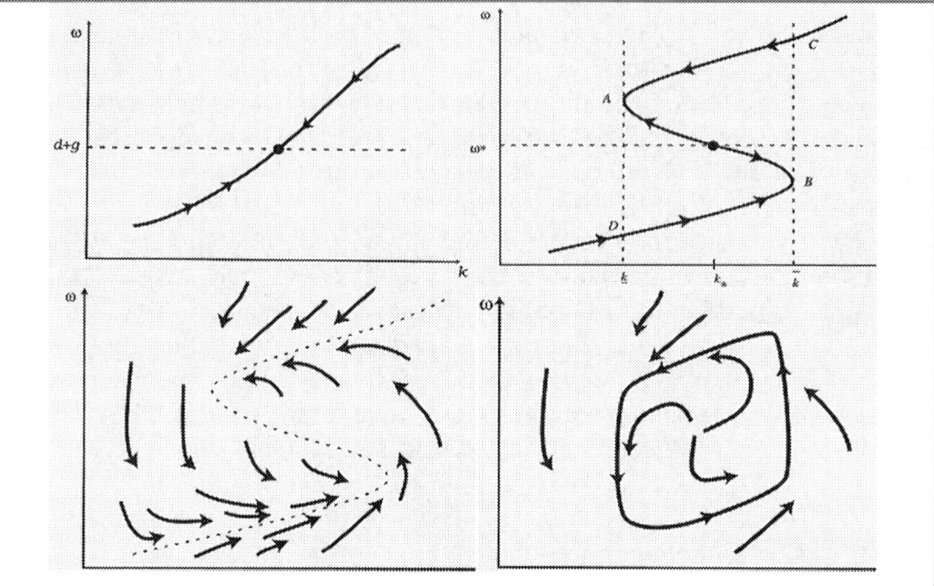

Fig. 40 –   Examples of behavior of a financial system when assuming the one-sector economy, which shows the economic growth and relaxation phenomena in the graph of k (capital/labor ratio) against ω (ratio of wage rate vs. rental rate): a) stable state with an equilibrium attractor positioned in the middle of monotonously increasing curve, b) unstable state with the s-shape curve (resembling the phase transition curve in physics, cf. Fig. 48) with two end-point attractors where a jump have got to occur (having thus apparent equilibrium in the middle, which location, however, is inaccessible from outside), c) unstable phase-plane picture when the capital/labor ratio in each sector must instantaneously rearrange itself in order to satisfy the constrains of full employment or balance competition for resources (reminding the quasi-steady state, which often witness a self-limiting growth of cells lacking interactions among the species) and finally d) closed-curve oscillatory state if both variables vary smoothly in time and when all trajectories tend toward this so called 'limit cycle curve'. Position marked ω^* is equilibrium-like location and 'k and k' are trajectory edge limits.

turned into the exact form of dW by adding an integrating factor T (representing the mean capital per person or the mean price level or the standard of living) so that $dK = dW - TdS$ where K reassumes the meaning of energy while entropy S retains its standard notion. It displays an important feature that besides work the main source of economic growth is entropy that hold up the enviable chance of diversification, the variety for independent economic systems, etc.

Let us now consider a separate production system for consumer goods (as bread) and capital goods (like electricity), each requiring its own supply of capital and labor. This leads to the neo-classical two-sector model of economic growth [338], which compares quantities in each sector providing thus wage rate (the value of labor) and capital depreciation (so called rental rate) which ratio is abbreviated by ω and is independent of prices. Introducing k as the ration of capital over labor we can see following two kinds of behavior for the function $k = f(\omega)$, see Fig. 40, i.e., steady (upper) having a robust toward-stability attractor and unsteady, which in limit can be predisposed to chaotic actions [335].

Most economic processes are periodic and we have to complete a closed cycle by interconnecting income Y with the cost (consumption) C. We again can recall to the classical thermodynamics via Carnot diagram in the $T - S$ plane. Let us recall the Carnot cycle [311,336,339,340] in Fig. 40. Manufacturing products like agriculture yield. cars, furniture or other innovative goods from a single input part requires comparable labor (assumed to represent a change in entropy) in any source (farm, business). A farm, for example, conducts itself as a creator of capital and thus operates like a generator (of electric energy) The farmer collets a high amount of energy (capital, Y) like grain, cattle, etc. from laborers (C). The farmer pays a lower amount of energy (capital, C) for food, costs, wages, etc. to the laborers. The profit (ΔQ) is thus Y-C with the efficiency (ef) of the system given by the ratio of profit to the invested capital. The farm, like all viable production systems, has to gain more than it spends (Y-C>0) in order to survive, which is also true for all biological, ecological or business systems because the price of merchandise depends on the quality of labor, tools and generally on the living, information and technology standards (T). This methodology approach enables us to replace the classical thermal power cycle by an assumed economy profit cycle [336,339]. The useful work extracted (originally from heat Q) may then represents money gained by producing in a cheap market and selling in a more affluent market at the difference ΔT (due to unequal economies). Dashed, non-equilibrium delay is then caused by transfer of matter, know-how, etc, representing variety of economic hindrance similar to dissipation of friction [311]. This implies that the rich become richer but, on the other hand, job creation or agriculture support may install equalization.

For farms, companies, traders, etc., at least two groups of people are involved: e.g., farmers and laborers (or owner and workers, capital and labor, two trading countries). Both groups together form an economic system and, accordingly, have to resolve how to divide the profit (ΔQ) of each cycle. This is negotiated periodically by workers and employers, by union and industry or even by the world trade conference. As a result the workers obtain percentage (p) of the profit (P) and the employer will take a rest (1-P). If both groups reinvest their shares $\{P(Y_1$-$Y_2)$ and $(1$-$P)(Y_1$-$Y_2)\}$ they will grow in time and we can arrive to equations C=$Y_1(t)$ and Y=$Y_2(t)$. For $P\neq1/2$ the solution of this set of differential equations [336] gives exponential relation of the type $Y(t) \approx P\Delta Y\{exp(Pt)$-$1\}/(1$-$2P)\}$. If all profit goes to the richer party at P=0, the income of the second group will grow exponentially while the income of the first party stays constant. At 10% profit for the poorer party both parties grow exponentially while for P=0.25 the richer party still growth exponentially but with lower rate linked with a less growth of the poorer party. This indicates well that a high rise in wages may weaken the economy leading to lower wages in a long run. At a split case of $P = 0.5$ the efficiency of the system stays constant with time leading to a linear growth only. Assuming $P= - 0.25$ the poorer side makes only losses going bankrupt while the richer party will still grow exponentially.

8.3. Thermodynamic laws versus human feelings

Recently there arose a remarkable idea that the entire human society can be regarded as a kind of many-celled super-organism [311], the cells of which are not cells, but rather we, the human beings, cf. Fig. 41. The Internet might represent a kind of embryonic phase of the neural system of our "as-if organism" in which a "global brain" may facilitate the linking up of partial intelligence of the individual users. Later on, perhaps, it may develop its own ideas, strategies and even a consciousness of an unknown order. From a systematic point of view, such a "cyberspace" [87] is in a similar category as language (the system), but differs in many important respects: such as the temporal scales of all relevant levels in such a cyberspace are similar, the system has a "body (neural network), some participants (system programmers as well as hackers) may deliberately influence the system at all levels, etc. It may become as common a property as "fire" and it may develop a systematic theory such as thermodynamics or thermal physics.

We just need to look for some basic links between the mathematical description of particles strictly controlled by the laws of thermodynamics and human beings affected by their feelings. Such correlation of thermodynamic ideas and rules applied to economics [336-342] was already made by *Lewis* [152] as early as in 1925, and, certainly, it was preceded by pure mathematical consideration rooted back to the time of *Newton*, and his experience while he was functioning as an accountant.

Viewing perceptible activities of the human population from a greater distance, it may be possible to observe and compare the behavior of societies as a system of thermodynamic-like partners. Metals, a similarly viewed society of different species of atoms, can be described by *functions* derived long ago within the field of thermodynamics, which so far well-determines the state of integration and/or segregation of resulting alloys. By analogy to such a vast variety of problems (known in the associated field of materials science) similar rules can be established to become useful for the application to various problems in segregated societies now governed not by mathematics with its well-defined functions, but by *human feelings* (sociology), as introduced by *Mimkes* in 1995) [339]. For example, important forces may be created by the integration of foreign workers in different states or integration of women in leading positions of politics, business or science. It may help to provide solutions to long-lasting

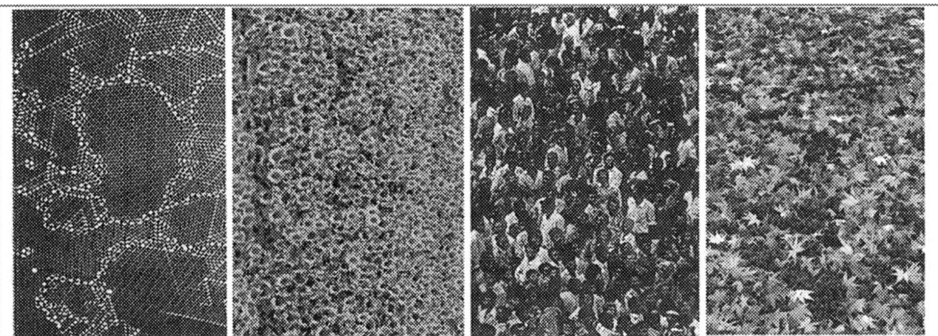

Fig. 41. – Photos exemplifying some aggregations of variously assumed disorder of particles: Left, microscopic picture of the arrangements of atoms, which are homogeneously ordered within microcrystalline domains (separated by whiter interfaces). Middle, macroscopic photo of the plantation field of sunflowers in a field at Giant mountains (Czechia) and the crowd of people somewhere in India. Right, windfall leafs in a Kyoto garden.

conflicts in binary societies such as Northern Ireland, Africa or the former Yugoslavia, etc. It may be extended to develop transitional terms of traditional thermodynamics to sociology, economy or even to some military ideas. Societies would exhibit different states depending on the degree of their development and organization. They can show characteristic quantities, which appear to have a certain similarity to those already known in thermodynamics.

Therefore, let us freely move to the matching area of our thermodynamic background and to the previously discussed field of *phase diagrams*. The same structure may arise when we try to describe the state changes within a society, whose state is connected with the formation of mixtures of independent individuals, i.e., the associations of the original constituent abbreviated as A ($x\%$) and B (($1-x$)$\%$). We can recall the chemical parallel described by the classical change in the Gibbs energy [339] repeating from the previous Chapter: ΔG^s_{mix} (= $\Delta H^s_{mix} - T \Delta S^s_{mix}$ = ΔG^{ex}_{mix} + ΔG^{id}_{mix}) which is fortuitously composed of two terms. For ideal (*id*) non-interrelated behavior, ΔH^s_{mix} is zero, and mixing is only governed by the *distribution statistic* of constituents. Such a simple model of the regular mixing of A-B associates is thus conveniently based on the logarithmic law of entropy and can be customarily described by:

$$\Delta G^{id}_{mix} = - T \Delta S^{id}_{mix} = RT \{x \, lnx + (1-x) \, ln(1-x)\}$$

For the interrelated behavior of constituents in the mixture, ΔH^s_{mix} is either positive or negative and can be conventionally depicted on the basis of previously explained *regular behavior*, where $\Delta G^{ex}_{mix} = \Omega x(1-x)$. Whereas the *fraction*, x (members, species, contents or generally concentration) can be directly measured, certain assumptions must be made regarding the *interaction parameter* Ω (which must evidently be zero for the mixture exhibiting ideal behavior because the components A and B behave equally).In addition of this excess term of non-ideality, which is usually dealt with in terms of *cohesive energy*, E, we introduce ε to express the *interactions* between the inherent pairs of components A and B, i.e., $(\approx E_{AB} + E_{BA} - E_{AA} - E_{BB})$. For example, an *A-B* mixture (typically alloy) will be stable if ΔG^{s}_{mix} is at its maximum, which results from either positive or negative interaction energy between the *A-A*, *A-B* and *B-B* neighbors. For a strong *A-B* interaction, ε will be positive while for a strong *A-A* and/or *B-B* attraction, it will lead to a negative sign resulting in a limited solubility of the mixture.

Surprisingly, *Mimkes* [339] found statistically well-compiled evidence that such a regular solution model can also be satisfactorily applied to describe the intermarriage data of binary societies consisting of *partners (= components)*, e.g., girls and boys in different *societies (= mixtures,* such as: immigrant foreigners and domestic citizens in the middle Europe, religious Catholics and non-Catholic inhabitants of British islands or African (black) and non-black citizens of the USA. The observed structural analogy is due to the general validity of mathematical laws of statistics of mixing, which allows one to translate the well-established laws of thermodynamics into the social science where the state of binary societies is determined by the above-mentioned feelings, i.e., maximizing mutual *happiness and/or satisfaction[1]*.

---

[1]* In order to change the interaction behavior of inorganic mixtures we can add admixtures, surface reactants, dopants, etc., so that the appropriately adjusted composition matches the lowest value of the system's Gibbs energy. For living organism it is far more complicated to intervene its "chemical factory" in order to achieve desired feeling of maximal satisfaction or happiness. The primary chemical in charge of the living function is surprisingly nitric oxide. It's a vascular traffic cop, activating the muscles that control the expansion and contraction of blood vessels. If the mind is in the mood the body appropriately responds. Alike the interaction of differently charged molecules the interaction between two heterosexual beings is controlled by hormones: male testosterone and female estrogen, which triggers desire by stimulating the release of neurotrasmitters in the brain. These chemical are ultimately responsible for our moods, emotions and attitudes. Most important seems be dopamine, partly responsible for making external stimuli arousing, among other things, from drug addiction or sport/work/hobby or other activity holism. Dopamin-enhancing drugs (antidepressants) can increase desire and erection as well as apomorphine. Another neurotransmitter almost certainly involved in the biochemistry of attraction and desire is serotonin, which, like dopamine, plays a significant role in feelings of satisfaction. Another imperative hormone is oxytocin, which is in charge of feel-good factor, e.g., for couples holding their hands, watching romantic movies or grouping with the same leisure time as well as melanocyte-stimulating hormone with a dual effect of giving men erections and heightening their interest in sex and attractiveness. Probably the most controversial issue on the chemistry of attraction and sexuality is the role of pheromones, so far only used by perfumemakers that have latched market by pheromone-based scents. Therefore the mutual

Case A – Order and sympathy, $\varepsilon > 0$,

Crystal: In rock salt, the attraction of the ion components A (sodium-Na) and B (chlorine-Cl) is much stronger than the attraction between similar pairs of Na-Na and Cl-Cl. The maximum of the (negative) Gibbs energy is given for equal numbers of both components at x = 0.5. It is well known that due to the strong Na-Cl attraction, rock salt crystallizes in a well-ordered "..*ABABAB..*" – ball-layered structure. This is associated with a negative enthalpy change and accompanied with heat liberation.

Society: A group of English-speaking tourists is visiting a Japanese fair. They are more attracted to English speaking sellers while domestic buyers prefer Japanese- speaking sellers. Mutual happiness (including economic gain) will be low if there are few sellers or few buyers. The maximum of mutual satisfaction will be at equal numbers of all kinds of buyers and sellers. Each group gets a certain degree of excitement from the shopping process when buying and selling; emotion possibly bears a certain analogy with (or relation to) the above-mentioned heat of mixing.

Ancient philosophy: water and wine, fire together with air

Case B – Disorder, integration, indifference and apathy, $\varepsilon = 0$,

Alloys: There are no interactions between the neighbors at all. The Gibbs energy is negative because of entropy disordering effects, only. Such an ideal solution can be illustrated by silicon (Si) and germanium (Ge), in which the arrangement is accidental (random).

Societies: Two kinds of solution can be found: *Indifference*: Equal partners are as attractive or repellent as different partners, which leads to ($\varepsilon =$) EAB + E BA -EAA – EBB = 0. *Apathy:* The attraction of all partners is zero so that ($\varepsilon =$) EAB = EBA = EAA = EBB = 0. In a downtown supermarket in Kyoto, we will find a random distribution of men and women. For busy shoppers, a short cashier line will be more important than male and/or female neighborhoods. This corresponds to indifference, and the society of shoppers is mixed by chance or is integrated.

Case C – Segregation and antipathy, $\varepsilon < 0$,

Alloys: Mixtures of gold (Au) and platinum (Pt) segregate into two different phases. The (negative) Gibbs energy exhibits two maxims, one for Au with few Pt atoms, the other for Pt with few Au atoms. The degree of segregation is not generally a full 100% unless the equilibrium temperature is close to absolute zero. On the other hand there is a $T_{max}(x)$ temperature needed to completely dissolve or integrate a given composition, x, of two components which can be determined from the derivation dG/dx to yield $T_{max}(x)$.

---

attraction and associated sexual desire is yet seen as a sort of magic, a phenomenon filled with delightful mystery thus for long outside of a rigorous description in any thermodynamic language.

240

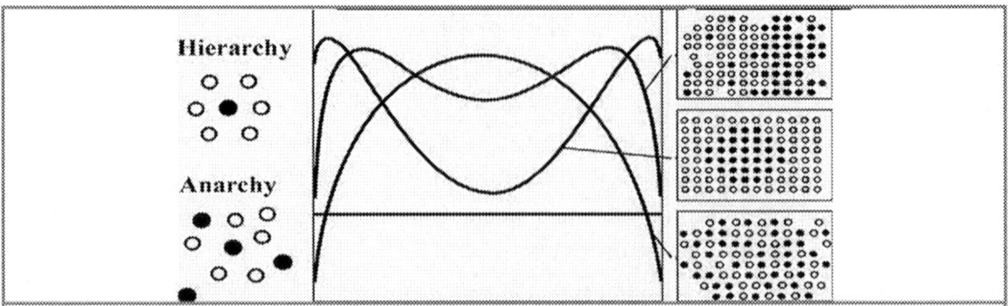

Fig. 42. – Dilution profiles of the Gibbs energy curves of binary-like blend akin to people blend (middle, black and white circles) under the assumption of various orders (arrangements) of their mixing in dependence on the character (intensity) of their mutual relations.

Societies: Mutual happiness of a society of, for example, black and white neighbors (or Moslim and Serbian people in the former Yugoslavia) would again show two maxims: one maximum is obtained if the percentage of black neighbors is low and the white people feel at home; on the contrary, the other maximum is obtained at a high percentage of black neighbors, where blacks feel at home. In order to attain a maximum of happiness, the town (society) will segregate into areas mostly white with just a few black renters and the areas mostly black with a few white renters. In this way both, black and white people, will *feel* mostly at home. In general, the degree of neighborhood segregation will not be 100%. This can happen only if the *tolerance* between different groups is close to zero; resulting in ghettos, e.g., the pure ethnic Moslem and/or Serbia homeland areas.

Ancient philosophy: water and oil, water in contradiction of fire.

The above grouping can further be discussed in more detail in terms of the extreme values of interactions, such as when the degree of mixing is much greater than zero – this will yield a *hierarchy*, and for highly negative values the situation results in an aggressive society, cf. Fig. 42. On simulating the classical (T-P) phase diagram of matter (solid-liquid-gas), we can define analogous states of societies, i.e., *hierarchy-democracy-anarchy* (see Table/Fig. 43.). Such thermodynamic-like considerations offer a wider source of inspiration: for example, relating P to *political pressure* and V to *freedom*, the constancy of their multiplication (similar to Boyle's law) shows that for a higher political pressure, the society generates a lower freedom and vice versa. Associating P with the pressure of political relations, then temperature can also be characterized as a measure of the extent of internal proceedings: the warmer the international proceedings, the lower the number of possible collisions. Two neighboring states developing at different rate would mutually interact (the quicker accelerating the slower and vice versa).

Table 1. Analogies between the Natural Sciences, Social Sciences, Economics and Military regarding thermodynamic analysis and functions.

Symbols binary	NaturalSciences	Social Sciences	Economics	Military
A-B	gold-platinum	Blacks-Whites	sell-buy	NATO-Russia
	silicon-germanium	Catholic-Protestant	rich-poor	North-South
	sodium-chlorine	male-female	Europe-USA	SEATO-China
	concentration (%)	minority size (%)	demand/supply	relation of forces
ordering				
$\omega > 0$	compound	sympathy	trade links	military treaty
$\omega = 0$	ideal solution	indifference	free market	neutrality
$\omega < 0$	segregation	antipathy	business/competition	military block
functions				
G	Gibbs energy	satisfaction/happiness	prosperity	security
T	temperature	tolerance to chaos	mean property	reconciliation
Q	heat	self-realization/health	money	weapons
E_{AA}	cohesive E	tradition/heritage	profit/earnings	friendship
$E_{AB} > 0$	binding E	curiosity/love	investment	joint activity
$E_{AB} < 0$	repelling E	distrust/hate	cost	hostility
$E = 0$	indifference	apathy	stagnation	neutrality

stages	
metallurgy	anthropology
disordering/solubility	integration
solubility limit	segregation
phase diagram	intermarriage diagram

$\Leftarrow$ states and state diagrams $\Rightarrow$		
ancient philosophy	pressure vs temperature	political pressure vs living standard
air	gases	anarchy
water	liquids	democracy
earth	solids	hierarchy

Fig. 43. – Analogies between various scientific, societal and economic relationships based on thermodynamics [9]. Courtesy of *Jürgen Mimkes*, Paderborn, Germany.

It certainly does not include all inevitable complications when assuming boundary conditions. A material property known in physics as viscosity can be renamed within international relations as *hesitation,* which can be determined as a function of population density, speed of the information flow and the distance between possible collision centers and the places of negotiations (past East-West Germany relations stirred via Moscow communistic administration or Middle East conflicts taking place on the Arab-Jews territories but negotiated as far away as in New York). Understanding the freeway as the mean distance between two potential collision (ignitions), we can pass to the area of transport phenomena observing *inner friction* as hindrance to the streaming forward characteristic, e.g., for migration of people from the East/South to the West. We again can describe such people migration as the problem linked with thermodynamics of diffusion. We can imagine a technology transfer which depends not only on the properties of state boundary (surface energy standing for the transferred meaning of administration obstacles) but also on the driving force which is proportional to the differences of technology levels (e.g., North and South Korea) bearing in mind that the slower always brakes the quicker and *vice versa*. As with chemistry, nucleation agents (ideas, discoveries, etc.) can help to formulate societal embryos capable of further spontaneous growth. Surfactants are of ten used to decrease the surface energy, like the methods of implantation of production factories or easing of customs procedures (well-developed Western Europe against post-communistic Eastern Europe). Human society often suffers disintegration, but the overall development can be seen in open loops (or spirals) which finally tend to unification upon increasing the overall *information treasure* of civilization. This is similar to a technological process where raw material first undergoes disintegration followed by separation, flotation and other enrichments to upgrade the quality for an efficient stage of final production of new quality material.

Such examples of brief trips to the nature of sociology could give us certain ideas to improve the art of both the *natural and humanistic sciences whose interconnections were natural until the past century* when the notion of heat was analyzed to become the natural roots of thermodynamics. Researching the analogy of physical chemistry to sociological studies of human societies is a very attractive area particularly assuming the role of thermodynamic links, which can be functional until the relation between inherent particles and independent people, is overcome by the conscious actions of humans because people are not so easy classifiable [331] as are mere chemicals. Such a feedback between the human intimate micro-world to the societal macro-state can change the traditional form of thermodynamic functions, which, nevertheless, are here considered only in a preparatory stage of feelings. Therefore this sociology-like contribution can be classified as a very first though rather simplified approach to the problem whose more adequate solution will not, hopefully, take another century as was the development of the understanding of heat and the development of the concept of early elements.

8.4. Rules of behavior: strategy for survival and evolution

In our previous ´physico-chemical´ approach, we assumed people as unvarying ´thermodynamic´ particles without accounting for their own ´human´ self-determination [331]. The interactions of such 'sedentary' particles are given by their inherent nature (internal state, charge), properties of the host matrices (neighboring energetic) and can be reinforced by their collaborative integration (collective execution, stimulated amplification) in the set-up vicinity. On the other hand, in the diluted intermediates, where a few impurity particles are scattered in a uniform lattice of the other host atoms, such visitants prefer to cooperate with each other, or align in a mutually favorable way, and the nearest-neighbor reciprocated 'orientation' can be positive or negative. That is, for example, the case of magnetic spins tending to be aligned antiprallel for the typical case of antiferromagnetic arrangement. Problems occur if an odd number of such aligned atoms are arranged in a loop, impossible to satisfy globally the desire of all spins to align in a definitely opposite way to one another (characteristic for a trimerous or pentamerous network). Unfulfilled compensation and resulting competition between the various types of interactions can lead to the effect of frustration, well known in the concrete case of 'spin glasses'. It may be generalized for a rolling landscape under-peopled by inhabitants representing thus a very complicated free-energy terrain showing the locations of the global maxims (peaks) and minims (valleys). How easily the occupants of neighboring valleys can communicate will depend on the 'towardness' of environment (e.g., temperature excitations or tolerability conditions needed to facilitate flipping across the separating barriers). It is clear that the nature is rich even when accounting insensible inorganic world.

Certainly, there are works concentrating on a plausible cooperation between people as (genetically) unrelated individuals aimed to show their reciprocal strategies of behavior that can emerge spontaneously as a result of the same blind driving forces of survival [333,343-345]. This subject has been investigated using a branch of mathematics that today is called *game theory* [332,346]. It aspires to determine the strategies that individuals (or their whole groups – organizations) should adopt in their search for premium valuation or loss compensation for their doing when the outcome is uncertain and depends crucially on what strategies the other counter-partners may adopt. *Von Neumann* is the founding father of this matter, which can be traced back to 1928. It involves the evaluation of risks and benefits of all strategies in games (as well as in realistic actions of wars, economics, endurance) or whatever we can account as describable in its ongoing competitive advancement. It is known to have important impact not only on economics but also on mathematics, which is enriched through the advances simultaneously made in combinatory (i.e., the theory of arrangements of sets of objects) [346]. The drawn-in principles are found applicable to human societies in better understanding how cooperation emerges within the world of egocentric self-seekers, whatever they are politicians or private individuals, superpowers or small societies, where there is no strictly applied central authority to rigorously police their actions[347]. Later in 1970s it was found functional in the sphere of biology [345] for understanding the group behavior and so-called *reciprocity cooperation*, which is concerned with symbiosis, often accounted even between different species.

The most instructive and widely exercised computer model (by mathematicians, social scientists or biologists) is a simple game called 'Prisoner's Dilemma' [332]. The idea of this approach is to simulate the conflicts that exist in a real situation between the selfish desire of each player to pursue the 'winner-takes-all' philosophy, and the necessity for cooperation and compromise to advance that selfsame need. It is similar to the previously mentioned problem of 'spin glasses' looking for the determination of the lowest energy state or, in general, the optimal solution for the traveling salesman's problem to find the most favorable scheme that must be solved in the presence of conflicting constrains. The two counter-partner actors can choose to cooperate with each other or not. If yes, they receive some bonus, if only one does ('bum- sucker') he receives nothing while the other ('defector') gets a reward though the bonus is smaller. If both defect, each gets a minor reward. Even though each player inevitably gains if both cooperate, there is always a temptation to defect, both to maximize the profit and avoid being betrayed.

The dilemma 'of what strategy is better' often results in the so-called 'tit-for-tat' ('blow-for-blow') policy, i.e., the signalization of willingness to cooperate at first and then retaliating to whenever the opponent defects. It includes the property of forgiveness, which is perpetually making available the opportunity of re-establishing the trust between the opponents; if the counter-partner is conciliatory, it forgives, and both reap the greater reward of cooperation. In conclusion, however, it does not seem too clever as any highly complex strategy seems to be incomprehensible; if one appears irresponsible, the other's adversary has no incentive to cooperate. Let us mention the biological context where one can interpret the reward and penalty offered in the game in terms of the ability of individuals of different species to survive through reproduction – the reward is simply the numbers of offspring produced during the breeding season. In terms of ecosystem, this strategy of retaliation in kind is vigorous and does well when playing a wide variety of other tactics. Though no strategy is evolutionarily stable, it turns out that 'tit-for-tat' cannot be invaded and displaced by all-out defectors in a long-term relationship. The detection that retaliation is a common strategy sends an optimistic message to all those who fear that human nature is founded on greed and self-interest alone, hopefully, nice people do not have to finish last.

Nevertheless, it is worth mentioning that evolutionary games generate irregular or better the regularly shifting mosaic, where strategies of cooperation and defection are both maintained. The outcome is milling chaos, with clusters expanding, colliding and fragmenting with both preeminent and nastiest persisting. The term of punctuated equilibrium is also mentioned saying that the evolutionary changes occurred usually in shorter bursts separated by longer periods of stability. Catalyzing any replication of a molecule is a form of giving a self-support, with each link feeding back on itself, which would be the earliest instance of mutual aid, and in this sense, cooperation could be older than life itself. Remarkably enough, the genetic algorithm use to evolve from a random initial point and became a dominant member whose rule base was just as successful as the 'tit-for-tat' strategy that won the tournament. Indeed, most of the rules are shared, but some are better. One is the kind of sexual reproduction, in which the chromosomes of two parent solutions are recombined. It means that the population still evolved towards rules that performed about as well as 'tit-for-tat', but they were only half as likely as in the sexual case to produce rules substantially better than that for retaliation. Besides the importance of sex, the strategic environment itself was changing and was taken to be the evolving population of chromosomes moving away from cooperation and subsequently reversing back again. Eventually, the reciprocators spread to dominate the entire population. The computer simulations showed that sex helped a population to explore the multidimensional space and so find the gene combinations with the greatest fitness; that there became a trade-off in evolution between the gains to be had from flexibility (advantageous for a long run) and the specialization (for immediate survival).

Sznajd's model is also intermittently cited [248] explaining certain features of opinion dynamics based on the saying "united we stand, divided we fall". The individuals involved in this contest are facing the choice between two opinions (parties, manners, products, etc.) and in each updated step of decision the pair of neighbors who share common opinion try to persuade their neighbors to join them to share the opinion. Therefore a definite modification from the standard models (Ising in physics or voter in politics) is achieved so that information does not factually flow from the neighborhood to the selected species (spins, voters) but conversely, from the selected cluster (of the same behavior) to its neighborhood. The results do not depend much on the spatial dimensionality or on the kind of chosen neighborhood providing thus no chance to escape from the homogeneous state , which thus serves as an attractor of (zero-temperature-like) dynamics. Solving the underlying diffusion equations an interesting analytical solution is obtainable for several dynamical properties including the inherent phase transition [349].

It is anticipated that economics straddles the divide between science and the humanities. The world's economies possess nonlinear features characteristic of complex dynamical systems, although the marketplace is rather associated with a form of financial survival of the fittest. There are objective measures of economical and financial success, whether of nations of companies, such as gross product, budget deficit, market share, revenues or stock prices, however, yet many factors may or will be seen ill defined. For decades, the central dogma of economics revolved around decayed equilibrium principles in a manner entirely analogous to the application of equilibrium thermodynamics. A recent Nobel Prize was awarded in the sphere of establishing game theory as a powerful tool with applications ranging from the industrial economics to international trade and monetary policy. It may be simply illustrated on the strategy, which a monopolizing company must consider in a bid to prevent a would-be competitor from encroaching on its market. Engagement in a price war may inflict losses on rivals and persuasion of a more welcoming strategy (cartel) may assure profits for both companies, but often agreeing illegitimately high prices. It may also help answering the trade-off dilemma between cheap, high-polluting transport and expensive green alternatives.

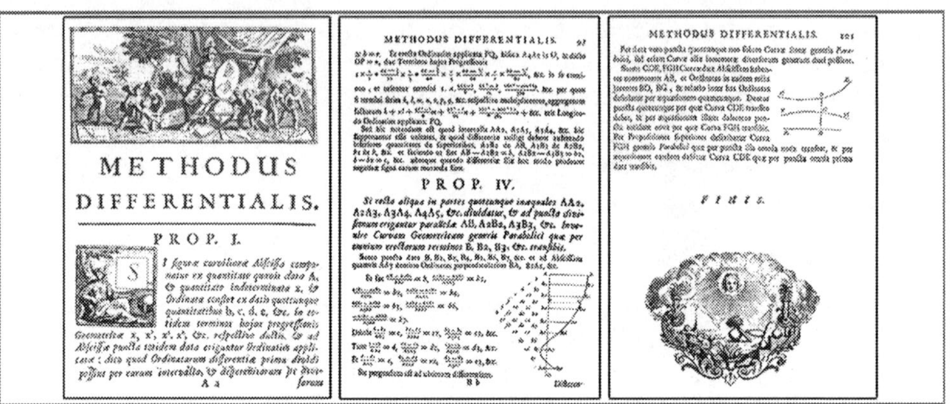

Facsimile of the *Newton* book "Methodus Differentialis" (1675) and the inherent subject of interpolation by means of formulas of finite differences, which was highly recognized by the members of the 8th International Congress of Actuaries (1927), whose members owe so much in their daily work on the subject of interpolation. It gave one of the earliest traces paid to the subject application in economics [e.g., D.C. Fraser, J. Inst. Actuaries, *lviii* (1919) 53

Certainly, we can draw yet other analogies between the crisis points associated with self-organization and chaos that occur in inanimate processes, from ordered chemical reactions through certain phenomena that arises within human societies (revolutions) and behavior of people. It is worth mentioning that there always remains an intriguing question concerning the distinction between computer simulation and reality. In many cases, a simulation of a physical or chemical process on a computer could not be confused with the process itself, i.e., a computer simulation of self-organized chemical waves is not the reaction in nature. For the scientists, who are deciphering the wide-ranging complexity, it places renewed emphasis on interdisciplinary research in the over and done Renaissance style, and helps to underline the symbiosis between the science, technology, economy and humanities.

8.5. Openness, interface and useful work as exergy in ecosystems

Other fascinating world of application is *bionetwork*. Any such ecosystem structure, presenting the transfer of energy from one species to another trapped in food, linked together by resource-versus-consumer (or better prey-predator) type relations [77], is known as the *tropic chain*. Such tropic level is considered as a horizontally structured biological community of rival populations (better competing community). The number of species varies in a wide interval, from single to hundreds, but they are all connected by a relation of competition and are typical dissipative processes producing entropy, which is exported from the system into its environment. It is interesting that possible interactions between two species all become more beneficial for both components due to the presence of the network that ensures constant cycling of matter, energy and information (symbiosis). The required export of entropy can be provided by either of processes, heat transfer, matter exchange or state transformation within the system. Phenomenological theory of fission of living cell can be a good illustration following somehow the philosophy of nucleation dealt with in solid-state physics as shown in the preceding Chapter.

We can assume that the entropy production is here proportional to the cell volume, and the entropy outflow is then comparative to its surface [350-352] with the proportionality coefficients, a_1 and a_2. The cells grows until the stationary state at $dS/dt=0$ with the cell radius r equal to the ratio of coefficients $(3a_2/2a_1)$. If the equilibrium radius r_{eq} becomes smaller than the actual radius r, the internal entropy production does not keeps compensating its outflow and the cell has to die. However, if the cell is divided, then the volume is conserved but the surface increases. As a result, dS/dt become negative and the entropy change ΔS (equal to $a_1 4\pi r^3/3$) provides that the radius of two new cells will be equal to $r'= r/(^3\sqrt{2})$. The entropy out-flow for these cells is $\Delta S = 8a_2\pi r^2/(^4\sqrt{4})$ and at $r \rightarrow 3a_2/a_1$ the negative increment of the total entropy is $36\pi(a_2/a_1)(1-\sqrt{2}) \approx -29.4(a_2^3/a_1^2)$. Certainly, such an illustration will become more complex for organisms which are more multifaceted and complicated but generally would display that the entropy change is determined by a combination of negative (differentiation) and positive (growth) terms, and that a simple monotonous dependence of the entropy change is not expectable.

Assuming the previously used correlation of the volume-to-surface entropy flux $dS/dt \cong 4a_1\pi r^3/3 - 4a_2\pi r^2$, we can rewrite it in a more general form of the proportionality of entropy on the biomass density ρ and the organism volume V through $S \cong \rho V$. If the volume V and surface area A are determinedly interconnected and if the $(1/V)dS/dt$ is the specific entropy production, we can arrive to a simply-connected geometric figures with central symmetry, where volume and surface area are determined by the single characteristic scale-size, $m=2/3$ because $A = f(V^m)$. By denoting a new variable $z = V/V_\infty$ we can write an asymptotic analogue for dV/dt in the form of $dz/dt \approx z\{1 - 1(z^{1-m})\}$, which reveals that the derivative of z is negative within the interval $<0,1>$ and which implies that the equilibrium with $z - 1$ is unstable. This means that the growth process do not tend to reach the final stationary state, where the constant rate of entropy production will be observable due to the interruption by the organism death showing the impossibility of an unlimited life.

If the energy content in an organism is $E \cong \rho V$ we can assume that the energy input is proportional to both the surface area of the organism and the metabolism, which is proportional to its volume (biomass) so that we may anticipate $dV/dt \approx V^m \, (const - rV^{1-m})$. For a stationary level of basic metabolism, we can estimate equilibrium at $V_\infty \cong {}^{(1-m)}\sqrt{(const/r_\infty)}$, which is stable. From this we can derive a general law of the allometric relationship (cf. paragraph 2.4.) $r_\infty \cong W^{m-1}$ where r_∞ may, for instance, the intensity of oxygen uptake per unit biomass and W is the total mass (or weight) of the organism, or their size versus generation time. Beside the respiration $(m-1)=0.8$ analogous allometric relations are feed consumption (0.65) or ammonian excretion (0.72), which has been generalized by *Odum* [353,354] and reexplained by *Jørgensen* [77,351,355].

We have illustrated that that the surface area of a species is a fundamental property as it quantitatively determines the interactive boundary with the environment. It is typically a heat exchange (lost), which would be a limiting factor even for any artificial network system, such an integrated circuit, which integration size will be determined by its large enough surface yet capable to keep the unit at the working (within limits – not overheated) temperature.

It is obvious to seek a possibility to include species' variation in the form of an additional form-factor. Allometric principles are legitimate on various hierarchical levels and the process rates, determining thus the need for the energy supply. It is worth noting that the so called degree of *openness* plays a significant role in the establishment of all necessary relations [77], which, again, maintains the critical surface-to-volume proportionality. Its consequence can be illustrated in the relationship between the numbers of species, NS of an ecosystem on islands they reside, with the area a, i.e., $NS \cong A^m$. The perimeter relative to the area of an island determines how exposed (open) the island is to immigration or to dissipative emigration from the neighborhood (other islands or adjacent continent There is a close relationship between energy flow rates and organism size, which is denoted by this allometric principle [58,356] discussed previously in paragraph 2.4.

Trends to apply the various extreme (teleological) principles in science have a sufficiently long history (mechanics, physics). In ecology, however, it has started only recently trying to adjust the population and ecosystems dynamics in the form of extreme principles as a consequence of the perennial teleleologicity of our thinking. It is obvious that in the process of its own evolution the ecosystem tends to increase the energy through-flow reaching thus a maximum of steadiness (certainly yet comprising all different constrains). The hypothesis of *Lotka* [358] went even so far as to suggest calling this statement the '*Fourth Law of Thermodynamics*' being close to the concept of *Jørgensen's* exenergy [77,351] and *Kauffmann* investigational thoughts [43]. It follows the sense of useful energy, which was made easy available by *Morowitz* [358] who expanded this bonus-law to the states, which ordered structure among possible ones survive selection. However, we are still cautious about the concept of usefulness because the import energy and the export of heat in ecosystems are non-spontaneous process, which are realized by means of some special impellent or better pump. If more useful energy is available, the system is moved further away from equilibrium as reflected in growth of gradients. If more than one pathway to depart from equilibrium is offered, the one yielding the most work under prevailing conditions and ultimately moving the system the farthest from equilibrium providing the most ordered structure, tends to be selected. Such active ecosystems with their own internal pumps possess a high degree of internal organization, i.e., exceedingly complex structure. Some share of the useful work must be used for the creation of this vigorous structure and the maintenance of its functioning as well as for the growth of biomass and upholding of metabolic and reproductive processes. It is adjacent to the original idea of *Odum's* formulation of maximum power principle [353,354].

One of the founders of modern thermodynamics, *Ostwald* [359] found the innovative term of entropy objectionable and tried to stay away from its using both as a word and a concept. He challenged to replace the latter by the concept of work assuming two sorts of work, namely work done by the system on its environment and work done by the environment on the system embedded within it. In view of that, *Jørgensen* [360] introduced a new term *exergy* defined as the amount of work (=entropy-free energy), which a system can perform when it is brought into thermodynamic equilibrium with its environment, We should remind that exergy is not a state variable and is not conserved unless entropy-free energy is transferred, which implies that the transfer is reversible. All processes in reality are, however, irreversible, which means that exergy is lost (and entropy produced). Loss of exergy and the production of entropy are two different descriptions of the same reality, namely that all real processes are irreversible and we always have some loss of energy forms, which can do work (exergy). The energy forms, which cannot do work are called anergy. So the formulation of the Second Law when using exergy may be altered [77] as, citing *'... all real processes are irreversible, which implies that exergy is inevitably lost...'* while energy is, needless to say, conserved by all processes according to the First Law. Therefore, it is of interest for all environmental systems to set up and exergy balance together with and energy balance keeping in mind that a first-class energy capable to do work is lost and replaced by *second-class* of energy (as heat at the temperature of environment), which cannot do work. Therefore, the energy can be represented as a sum of two items, exergy + anergy, and in accordance with the Second Law, anergy is always positive. Any ecosystem (due to the through-flow of energy) have the tendency to move away from thermodynamic equilibrium gaining exergy $dE_{ex}/dt \geq 0$ (and information) so that we can put forward a proposition of the ecosystem relevance: *"...ecosystem attempts to continuously develop towards a higher level of exergy and during its process of evolution towards its climax state with maximum of its own exergy...".*

The exergy-storage hypothesis might be taken as a generalized version of LeChatelier's principle: when the energy is inserted into a reaction system it forceshifting its equilibrium composition in a way to counteract the input change. So that when habitually applied to biomass synthesis it is understood as a chemical reaction, which starts with energy and nutrients ending with exergy and organization along with dissipated energy as dealt with under various alterations elsewhere [43,77,360-364].

Nature and man, laws and feelings,
how to make it feasible?

Chapter 9

9. THERMAL PHYSICS OF PROCESS DYNAMICS

9.1. Phase transitions and their order

The equality equation given in the preceding Chapter 7 can be used for the description of *phase transition*, i.e., the point where the potential, Φ, have the same values on both sides of the boundary between the phases *1* and *2* in question. Such a system must not only be in thermal, concentration and mechanical equilibrium, but must also meet further conditions of electromagnetic equilibrium. In other words, a continuous change in temperature must be accompanied by continuous changes in the other intensive parameters, *I*. However, the conditions for such a thermodynamic equilibrium place no further limitations on changes in the derivatives of Φ with respect to *I*, which can have various values in different phases. These phases must differ at least in one value of a certain physical quantity characterizing some property such as the volume, concentration, magnetization or also the specific heat, magnetic susceptibility, etc.

The discontinuity in the first derivatives of function Φ thus appears as the most suitable for an idealized classification of phase transitions [3,297,365]. The characteristic value of a variable, at which a phase transition occurs, is termed the *phase transition point* (T_{eq}, I_{eq}). The changes in the derivatives can be then expressed according to *Ehrenfest* classification and give the limit for the *first-order phase transitions*

$$\Delta(T) = (\partial\Phi^1/\partial T) - (\partial\Phi^2/\partial T) \neq 0 \Rightarrow \Delta S\,(\Delta H) \neq 0,\ \text{at}\ T_{eq},$$
$$\Delta(I) = (\partial\Phi^1/\partial I) - (\partial\Phi^2/\partial I) \neq 0 \Rightarrow \Delta X \neq 0,\ \text{at}\ I_{eq}.$$

From the viewpoint of thermoanalytical measurements in which the temperature is the principal parameter, a *certain amount of heat is absorbed and/or liberated* during the transition. For T_{eq} it holds that $\Delta H - T\Delta S = 0$, where ΔH is the latent heat of the phase transition, which must be supplied for one mole of a substance to pass from the phase *1* to phase *2*. In the endothermic process, $\Delta H > 0$, and the heat is absorbed on passage *1→2* while in the exothermic process, $\Delta H < 0$, and the heat is liberated. The set of equations specified in the previous Table 6.I. then enables monitoring of the effect of a change in the external field parameters (T and I) of the process, on the basis of a step change in measurable extensive quantities of the material under study.

Another limiting case is attained when $\Delta(T)$ and $\Delta(I)$ equal zero, i.e., such a phase transition which is not accompanied by a step change in the enthalpy or the other extensive quantity and which is called a *second-order phase transition*.

These phase transitions are classified on the basis of discontinuity of the second derivatives of the general thermodynamic potential, or

$$\Delta(T,T) = (\partial^2\Phi^1/\partial T^2) - (\partial^2\Phi^2/\partial T^2) \neq 0 \Rightarrow \Delta C_p \neq 0 \text{ at } T_{eq} \text{ where } \Delta S = 0.$$

The same consideration holds for $\Delta(T,I)$ and $\Delta(I,I)$ (and the value of an uncertain expression of the 0/0 type is then found by using the *l'Hospital* rule indicated in the preceding Table 6.I.). Consequently, the higher-order phase transitions can be predicted, i.e., a third-order phase transition, where $\Delta(I,I)$'s are zero and only the third derivatives differ from zero – however, experimental evidence of such processes is yet absent.

For thermal analysis experiments it is important to compare the main distinctiveness of these first- and second-order phase transitions. With a first-order phase transition, each of the thermodynamic potentials can exist on both sides of the boundary, i.e., the functions Φ^1 can be extrapolated into the region of Φ^2 and vice versa, denoting thus a *metastable state* of the phase (*1*) above T_{eq} and vice versa for (*2*) below T_{eq}. From this it follows that the phase Φ^1 can be *overheated* at $T>T_{eq}$ (less common) and phase *2 undercooled* at $T<T_{eq}$ (quite frequent).

On the contrary, the feature similar to overheating and/or to undercooling is *impossible* with the second-order phase transitions because each phase, *1* and *2*, is *restricted* to exist only on its own side of the phase boundary. This is caused by the fact that there is a break in the first derivatives of the entropy but the entropy itself remains continuous ($S^1 \Rightarrow S^2$). Therefore the function Φ has a certain singularity causing different behavior on the two sides of the phase transition.

Landau [366] has shown that a general theory can be developed using certain assumptions without a more detailed knowledge of the singularity of Φ. For a generalized phase transition, the function Φ can be written as the sum of contributions supplied by the 1st- and 2nd-order phase transitions, multiplied by appropriate distribution coefficient whose sum equals unity. In view of the non-ideality, the degree of the closeness to a first-order phase transition is denoted as *p*, see Fig. 44., showing the limits, *p*=1 and *p*=0, and the intermediate cases.

$p = 1$	$\Delta(T) \neq 0,\; \Delta(T,T) = 0$	ideal 1st order
$p \leq 1$	$\Delta(T) \;>>\; \Delta(T,T)$	real 1st order
$p \cong 0.5$	$\Delta(T) \;\cong\; \Delta(T,T)$	broadened or lambda PT
$p \geq 0$	$\Delta(T) \;<<\; \Delta(T,T)$	real 2nd order
$p = 0$	$\Delta(T) = 0,\; \Delta(T,T) \neq 0$	ideal 2nd order

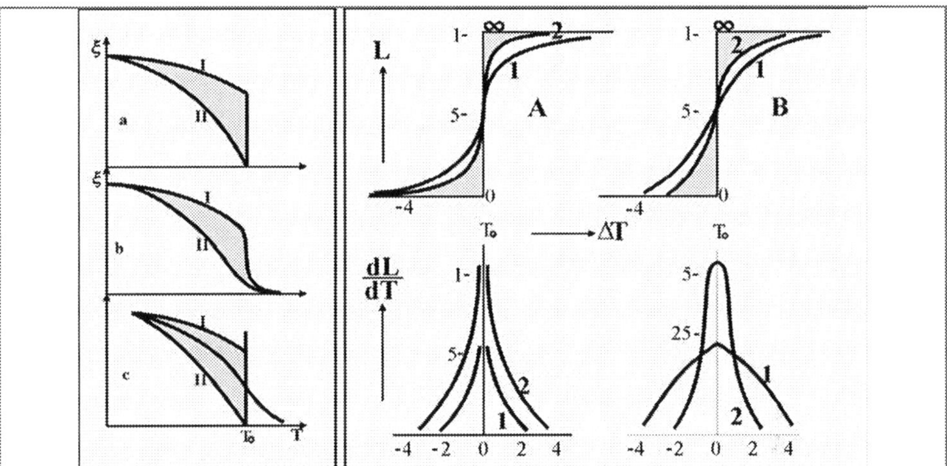

Fig. 44. – Left: The dependence of the degree of disorder, ζ, on temperature, T, for (a) ideal limiting phase transitions of the Ist and IInd order, (b) diffuse phase transition of the Ist and IInd order and (c) an approximate function – the degree of assignment of $Z(I)$ ands $Z(II)$. Right: the approximation of phase transitions using two stepwise *Dirac* δ-function designated as L and their thermal change dL/dT. (A) Anomalous phase transitions with a stepwise change at the transformation point, T_o (where the exponent factor n equals unity and the multiplication constant has values -1, -2 and ∞). (B) Diffuse phase transition with a continuous change at the point, T_o, with the same multiplication constant but having the exponent factor 1/3.

For a mathematical description of generalized course of phase transitions a complementary parameter is introduced [3,366] and called the *ordering transition parameter,* ξ , which can reflect a regular distribution of atoms, spontaneous magnetization or polarization, etc. For the initial parent phase, *1,* this parameter is assumed to be zero ($\xi = 0$ for the ordered phase) and for the resultant phase, *2,* nonzero ($\xi \neq 0$ for the disordered phase). It is obvious that ξ changes stepwise for the first-order transition and is continuous for the second-order or broadened phase transitions. The main significance of the *Landau* theory lies in the fact that the principal physical quantities can be expressed as functions of the instantaneous value of ξ. This approach can also be taken to express the general thermodynamic potential, Φ, which can be expanded in a series using the integral powers of ξ. From the condition of equilibrium it holds that $\xi^2 \cong a_o \, (T_o - T)/\, 2b_o$ where the constant a_o and b_o are the coefficients of the expansion. It gives further possibility to approximate the entropy, $S = a_o^2 \, (T_o - T)/\, 2b_o$, as well as the heat capacity, $C_p = = a_o^2 \, T_o \, / \, 2b_o$ .Therefore, the transition from the ordered phase to the disordered phase ($\xi=0 \rightarrow \xi\neq0$) is always accompanied by a finite increase in the heat capacity.

Let us now consider the behavior of an ideal first-order transition exposed to a thermoanalytical experiment, in which the temperature of the system continuously increases and when the kinetics of the process does not permit the phase transition to occur infinitely rapidly, i.e., at a single point of the discrete equilibrium temperature. Therefore it is often important to find the temperature shift of the ΔH value including the possible effect of a complementary intensive parameter, I, which could act in parallel with T, i.e. the dependence of $\Delta H = \Delta H \{I, \Delta S (I, T)\}$ type. Then $d\Delta H/dT = \Delta C_X + \Delta H/T - d \ln \Delta X/dT$. ΔH and ΔX can be replaced by corresponding differences in ΔV or ΔM. It is worth mentioning that if only one of the externally controlled parameters become uncontrolled, the transition process become undefined. This is especially important for the so-called 'self-generating' conditions, during the work at a non-constant pressure (i.e., under static atmosphere during decomposition reactions carried out at a sealed chamber) or which can happen under the other hesitating force fields.

9.2. Broadened phase transitions

In an experimental study of phase transitions, the equilibrium dependence of measured quantity, $Z(T)$, on temperature, which is representing the given system can be organized as follows [3,365]:

a) The phase transition point is simply situated in the inflection point of an experimental curve

b) The magnitude of phase transition broadening occurs as the difference of the temperatures corresponding to the beginning (onset) and the end (outset).

c) The degree of closeness of the experimental curve to the I^{st}- and IInd-order theoretical curve can be expressed by a suitable analytical formula.

The viewpoint of curves closeness between limiting cases of $p_I(T)$ and $p_{II}(T)$ can be approximated by $p_I(T) + p_{II}(T) = \{Z(T) - Z_{II}(T)\}/\{Z_I(T) - Z_{II}(T)\} - \{Z_I(T) - Z(T)\}/\{Z_I(T) - Z_{II}(T)\} \cong 1$, while the individual experimental curve can be approximated by $Z(T) = p_I(T) Z_I(T) - [1 - p_I(T)] Z_{II}(T)$ or by means of an *arctan* function. Of practical merit is the approximation of the C_p values as a power function of the temperature difference from the beginning, T_o, e.g., $C_p(T) = a_o + a_1 (T - T_o)^{a3}$. There is frequent use of the so-called *Dirac* δ-function $L(T) = {}_o\!\int \delta (T - T_o) dT$, see Fig. 44., so that a general thermodynamic potential of a broadened phase transition of the I^{st} order can be expressed as $\Phi(T) = \Phi_I(T) + \Delta\Phi(T) L(T)$, where $\Phi_I(T)$ is the normal part of the potential corresponding to the initial phase, *1*, and $\Delta\Phi(T)$ is the anomalous part connected with character of phase transition. If the analytical function of $L(T)$ is known, the corresponding thermodynamic quantities can also be determined, e.g., regarding entropy it holds that $S = - \partial\Phi/\partial T = S_I + \Delta S L(T) + \Delta\Phi(T_o) \{\partial L(T)/\partial T\} = S_I + S_{anom}$ with analogy for $C_p = C_{pI} + Cp_{anom}$. Using the *Clapeyron* equation (cf. Table 6.I) we can find anomalous parts of entropy and heat capacity on the bases of volume as $(\Delta S/\Delta V) V_{anom}$ or compressibility as $(\Delta C_p/\Delta\alpha_V) \alpha_{V\ anom}$. For the functional determination of $L(T)$ the *Pippard* relation is often used [367] in the general form, $L(T) = \{1 + exp\ [a_o(T - T_o)^{a1}]\}^{-1}$, often modified for a broadened phase transformation as $L(T) = \{1 + exp\ [3.52\ [(T - T_o)/(T_{init} - T_{fin})]\ ^{a1}]\}^{1/2}$, with the exponent $a_1 < 1$. For non-symmetrical anomalous phase transitions the asymmetry is included by adding the temperature differences in assorted powers, a_1 and a_1+1, i.e., $L(T) = \{1 + exp\ [a_o(T - T_o)^{a1} + b_o(T - T_o)^{a1+1}]\}^{-1}$.

For the anomalous case of non-equilibrium glass transition we can predict the behavior of the heat capacity changes according to the above models, i.e., stepwise, diffuses and linearly decaying and lambda shaped. Introducing the concept of normalized parameters in the form conveniently reduced by the temperature of melting, e.g., $T_{gr} = T_g/T_{melt}$, $T_{or} = T_o/T_{melt}$, $T_r = T/T_{melt}$ or even $\Delta C_{pr} = \Delta C_p/\Delta C_{melt}$, we can assume a certain limiting temperature where the difference between the thermal capacities of solid and liquid/melt vanishes. Using $T_{gr} = 2/3$ and applying some simplified manipulations [368,369], such as $C_p^{liq} - C_p^{sol} = \Delta C_p^{liq\text{-}sol} \approx constant$, we can assume that $\Delta C_p^{liq\text{-}sol}/\Delta S_{melt} \cong \Delta C_{po} \cong$ 1.645 at $T_r > T_{or}$ and zero at $T_r = T_{or}$ $(= (2/3)^{2/3} = 0.54)$. Using molar quantities, $\Delta S_r = \Delta S^{liq\text{-}sol}/\Delta S_{melt} \cong 1 + C_{po}\ ln\ T_r$, possibly to be approximated at $T_r > T_{or}$ by a constant value of 1/3 as the entropy difference frozen-in due to the liquid vitrification. It provides the related values of $\Delta H \cong 1 - C_{po}\ (1 - T_{gr}) \cong 0.452$ and of $H_o \cong 1 - C_{po}\ (1 - T_{or}) \cong 0.243$. It can be extended to the limiting case of T_g of 1/2 and to a wider range of ΔC_{pr}'s (down to 0.72) to calculate not only approximate values of H's but also μ's [6,306].

9.3. Equilibrium background and the kinetic degree of a phase transition

In practice, a transition under experimental study is defined as the development of states of a system in time, t, and at temperature, T, i.e., a time sequence of the states from the initial state (at starting time, $t = t_o$) to the closing state (at final time, $t - t_F$, corresponding to the temperature T_F). It is associated with the traditional notion of *degree of conversion*, α, normalized from 0 to 1, i.e., from the initial stage at the mole fraction $N = N_o$ to its final stage at $N = N_F$. Then it holds that $\alpha = (N - N_o)/(N_F - N_o)$ or, alternatively, as $(Z - Z_o)/(Z_F - Z_o)$, if we substitute the thermodynamic parameter, N, by the *experimentally determined*, Z, which is a particular measured value of the observed property. This expression is appropriate for strict *isothermal measurements* because for non-isothermal measurements we should introduce a more widespread expression [370], called the *non-isothermal degree of conversion*, λ, related to the ultimate value of temperature, where $T_F \rightarrow T_{F\infty}$. It is defined as [3,370,371] $\lambda = (N - N_o)/(N_{F\infty} - N_o)$, or $- (Z - Z_o)/(Z_{F\infty} - Z_o)$, where the value at infinity (indexed by $F\infty$,e.g. $Z_{F\infty}$) expresses the maximum value of Z that can be reached at the concluding temperature, $T_{F\infty}$. Introducing the previously defined *equilibrium advancement of transformation*, λ_{eq}, discussed in Chapter 7, we can

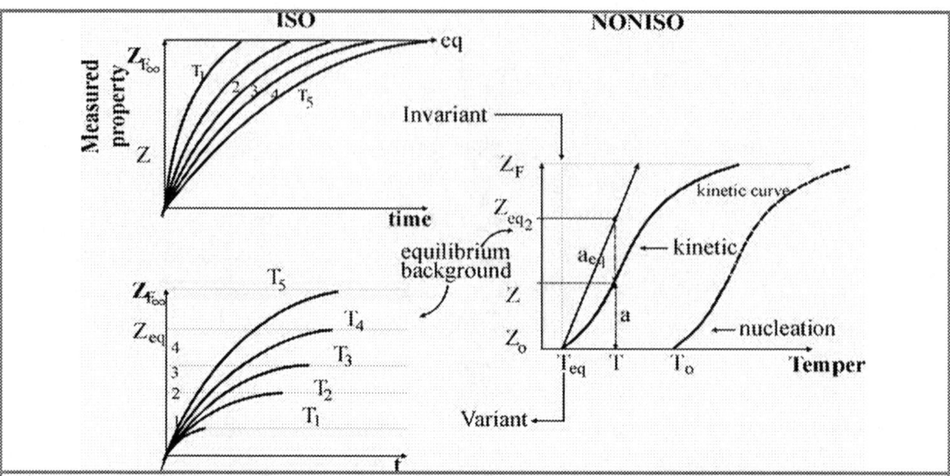

Fig. 45. – Detailed specification of possible variants of the degree of reaction progress based on the experimentally measured property, Z, which can reach the instantaneous value $Z_{F\infty}$ or the intermediate value, Z_{eq}. We can distinguish two limiting courses of invariant (sudden change) and variant (continuous change) processes, their equilibrium background and actually measured (kinetic) response.

see that there is a direct interrelation between α and λ , i.e., through the direct proportionality $\lambda = \alpha \, \lambda_{eq}$.

It follows that the isothermal degree of conversion equals the non-isothermal degree of conversion only for invariant processes, where $\lambda_{eq} = 0$. The difference of these two degrees of conversion is well revealed in Fig. 45., where we easily observe that in the isothermal measurement the system reaches a final, but incomplete value of Z (= Z_{FT}), which corresponds to the given temperature, T_F , whereas on continuous increase of temperature the system always attains the ultimately finishing value of measured property, $Z_{F\infty}$, corresponding to the definitive completion of reaction (Fig. 45.). Practical application of λ_{eq} is shown in the functional case of non-stoichiometry interactions of Mn_3O_4 (see Fig. 46.).

This description is closely related to the preceding definition of the *equilibrium advancement of transformation,* λ_{eq}, which, however, is not exhaustive as more complex processes can be accounted for by multi-parameter transitions, such as segregation in solid-state solutions or solid-spinodal decompositions. Here we ought to include new parameters such as the *degree of separation,* $\beta = (N_M - N_N)/N_T$, together with the complex *degree of completeness of the process,* $\zeta \{= (N_T - N_M)/(N_N - N_M)\}$, based on the *parallel formation* of two new phases, whose graphical representation leads to four-dimensional space [3,371], see Fig. 47. The degree of conversion of such a process is then characterized by three measurable parameters, i.e., by the composition of the initial phase, N_T , and the two newly formed phases, N_M and N_N. The composition of newly formed phase can be then expressed as $N_M = N_T$ *(1 − $\zeta \beta$)* or $N_N = N_T$ *(1 − $\beta − \zeta \beta$)* and the overall equilibrium degree of conversion of the process, λ_{eq} , is then represented by a spatial curve in the coordinates of ζ , β and T, pretentious that the degree of separation is normalized to vary from zero through a maximum (at one) back to zero. Such a complex course is called the *multiparametric* process and can become useful in the description of special cases of diffusionless mechanism of martensitic transformation [3,9,306,371]. If two such simultaneous processes have the identical initial states, the resultant curve can be given by a superposition of two partial complex curves, which may require another complication, newly defined degree of separation, $\delta = (N_T - N_M)/(N_N - N_M)/(N_N - N_M)^2$ making experimental evidence almost impractical.

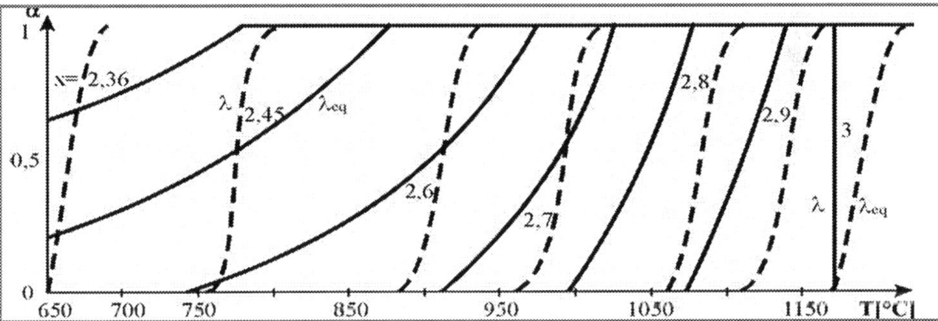

Fig. 46. – Practical cases of the course of equilibrium background, λ_{eq} , and actual kinetic degree of transformation, λ (dashed), for the variant processes taking place during the tetragonal-cubic transformation of manganese spinels of the composition $Mn_xFe_{3-x}O_4$ for the various levels of manganese content, see the inserted values, x, in the region from 2.3 to 3.0.

9.4. Aspects of invariant and variant processes

The passageway of a system through the phase diagram (cf. Fig. 46.) is best defined at the *invariant points*, where the system has no degree of freedom and where an ideal phase transition should occur at a single temperature [3]. This, however, is real only with completely pure compounds or for systems with well-defined eutectic compositions. Taking a hypothetical phase diagram, which represents the dependence of the composition on temperature, the principal kinds of *equilibrium passage* through the phase boundaries can be distinguished on basis of the so called *equilibrium advancement of transformation*, λ_{eq} [370]. Using the relation for Z, as the *particularly measured property* (such as concentration, volume, mass, magnetization, etc.), we can normalize the equilibrium advancement within the standard interval from 0 to 1, i.e., $\lambda_{eq} = (Z_{FT} - Z_o)/(Z_{F\infty} - Z_o)$, where Z_o , Z_{FT} and $Z_{F\infty}$ are respectively the limiting values of measured property at the final (index, F) and initial state (o), instantaneous (T) observed for given temperature, T, as well as its ultimate value function of temperature alone, as it actually expresses the shift of equilibrium with changing temperature. It is attained only if we assume a completed equilibration at any

258

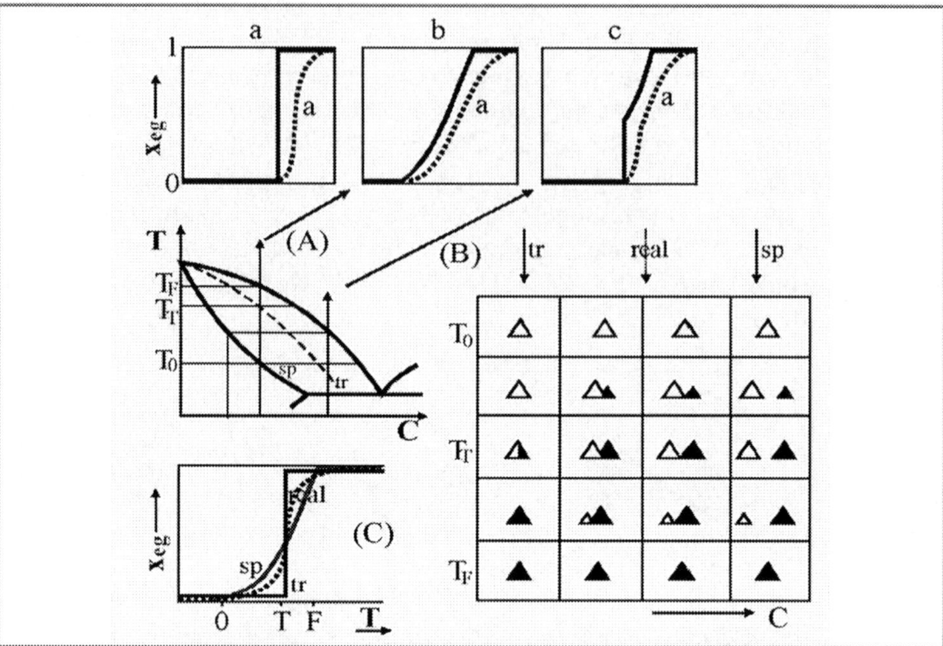

Fig. 47. – Illustration of a hypothetical multiparameter (mixed) process obtained by assuming a competition of following reactions: an invariant phase transformation (tr) and variant phase separation (sp) [1,3,371]. Upper row shows the equilibrium background (solid line) and kinetic curve (dashed line, where a denotes the normalized progress of transformation) for classical cases of invariant (vertical), variant (sloped) and shared (combined) reactions connected in this diagram through arrows to the actual compositional cuts in a binary phase diagram [370]. While (A) gives characteristics of transformation in the standard interval T_o to T_F, (B) is a yet uncommon schematic representation of the possible sequences of the system states, where the open triangles denote the initial (disappearing) phase and the black triangles the resultant (appearing) phase. The left column in the right block shows the normal case of an authentic transformation while the right column stands for the extreme linked case of variant phase separation. Bottom curves correspond to the equilibrium (tr and sp) and real degrees of completeness of the overall process. Courtesy of *Pavel Holba*, Praha, Czechia

given moment (ideally assured by infinitely slow heating). According to the shape of the $\lambda_{eq} = \lambda_{eq} (T)$ dependence, we can expect three types of dependence exemplified in the following Fig. 47.

For the process with the stable initial states, we can distinguish *invariant* process, where λ_{eq} changes in sudden break from 0 to 1, while when the change proceeds gradually over a certain temperature interval, from T_o to T_F, we talk about the *variant* process of a continual character. If the transformation consists of both the stepwise (invariant) and continuous (variant) changes, we deal with the so-called *combined* process [370]. It is necessary to stress out again that these three types of changes are valid only under ideal accession to phase transition whereas under real experimental conditions a certain temperature distribution (broadening) is always presented due to the inherent temperature gradient or other types of fluctuations possibly involved (concentration). Its practical use is best illustrated on the hypothetical sections through a phase diagram, exemplified in Fig. 48. For pure components, eutectics and distectic compositions, it exhibits invariant character while for two-phase regions it shows invariant course, which accounts the coexistence of two components. Under peritectic regions its change exhibits a combined character. Certainly it is valid only for processes with a stable initial state, not accounting yet the highly non-equilibrated states of metastable and unstable disequilibria.

In some cases the situation gets mixed up by the insertion of simultaneous effects of *phase separation,* β, which becomes actual due to the diffusion-less or nucleation-free progression of a transition. It is seldom appreciated within the current phenomenological kinetic models because it is not only complicated mathematically but also particularly plagued by the experimental difficulty in the parallel determination of numerical data of both degrees involved (α and β, which is analyzed in details elsewhere [3,9,278]).

Certainly, a more complete and accurate description of a generalized transition would be associated with analyzing its microscopic measures, which solution associates strong non-linearity mathematically exploitable with respect

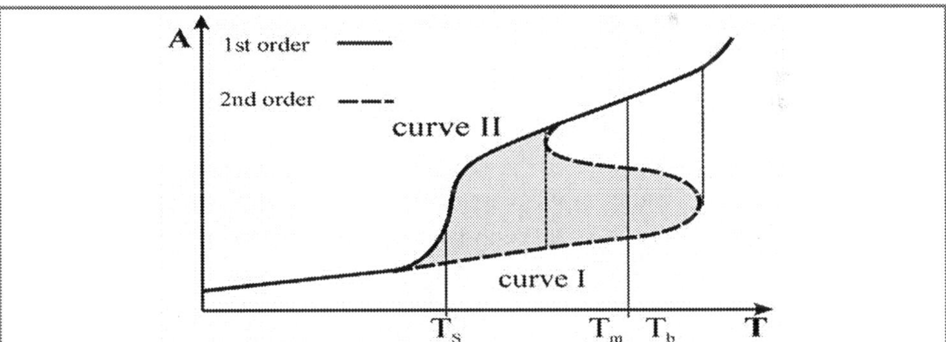

Fig. 48. – Schematic illustration of phase traditions [9,388] in the view of amplitude rise shown for a case of strongly damped non-linear oscillator (curve II) as the function of temperature. The second-order-like area for T_g is characterized by very high viscosity and any amplitude enlargements proceeds very slowly. Curve I displays the case of the first order transition (melting at T_m). This portrait is characteristic for the phenomenon known as 'hysteresis' when the value of the bifurcation parameter (cf. Fig. 9, here read as T) initially grows and afterward diminishes. In particular, if the system is in the stationary state (coupled with the lower branch) it stays in it even if T is increased (overheating), but at the moment of $T - T_{melt}$, the system suddenly bounds to upper branch. On contrary, when undercooled it jumps down from the upper branch (which is a more frequent phenomenon bearing a more extensive character due to the associated 'geometrical' effects of nucleation, cf. Chapter 7). We can come across this effect in various remote instances of agglomerated multi-particle systems such as physics of lasers, biological membranes; econophysics (cf. Fig. 40).

to an amplitude switch in localized spots, where the centrally inspected micro-particle is able to dislodge by pushing aside the neighboring particles creating thus in its vicinity a highly expansive spot. Such a created cavity or void can be responsible for the higher expansion coefficients identified for viscous media close to the transition. The entropy contribution connected with such a 'semi-evaporated' state, which is created inside the system in vacant regions, can be eventually contemplated as well, providing a wide-ranging relation for entropy. Because the vacant regions (thanks to the possibility of direct experimental observations) have a well-defined size (larger than the *Van der Waals* volume), which, however, is smaller than the critical volume of the particles involved. Thus we can estimate the change of enthalpy connected with the semi-evaporated state, which can be taken as a fraction of evaporation enthalpy, see the above figure. Though the solution of the associated non-linear equations is complex it can provide illustrative examples, which can be easily visualized by the transformation into two separate, first-order differential equations. This procedure is usually performed in the determination of chaos theories [59] as well as in studies of self-organized structures considered in non-equilibrium thermodynamics, and analyzed in more detail in [9,388] (cf. Fig. 48.).

9.5. Kinetic phase diagram

The theory of phase equilibrium and the development of experimental methods used for such investigations gave the impetus to the extensive physico-chemical program aimed at the compilation, tabulation and interpretation of phase diagrams of substances in practically all the fields of natural science and technology. It started by the construction of basic metallurgical phase diagram of the system Fe-C, whose first version was completed in the middle of nineteenth Century. The theoretical foundation of thermodynamic analysis of the metastable equilibrium occurring due to martensitic transformation was laid in late thirties of the twentieth Century when transformation of austenite to martensite was calculated for the given content of carbon (to include strain energy of the crystal network rearrangement, interfacial energy, undercooling (see Fig. 49), as well as under the use of activities instead concentration).

It changed the classical view to the 'true equilibrium' approach by admitting that the shape of experimentally determined phase diagrams can be affected by the 'degree of equilibration' shift during the experiment itself. The best example is the phase diagram of $SiO_2 - Al_2O_3$ which has been studied almost for a hundred years. The interpretation of its high alumina regions varied from experiment to experiment according to the duration of the system annealing (from hours to weeks) and was indicated as having both the incongruent and congruent melting point. Other disparities were reported in the composition range of mullite solid solution.

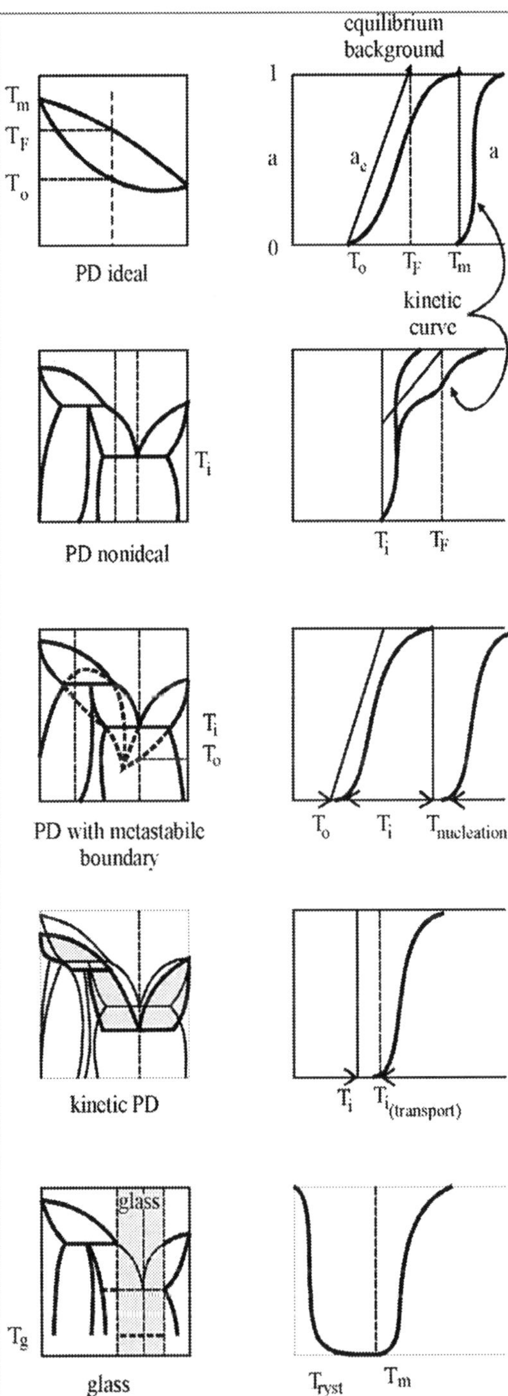

Fig. 49. – Left column shows a global view to the gradual development of equilibrium (upper two) and non-equilibrium (lower three) phase diagrams (PD) screening in the upper two rows the PD shape progress from the ideal case to interactive non-ideal case both being only dependent on the character of the material itself (cf. Fig. 36.). The three PD below exhibit the consequence of intensified experimental processing, i.e., the gradual impact of temperature handling, which, for low cooling rates, results to the slight refocus of phase boundary only, i.e., the implementation of curves extension (dotted extrapolation) to metastability (yet thermodynamically candid or loyal) regions. Faster cooling, however, forces the system to enter 'kinetic status', controlled by the rates of transport processes so that whole boundary (thick lines, new shadowed PD) are shifted away from the originally equilibrium pattern (thin lines). Finally, at the extreme temperature changes (bottom PD), the entire state is relocated to the highly non-equilibrium condition, where the material set-up cannot match the transport processes, so that a kind of forcefully immobilized ('frozen' or amorphous) state is established, where the actual phase boundary are missing (or being featureless, shaded), and the region is factually characterized by glass transformation. The right-hand column depicts the system behavior under the experimental observation of a given property (normalized within $0 \leq a \leq 1$) where thin straight lines represent the development of equilibrium background (balanced conditions, where the vertical lines stand for invariant and sloping lines for variant processes). The thick s-shaped curves endure for the practically measured trace, i.e., for the actually observed 'kinetic' curve.

It can be generalized that the resulting data can be prejudiced by the experimentalists' view on what is appropriate experimental arrangement and adequate processing of the system resonance in order to attain equilibration in the reasonable time and, in fact, a really true equilibrium is ideally achievable only during almost limitless 'geological-like' processes. Actually available rates of experimental measurements fall, however, within the boundary area of so-called 'thermodynamically imposed' (somehow extreme) conditions. The introduction of real conditions to the existing thermodynamic description must refer to the state of local equilibrium, whether the rates of changes of state (macroscopic variables) are comparable with the rates of elementary (molecular) processes, which determine the state of the system at the microscopic level. It is often related to the ratio of $\Delta T/\Delta t << \langle T \rangle/\tau_T$ and/or $\Delta T/\Delta x << \langle T \rangle/\lambda$ where ΔT is the variation of temperature at macroscopic level during the time interval, Δt, (or over the distance, Δx), where τ_T is the matching period of the elementary thermal motion of molecules or their mean free path, λ, at the given average temperature, $\langle T \rangle$.

As already noticed, any new phase can be formed only under certain non-equilibrium conditions of a definite driving force, $\Delta\mu > 0$, and *at the minute of absolute equilibrium no new phase can ever originate*. The conditions of particular points at which the phase transformation takes place are further deter mined by the transport of heat (energy) and mass. For the probability of formation, P_{cs} , of the first critical nucleus, concentration C_{sol} , in a small time interval, $V\tau$, holds within the framework of linear approximation, $P_{cs} = p(T, C_{sol}, C_{liq})\ V\tau$, where $p(T, C_{sol}, C_{liq})$ is the probability density equal to the product between the kinetic coefficient of homogeneous nucleation, K_n , and the unit time probabilities of a cluster realization ($p_1 \sim$ concentration difference), and its transition from liquid to solid phase ($p_2 \sim$ formation work). Thus we must study the temperature and concentration distribution in the system and their time changes in connection with the initial state and boundary conditions. So there arises a question how to correctly interpret the experimentally determined phase diagrams which do not fully comply with equilibrium conditions [6,7,305], the divergence being dependent on the proximity to the actually applied and thermodynamically required conditions. In determining the phase boundaries one can find experimental factors of different value and order, cf. Fig. 49.:

- arising from the nature of the material under study,
- associated with physical-chemical processes of the actual preparation of a representative sample from raw input materials,
- affected by the entire experimental set up and
- dependent on the kinetics of the phase transition (time hindrances of the new phase formation).

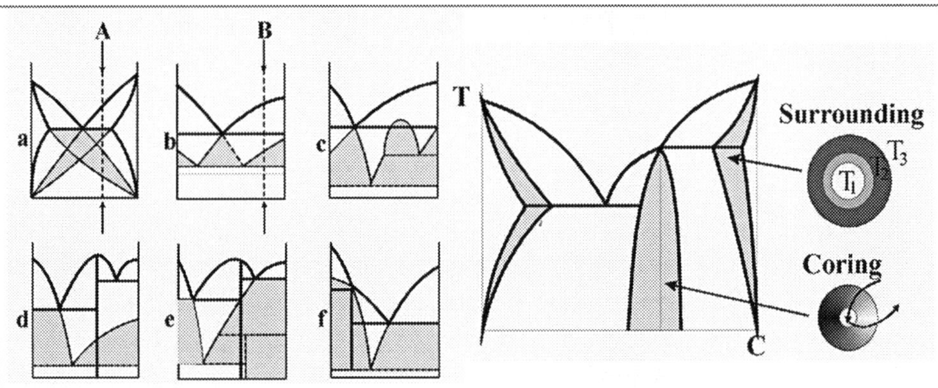

Fig. 50. – Upper two rows show model phase diagrams with the phase boundaries conventionally extrapolated into the below metastable regions (dashed lines), such as (a) liquid-solid curves indicating metastable point of melting as their bottom intercept, (b) shift and splitting of the eutectic point, (c) formation of a deep eutectic with a metastable region of limited solubility, (d) formation of simple eutectic point under the distectic and/or peritectic points (f) and finally (e) common experimental case of the observed change of congruent melting to incongruent. Below, the coring (along the vertical peritectic line) and surrounding (taking place at the both side curves near the hypoeutectic points of PD) are depicted at the bottom diagram. Both effects belong to true kinetic phenomena, common and long-known (and troublesome) in the field of metallurgy where they yield the uneven stratification of solidified material having experimental evidence from cross gradients of grown grain. In the case of coring it is caused by the opposite character of peritectic (liquid-solid) reactions requiring to re-form a new compound by inward diffusion of one component, while the other one moves outwards, so that the completeness of the conversion is governed by the inherent ease of reaction kinetics. Surrounding is little simpler in its mere transport characteristics, so that it is even more frequent as it accounts for the formation of layers across the growing grain due to the insufficient surface instantaneous equilibration at each temperature, T_i , as a rule caused by the carrying delays.

Hence it follows that the phase diagrams can generally fall into three, equally important, all-purpose groups regarding their practical validity and applicability. There are those having the effectiveness of unalterable relevance in order of :
- scientific age (suitable for tabulation of equilibrium data),
- livelihood (good enough for scientific generation in the sequence of years and useful to prepare materials durable long enough to be defined as stable) and
- for a given instant and experimental purpose (respecting particularity of a given arrangement and functionality to get ready-to-use materials of unusual properties to last for a certain limited period).

The latter requires deeper knowledge of metastable and unstable phase equilibria, which would also promote the application of the thermodynamics of irreversible processes in new branches of science and technology and may bring about the discoveries of new phases and of the yet unknown properties.

Current metallurgy bears the first accounts of the consequences of non-equilibrium solidification, i.e., phenomena known as *'coring'* and *'surroundings'* ndubitably to occur in the vicinity of all characteristic (invariant) points [306,372] , see Fig. 51. In the first case of coring, the melt under solidification does not encompass sufficient time to follow equilibration process along the balanced curve of solidus, which is caused by insufficient mass transport towards the phase interface. The precipitated grains, whose centers are closer to the equilibrium composition than their core-layers grown later, are the typical result. In the second case of surroundings, the originally precipitated phase starts to react with the remaining melt on reaching the peritectic temperature to form a new peritectic phase. It requires the atoms of one component to diffuse out from the melt to reach the phase interface. The thicker is the layer to cross, the slower the diffusion proceeds, particularly if the atoms of the second component must diffuse in the opposite direction through the solid layers to counterpart the reaction. This, evidently, must imply certain non-equilibrium conditions of the solidification and the gradual modification of the phase 'equilibrium' composition. Supplementary elimination of such non-equilibrium concentration gradients is usually accomplished by subsequent thermal treatment to allow equilibrating rearrangement. Similar phenomena are often accoutered when single crystals are grown from foreign melts, their concentration gradients being dependent on the concentration changes in the matrix melt. These gradients can be again removed by specific thermal treatment or by melt agitation. There is increased interest in more detailed mathematical analysis, description and synthesis of such 'dynamic-like' phase diagrams, which we coined the name *'kinetic phase diagrams'* [6], cf. Fig. 49. First, the *Gibbs* energies must be assumed for metastable phases, which occur at higher energy levels and non-equilibrium concentrations, because the *kinetic hindrance of nucleation* makes centers, suitable for the new phase growth, unattainable. For the practical use, metastable boundaries can be simply predicted by a straightforward extrapolation of the coexistence lines for the stable states into these non-equilibrium metastable regions, usually down to lower temperatures, Fig. 50. Alternatively, the preliminary shapes of metastable lines can be estimated from a superposition of two corresponding (usually simple-eutectic) phase diagrams. This is of considerable importance for all dynamic methods (from instrumental thermal analysis to laboratory quenching methods) to attain a correct interpretation of the phases on basis of observed effects.

For a system that cannot follow the experimentally enforced (usually strong) changes, even by establishing the previously discussed metastable phase equilibria due to backward nucleation, the boundary lines shift almost freely along both the concentration and temperature axes. Thereby the regions of unstable phases are formed to be described in terms of the *kinetics of the physical-chemical processes* (especially fluxes of mass and heat).

Quantitative dependences of the growth parameters on the kinetic properties of solidification

Change of values Increase of values	T_f	ΔT	C_{liq}	$\Delta \sim C_{liq} - C_{solid}$	growth rate
Cooling rate	intensive				
$10^3 \to 10^6$ K/s	increase	increase	increase	decrease	increase
Diffusion coeff.					
$10^{-7} \to 10^{-3}$ m^2/s	decrease	increase	increase	decrease	increase
Kinetic const.	increase	decrease			
$10^6 \to 10^1$ 1/s	to equilib.	to zero	decrease	increase	increase

where T_f, ΔT, C_{liq}, $\Delta \sim C_{liq} - C_{solid}$ are respectively the temperature on the solidification front, undercooling, concentration of the liquid phase and the concentration difference on the solidification front, and G the growth rate of solid phase formation [6].

Such a truly kinetic phase diagram is fully dependent upon the experimental conditions applied (cooling rate, sample geometry, external fields, measuring conditions) and can portray materials, which state become fixed at room temperature by suitable freeze-in techniques. It is best treated mathematically in the case of a stationary process conditioning (e.g., *Focker-Planck* equation, *Monte-Carlo* methods or stochastic process theory [6]). It is complicated by the introduction of local cooling rates and degrees of undercooling in bulk and at interfaces and a mathematical solution requires very complicated joint solution of the equations for heat and mass transfer under given boundary conditions as well as that for the internal kinetics associated with phase transition on the solidification front. Evaluation yields interesting dependences between the undercooling, concentration, linear growth rate and cooling rate, see the following Table.

9.6. T-T-T and C-T diagrams

During recent decades a considerable utilization has been made of the so-called *T-T-T diagrams (Time-Temperature-Transformation)* as well as their further derived form of C-T diagrams (*Cooling-Transformation*) particularly in assessing the critical cooling rates required for tailoring quenched materials [6] (initially metals) and their disequilibrium states (glasses). An important factor is the critical cooling rate, ϕ_{crit}, defined as the minimum (linear) cooling at which the melt must be quenched to *prevent crystallization*. Certainly, the ϕ_{crit} determination can be performed only in limited, experimentally accessible regions, ranging at present from about 10^{-4} to as much as 10^7 K/s Critical cooling rates are an integral part of glass-forming theories, which view all liquids as potential glass-formers if, given a sufficiently rapid quench, all crystal-forming processes are essentially bypassed.

In this prospect a simplified kinetic model was developed based on the JMAYK equation (see next Chapters 10 and 11) assuming the limiting (already indistinguishable) volume fraction of crystalline phase, traditionally assumed to be about 10^{-6} % [6,373]. Neglecting transient time effects for the steady-state rate of homogeneous nucleation, the ϕ_{crit} is given by a simple proportionality $\phi_{crit} = \Delta T_{nose}/t_{nose}$, where $\Delta T_{nose} = T_{melt} - T_{nose}$, and T_{nose} and t_{nose} are temperatures and time of the nose (extreme) of the time-temperature curve for given degree of transformation. The actual application of this kinetic analysis to wide variety of materials led to the development and construction of *T-T-T* diagrams [374]. It was found [373] that the ratio T_{nose}/T_{melt} falls in the range from 0.74 to 0.82 and that the ϕ_{crit} decreases with decreasing T_{gr} – for its value of 0.7, metallic glasses could be formed even under such a mild cooling rate as 10 K/s. If for the given fraction, T_{nose}/T_{melt} is moved to higher values, the *T-T-T* boundary is shifted to lower time [6,375]. Practically ϕ_{crit} implies the maximum sample thickness, δ, that can be wholly vitrified in one quenching stage is $\delta_{crit} \approx \sqrt{(\lambda\, t_{nose})}$ and will increase with decreasing ϕ_{crit} and increasing T_{gr} , certainly dependent on the sample thermal property (λ). In addition, a simple relation between ϕ_{crit} and undercooling, ΔT_{crit}, was devised in the logarithmic form *log* $\phi_{crit} = a - b/\Delta T^2_{crit}$ [375], where a and b are constants. It can further provide an estimate for the change of fraction crystallized, α, with undercooling, so that $(d\alpha/d\Delta T)_{\Delta Tcrit} \cong 3b/(\Delta T\ T^3_{crit})$ being useful for various numerical computations. For silicate ma terials the important factor, viscosity η, may also be fitted within about an order of magnitude by the traditional *Vogel-Fulche* type of logarithmic equation [393] in the form, *log* $\eta = -1.6 + 3.12/(T/T_g - 0.83)$.

A more realistic estimate of ϕ_{crit} can be obtained when constructing the continuous-cooling-curves, often called *C-T* diagrams [6,374,338]. It is based on the additivity following from the assumption that the momentary value of rate, α', which depends merely on α and T and not on the temperature history of the transformation process. In comparison with the classical *T-T-T* diagrams the same α is reached at relatively lower temperatures and in longer times. During the construction of *C-T* curves we can assume that α crystallized from temperature T_1 to T_2 (corresponding to t_1 and t_2) depends on the related values from the *T-T-T* diagram read out for the mean temperature $(T_1 + T_2)/2$. Similarly α formed during cooling up to the point given by the intercept of ϕ and *T-T-T* curve is found to be negligible in comparison to α corresponding to isothermal condition of this point, see Fig. 51. It should be noted that the value of ϕ_{crit} found by the nose method are typically larger than those computed by other techniques [378-383] because appreciable crystallization does not occur at temperatures below the nose, and thus the basis the basis for the nose method theory needs additional inquiry [380].

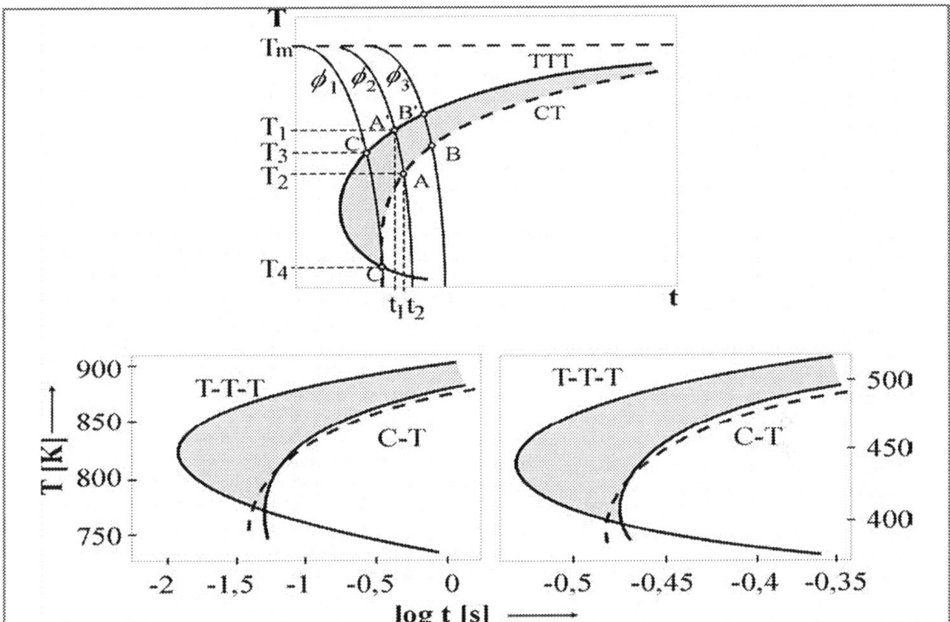

Fig. 51. – Scheme of the construction of the nonisothermal C-T curves (upper) from the known shape of the isothermal T-T-T diagram in the coordinates of temperature (vertical) and logarithm of time (horizontal). Shade areas put on view the difference between both treatments. Below there are shown two real cases, C-T curves calculated by the approximate method of *Grange* and *Kiefer* [385] (dashed line) and by the direct method suggested by *MacFarlane* [374] (solid lane). The curves are drawn in the comparison with their associated, classical T-T-T curves derived for the following sample alloys: Left figure – $Pd_{82}Si_{18}$ using the data for critical cooling rate, $\phi_{crit} = 2 \times 10^7$ [K/s] and right – $Au_{72}Ge_{14}Si_9$ with $\phi_{crit} = 7 \times 10^3$ [K/s] [6].

There is a close link between the particular crystallization mode and the general resolution of the *T-T-T* or *C-T* diagrams. The time needed to crystallize a certain fraction, α, at a given temperature, T, can be conveniently obtained by the DTA/DSC measurements, see Chapter 10. On the basis of nucleation-growth theories the highest temperature to which the kinetic equation can reasonably well reproduce the *T-T-T* or *C-T* curve was estimated to be about 0.6 $T_{melt.}$ There are certainly many different methods used for numerical modeling [374,380, 385], which detailed analysis [6,375] falls, however, beyond the scope of this chapter.

9.7. Thermodynamics of non-equilibrium glass transition

Special cases of non-equilibrium phase transitions are so-called *'kinetically enforced'* transformations, which are best represented by the *glass transformation,* abbreviated as T_g. If we measure some macroscopic property, Z, of a system, which does not undergo easy crystallization during its solidification, then its temperature dependence exhibits a characteristic change in the passage region from an undercooled, yet fluid, melt to highly viscous, rigid substance (solid). This rather broad process of glass transition is often characterized by single temperature, T_g. This process accounts for the course of liquid *vitrification* and the resulting amorphous solid is called *glass.*

Interestingly the topic of glass transition seemed clear in the 1970s, i.e., viscosity increases with shortage of free volume, and molecular cooperativity must help to save fluidity at low temperatures. Consequently an unexpected and previously hidden complexity emanated, which persists until now. Namely, it was the mysterious crossover region of dynamic glass transition at medium viscosity and two independent courses of a nontrivial high-temperature process above and a cooperative process below the crossover providing a central question whether is there any common medium, such as the spatial-temporal pattern of some dynamic heterogeneity. Certainly this is the composite subject of specialized monographs published within various field series [386-388].

The standard observations, based on measuring crystallographic characteristics and the amount of crystalline phases, such as typical XRD, are capable to detect the crystalline phase down to about 2-4 % within the glassy matrix. Indeed, we are not assuming here the delectability of minimum crystal size (peak broadening), nor we account for a specialized diffraction measurement at low diffraction angles. The critical amount of crystalline phase in the glassy sample became, however, the crucial question: how to define the limit of 'true glassiness'. Several proposals appeared but the generally accepted value is 10^{-6} vol. % (less common 10^{-3}), of crystallites to exist in glass not yet disturbing its glassy characteristics. The appropriateness of this value, however, is difficult to authorize. Nevertheless, the entire proof of the presence of glassy state is possible only on the basis of thermal measurements, i.e., upon the direct detection of glass transformation, which is the only characteristics for the glassy state alone without accounting for the interference of crystal counterparts.

Though this transformation exhibits most of the typical features of a second-order phase transition according to the *Ehrenfest* classification, it is <u>*not a true*</u> second-order phase transition due to the inherent irreversibility.

Let us employ the state function of affinity, A, defined previously in the preceding Chapter 7, together with the parameter, ζ, analogous to the above mentioned order parameter [389]. For an undercooled liquid (subscript, A), $\zeta = 0$, and for a glassy state (subscript, ζ), ordering arrives at constant value. For a second-order phase transition at T_E, it would hold that $(d\zeta/dT)_E = 0$, i.e., illustrated by the shift to the dashed line in Fig. 53. However, because ζ also changes to $(\zeta + d\zeta)$ for a glass, the equilibrium can be written as,

$$liquid \leftarrow dV_{E,A}\,(P,T) \;=\; dV_{E,\zeta}(P,\,T,\,\zeta) \rightarrow glass,$$

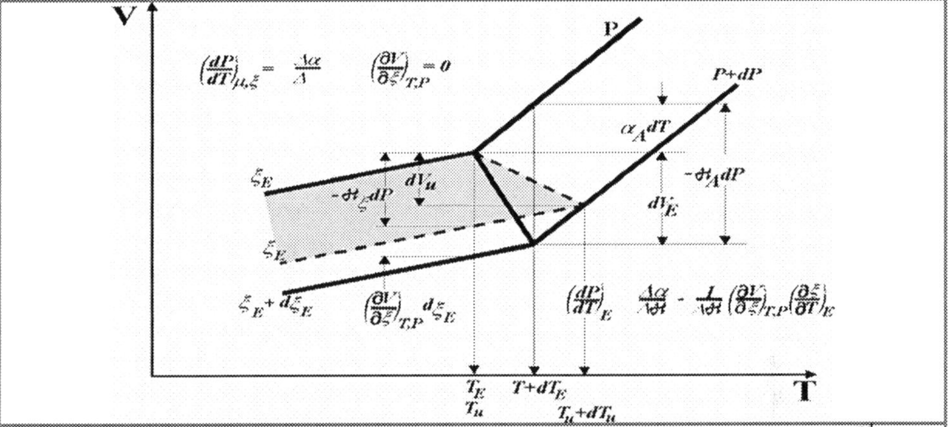

Fig. 52. – Graphical representation of the *Ehrenfest* relationship [3] for an ideal 2nd-order transition (point E) designated by the upper solid line and the middle dashed line, compared with glass transformation (point U, ordering parameter ξ) denoted by solid lines in the volume versus temperature diagram, where Δ denotes the change of volume compressibility, Ξ, volume expansion, α_v , and heat capacity, *Cp*.

so that it holds $- \Xi_A\, dP + \alpha_A\, dT = \Xi_\zeta\, dP + \alpha_\zeta\, dT + (dV/d\zeta)_{T,P}\, d\zeta$ from which it follows that $\Delta\Xi_A V = (dV/d\zeta)^2_{T,P}/(dA/d\zeta)$, $\Delta\alpha_V V = (dV/d\zeta)_{T,P}\,(dS/d\zeta)_{T,P}/(dA/d\zeta)_{T,P}$ and $\Delta C_P/T = {}_P\,(dS/d\zeta)^2_{T,P}/\,(dA/d\zeta)_{T,P}$ where $\Delta\alpha_V$ is the difference between the expansions of the equilibrium undercooled liquid and that of frozen glass (the same valid for the difference in volume compressibility, $\Delta\Xi_A$ and heat capacity, ΔC_P). In practice, it follows that the classical form of *Ehrenfest* equation is supplemented by another term dependent on the degree of ordering, which states that if the properties of an undercooled liquid would be known, the properties of the glass close to this state can be derived, i.e., $\Delta(\Xi\, C_V/\alpha_V)\Delta(1/\alpha_V) = \Delta(C_P/\alpha_V)\Delta(\Xi/\alpha_V)$, where T_o is the thermodynamic limit of glass transformation at $T_g \rightarrow T_f$.

The degree of irreversibility, π, can also be settled on the basis of a modified *Ehrenfest* relation established from statistical thermodynamics [390], or $\pi = \Delta C_P\, \Delta\Xi\, (T_g\, V_g\, \Delta\alpha_V^2)$, where π denotes a ratio number to follow the concept of free volume, V_f, that decreases to a minimum at the glass transformation point and which is close unity for a single ordering parameter. For more ordering parameters it exceeds unity, which may lead to a concept of several T_g values as a result of gradual freezing of the regions characterized by the individual ordering parameters.

Geometrically, this means that the thermodynamic surface, which corresponds to the liquid, is formed by the set of characteristic lines of the single-parameterized glass characterized by the constant, ζ. The shape of the surface at constant ζ in the vicinity of the equilibrium point is thus determined by two independent parameters, specifying the slope and fixing the curvature radius. Unfortunately, all the coefficients remain numerically undefined unless ζ is completely specified. To solve more complicated systems arising from penetration framework of multiple glasses, several ordering parameters (sub-systems) must be introduced. A number of workers tried to approach thus problem by introducing experimentally measurable, the so-called *fictitious temperature*, T_f [391,392], which are sensitively dependent to the cooling rates applied, experimental conditions, etc., see Fig. 54. If a glass is annealed at a temperature close above T_g, an equilibrium-like value, corresponding to the state of undercooled liquid, is routinely reached at this temperature T_f. It can be geometrically resolved for any point on the cooling curve as the temperature corresponding to the intercept of the straight line passing through this point and parallel with heating line, with the equilibrium curve of the undercooled liquid.

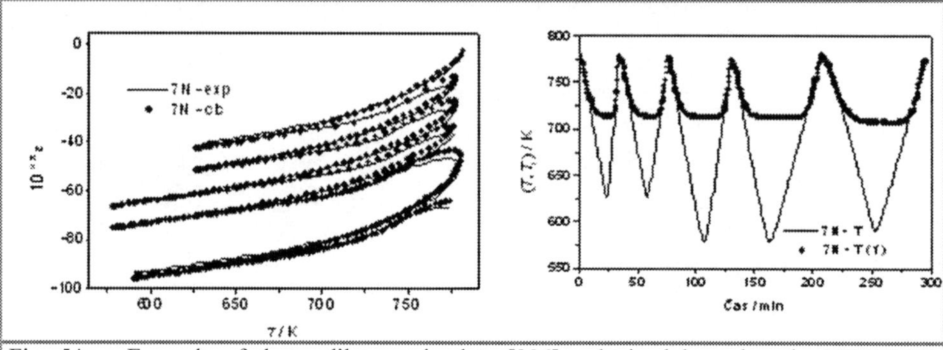

Fig. 54. – Example of thermodilatometric data [396] obtained by using the NETZSCH instrument TMA 402, which shows the actually measured curve (thin solid line) and simultaneously calculated data of the relative deformation (solid points) by means of the Narayanaswam's model [400]. Right, there is the graph of the time dependence of thermodynamic temperature (thin solid line) and calculated fictitious temperature (solid points). Courtesy of *Marie Chromčíková* and *Marek Liška*, Trenčín, Slovakia.

The temperature coefficient at the temperature $T^{\#}$ is then given by $(dT_f/dT)_{T\#} = [(dZ/dT) - (dZ/dT)_g]_{T\#} / [(dZ/dT)_{eq} - (dZ/dT)_g]_{Tf} = (C_P - C_{pg})/(C_{Pe} - C_{pg})$. The structural relaxation can be likewise described by using the parameter, ζ, upon the combination of above equation with the relationship $(dZ/dt)_T = (dZ/d\zeta)_T (d\zeta/dt)_T$ that yields $(dT_f/dT)_T = \{[(dZ/d\zeta)_T (d\zeta/dt)_{eq}]/[(dZ/dT)_{eq} - (dZ/dt)_g]\}[(T_f - T)/\tau]$, where τ is the *relaxation time* corresponding to the given degree of ordering, which bears mostly an exponential form.

It was also shown that many transport properties, Z, such as diffusion coefficient, viscosity or electric conductance of undercooled liquids obey a general *Vogel-Fulcher* equation [393] $Z = Z_o T^{1/2} \exp [a_Z/(T - T_o)]$.

One of the most convenient tools for practical determination of fictitious temperatures is thermomechanometry [396] see Fig. 54, where the time dependence of fictitious temperature can be obtained on the basis of the *Tool-Narayanaswami* relation [391,396,400] by the optimization of viscosity measurements *(logη{T,T$_f$}* versus temperatures) using the *Vogel-Fulcher* equation again.

In general, when a glass is equilibrated at a certain temperature, T_o, and then subjected to a temperature down-jump, $\Delta T = T_o - T$, one observes an instantaneous change of structure related properties due to vibrational contributions, which is followed by a slow approach to their equilibrium values corresponding the temperature, T. This structure consolidation is usually monitored by volume or enthalpy relaxation experiments and can be described by means of relative departures, δ_V or δ_H, respectively understood as $\{\delta_Z = (Z - Z_\infty)/Z_\infty \}$. Initial departure from equilibrium can be related to the magnitude of the temperature jump and follows $\delta_V = \Delta\alpha \, \Delta T$ and/or $\delta_H = \Delta C_p \Delta T$ where $\Delta\alpha$ and ΔC_p are the differences of the thermal expansion coefficient and heat capacity between the liquid during its equilibrium-adjacent undercooling and the as-quenched glassy liquid (solid). An attempt to compare both volume and enthalpy relaxation rates was made by *Hutchinson* [394] using the inflections slopes of $\delta_V(t)$ and $\delta_H(t)$ data sets plotted on the logarithmic time scale. Seemingly this data are not fully comparable quantities because $\Delta\alpha$ and ΔC_p may be different for various materials. *Málek* [395] recently showed that the volume and enthalpy relaxation rates can be compared on the basis of time dependence of the fictive temperatures, $T_f(t)$, obtained using relations $T_f(t) = T + \delta_V/\Delta\alpha$ and/or $= T + \delta_H/\Delta C_p$. It keeps relation to the fictive relaxation rate $r_f = - (dT_f / d \log t)_i$ identifiable as the inflectional slope of the stabilization period.

9.8. Use of temperature-enthalpy diagrams for a better understanding of transition processes in glasses

We again note that *glasses* are obtained by a suitable rapid cooling of melts (which process exhibits a certain degree of reproducibility) while *amorphous solids* are often reached by an effective (often unrepeatable) disordering process [6,9]. Certainly, there can be further classification according to the glass origin, distinguishing thus liquid glassy crystals, etc., characterized by their own glass formation due to the specific arrestment of certain molecular movements. The non-crystalline state thus attained is in a constrained (unstable) thermodynamic state, which tends to be transformed to the nearest, more stable state on a suitable impact of energy-bringing warmth (reheating). The suppression of nucleation is the most important factor for any process of melts vitrification, itself important in such diverse fields as metglasses or cryobiology trying to achieve non-crystalline state for apparently non-glass-forming alloys or intra- and extra-cellular vitreous solutions needed for cryopreservation. Characteristic processes worthy of specific note are the sequence of relaxation-nucleation-crystallization phenomena responsible for the transition of metastable liquid state to the non-crystalline state of glass and reverse process to attain back the crystalline state. Such processes are always accompanied by a change of the content of system enthalpy, which in all cases is sensitively detectable by thermometric and calorimetric measurements.

A customary plot can be found in the form of enthalpy versus temperature [3,370,397], which can be easily derived using an ordinary lever rule from a concentration section of a conventional phase diagram, see the preceding section 6.6. The temperature derivative ($d\Delta H/dT$) of this function resembles the DTA/DSC traces (for DTA see the subsequent chapter 12), each thermal effect, either peak or step, corresponding to the individual transformation is exhibited by the step and/or break in the related H vs. T plot [397,398], cf. Fig. 54. Such diagrams well illustrate possible routes of achieving a non-crystalline state. It shows the glass ordinarily prepared by liquid freeze-in via metastable state of the undercooled liquid (RC). Another state of amorphous solid can be prepared by the methods of intensive disintegration (milling) applied on the crystalline state of solids, which results in submicron grain–sized assemblage (MD, upwards from solid along the dotted line). Deposition from the gaseous state (VD, to thin sloped line) can be another source for amorphous solids. In the latter cases, however, the glass transformation is accelerated upon heating and its characteristic region turns out so early that is usually overlapped by coexisting crystallization, which can be only distinguished through the careful detection of the baseline separation, which always occurs due to the change of heat capacity. Such cases exist in oxide glasses but are even more enhanced in non-crystalline chalcogenides. For most metallic glasses, however, there is such negligible change of C_P between the glassy and crystalline states that the enthalpy record does not often show the expected baseline displacement, which even makes it difficult to locate T_g at all.

During any heat treatment of an as-quenched glass, relaxation processes can take place within the short or medium range of its network ordering, which covers *topological* movements of constitutional species, *compositional* rearrangements (when neighboring atoms can exchange their position) and *diffusional* reorganization connected with relaxation of structural defects and gradients. Such processes are typical during all types of annealing (such as isothermal, flush or slow heating) taking place below and/or around T_g and can consequently affect the glass transformation region. After the glassy state

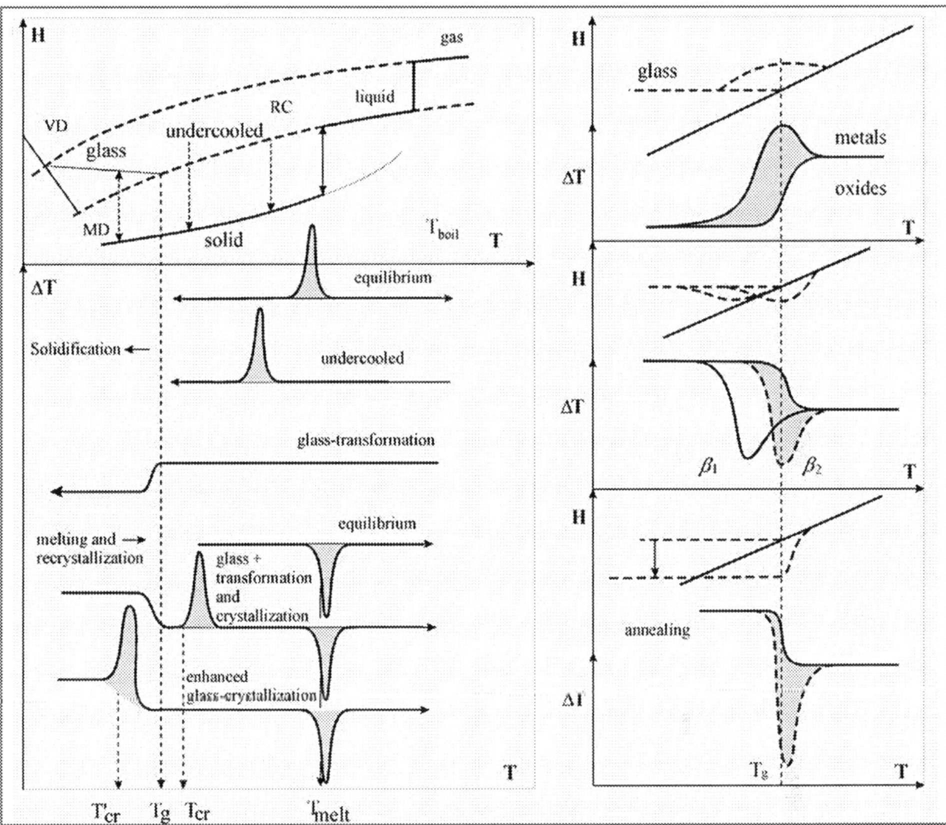

Fig. 54. – Left: a schematic diagram of enthalpy, *H,* versus temperature, *T,* (upper) and its derivative (*dH/dT ≅ ΔT,* bottom), which is accessible and thus also characteristic in its actual form upon the reconstruction of DTA /DSC recording. In particular, the solid, dashed and dotted lines indicate the stable (gas, liquid, solid), metastable (undercooled) and unstable (glassy) states. Rapid cooling of the melt (RC) can result in equilibrium and non-equilibrium (instantaneous and delayed, characterized by peak) solidification and, in extreme, also glass formation (characterized by stepwise *Tg,* which on reheating exhibits exothermic recrystallization below T_m). On the other hand, an amorphous solid can also be formed either by deposition of vapor (VD) against a cooled substrate (thin line) or by other means of disintegration (e.g., intensive grinding, high-energy disordering) of the crystalline state (moving vertically to meet the dotted line of glass). Reheating such an unstable amorphous solid often results in the early crystallization, which overlaps (or even precedes) *Tg* region and thus remain often unobserved. The position points are serviceable in the determination of some characteristics, such as the reduced temperature (*Tg/Tm*) or the *Hrubý* glass-forming coefficient (*Tc-Tg*)/(*Tm-Tg*). On the right-hand side the enlarged part of a hypothetical glass transformation region is shown in more detail and for several characteristic cases. These are: cooling different sorts of materials (upper), reheating identical samples at two different rates (middle) ands reheating after the sample prior reheating (often isothermal annealing, bottom).

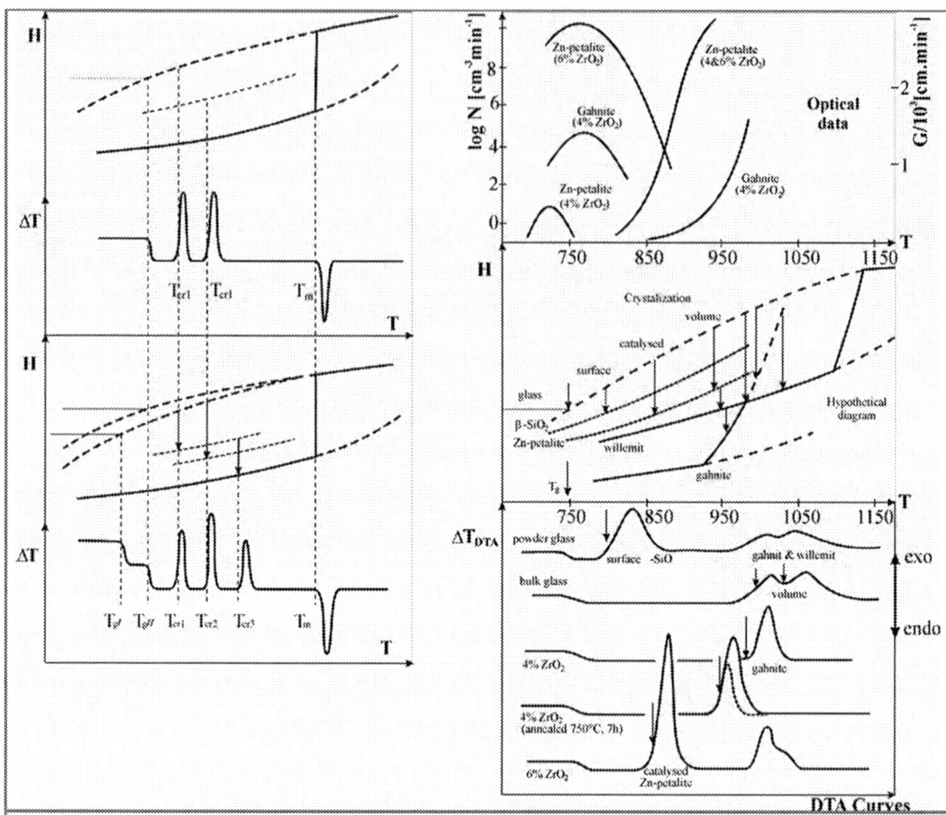

Fig. 55. – Left: more complex cases of crystallization showing a hypothetical metastable phase formation (dotted) for one (T_{cr1} upper) or for two (T_{cr1} and T_{cr2} , bottom) phase formation. Bottom example also includes possibility of the interpenetration of two glasses yielding the separate glass-formation regions (T_{g1} and T_{g2}) as a possible result of preceding liquid-liquid phase separation. Right: a complex view to the real crystallization behavior of the 20ZnO-30Al$_2$O$_3$-70SiO$_2$ glassy system [401]. Upper curves show the nucleation (N) and growth (G) data obtained by the classical two-step optical observation while bottom curves are the actual DTA traces measured on powdered and bulk glassy samples (cast directly into the DTA cell – 200 mg under heating rate of 10 C/min). The middle graph gives a schematic view of the enthalpy versus temperature exhibit in the presence of two metastable phases (β-quartz and Zn-petalite, dotted) and two stable phases (willemit and gahnite, solid). The mutual crystallization preference (arrows) is given by the sort of pretreatment of the glass (sample powdering, addition of ZrO$_2$ catalyst) and upon the controlled crystallization of Zn-petallite the low-dilatation glass-ceramics can be tailored for the technological applications.

undergoes the glass transformation, which may be multiple due to the existence of two (phase-separated, interpenetrated) glasses, it is followed by precipitation of the closest, usually metastable, crystalline phase (dot-and-dashed lines in Fig. 55). Consider a more complex case, e.g. the existence of a second metastable phase: the sequences of processes is multiplied since the metastable phase is probably produced first, to precipitate later into a more stable phase or catalyze a simultaneous or subsequent crystallization of the second phase. It may provide a complex pattern of heat effects, the sequence of which is difficult to analyze; nevertheless, it is always helped by means of this hypothetical itemization.

Hrubý [399] attempted to give a more practical significance to glass-forming tendencies using easily available values of the characteristic temperatures determined conveniently by DTA/DSC for reheating of the already prepared glassy sample. The so-called *Hrubý* coefficient, K_{gl}, is then given by the ratio of differences $K_{gl} = (T_{cryst} - T_g)/ (T_{melt} - T_{cryst})$. The greater the value of K_{gl} the better the glass-forming ability is approached. Despite the strong dependence of K_{gl} upon the method of glass preparation and measurement this coefficient exhibits a more sensitive interrelation to the glass formation peculiarities than a simple ratio T_g/T_{melt}. Best utilization of K_{gl} is in the comparison of glass-forming ability of different materials under different conditions and thermal treatments [399] cf. Fig. 55 again, but, regrettably, it is only assured by the entire use of material prepared beforehand.

The mechanical beginning and pre-treatment of glassy samples affect thermal measurements. The bulk and powdered sample can yield different enthalpy recordings and their identity is best verified by the measurements where their properties are directly compared. For example, we can use a DTA set-up where the reference sample is replaced by either of them, i.e., the powder-filled cell is compensated against the cell with the in-cast glass. Powdered sample may initiate surface-controlled crystallization at lower temperatures whilst the temperature region for the growth of nuclei in bulk glass is surface-independent and can be controlled by the gradual pulverization of sample [401,402] cf. Fig. 55. Thus for larger particles, the absolute number of surface ready-to-grow nuclei is smaller, owing to the relatively smaller surface area in the particle assembly resulting in predominant growth of nuclei in bulk. With decreasing particle size the relative number of surface nucleation sites gradually increases, becoming responsible for a better-shaped thermal effect. The particle-size-independent formation energy can be explained on the basis of the critical nucleus size being a function of the curved surface [383,397,402]. Concave and convex curvature decreases and increases, respectively, nucleus formation depending on the shape of glassy particles. It is often assumed that as-quenched glass has an amount of accessible nuclei equal to the sum of a constant number of quenched-in nuclei and that, depending on the consequent time-temperature heat treatment, the difference between the apex temperatures of enthalpy effects for the as quenched and purposefully nucleated glass $(T_{apex} \quad T^o_{apex})$ becomes proportional to the number of the nuclei formed during the thermal treatment. A characteristic nucleation curve can then be obtained by simple plotting $(T_{apex} - T^o_{apex})$ against temperature (or just using DTA-DSC peak width at its half maximum, for details see next section).

10. MODELING THE REACTION MECHANISM: THE USE OF EUCLIDIAN AND FRACTAL GEOMETRY

10.1. Constitutive equations applied in chemical kinetics

Chemical kinetics is based on the experimentally verified assumption that *the rate of change, dx/dt, in the state of a system (characterized by) x is a function, f, of the state alone*, i.e., $dx/dt = x' = f(x)$. Using this traditional postulation, the appropriate constitutional approach to inaugurate the desired constitutive equation can be written in the principal form of the dependence of the reaction rate, expressed as the time development of the degree(s) of change (transformation) on the quantities that characterize the instantaneous state of the reacting system (chosen for its definition).

In a simplified case, when $\lambda_{eq}=1$ (i.e., $\lambda=\alpha$), and under the unvarying experimental conditions (maintaining all intensive parameters around the sample constant, e.g., $T' = 0$, $P'= 0$, $\mathcal{H}' = 0$, etc.), we can write the set of *constitutive equations* with respect to all partial degrees of conversion, α_1 , α_2 .. α_i , thus including <u>all</u> variables in our spotlight. In relation to our earlier description we can summarize it as [3,402] $\alpha' = f_\alpha(\alpha,\beta,T)$ {possibly including other $\alpha_i' = f_i (\alpha, \beta, T)$} and $T' = f_T (\alpha, \beta, T)$, where the apostrophe represents the time derivative. Such a set would be apparently difficult to solve and, moreover, experimental evidence usually does not account for the particularity of all fractional degrees of conversion, so that we can often simplify our account just for *two basic variables*, i.e., α and T, only.

Their change is traditionally depicted in the form of a set of two basic equations:

$$\alpha' = f_\alpha (\alpha, T) = k(T) f(\alpha) \; and$$
$$T' = f_T (\alpha, T) \cong T_o + f(t)_{outer} + f(t)_{inner.}$$

The analytical determination of the temperature function, f_T, becomes the subject of both thermal effects so that let us see it firsts:

(1) The *external temperature*, whose handling is given by programmed experimental conditions, i.e., thermostat (furnace) control when $f_{T'} (T) = T_o + f(T_T)_{outer}$, and

(2) The *internal temperature*, whose production and sink is the interior make up of the process investigated, $f_{T'}(\alpha) = T'_o + f(T_\alpha)_{inner}$.

In the total effect these quantities govern the *heat flows outwards and inwards* the reaction interface. Their interconnection specifies the intimate reaction progress responsible for the investigational quantity, α', which is derived from the measured properties of the sample under study (cf. previous Chapter).

Here we can specify the externally enforced program of time-temperature change of thermostat, i.e., $dT/dt = f(t) = at^m$, or $T = T_o + at^{(m+1)}/(m+1)$, which can have the following principal forms:

Program:	$f(t)$ function:	a:	m:
Constant	0	0	0
Linear heating	ϕ	ϕ	0
Exponential	at	a	1
Hyperbolic	bt^2	b	2
Parabolic	$1/ct$	$1/c$	-1

Even more important is to see the properties of the first function, f_α, which is liable for an appropriate match of temperature progression of reaction mechanism under study. The long-lasting practice accredited the routine in which the function f_α (α,T) is divided into the two mutually independent functions $k(T)$ and $f(\alpha)$. This implies that the *rate of change, α', is assumed to be proportional to the product of two separate functions, i.e., the rate constant $k(T)$, dependent solely on the temperature, and the mathematical portrayal of the reaction mechanism, $f(\alpha)$, reliant on the variation of the degree of conversion, only.*

The entire temperature progress, f_T (α,T), is the subject of external and internal flows of heat and their contributions are usually reduced to that which is simply provided by externally applied temperature program, $T_o + f(t)_{outer}$, specifying the external restraint of the reaction progress in relation to the reaction rate, $\vec{r}$, derived from the factually measured properties of the sample.

When accounting for $\lambda_{eq} \neq 1$, the constitutional equation holds as, $\lambda = f(\lambda, \lambda_{eq}, T)$, and the true non-isothermal rate of chemical reaction, $d\lambda/dt$, becomes dependent on the sum $\{d\alpha/dt\ \lambda_{eq} + \alpha\ d\lambda_{eq}/dt\}$. Then the modified rate of a non-isothermal process, r, becomes more complicated [3,370,371,403] $\vec{r} = [d\lambda/dt - \lambda\ dT/dt\ (dln\lambda_{eq}\ dT]/\lambda_{eq} \cong k(T)\ f(\alpha)$. This expression, however, is of little practical use as the term, $d\lambda_{eq}/dt$, tends to become zero for all near-equilibrium conditions.

In the region of significant contribution of *reversible reaction* [403] (i.e., if the measurement is performed at the temperatures below or close to the thermodynamic equilibrium temperature, $T < T_{eq}$), another solution can be found beneficial to account for the instantaneous value of equilibrium constant, ΔG, of both reversal reactions, or $\alpha' = k(T)\ f(\alpha)[1 - exp\ (-\Delta G/RT)] \cong k(T)\ f(\alpha)\ [1 - (\lambda/\lambda_{eq})^v]$ where the ratio λ/λ_{eq} can factually be the same as the thermodynamic yield of reaction. We can assume that the change of *Gibbs* energy, $\Delta G = -\nu RT \ ln \ (\lambda/\lambda_{eq})$, become significant at the end of reaction reaching, at most, the value of the conversional degree equal to, λ_{eq}, at the given temperature, T. We can even continue further on applying the modified *van't Hoff* equation, introducing thus the proportionality $ln \ \lambda_{eq} = (\Delta H/\nu R)(1/T - 1/T_{eq})$, when $T = T_{eq}$ and $\lambda_{eq} = 1$, which, however, falls beyond this simplified account.

Another extension would be the insertion of the effect of pressure, P, into the set of constitutive equations [3], i.e., $\alpha' = f\ (\alpha,\ T,\ P) = k(T,\ P)\ f(\alpha) = k(T)\ K(P)\ f(\alpha)$. The dependence of the reaction rate on the pressure can than be expressed by an asymptotic function of the *Langmuir* type, e.g., $\alpha' = k(T)[a_oP/(1+a_1P)]\ f(\alpha)$, or by a more feasible, general *power type* function, $\alpha' = k(T)\ P^m\ f(\alpha)$, where $\underline{a}$ and $\underline{m}$ are constant.

Often misguided concept of an erroneous application of constitutive equations in chemical kinetics is worth of a special note. This is the famous 'puzzle' of the apparently blindfold misinterpretation of the role of degree of conversion, α, which is insensibly understood as the state function of both parameters apparently involved, i.e., the temperature and time, such as $\alpha = f(T, t)$. It was introduced as early as in the turn of 1970's [404] and challenged to the extensive discussion whether the consequential total differential, $d\alpha$, exists or not [405]. Curiously this query has been surviving until the present time [406,407] and is aimed at the inquiry whether the following equation is valid or whether it is only a mathematical fudge, i.e., $d\alpha = (\partial\alpha/\partial T)_t\ dt + (\partial\alpha/dt)_T\ dT$ and thus $d\alpha/dT = (\partial\alpha/\partial T)_t\ /\phi + (\partial\alpha/dt)_T$. The downright analytical solution for $d\alpha/dT$, assumes (for the simplest case of controlled linear heating, i.e., $dT/dt = constant = \phi$) the direct substitution of the standard reaction rate, α' for the 'isothermal' term $(\partial\alpha/dt)_T$, which thus can mathematically yield a newly modified rate equation in the extended form [370,407], such as $(\partial\alpha/dt)_T = k(T) [1 + E(T - T_o)/RT^2]\ f(\alpha)$. There immediately arises a question whether this unusual alteration is _ever_ applicable (and thus more principally useful) if it puts on a display the extra temperature term, $[1 + E(T - T_o)/RT^2]$, see Fig. 56. We have to assume two class of processes, i.e.,

isothermal:	nonisothermal:
$\alpha = \alpha_T\ (t,\ T)$	$\alpha = \alpha_\phi\ (t,\ \phi)$
$\alpha' = (\partial\alpha_T/\partial t)_T + (\partial\alpha_T/dT)_t\ T'$	$\alpha' = (\partial\alpha_\phi/\partial t)_\phi + (\partial\alpha_\phi/dT)_t\ \phi'$
because $T' = 0 = \phi$	because $\phi' = 0 = T''$

$$\rightarrow \quad then \quad \leftarrow$$

$$(\partial\alpha_T/\partial t)_T = \alpha' = (\partial\alpha_\phi/\partial t)_\phi$$

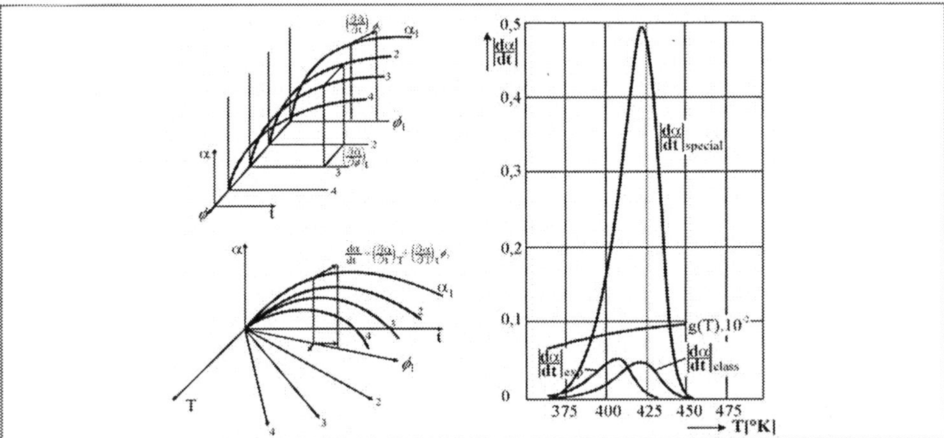

Fig. 56. — Graphical representation of the persisting query for the correct selection of variables in determining constitutional equation [370]. Left, diagrammatic solution of the given differential equations (reaction rates in the two coordinate system) which elucidation depends on the selected temperature program, i.e., $\alpha = \alpha\,(t,\,\phi) = \alpha\,(t,\,T/t) \equiv \alpha\,(t,\,T)$ using isothermal $(\alpha,\,t,\,T)$ (upper) and linear-nonisothermal $(\alpha,\,t,\,\phi)$ (bottom) modes. Right, graphical confrontation of the experimentally obtained reaction rate (exp) with the two theoretical evaluation based on both the classically standard (class) and specially proposed (spec), the latter including the extra multiplication term $[1 + E/RT]$.

This somehow curious version has both the recently supporting [406,407] as well as early abandoning [3,370,405] mathematical clarifications. Avoiding any deeper mathematical proofs let us show its absurdity using a simple example of the change in the water content in an externally heated reaction chamber. The process, taking place inside the chamber, is under our inspection and can be described on the rate of water content change (denoted as $d\alpha$), which just cannot be a mere function of the instantaneous temperature, T, and time, t, elapsed from filling the chamber with water, regardless of whether any water is remaining in the chamber. Instead, the exactly and readily available constitutional equation says that the instantaneous rate of water removal, represented by the change of water content, $d\alpha/dt$, must be a function of the water content itself, α, and its temperature, T, at which the vapor removal occurs. This reasoning cannot be obviated nor approved by any re-derivation of the already integrated form of basic kinetic equation [307] although the reverse treatment, $g(\alpha) = {}_0\!\int^{\alpha} d\alpha/f((\alpha)$, can have some (in the first sight almost reasonable) impact because such a mathematical procedure is next to leap-frog of the physical sense of a chemical reaction [405].

10.2. Modeling used in the phenomenological description of a reaction path

In chemical kinetics, the operation of $k(T)$ function is customarily replaced with the known and well-defined analytical form of the *Arrhenius* exponential function, $k(T) = A \exp(- E/RT)$, thoughtfully derived from the statistics of evaporation. The inherent constants are the so-called *activation energy, E,* identified as the energy barrier (or threshold) that must be surmounted to enable the occurrence of the bond redistribution steps required to convert reactants to products. The pre-exponential term, or *frequency factor, A,* provides a measure of the frequency of occurrence of the reaction situation, usually envisaged as incorporating the vibrating frequency in the reaction coordinate. There are also alternative depictions related to the activated complex theory and the partition functions of the activated complexes to give limited freedom along the reaction coordinate which details can be found elsewhere [3,408-410].

The validity of the Arrhenius equation is so comprehensively accepted that its application in homogeneous kinetics normally requires no justification. Although the Arrhenius equation has been widely and successfully applied to innumerable solid-state reactions, this use factually lacks a theoretical justification because the energy distribution, particularly amongst the immobilized constituents of crystalline reactants, may not be adequately represented by the *Maxwell-Boltzmann* equation. Due allowance should be made for the number of precursor species within the reaction zone as well as for the changing area of the reaction interface. It can affect the enhancement of reactivity at the active sites (reactional contact), exaggerated by strain (between the differently juxtaposed networks), defects, catalytic activity of newborn sites, irreversible re-crystallization, or perhaps even by local volatilization (if not accounting for the thickness of reacting zone contoured by local heat and mass fluxes across and along the interfaces). It brings multiple questions about its applicability and we can humorously cite from ref. [411] *"Everybody believes in the exponential law of energy distribution, the experimenters because they think it can be proved mathematically, and the mathematicians because they believe it has been established by observations"*.

Nevertheless, it was shown [3,410,412] that energy distribution functions of a similar form arise amongst the most energetic quanta. For values significantly above the Fermi level, both electronic energy (*Fermi-Dirac* statistics) and phonon energy (*Bose-Einstein* statistics) distribution approximate to the same form as that in the Maxwell-Boltzmann distribution. Also the interface levels, capable of accommodating electrons, are present within the zone of imperfections where chemical changes occur. These are analogous to impurity levels in semiconductors, imperfections levels in crystals (F-centers) and, similarly, can be located within the forbidden range between the energy bands of the crystal. Such levels represent the precursor energy states to the bond redistribution step and account for the increased activity relative to similar components in more perfect crystalline regions. Occupancy of these levels arises from activation through an energy distribution function similar in form to that characteristic of the Maxwell-Boltzmann equation and thereby explains the observed fit of $k(T)$ data to an Arrhenius-type equation.

However, for the resolution of the second function, $f(\alpha)$, we must act in a different way than above because the exact form of this function [3,9] is _not a priori known_ and we have to determine its analytical form (taking, in the same time, its diagnostic potential into our account). The specification of $f(\alpha)$ is thus required by means of the substitution of a definite analytical function derived on basis of modeling the reaction pathway usually by employing some physical-geometric suppositions. In contrast to the determination of predefined characteristic parameters of the $k(T)$-function, the true purpose of kinetic studies is to hit upon this pathway, i.e., to find our imaginative insight into the reaction mechanism. Thus it is convenient to postulate a thought model visualizing the reaction, usually splitting it into a sequence of possible steps and then trying to identify the slowest step, which is considered to be the rate-determining one. Such models usually incorporate (often rather hypothetical) description of consequent and/or concurrent processes of interfacial chemical reactions and diffusion transport of reactants, which governs the formation of new phase (nucleation) and its consequent (crystal) growth. Such a modeling is often structured within the perception of simplified geometrical bodies, which are responsible to depict the incorporated particles and such visualization exemplifies the reaction interfaces by _disjointing lines_. Such a derived function $f(\alpha)$ then depends on all such physical, chemical and geometrical relations, which are focused to the interface between the product and the initial reactant.

When not accounting on the interfaces or other inhomogeneity phenomena we deal with _homogeneous reactions_ determined by an averaged concentration in a whole reacting volume and thus the $f(\alpha)$ function is paying attention to a most simplified description only, which is associated with the so-called _reaction order_. However, for solid-state reactions the reactants are neither mixed on an atomic level nor equally distributed in the whole reactant volume and must, therefore, penetrate, flow or diffuse into each other if the reaction is to start and propagate within the given volume. Accordingly the space co-ordinates become a controlling element, which creates heterogeneity effects inevitably to be included and which are actually accounted for creating interfaces by means of '_defects_' conveniently symbolizing a pictographic contour (borderline, curve). Hence the mathematical description turns out to be much more complicated due to the fact that _no mean' bulk concentration' but the spot 'phase interfaces' carry out the most significant information_ undertaking the true controlling role for the reaction progress, as illustrated in Fig.57.

The most common models are derived isothermally and are associated with the shrinking core of a globular particle, which maintains a _sharp_ reaction boundary [3,413,414]. Using a simple geometrical representation, the reacting system can be classified as a set of spheres where each reaction interface is represented by a curve. We assume that the initial reactants' aggregation must be reached by (assumingly) well distributed (reacting) components (often through diffusion). Any such an interfacial (separating) layer, y, bears thus the role of *kinetic impedance* and the slower of the two principal processes, diffusion and chemical interface reaction, then become the *rate-controlling process* responsible for the over-all reaction progression.

Reaction rate is often proportional to the interface area, s, i.e., $\alpha' \cong k(T)s/V_o$, where V_o is the initial particle volume. Assuming a unidirectional process, the rate of thickness growth of product layer, y, is given by $dy/dt = D/y$ where D is the diffusion constant independent of time and the area of contact. For a real particle diameter, r, the mathematical manipulations yields the most famous model of simplest three-dimensional diffusion called the *Jander* Parabolic Law [415] where Dt/r^2 became proportional $\{1 - 1/(1 - \alpha)^2\}^2$. It has been widely adapted by modifying the growth rates to became inversely proportional on time [416], i.e., $dy/dt = 1/yt$ (*Kroger-Ziegler*) or to the fraction reacted [417], i.e., $dy/dt = (1 - \alpha)D/y$ (*Zhuralev-Lesokhin*). *Ginstling* and *Brounshtein* arrived at an another model [418] assuming that the parabolic law asserted that the reaction surface does not remain constant but actually decreases when the reaction proceeds so that the volume consumed becomes proportional to the actual particle diameter so that Dt/r^2 is proportional to $\{1 - 2\alpha/3 - (1 - \alpha)^{2/3}$. *Carter* further improved this model [419] by accounting for differences in the volume of the product layer with respect to the volume of the reactant consumed. *Komatsu* [420] made yet further modification by using the inverse proportionality between D and t accounting also for the mixing ratios and the radii differences of the two starting components (i.e., the manner of powders packing) as well as for a certain number of contact points existing between the reacting particles arriving thus to the so called *counter-current diffusion*. The majority of these already classical models are collectively illustrated in Fig. 58.

As a rule, the processes controlled by the chemical event and diffusion are respectively dependent on $1/r_o$ and $1/r_o$ where, r_o, is the initial radius of incorporated particles. Most of individual equations derived for particular cases [1,3,413,414,421-426] are summarized in Table 10. Together with their integral forms they are useful for their further experimental assessment (see below in this Chapter).

The above solid-state-reaction models applied to powdered compacts assumed that the particle surfaces are instantly coated by primary (easy-gathering) thin reactant blanket, which is capable of further growth under the previously specified conditions. There is, however, another important way of looking at the initial product formation and subsequent growth. This approach considers the nucleation at the active sites distributed over the interface. The rate at which this embryonic spots become escalating and thus turning out to be growing nuclei gradually determines the consummation of the reactant particles.

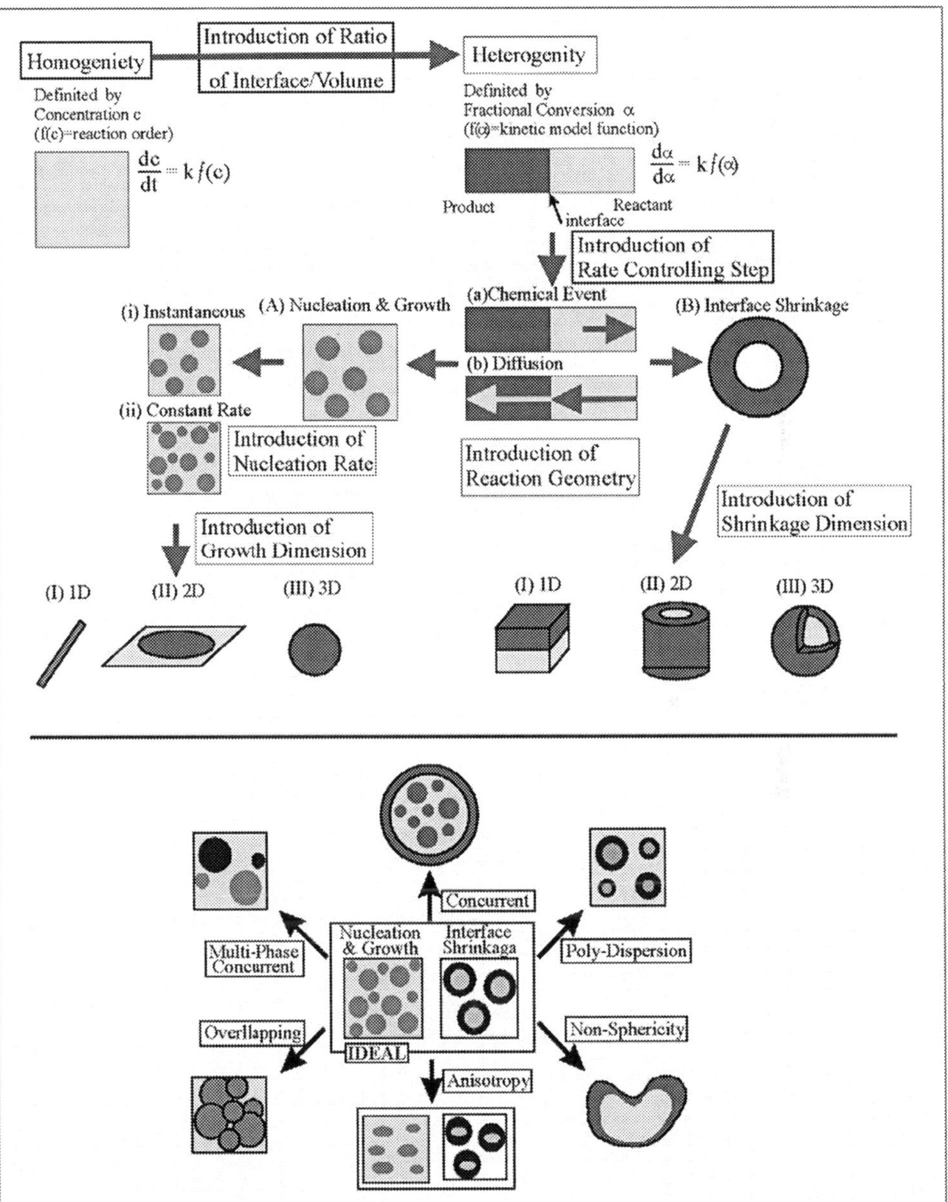

$$\frac{dc}{dt} = k f(c)$$

$$\frac{d\alpha}{d\alpha} = k f(\alpha)$$

Fig. 57. – Illustrative portrayal showing the development of the heterogeneity concept (upper) from the homogeneous description based on the uniform concentration to the introduction of interfacial separation of reacting phases, including thus the aspects of dimensionality. Bottom picture exhibits possible causes of non ideality, which are not commonly accounted for in the standard physical-geometrical models, but which have actual qualifications observed during practical interpretations [423]. Courtesy of *Nobuyoshi Koga*, Hiroshima, Japan.

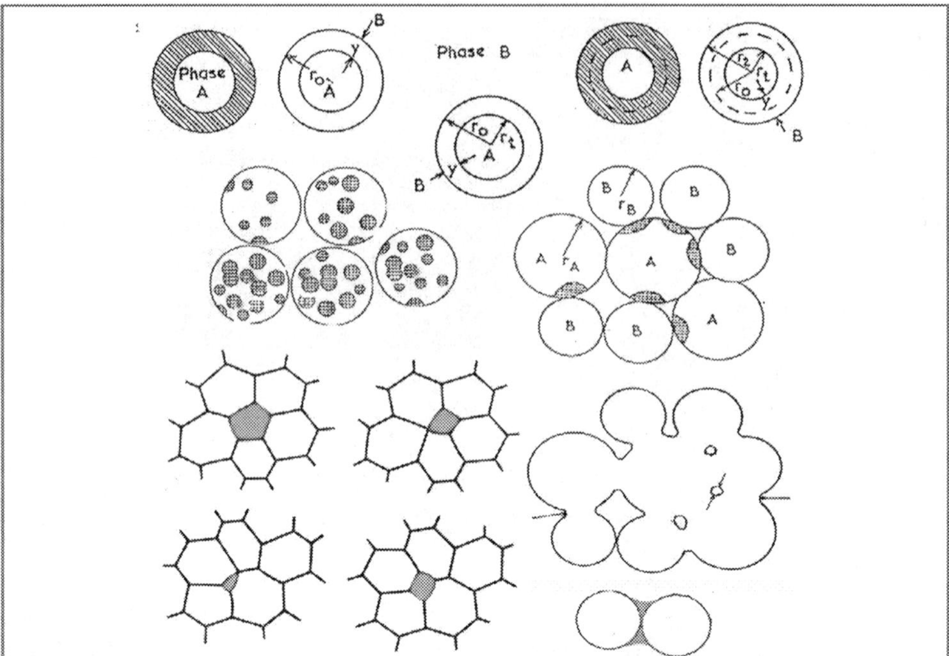

Fig. 58. – Upper raw, simplified circular models of diffusion from the surrounding tiny particles of the phase B into the enveloped phase A, which surface is completely and instantly covered by the product (shaded). It is worth of attention that all borderlines in this (and all other) figure factually represent the reaction interface. Left, the model of *Jander* parabolic law where the product layer, *y*, is formed on the surface of *A* (radius r_o). Middle, the *Ginstling-Brounshtein* model accounting, in addition, for changing the surface area (r_t) and Carter symmetry (right) accounting for initial (r_o), unreacted (r_t) and overlapping (r_2) material. Middle raw shows the *Komatsu* model of diffusion (right) that combines contact-point-geometry (shaded) with contracting-sphere of phase-boundary models while left is the schematic representation of nuclei-growth model (cf. Fig. 59.) and bottom right is the related portrayal when accounting for sintering, in which the decisive process if the formation (curvature, arrows) and growth of connecting necks (shaded) between point-connected particles. Real grain structures, however, are far from simple circular (porous) geometry and all dense agglomerates exhibit certain instability, which depends on the interactions between the topological requirements of space-filling and the geometrical needs of the surface tension balance. Grain boundary migration (similar to flows in sintering) then tends to occur because of curvature and the convex grains with sides less than 6 (< hexagon) incline to shrink to 4 and 3 sides finely to disappear (bottom left, clockwise).

Many mathematical models have bees advanced relating nucleation and nuclei growth rates to the overall kinetics of phase transformation, such as *Johnson* and *Mehl* [427], *Avrami* [428], *Yerofyeyev* [429], *Kolmogorov* [430] as well as *Jacobs-Tompkins* [431] or *Mampel* [432] and were agreeably summarized elsewhere [1,3,413,144, 421,422,423,426,431].

Table 10. I. Typical physical-geometric (phenomenological) model functions *derived for kinetic description of some particular solid-state reactions*

Model:	Symbol:	function $f(\alpha)$:	integral form $g(\alpha) = {}_o\!\int^\alpha d\alpha/f(\alpha)$:
Nucleation-growth (JMAYK eq.) (Johnson-Meal-Avrami-Yerofeev-Kolmogorov)	A_m	$(1-\alpha)[-\ln(1-\alpha)]^{1-1/m}$ $m = 0.5, 1, 1.5, 2, 2.5, 3$ and 4	$[-\ln(1-\alpha)]^{1/m}$
Phase boundary n-Dim.: (equivalent to the concept of reaction-order n)	R_n	$n(1-\alpha)^{1-1/n}$ $n = 1, 2$ and 3	$[1 - (1-\alpha)^{1/n}]$
Diffusion controlled	D_i		
1-dim. diffusion	D_1	$1/2\alpha$	α^2
2-dim. diffusion	D_2	$-[1/\ln(1-\alpha)]$	$\alpha + (1-\alpha)\ln(1-\alpha)$
3-dim. diffusion (Jander)	D_3	$(3/2)(1-\alpha)^{2/3}/[(1-\alpha)^{-1/3} - 1]$	$[1 - (1-\alpha)^{1/3}]^2$
3-dim. diffusion (Ginstling-Brounshtein)	D_4	$(3/2)/[(1-\alpha)^{-1/3} - 1]$	$1 - 2\alpha/3 - (1-\alpha)^{2/3}$
3-dim.counter dif. (Komatsu-Uemura, sometimes called 'anti-Jander' equation)		$(3/2)(1+\alpha)^{2/3}/[(1+\alpha)^{-1/3} + 1]$	$[(1+\alpha)^{1/3} - 1]^2$
Normal grain-growth (Atkinson – long-range diffusion where r_o is initial grain radius)	G_n	$(1-\alpha)^{n+1}/(n\,r_o^n)$	$[r_o/(1-\alpha)]^n - r_o^n$
Unspecified – fractal approach (extended autocatalytic)			
fractal-dimension (Šesták-Berggren)	SB	$(1-\alpha)^n\,\alpha^m$	no particular analytical form

When deriving nucleation-dependent kinetic models [427-432] we assume that the rate of nuclei formation depends on the number of energetically favorable sites, N_o. If the number of nuclei at time t is N, then it holds that $dN/dt = k_N(N_o-N)^n$ with an optimal case of ideal distribution at $n=1$. On the integration we obtain a standard exponential relationship. When k_N is small, the relationship can be approximated by a linear dependence $N = k_N N_o t$, which can be supplemented by the power exponent n (related to nucleation veracity) and by the relaxation term $exp(\tau/t)$ responsible for a time delay τ (at the beginning of nucleation). Nucleation is followed by the growth, G, of tiny nuclei to substantial grains of the volume $V(t)$, which process can be collectively described by the general form of the joined nucleation-growth equation $V(t) = {}_o\!\int \{{}_\tau\!\int(dG/dt)dt\}(dN/dt)d\tau$. The growth rate dG/dt can bear the character either of surface chemical reaction (R) or of diffusion (D).

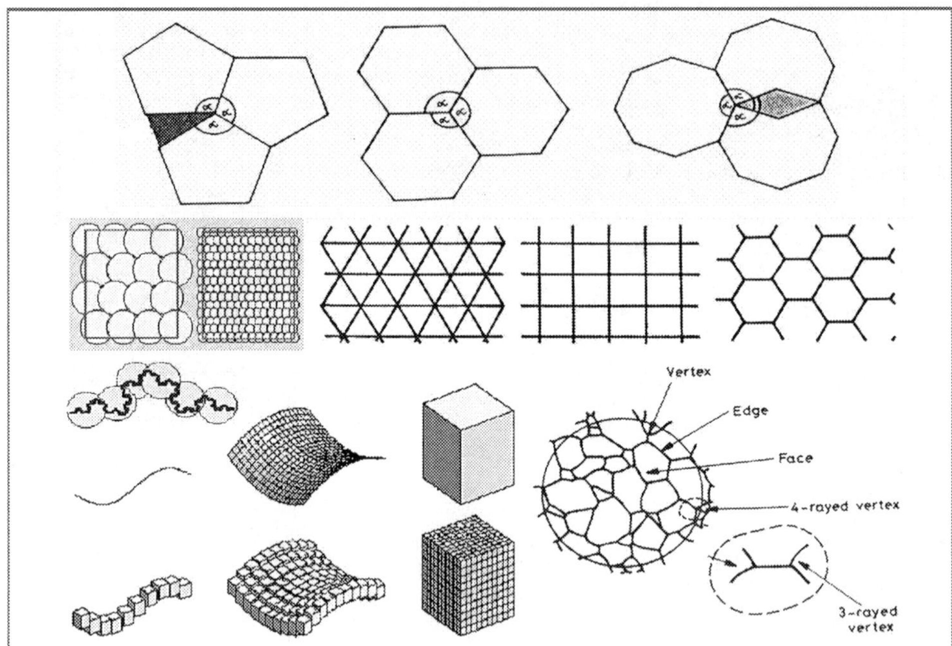

Fig. 59. – Building faces (blocks) available for modeling. Upper row shows unsuitability of pentagons and heptagons to form a continuous web because of respectively uncovered and overlapping areas (shaded), which can only be matched by curving their edges or adjusting angels (5-sided→convex and 7-sided→concave). Symmetrical network can only be satisfied with a collection of trigonal, tetragonal and hexagonal faces (middle) and their combinations. For an array of equal balls and/or cubes (even when crimped) we always face restrictions due to the strict Euclidean dimensionality, which we do not find in any actual images, often characterized (or observed) by typical 2-D cross-sections (bottom right). Irregular grain structure possesses distinctive faces, which can best be characterized by the degree of vertices (4-reyed vertex decomposing to 3 as a spontaneous growth occurs).

The procedure for the determination of the whole course of conversion in the overall time t depends on the analytical expression of the instantaneous volume of the crystallized phase $V(t)$ as a function of the rates of nucleation dN/dt and growth dG/dt. This volume of nuclei capable of further spontaneous growth at time τ is, e.g., $G_{(t)} = (4/3)\,\pi k_R(t-\tau)^3$ for the case of 3-D isotropic growth controlled by the surface reaction (k_R). Introducing the normalized degree of crystallization α as the ration $V(t)/V_\infty$ and employing the simplest form of constant nucleation $dN/dt = k_N N_o$ we can obtain $\alpha \cong \pi N_o k_R^3 k_N t^4$, which was derived by *Mampel* [432] for the initial stages of decomposition of solids and converted by *Yerofyeyev* [429] for the instantaneous nucleation with a constant number of ready-to-grow nuclei $N=N_o$ as $\alpha \cong \pi N_o k_R^3 t^3$. *Avrami* [428] introduced simplifying relationship $d\alpha = (1-\alpha)d\alpha_{max}$ valid for a maximal attainable (ideal) degree of crystallization assuming that the growing nuclei do not overlap.

We can conclude that the main idea of the nucleation-growth approach, which is hidden in a simplified relation where the fraction of reactants consumed *(1 -α)* is put in a general proportionality to a combined effect of the space and time dependent outcome of interrelated measures of nucleation, $N_{(T)}$, diffusion, $D_{(T)}$ and interface chemical reaction, $k_{(T)}$, which is regularly written in a compact form of *-ln(1 − α)* $= k_T t^r$. The invariable but temperature dependent k_T stands here for the inclusive (overall) kinetic constant and the power exponent r provides a condense representation of the processes interconnected heterogeneity. This concise relationship is called the JMAYK equation [3] after the authors [427-430] behind its derivation. It, however, is worth repeating that crystal (nuclei-aided) growth, $G_{(T)}$, can be controlled by either (i) anisotropic (unidirectional) chemical (interface) reaction, $k_{(T)}$, if the diffusion is so rapid that the reactants cannot combine fast enough at the reaction interface to assure the adjustment of equilibrium (phase-boundary reactions) or by (ii) isotropic (directional) diffusion $D_{(T)}$ if the chemical reaction is fast enough so that the grows move becomes dominating the transformation. In this case either of its 1-, 2- or 3-dimensionality steps forward into calculation and the growth function is replaced by the parabolic proportionality of the kind $D^{1/2}t^{1/2}$, which detailed manipulation is shown elsewhere [1,3,413,421,425,433,434].

10.3. Idealized models contrary to the real process mechanisms and morphologies

It is worth repetition that the above mentioned mathematical modeling can be broken into three major but reduced designs according to the downgraded reaction geometry. We can envisage either elementary process (i) diffusion of reactants to the reaction interface through a continuous product layer, (ii product nuclei formation and their consequent growth and (iii) chemical reaction restricted to the phase boundary. These events can roughly concur to what can specifically happen when we try to measure and study the overall reactio path. There are no assured means that suchlike aggregated form of reactants would conduct themselves within our simplified picture and that the selection of a particular model is acceptable; however, a nuclei growth analysis shows as a rather useful tool for a preliminary screening of reactions within powdered compacts [421-426]. Reaction dynamics of solid-state processes are extensively studied by methods of thermal analysis and there is a vast amount of data published on such a kind of "non-isothermal" kinetics [422-423], frequently treated on this focal but figurative point of oversimplified modeling revealed above.

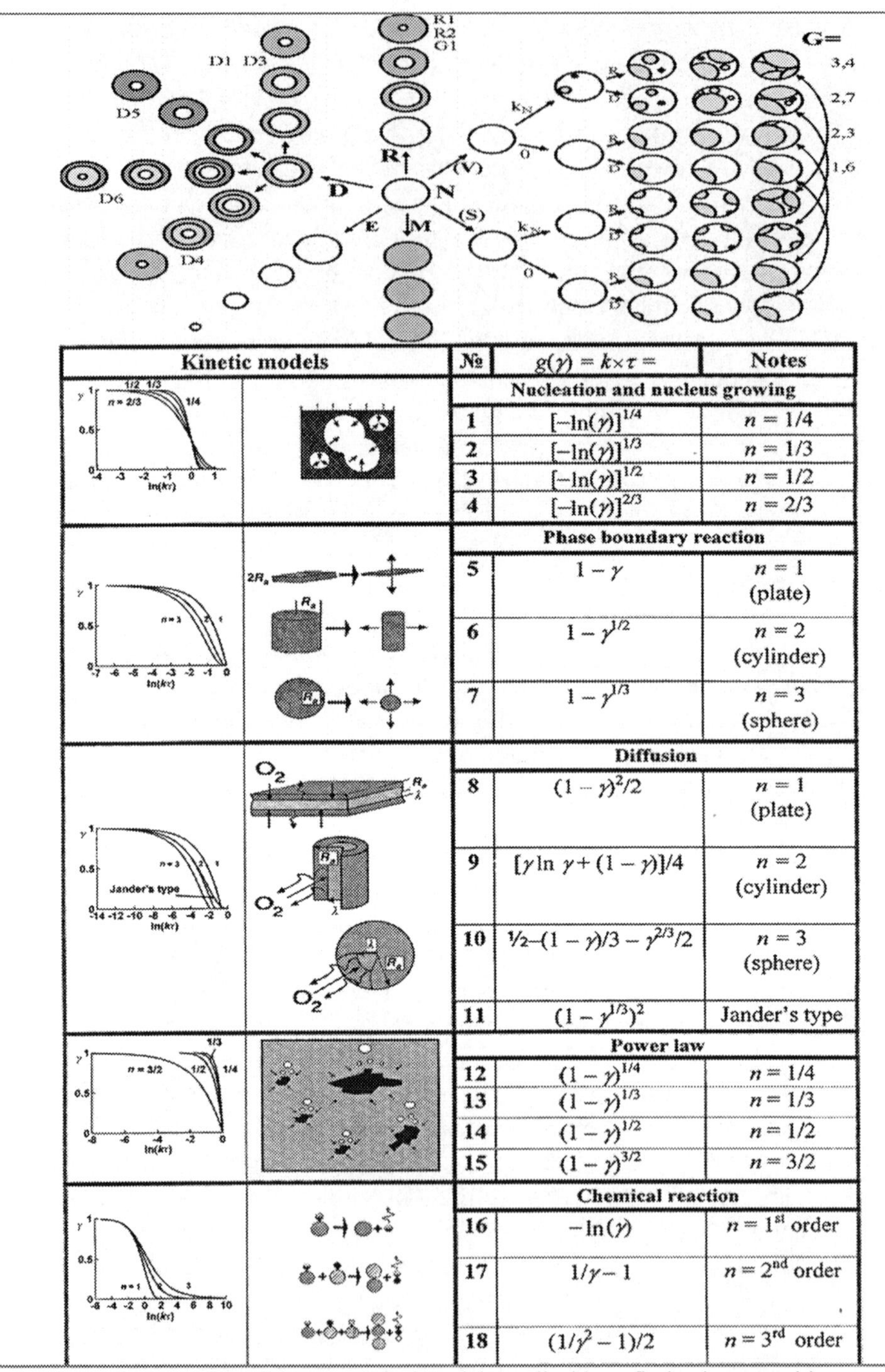

Kinetic models	№	$g(\gamma) = k \times \tau =$	Notes
		Nucleation and nucleus growing	
	1	$[-\ln(\gamma)]^{1/4}$	$n = 1/4$
	2	$[-\ln(\gamma)]^{1/3}$	$n = 1/3$
	3	$[-\ln(\gamma)]^{1/2}$	$n = 1/2$
	4	$[-\ln(\gamma)]^{2/3}$	$n = 2/3$
		Phase boundary reaction	
	5	$1 - \gamma$	$n = 1$ (plate)
	6	$1 - \gamma^{1/2}$	$n = 2$ (cylinder)
	7	$1 - \gamma^{1/3}$	$n = 3$ (sphere)
		Diffusion	
	8	$(1 - \gamma)^2/2$	$n = 1$ (plate)
	9	$[\gamma \ln \gamma + (1 - \gamma)]/4$	$n = 2$ (cylinder)
	10	$\frac{1}{2} - (1 - \gamma)/3 - \gamma^{2/3}/2$	$n = 3$ (sphere)
	11	$(1 - \gamma^{1/3})^2$	Jander's type
		Power law	
	12	$(1 - \gamma)^{1/4}$	$n = 1/4$
	13	$(1 - \gamma)^{1/3}$	$n = 1/3$
	14	$(1 - \gamma)^{1/2}$	$n = 1/2$
	15	$(1 - \gamma)^{3/2}$	$n = 3/2$
		Chemical reaction	
	16	$-\ln(\gamma)$	$n = 1^{st}$ order
	17	$1/\gamma - 1$	$n = 2^{nd}$ order
	18	$(1/\gamma^2 - 1)/2$	$n = 3^{rd}$ order

Fig. 60. – Diagram of possible geometrical models applicable to fit the decomposition mechanism [426] (notice that γ is used instead α). Courtesy of *V. Mamleev*, Almaty, Kazachstan

It has become a subject of criticism and discussion, which we do not want to repeat, other than stressing that real solid-state reactions are often too complex to be described in terms of a traditional set of simple circles and balls, see above Fig. 60, and based on the plain reaction orders (represented as integrals) and the associated numerical pairs of Arrhenius parameters. The ensuing data are treated in terms of customary and almost 'religious' constants [3,278,424], mostly linked with the activation energies that never express the ease of reaction (to be desirably related to the reactivity as a kind of 'tolerance' and to the reaction mechanism as a kind of 'annexation'). High values of activation energies are often misleading when determining the character of the process investigated, because high values do not mean difficult reactivity (typical for spontaneous and rapid exothermic crystallization) and low values do not imply easy reactivity (e.g. habitual for slow diffusion-controlled processes). The traditional interpretation is exactly the opposite, the best example being the repeatedly studied case of the reversible $CaCO_3$ decomposition, which is strongly mass-flow and heat-flow dependent (CO_2 partial pressure and through-out diffusion), creating the concentration gradients within the solid samples. Gradient-insensitive kinetic evaluation generates the numberless figures of somewhat insignificant numbers (still habitually called as the activation energies), which are thus strongly dependent on the experimental conditions applied (but often not refined or adjudicated).

However, the modern mathematical tools of thermal physics [9,437-439] make available powerful theoretical models employing nonparametric evaluations or neural networks [438] to be associated with the true reality of natural processes that are never at equilibrium neither without gradients by appreciating the decisive role of thermal fluxes [5,6,278,398,440-442]. In the scientific intent (often directed to generate publications based on non-isothermal kinetics), this approach has not yet been applied, but in the more urgent technological processing, such as industrially significant cement firing, ingot casting, arc melting or welding, which adoption became a real necessity to overcome manufacturing difficulties. In particular, we can generally assume that at some distance from the reaction zone where the phase transformation (solidification) is taking place, the viscous (molten) material undergoes irregular (turbulent) motion. It creates a mushy zone consisting of cascade of branches and side branches of crystals and interspatial melts, that remains lying between the original reactant (fluid) and the product (fully solidified region). Some chemical admixtures (as the alloy solution) are concentrated in the interspatial regions and ultimately segregated in the resulting micro-texture pattern. Such a highly irregular microstructure of the final solid can become responsible for alternative properties, e.g., reduced mechanical strength that is a costly factor thus worthy of active search as to resolve the intricacy of the processes involved. It follows that small changes in the surface tension, microscopic temperature fluctuations or non-steady diffusion may determine whether the growing solid looks like a snowflake, seaweed or spongy sinters. The subtle ways in which tiny perturbations at the reacting interface are amplified (cf. Fig. 62.) then become important research topics bringing necessarily into play either higher mathematics or non-Euclidean geometry of fractals. The challenge of theorists turns out to be the prediction of spacing of the final crystalline array [9,278,438,440-442], which requires computations:

(i) How the initially stationary flat interface accelerates in response to the moving temperature gradient, how local concentrations (e.g., impurities) adjust to this motion,

(ii) How the flat interface destabilizes, fluctuates and becomes branched,

(iii) How the resulting crystalline twigs interact with each other and

(iv) How the branched array coarsens and ultimately finds a steady-state configuration. In every detail [9], it is not an easy task at all.

A real solid-state reaction under thermophysical investigation, even those most ideal one, is intrinsically more complicated than most of us would like to believe requiring often coping with all these complications caused by actual localization of generated heat and fluids with respect to actual geometry necessary to achieve new levels of performance. The conceptual underpinnings for much of our more advanced perception of phase transformations have thus to use more complicated mathematics of non-equilibrium thermodynamics that is curiously employed to describe both the pattern formation in crystal growth and the so-called symmetry breaking (also known when describing the origin and distribution of elementary particles in the early Universe). Therefore, such an intricate approach is not too welcome in the ordinary practice of chemical kinetics and its further application to daily kinetic evaluations has not been assumed yet and neither is foreseen in the near future.

We, at least, can indicates that that the above discussed kind of `as-belief´ models depict both the ideal situation of only single reaction-controlling mode as well as the spherical representation for _all_ reacting particles. Though this simplification has no any investigational authorization such theoretical models sometimes (and from time to time even routinely) provides surprisingly good fitting for experimental measurements. It can be rationalized by certain model coincidence when assuming an improved geometrical fitting if we incorporate some additional symmetry features such as an adjustment of some regularity of pattern-similar bodies (globe-prism-cube-block-hexahedral-dodecahedral etc.). It somehow helps us to authorize the relations truthfulness and applicability of the above discussed oversimplified models when applied to more irregular structures, which we often are anxious to observe visually. Even symmetry generalization does not facilitate above modeling to the full-scale matching the real morphologies as witnessed in practice. Therefore, we ought to adapt another

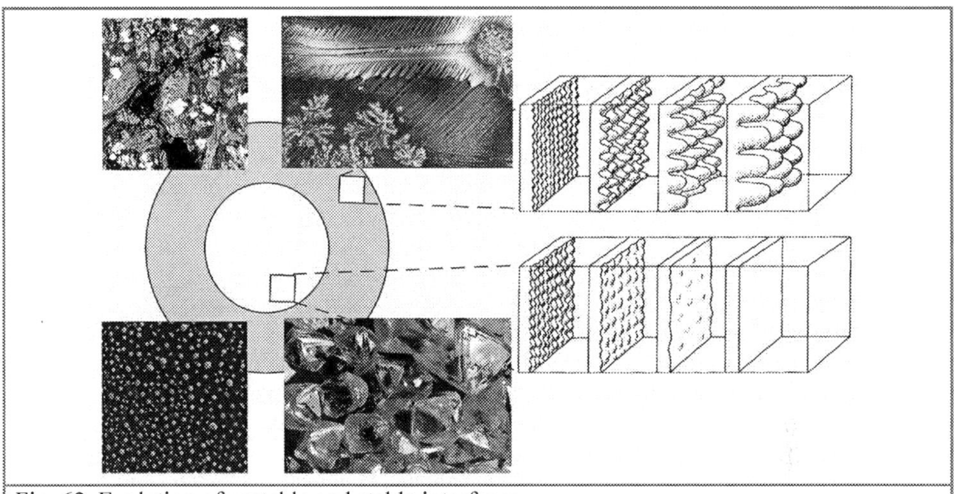

Fig. 62. Evolution of unstable and stable interfaces.

philosophy of modeling the reaction mechanisms either surviving with a simplified modelfree description using a blank fractal pattern or learning how to employ more complex mathematics providing functions instead of numerical value. The development may be different from what we presume, see above – Figure 62. Our theory proposed above in the Chapter 9, which is based on near-equilibrium thermodynamics [9,438], is applicable to thermal treatment and analysis only if a truly constant heating is assured, which assumes straight-forward heat interaction between the sample and a regulated thermostat or furnace. It does not involve the actual effect of heat liberated and/or absorbed by the reacting sample itself. Hence, it is necessary to extend it to the areas so far not commonly applied in the traditional domain of thermal analysis, although it is most pertinent to its feature of "real heating and/or cooling" phenomena (where the second derivatives can often become non-zero). Moreover, the classical sphere of thermodynamic definitions of stability is inapplicable to the determination of the morphology of growing interfaces, and current extensions have not yet furnished a fully acceptable alternative. The simplest assumption made is that the morphology that appears is the one which has the maximum growth rate and/or minimum undercooling (or, less commonly, overheating). Disregarding the initial process of new phase formation (nucleation), the kinetic models are described in terms of the overall atom attachments to the reaction interface due to either chemical reaction (bond redistribution steps) or interfacial diffusion (reactant supply). A stabilized (steady) state is taken for granted neglecting, however, directional changes (fluctuations). The physical-geometrical models also neglect other important factors such as interfacial energy (immediate curvature, capillarity) and, particularly, the internal and/or external transport of heat and mass to and from the localized reaction boundary, which may result in the breakdown of the planar reacting interface, and which anyhow, at the process termination, are responsible for complex product topology. Various activated disturbances are often amplified until a marked difference in the progress of the tips and depressions of the perturbed reacting interface occurs, making the image of resultant structures irregular and indefinable, see Fig. 62. This creates difficulties in. correlating traditional morphology observations with anticipated structures that are usually very different from the originally assumed (simple, planar) geometry. Depending on the directional growth conditions, so-called *dendrites* (from the Greek *'dendros'* = tree) develop, their arms being of various orders and trunks of different spacing due to the locally uneven conditions of heat supply. This process is well known in metallurgy (quenching and casting of alloys [306]), water and weather precipitates [9,440] (snow flakes formation, crystallization of water in plants) but also for less frequent types of other precipitation, crystallization and decomposition processes associated with dissipation of heat, fluids, etc.

It is always interesting to see how far the use of the above-mentioned classical methods can be extended into this non-equilibrium situation [278,438-441]. Rates of flows and grows, which associate undercoolings and supersaturations, are closely related by the functions whose forms depend upon the processes controlling transformation (atomic attachment, heat and electrical conduction or mass and viscous flow). In each case, the growth rate increases with increasing degree of undercooling and supersaturation. Perturbations on the reaction interface can be imagined to experience a driving force for such an accelerated growth that is usually expressed by the negative value of the first derivative of the Gibbs energy change, ΔG, with respect to the distance, r. For small undercooling, we can still adopt the concept of constancy of the first derivatives, so that $d\Delta G$ equals to the product of the entropy change, ΔS, and the temperature gradient, ΔT, which is the difference between the thermodynamic temperature gradient (associated with transformation) and the heat-imposed gradient at the reaction interface as a consequence of external and internal heat fluxes. Because ΔS is often negative, a positive driving force will exist to allow perturbations to grow, only if ΔT is positive. This pseudo-thermodynamic approach gives the same result as that deduced from the concept of zone constitutional undercooling [6,442] and its analysis is important for the manufacturing of advanced materials such as fine-metals, nano-composed assets, formation of quantum low-dimensional possessions, composite whiskers, tailored textured configurations, growth of oriented biological structures, processes involving water freeze-out (in, e.g., cryopreservations), all worth more detailed examination but, however, beyond the scope of this chapter [9].

We should also focus our attention to a specific case often encountered when an experimentalist faces chaotic trends in his resulting data [9,252,256] while studying chemical reactions in an apparently closed system. Such results are frequently refused by reasoning that the experiment was not satisfactorily completed due to ill-defined reaction conditions, unknown disturbing effects from surroundings, etc. This attitude has habitual basis in traditional view common in classical thermodynamics that the associated dissipation of energy should be steadily decelerated to reach its minimum (often close to zero) at a certain stable state (adjacent to equilibrium). In such a case, we are examining the reaction mechanism as a time-continuous development of regularly successive states, as shown above. In many cases, however, the reaction is initiated to start far away from its equilibrium or external contributions are effectual (in a partly open system) or reaction intermediates play a role of doorway agents (i.e., feedback catalysis). In such a case, the seemingly chaotic (oscillatory) behavior is not an artifact but real scientific output worth of a more detailed inspection where the reaction mechanism should not be understood in its traditional terms of time-continuous progress but also as a reflection of reaction time-rejoinder which feedback character yields rather complex structure of self-organization. Statistics show that the stability of such a non-equilibrium steady state is reflected in the behavior of the molecular/atomic fluctuations that became larger and larger as the steady state becomes more and more unstable, finally becoming cooperative on a long-range order. In many cases this effect is hidden by our insensitive way of observations. Particularly it becomes apparent for those reactions that we let start far from equilibrium; which first exhibit non-equilibrium phenomena but later they either decay (disappear) close to their steady state or are abruptly stopped (freeze-in) by quenching phenomena (often forming the reinforced amorphous state of non-crystallites).

Inorganic solid-state reactions are often assumed to proceed via branching [3,413,425] which may ultimately reveal a repetitive order. Let us assume a simple case of synthesis customarily identified in the manufacturing of cement. There are ideal and real reactions, which can be practically and hypothetically supposed to follow processes taking place during silicate formation [3,9,443]. There are two starting solid reactants A and B (e.g., CaO and SiO_2) undergoing synthesis according the scheme shown below (left) to yield the final product AB ($CaSiO_3$) either directly or via transient products A_2B ($CaSi_2O_4$) and A_3B ($CaSi_3O_5$). The formation of these intermediate products depends, beside the standard thermodynamic and kinetic factors, on their local concentrations (the degree of mutual admixture). If A is equally distributed and so covered by the corresponding amount of B, the production of AB follows standard kinetic portrayal (left headed arrow). For a real mixture, however, the component A may not be statistically distributed everywhere so that the places rich in A may affect the reaction mechanism to prefer the formation of A_2B (or even A_3B). The later decomposition of A_2B is due to the delayed reaction with the deficient B that is becoming responsible for the time prolongation of reaction completion. If the component A tends to agglomerate, the condition of intermediate synthesis becomes more favorable, undertaking thus the role of a rate-controlling process [3,9,443], see Fig. 63.

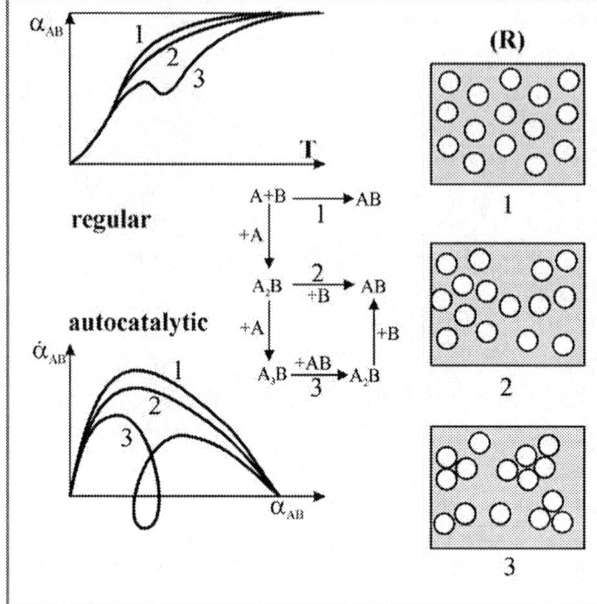

Fig. 63. – Ideal and the actual course of a potential solid-state reaction [3] where two reactants, A and B, undergo synthesis to product, AB, via transient products, A_2B and A_3B. The formation of the intermediates depends, besides the standard thermodynamic and kinetic factors, on the local concentration (particle closeness) dependent to the degree of reactant segregation. If mixed ideally the models discussed previously (Chapter 8 and 9) are applicable. If the agglomeration is effective the synthesis becomes favorable to produce intermediates and the entire course of reaction becomes self-catalyzed and may even exhibit oscillatory character.

The entire course of reaction can consequently exhibit an oscillation regime due to the temporary consumption of the final product AB, which is limited to small neighboring areas. If the intermediates act as the process catalyst, the oscillation course is pronounced showing a more regular nature. Their localized fluctuation micro-character is, however, difficult to be detected by direct physical macro-observations and can be only believed upon secondary characteristics read from the resulting structure (final morphology). Similarly, some glasses may exhibit a crystallization pendulum: after proceeding very fast in certain direction(s) the growth often stops due to the changes in concentration and converts into dissolution while in the other direction(s), where the growth rate was initially lower, it never becomes negative even if it decelerates effectively. Hence, a competition between several simultaneous processes takes often place, typical for such a non-equilibrium system and leading to curious morphology (plate or needle-shaped crystals [444]).

One of the most common phenomena is the self-organization due to the diffusion-controlled processes schematically given as follows.

$$dA \rightarrow \quad \text{diffusion} \quad \rightarrow dA$$
$$\downarrow \text{react} \qquad \Downarrow \qquad \downarrow \text{react}$$
$$dB \rightarrow \quad \text{diffusion} \quad \rightarrow dB$$
$$\textit{reaction} \Downarrow \textit{interface}$$

The local effect of counter-diffusion would become an important factor that may not only create but also accelerate above-mentioned oscillations, which is often an observable fact coming mostly from the interface reactions. As a result, many of peritectic and eutectic reactions turn out to pass an oscillatory regime providing regularly layered structures. For example, the directional solidification of the $PbCl_2.AgCl$ eutectic [9,35,372,447] is driven by temperature gradient and provides lamellar structure separated repetitively at almost equal lamella partition (see the previous shown Fig. 32.). Solidification under microgravity starts with higher undercooling compared with that observed in terrestrial condition, obviously due to the lack of convection. Typically, gravity-enhanced mass transfer leads to the effect of coarsening experienced at prolonged time and often at increased temperature [35].

Equally important is the domain of oscillatory processes common in solution chemistry, particularly known as the *Belousov-Zhabotinsky* (further abbreviated as BZ) reaction [9,263,264,445,446,448], cf. earlier Fig. 10. These processes were successfully simulated by the use of computers. Most famous is a simple scheme known as *"Brussellator"* [9] describing autocatalysis of the type $2X + Y \Leftrightarrow 3X$.

$$
\begin{array}{ccccccc}
reactants & & \rightarrow A & & products & & \\
& \downarrow & & \uparrow & & & \\
reactants & \rightarrow B & + X & \rightarrow & Z & + & Y \\
& \uparrow & & \downarrow & & & \\
products & \leftarrow P & \leftarrow 3X & \Leftrightarrow & 2X & + & Y \\
\end{array}
$$

A more complex case of so-called cross-catalytic reactions may involve two reactants A and B and two products Z and P. The intermediates are X and Y and the catalytic loop is caused by multiplication of the intermediates X, see the scheme above. Figure 63 above may well illustrate the input effect of reactant concentration within the given reaction mechanism (at the threshold concentration of A the steady sub-critical region changes from the sterile to the fertile course of action capable of oscillations in supercritical region. Although first assumed hypothetically, it enabled to visualize the autocatalytic nature of many processes and gave to them the necessary practical dimension when applied to various reality situations:

This scheme is typical for many biological systems such as the glycolytic energetic cycles where the oscillatory energy-intermediates are adenosintriphosphate (ATP) and adenosindiphosphate (ADP). It is also likely to explain the functioning of periodic flashes of the biogenic (cold) light produced by some microorganisms where the animated transformation is fed by oxygen whose energy conversion to light exhibit efficiency over 90%.

10.4. Accommodating non-integral power exponents

Let us first note that there are sometimes fussy effects of particle radius, r, which often encompass a wide range of reacting compacts. The previously derived relations stay either simply reciprocal $(\sim 1/r)$ if the whole reacting surface is exposed to ongoing chemical events or is inversely proportional to its square $(\sim 1/r^2)$ if the diffusion across the changing width of reactant/product layer became decisive. It is clear, however, that for a real instance we can imagine such a situation when neither of these two limiting cases is truthful so that the relation $1/r^n$ becomes effective and a new non-integral power exponent, n, comes into view to lay in the fractal region $1 \leq n \leq 2$.

It is somehow similar to the case of heat transfer across the layer d, which can fall in between two optimal cases limited by $1/d$ and $1/d^2$. The associated cooling rate ϕ is essentially influenced by the heat transfer coefficient, Λ, and the thickness of cooled sample, d, and relatively less by its actual temperature, T. At the condition of ideal cooling, where we assume infinitely high coefficient of heat transfer, the cooling rate is proportional to $1/d^2$, while for the Newtonian cooling controlled by the character of phase boundary, ϕ correlates to $1/d$, only. In practice we may adopt the power relation $\phi = 1/d^n$ (where n is an experimental constant lying again within $1 \leq n \leq 2$).

More complications can be anticipated when the involved particles undergo simultaneously the process of densification through shared sintering. Such kinetic description is often based on an analytical expression of the mass flux, which occurs between sites with positive and negative curvature along the connecting *necks* between two coupled particles. Such assessment respected both the Newtonian and Bingham material characteristics and provided linear contraction in the form of a power law [3] $\alpha = \Delta l/l_o = kt^m$, which can be mathematically transformed to a differential equation $d\alpha/dt = k\alpha^n$. The values of exponents varies from $m=1/2$ $(n=-1)$ for viscous and plastic flow to $m=2/5$ $(n=-2)$ for diffusion at the grain boundaries with variety of intermediate states $(m=1, 1/3, 2/7)$.

Discrepancies between the theory and the experimental results have led to the development of a sintering model, in which diffusion of vacancies from places with a negative surface curvature towards the particle center is the controlling process [449]. Survey of methods was given in [450] and the most known is the *Kingery-Berg* equation [451], $\alpha = \Delta l/l_o = r^{-6/5} k^{2/5} t^{2/5}$, mostly applicable for initial stages of particle sintering with the mean radius r [3,421].

We cannot omit the classical theories by *Burke-Turnbull* [452] who deduced parabolic relationship for grain growth kinetics or *Greenwood* equation

[453] relating surface energies and saturated concentrations between planar and curved surfaces. The theories for normal growth in pure single systems were reviewed by *Atkinson* [454] and results are mostly intended for the case of crystallization of finemetals, see next Chapter.

Smith [455] proposed at the early 1950s a classical approach emphasizing "*normal grain growth result for the interaction between the topological requirements of space-filing and the geometrical needs of surface tension equilibrium*". We can distinguish that in both 2-D and 3-D arrangements, the structure consists of vertices joined by edges ('sides'), which surround faces and in the 3-D case, the faces surround cells. The cells, faces, edges and vertices of any cellular structure obey the conservation law (Euler's equation), i.e., F − E + V = 1 (for 2-D plane) and F − E − C + V = 1 (for 3-D space). Here C, E, F and V are respectively the number of cells, edges, faces and vertices. Moreover the number of edges joined to a given vertex is its coordination number, z. For a topologically stable structure, i.e., for those in which the topological properties are unchanged by any small deformation, z = 3 (2-D) and z = 4 (3-D) everywhere. This can be best illustrated for 2-D structure by a 4-rayed vertex, which will tend to be unstable and to decompose into a two 3-rayed vertices, which is often termed as 'neighbor-switching'. For a 2-D structure, in which all boundaries have the same surface tension, the equilibrium angels at a vertex are 120°. The tetrahedral angle at 109°28' is the equilibrium angle at a four-edged vertex in 3-D having six 2-D faces.

The grain growth in 2-D is inevitable unless a structure consists of an absolutely regular array of hexagons. If even one 5-sided polygon is introduced and balanced by a 7-sided one then the sides of the grains must become curved in order to maintain 120° angles at the vertices. Grain boundary migration then tends to occur because of the curvature operational to reduce boundary surface tension so that any grain with number of edges above six will tend to grow because of concave and below six will incline to shrink because of convex sides.

It is clear that any reaction rate, particularly at the beginning of its acting-ion-exchange, must depend upon the size of the solid grains, which undergo transformation, growth or dissolution. Reaction rate, $\vec{r}$, should thus be inversely proportional to the particle size, r, in the form of a certain power law, i.e., $\vec{r} = r^{Dr-3}$, where D_r is the characteristic reaction dimension, which can be allied with fractal [456]. It is obvious that a mere use of integral dimensions, such as r^1 and r^2, is an apparent oversimplification. Moreover, we have to imagine that the initial rate is directly proportional to the extent (true availability) of ready-to-react surface as well as to its roughness (a kind of characteristic dimension, again). It seems that such a concept can be distinguished rather useful to describe the responding behavior of a reacting object towards the reaction while the characteristic fractal dimension relates in self the sum of all events occurring during the overall heterogeneous process.

There, however, is not a regular polyhedron with plane sides of exactly the tetrahedral angle $109^{\circ}28'$ between the edges. The nearest approach to space filling by a regular plane-sided polyhedron in 3-D is obtained by Kelvin ideal tetrakaidecahedra spaced on a body centered cubic lattice, but even that the boundaries must become curved to assure equilibrium at the vertices and the grain growth occur. It can be best illustrated by beer frost, which can be of two kinds (at-once draft beer with more fluid characteristic and the one of already aged beer with a more rigid structure) and which apparently differentiate in experts' taste being capable to self-adjust by boundary migration and gas permeate through the cell membranes to equalize pressure of adjacent bubbles.

The mobile-size of propagating particles in some highly viscous (such as 'gummy') system may insert another significant contribution, which are the relative slip velocities of the particle surroundings, particularly influential in the fluid media. For example, the growing particles can be observed as compact Brownian particles (for $r < 10$ μm, with the slip velocity naturally increasing with the decreasing size, $D_r \geq 1$) or as non-Brownian particles (where the slip velocity increases with the increasing size but only for mechanically agitated systems, $D_r \geq 2$) not accounting on the clustered Smoluchowskian particles ($r < 100$ nm, as 'perikinetic' agglomerates induced by Brownian motion and/or $r > 100$ nm, as a result of 'orthokinetic' agglomerations induced by shear fluid forces, both exhibiting $D_r \geq 0$). However, such a 'multi-fractal' behavior is often involved, which yields more than one fractal dimension at different levels of resolution or sizes of objects.

It can be summarized in the three major categories: boundary (B) fractal dimension, which characterize the boundaries of objects (with the values $1 \leq D_B \leq 2$); surface (S) fractal dimension, which describes the quality of surfaces of objects (with values $2 \leq D_S \leq 3$); and the mass (M) fractal dimension, which gives a measure of the way in which the individual sub-units of a whole are packed into the structure (falling in the range of $1 \leq D_M \leq 3$) [457]. To ease the fractal interpretation a certain scaling rule must be applied, which defines the specific overall linear growth rate, $\vec{g}$, as the ratio of the true linear growth rate, G, to the particle size, L. Particles with variable size, $L_1 \leq L \leq L_2$, are considered self-similar within the lower (1) and upper (2) cut-offs so that a range of D_R's is obtained ranging from 0 to 5 [458].

Ozao and *Ochiai* published and important treatise [481,482] assuming an arbitrary position of the particle fractal size (r) in the D-dimension (fractal) agglomerates where the diffusion flux (dN/dt) may become constant irrespective of r, i.e., $\approx \Delta N \{-ln (r/r_o)\}^{-1}$ or $\approx \Delta N \{r^{(2-D)} - r_o^{(2-D)}\}^{-1}$ for $D=2$ and $2<D\leq3$, respectively. This provides modified models of diffusion such as D_4 in the form of $d\alpha/dt \approx \{(1-\alpha)^{(2-D)/D} - 1]$ or can apply to the growth (JMAYK) models with the fractal exponent $(D-1)/D$ as $(1-\alpha)^1 \{- ln (1-\alpha)\}^{(D-1)/D}$. By a simple

substitution of D's with the integers 1,2 and 3 we match the standard equations listed in the previous table 10.I.

Specifically, the process of diffusion may not be as simple as a straightforward shift across the reactant or product layer as shown in the previous paragraph but it may be obstructed by the layer inherent structure, i.e., its internal make-up capable of gradual disordered shipping of ready-to-move species. In such a case the customary model of a *random walker* (often treated in terms of the self-styled ritual of a nicknamed "drunken ant") is worth noting as a useful illustration [9,459,460]. With the advancement of time, t, we may see that the walker progresses its wandering in such a way that the average of the square of his displacement increases monotonically. The explicit form of this increase is contained in the law, which concerns the mean square of displacement, i.e., $\{x^2\}_t = t$, or better, $\{x^2\}_{t+1} = t + 1$. Additional information can be found in the expectation values of higher powers of x and we can assume that $\{x^k\}_t$ equals zero for all odd integers, k, while $\{x^k\}_t$ is nonzero for the other even integers. The displacement can be identified either with certain length, say L_2, which is given by $\sqrt{\{x^2\}} = \sqrt{t}$ or with another length, $L_4 \equiv \sqrt[4]{\{x^4\}} = \sqrt[4]{3} \sqrt{t} [1 - 2/3t]^{1/4}$. The both distances display the same asymptotic dependence on time regardless of the definition of L's. We usually call the leading exponents as the scaling exponents, while the non-leading exponents associate with a correction-to-scaling, which is easy to generalize for $\sqrt[k]{\{x^k\}}$ as $A_k \sqrt{t} [1 + B_k t^{-1} + ...]^{1/k}$.

We can see that several different definitions of a certain, universally characterisable length, ξ, are scaling correspondingly to the value of $\sqrt{t}$, which naturally involves the above-mentioned interrelation between the general behavior of power laws and a symmetry operation. The scaling symmetry [461] is thus a natural root of the wide applicability range of fractal concepts in physics. The relation, $t \sim \xi^d$, gives the scaling exponent, d, which explicitly reflects the asymptotic dependence of a characteristic mass on a characteristic length. The most important consequence is related to diffusion limited aggregations, which are in the centre of interest of many physical systems such as electrochemical deposition, dendritic solidification and growth [9,440,441], viscous fingering, chemical dissolution, rapid crystallization or dielectric and other kinds of electric breakdowns. It covers as many as thousands of recognized fractal systems known today in nature, which even includes diffusion-limited aggregations of aficionados (neuronal outgrowth, augmenttation of bacterial colonies, etc.).

We can presuppose that transport properties due to fractal nature of percolation change physical laws of dynamics. For a sufficiently randomly diluted system, even the localized modes occur for larger frequencies, which can be introduced on basis of bizarrely called *'fractons'* [462]. Their density of states then shows an anomalous frequency behavior and, again, the power laws can characterize the dynamic properties. On fractal conductors, for example, the density is proportional to L^d and approaches zero for $L \to \infty$. If we increase L, we increase the size of the non-conducting holes, at the same time decreasing the conductivity, σ, which, due to self-similarity, decreases on all length scales, leading to the power law dependence, thus defining the critical exponent, μ, as $\sigma \sim L^{-\mu}$. Due to the presence of holes, bottlenecks and dangling end, the diffusion is also slowed down on all length scales. Classical Fick's Law, assuming that the 'random walker' has also a probability to stay in place (using the standard relation, $x^2(t) = 2dDt$, where D is the diffusion constant and d is the dimension of lattice) loses here its validity. Instead, the mean square displacement is described by a more general power law, $x^2(t) \cong t^{2/dw}$, where the new exponent, d_w, is always greater than two. Both exponents can be related through the Einstein relation, $\sigma = e\, n\, D/(k_B T)$, where e and n denote respectively the charge and density of mobile particles. Simple scaling arguments can be used to interrelate, dw and μ, since n is proportional to the density of the substrate and thus the whole right-hand side can be taken proportional to the term, $L^{d'-d}\, t^{2/dw-1}$. As a result, $dw = d' - d + 2 + \mu$, where d' can be substituted by the relation, $log3$-vs-$log2$, so that dw becomes proportional to $log5/log2$, which, however, is not easy to ascertain in general cases.

For auxiliary modeling, we start with a model of a walker again. Assume that the walker has four perimeter sites to enter with an equal probability, followed by two walkers with six growth sites and their corresponding (but non-identical) probabilities, and so on. The most that we can say about the walker's forming cluster is to specify the probability distribution of growth sites. If the growth rule of diffusion limited aggregation is simply iterated we obtain a larger cluster characterized by a range of growth probabilities that span several orders of magnitude, possibly stretching out literally from tips to fjords. Aggregation phenomena based on random walkers correspond to the Laplace equation, $\nabla^2 \Pi(r,\, t)$, which stands for the probability, Π, with which a walker is at the position r and at time t. However, a range of circumstances can turn up that have nothing to do with random walkers, but have a crucial effect, such as that experienced in viscous fingering phenomena.

Consider a random walker on the infinite cluster under the influence of a bias field. This can be modeled by giving the random walker a higher probability of moving along the unrestrained direction. In a Euclidean bias field, the walker gets a velocity along the direction of the field. In a topological bias field, the walker can get stuck in loops and kept in dangling ends so that both act as random delays on the motion of the walker. The length distribution of the loops and dangling ends in the fractal structure determines the biased diffusion. At the critical concentration, due to self-similarity on all length scales, the length, L, of the teeth is expected to follow a power distribution and the time, τ, spent in a tooth increases exponentially with its length. A random walker has to wait on the average time steps, τ_i , before he can jump from a site i to one of the neighboring site, $(i +1)$. The singular waiting time changes the asymptotic law of such diffusion drastically, from the power law to the logarithmic form. We often have a paradoxical situation that on the fractal structure of the percolation cluster the motion of a random walker is slowed down by a bias field.

In general, the experience with a simple random walk provides us familiarity, whenever entropy literally prevails over energy; the resulting structure will be dominated by randomness with disorder rather than by a strict Euclidean order. Therefore, we may well expect to find fractal structures with a scaling geometry analogous to that of the simplest unbiased random walk discussed above. Of course, the real structures such as dendrite growth patterns, which are most familiar in the case of familiar snowflakes, do not often occur in the acquiescent environment of periodic fluctuations. Rather, their curious shape arises from the local asymmetry of the constituent water molecules and their capability to bias random walk by mutual limiting the inward and outward transport from the growing crystals (latent heat released away and vapor or liquid water moved inside). Such fractal models achieve practical use for the description of roughness of materials' surfaces (even the texture of fabric [463]) as well as in the portrayal of thermal waves (hyperbolic heat conduction) under the application of global space-time modeling [464].

For many multifaceted (i.e., complex fractal-like) systems, the instantaneous rate constant must be further specified as an evolutionary (time-dependent) coefficient, $k(T,t)$, conventionally represented by the product, $k(T)$ with t^{-h}. The exponent-parameter, h, expresses thus the system 'fractality' , which is for strictly homogeneous conditions (under vigorous stirring and without accounting for the effect of initial conditions) equal to unity. For a simple case of [A+A] reaction in the one-dimensional space, h is ½, but is actually given by ($h = 1 - d_s$), where d_s is the so-called random-walk occurrence (spectral) dimension. Thus for both the [A+A] reaction and the geometrical construction of porous aggregated (based, e.g., on triangles called *Sierpinski* gasket [465], h is around 0.32, because the typical values for d_s are 1.36 (gasket) or 1.58 (percolation). Assuming a more general exponent, n, for $[A]^n$, we can find for the ordinary diffusion-effected case $n = 1 + 2/d_s = 1 + (1 - h)^{-1}$ so that the expected value for n is 2.46 for a gasket, 2.5 for a percolation cluster and 3 for the standard homogeneous one-dimensional reaction, still retaining its bimolecular character in all above cases. Practically, we have to distinguish a connected fractal, such as reactants in the form of 'dust' or 'colloids' where the effective dimension is $h \cong 1/3$ (i.e., $0 < d_s < 1$, providing $3 < n < \infty$), and segregated reactants, which often conjure self-organized reactions, which are more complex for the description [9,252,258]. The effect of segregation is in concurrence with the almost unknown *Wenzel's Law*. Already in 1777 Wenzel stated [466] that *for heterogeneous reactions holds that: the larger the interface, the faster the reaction*, that is, the rate per unit surface is reconcilable and the interface (and its character) thus backs modeling with self-similar fractals.

Hidden heterogeneity of even apparently homogeneous reacting systems (unseen effect of impurities, surface structure of a container, external influences of undetected fields or even rays) can change the simple reaction path to a more complex course of action where a customary analysis in terms of the integral reaction orders may fail. It was already noticed and analyzed in more details by *Kopelman* [477,478] who introduced fractal orders even for homogeneous reactions. In addition, anomalous diffusion and associated 'fracton' or spectral dimension became a progressive tool in the characterization of fractal properties of variously structured materials such as porous glasses, organic membranes or filter papers [479]. Habitual practice reveals that the diffusion experiments require microscopic measurements below the optical diffraction limit. However, the anomalous reaction kinetics, which are a direct consequence of anomalous diffusion, can be studied via macroscopic measurements by exciton (triplet) recombination characteristic [479], e.g., for naphthalene-doped microporous materials or other permeable stuff (e.g., Vycor etched-pores glass). The temperature studies enable separation of energetic and geometric features of the pore space and can also be approved by simulations.

The general meaning of nonintegrated exponents in solid-state kinetics was discussed at the turn of 1970s [1] and explored in details by *Šesták* and *Berggren* [480] who proposed the two-parameter power law capable for mathematical fitting for almost all types of processes. Though this generalized relationship does not depict any particular form of an analytical function for its integral representation, see the previous table 10.I., it has become the most widely applied model-free relation for the description of majority of processes where the traditional (but simplified) models based on regularly-shaped particles failed. This concept is factually close to the novel kinetic approaches based on model-free descriptions.

As shown above, the correspondingly motivated concept of self-similar size distribution in powdered compacts was commenced by *Ozao* and *Ochiai* in the turn of nineties [481,482] but has remained somehow underestimated in the recent literature and kinetic practice. It, however, is obvious that a self-similarity of fractal-like nature within the particle size distribution is naturally observed on all sorts of grain agglomerations, which are obtained by size reduction using common methods of pulverization such as crushing, milling, levigation, etc. Macroscopic feature of such a powdered sample (particle-size distribution, shape and its thermal and mechanical pre-treatment and consequent behavior during the structure breakdown) affects thermal activities of residue. Based on a formulation applied in the theory of stochastic processes a kind of master equations can be derived and introduced for the size distribution, which thus characterizes the size reduction process. Using a scaling concept a generalized form of both the Gaudin-Schubmann and Rosin-Rammler equations became functional, showing their intimately mutual relationship [481,482].

10.5. Alternative creation of geometrical models, significance of limits and self-similarity, Koch fractal curve

We learned in Chapter 1 that we live in an apprehended world that provides us a limited view within a certain horizon of knowledge. The associated limits are inescapable and always involve something mysterious about them so that it would be felt insufficient not to deal with them in more details although it somehow falls beyond the entire scope of heat but it links, however, with the associated domain of modeling and the related problems of order and disorder. Such limits create and characterize new quantities and new objects and the study of these unknowns was the pacemaker in the early mathematics and consequently has led to the creation of some most beautiful mathematical inventions, which involves the analysis of *patterns* and *forms*.

Pattern	Dimension, D	No. of piece, a	Reduction factor, s	Dimension
Line segments	1	$3 = 3^1$	1/3	$L(R) = N * r$
Square	2	$9 = 3^2$	1/3	$S(R) = N * r^2$
Cube	3	$27 = 3^3$	1/3	$C(R) = N * r^3$
Replicate symmetry	D	$n = 3^D$	1/3	$D(R) = N * r^D$

When building geometrical models we have to assume that there is a way to measure the degree of their complexity by evaluating how fast the body length, or surface, or volume, increases if we measure them with respect to smaller and smaller scales. The fundamental idea is that the two quantities (length-surface or surface-volume) on the one hand and the scale, on the other, do not vary arbitrarily but rather are related by a strict law, which allows us to compute one quantity from the other. The relevant relationship is the *power law* again, which turns out to be very useful for the discussion of true dimension. The best way to regulate the dimension is a process of transformation which we obtain from a photocopier with a reduction feature. For a reduction factor s shown in the table below it follows that there is a nice power law relation with the number of pieces a, i.e., $a = 1/s^d$, where the exponent agrees exactly with those familiar as the topological dimension of customary modeling. We can equivalently write [483] $D - log(a)/log(1/s)$ where D is the self-similarity dimension (which is for the special cases non-integral such as for Koch curve equals to 1.2619 $= log(4^k)/log(3^k)$ and for Sierpinski gasket to *1.5850* $= log(3^k)/log(2^k)$). Another question would be to see another relation between the power law of the length measurement (using different compass setting) and the self-similarity dimension of a general fractal curve, $D_s = 1 + d$, where d denotes the slope in this alometric (*log/log*) diagram of length measurement u versus its precision (compass setting) 1/s, i.e., *log(u) = d log (1/s)*.

There is yet another dimension called the correlation dimension D_{cor}, which is related to the number of point available to measurement, i.e., *log(C_r)/log(R)* = D_{cor} where C_r is the number of points having smaller distance that a given distance *r*. For a standard symmetry pattern used in the above reveled spherical models we can rephrase the standard (fractal) dimension for the array of spheres consisting of *n* linked balls with their diameter *r*, which cover the given (geometrical) shape. It ensues the relation *(r/n)(dn/dr)*, which provides the dimension proportionality always equal one for any of two compared arrays *log(N/n)/log(r/R)*.

Besides the traditional arranging of balls (circles) we can include to our modeling patterns build upon the simple geometrical gathering and multiplication of other shapes. Let us mention the discovery already ascribed to *Pythagoras* regarding the theme of *incommensurability* of side and diagonal of a square, i.e., to the ratio of the diagonal and the side of a square, which is not equal to the ratio of two integers. The computation of square roots is an interrelated problem and has inspired mathematicians to discover marvelous geometrical constructions.

A most common construction yields the family of Pythagorean trees when encompassing continuous attachment of a right triangle to one hypotenuse side of the initial square continuing by the attachment of two squares and so on, see Fig. 63. It rosettes broccoli and such a (fractal-like) branching construction is of a help to various branches of science (most common in botany). Even more allied is the construction passing from equilateral triangles to isosceles triangles with angles exceeding the right angles. It is worth noting that all these beautiful constructions are *self-similar*. The computed situation is, however, different to natural image of a broccoli, or better a tree, where the smaller and smaller copies accumulate near the leaves of the tree so that the whole tree is not strictly self-similar but remains just *self-affine,* as seen in Figure 63 above.

When *Archimedes* computed π by his approximation of the circle by a sequences of polygons, or when the *Sumerians* approximated √2 by an incredible numerical scheme, which was much later rediscovered by *Newton*, they all were well aware of the fact that they are dealing with the *unusual numbers* [484]. As early as in the year 1202, the population growth was evaluated by the number of immature pairs, i.e., $A_{n+1} = A_n + A_{n-1}$ with $A_o = 0$, $A_1 = 1$, continuing 1,2,3,5,8,13,21,34,55,89,144,... (meaning that the state at time n+1 requires information from the both previous states, n and n-1, known as two-step loops) called the *Fibonacci* sequence. The ratio of A_{n+1}/A_n is steadily approaching some particular number, i.e., 1,618033988..., which can be equaled to $(1+√5)/2$ and which is well-known as the famous golden mean ('*proportio divina'*). In fact it is a first member of a general series $x = a + (1/a + (1/a + (1/a + ...$, where $a = 1$, while the second member for $a = 1/2$ represents the limit of $\sqrt{2}$. This golden mean characterizes the side ratio of a rectangle, i.e., $x/1 = (x+1)/x$, which over several centuries inspired artists, architects and even scientists to speculate about such a lovely asymmetrical manifestation.

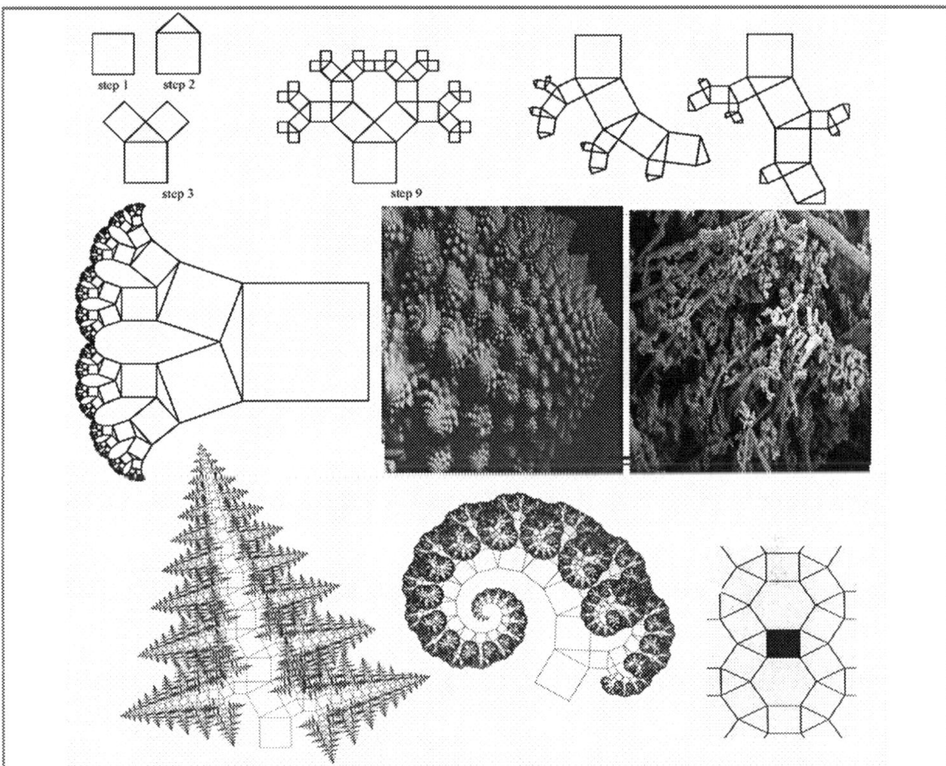

Fig. 63. – The popular construction of Pythagorean tree starts by simple children-like drawing of a square with an attached right triangle. Then two squares are attached along the free sides of triangle followed, by gradual repeating, attachment of squares and triangles, see upper left. It certainly can be modifies in various ways, the right triangles need not be isosceles triangles providing another degree of freedom, see upper right. After as many as 50 iterations, the results cannot look more different: In the first case, when the applied angle is greater than $90°$ (see middle), we can envisage the structures as some kind of broccoli-vegetable (photo left) or broccoli-like electrodeposited metallic tantalum (photo right) or a fern or even a pine tree (bottom left). Otherwise it can remind us of a spiraling leaf or decorated coiled shell (bottom right), worth noting that the size of triangles in the bottom are the same in both figures.

It is almost ironic that physics, at its most advanced level, has recently taught us that some of these conjectures, which even motivated *Kepler* to speculate about the harmony of our cosmos, have an amazing parallel in our modern understanding of nature. It literally describes the breakdown of order and the inquisitive transition to disorder, where the golden mean number is factually characterizing something like the 'last barrier' of order before chaos sets in (cf. previous Fig. 10). Moreover, the *Fibonacci* numbers occur in a most natural way in the geometric patterns, which can occur along those routes [484].

Instructive case is the geometry of spirals, which is often met and admired in nature as a result of self-organizing processes to take place in the distinctly reiterating steps. One of them allow us to construct $\sqrt{n}$ for integer n, which can be called the *square root spiral,* best exhibiting the so-called *geometric feedback loop,* see Fig 64. Another, but yet traditional *Archimedes* spiral is related to the arithmetical sequences when for an arbitrary but constant angle, ϕ, and points on the spiral r_1, r_2, r_3, ... each consequent number r_n is constituted by an arithmetic sequences, i.e., $r_3 - r_2 = r_2 - r_1$ or $r \approx \phi$. Replacing the arithmetic mean, $(r_1 + r_3)/2$, by the geometric mean, $\sqrt{(r_1\ r_3)}$, the other spiral is attained, which is known as famous *Bernoulli* logarithmic spiral, where $r_1/r_2 = r_2/r_3$ or $ln\ r \approx \phi$. Most amazing property is that a scaling of the spiral with respect to its center has the same effect as simply rotating the spiral by some angle, i.e., it shows a definite aspect of self-resemblance. Analogously we can generate an infinite polygon, which can easily be used to support any smooth spiral construction where the radii and length of the circle segments are a_k and $s_k = (\pi/2)a_k$ respectively. By adjusting a_k we can achieve the total length either finite ($a_k = 1/k$) or infinite ($a_k = q^{k-1}$ if q is a positive number smaller one). If for q we choose the golden mean we acquire a golden spiral which can also be obtained by a series of smaller and smaller reproducing rectangles (cf.previous Fig. 63).

As shown above for symmetrical bodies we can see that _any_ structure can become self-similar if it can be broken down into arbitrarily small pieces, each of which is a small replica of the entire structure. Here it is important that the small pieces can, in fact, be obtained from the entire structure by a *similarity transformation* (like cited photocopying with a reduction feature). It has a widespread applicability ranging from solid-state chemistry (sintering) to physiology. Already *Galileo* noticed that a tree cannot grow to unlimited height as the weight of a tree is proportional to its volume.

Scaling up a tree by a factor, s, means that its weight will level by s^3 but the cross-section of its stem, however, will only be scaled by s^2. Thus the pressure inside the stem would scale by $s^3/s^2 = p$ not allowing p to grow beyond a certain limit given by the strength of wood. From the same reason we cannot straightforwardly enlarge a match-to-sticks model to an authentic construction from real girders because each spill-kin must be *allometrically* full-grown to the actual building beam. Another case is the body height of beings, which is growing relatively faster than the head size. These features are quite common in nature and are known as allometric growth.

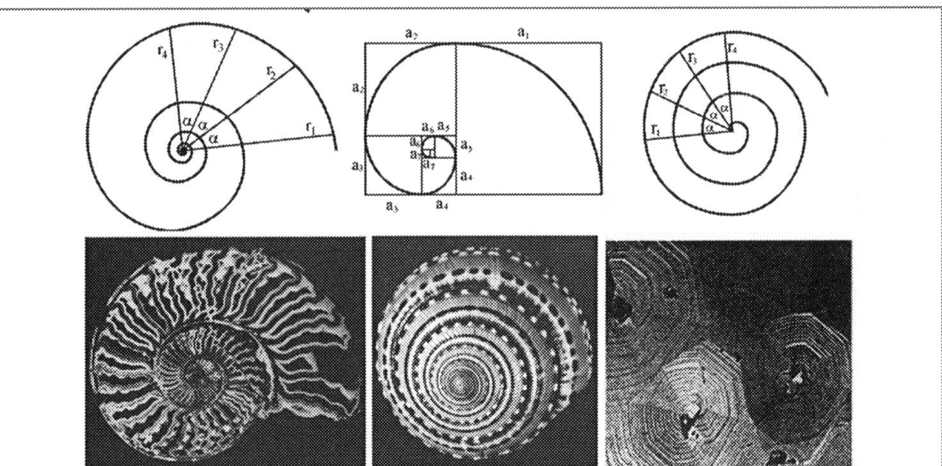

Fig. 64. – The previously shown structures responsible for yielding the family of Pythagorean trees are very much related to the construction of the square root spiral depicted here. It is much related to the construction of square root, $\sqrt{n}$, for any integer, n, (maintaining the sides adjacent to the right angle to have the starting length 1 and $\sqrt{2}$, $\sqrt{3}$, $\sqrt{4}$ and so on, upper left). This construction belongs to the class of objects, which defies length measurements, the most famous model is the Archimedes spiral exhibiting the winding of a rolled carpet, which maintains the distance between its winding invariable (constant angle, α, arithmetic sequences of radii, $r_2 = (r_1+r_3)/2$, see upper right). Stepping along the logarithmic spiral in steps of a constant angle yields the geometric sequence of radii (often known as *wonderful spiral* holding $r_2 = \sqrt{(r_1 r_3)}$ where $r_n/r_{n+1} = a$). Another case is the *golden spiral,* which starts from the a rectangle with sides, a_1 and a_1+a_2, where the ratio, a_2/a_3, is proportional to the golden mean, $(1 + \sqrt{5})/2$. The rectangles break down into squares and even smaller rectangles and so on. In general, the spiral growing has, in general, a very important consequences in nature, such as life-grown structures of shells (bottom middle) or even inorganic objects are attracted to exhibit logarithmic intensification akin to the crystal growth of ammonite safeguarding thus a law of similarity (as shown bottom left) and nucleation-growth pattern of plate-like formation of crystals of high-temperature superconductors (cf. Fig. 35.).

One of the most common examples of self-similarity are the space-filling curves, i.e., the derivation of a precise construction process, best known as the *Koch curve* [485], see Fig. 65., often serving to model the natural image of a snowflake upon a derived figure known as *Koch* island. The curves are seen as one-dimensional objects, because they fill the plane, i.e., the object, which is intuitively perceived as two-dimensional. This view envelops the sphere of *topology*, which deals with how shapes can be pulled and distorted in a space that behaves like a 'rubber' sheet. Here we imagine that a straight line can be bent into a curve, colligated to form circles, further pulled out to squares or pinched into triangles and further continuously deformed into sets, and finally crooked into a Koch island, cf. Fig. 65. The contour, which is commonly taken

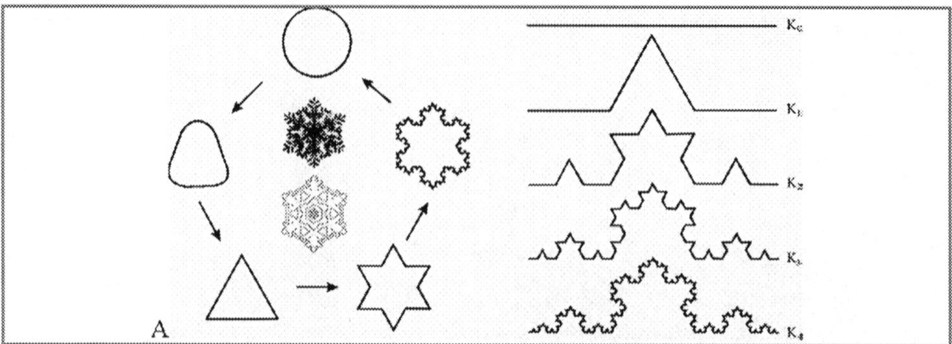

Fig. 65. – A circle can be continuously distorted, first into a triangle and hexagon continuing its deformation to the Koch island (left), which is composed of three congruent parts, each of them being a *Koch curve* (right). Here the each line segment of the middle third is replaced by two new segments forming a tent-like angle. All end-points of the generated line segments are part of the final curve. The circle (characterized by non-integral π) and its deformed constructions are topologically equivalent. Koch island has a typical fractal (non-integral) dimension exhibiting infinite length but enclosing a finite area similar to that of the circle, which surrounds the original triangle. Moreover, such a famous pattern of this artificially created Koch 'snowflake' obviously bears the similarities with the real snowflakes found in nature (see the inset). These formations have typically many branches at all levels of details and belong to the comminuting objects called *dendrites,* cf. Fig.61. (more details in ref. [9]).

as a linear representation of a reacting interface can be variously shaped and fragmented showing the great variability of its possible geometry.

As with any type of modeling we can define the primary object used for such a construction as an *initiator*, in our case this is a *straight line*. We can partition it into three equal parts and replace the middle third by an equilateral triangle taking away its base. As a matter of fact it completes the basic construction step forming the initial figure which is called *generato*r. By repeating and taking each of the resulting line segments, further partitioning and so on, calls for the Koch curve in the requested details. A different choice of the generator, usually a polygonal line composed of a number of connected segments, produces another fractal with self-similarity, such as three equal lines symmetrically pointing toward the central point will provide continual assembly of joined hexagons, which is the routine honeycomb structure common in nature. When each step in the construction is performed the length of the curve increases by a factor 4/3, so the final curve being the result of limitless number of steps is infinitely long. The similarities between this curve and the path of a quantum-mechanical particle become evident when we consider viewing the Koch curve with a finite spatial resolution [486]. In this case, the infinitely many wiggles in the curve, which are smaller than some minimum length, say Δx cannot be detected and the measured curve length will be infinite. However, this length will depend on Δx and will increase without limit as $\Delta x \rightarrow 0$. The conventional definition of length gives a quantity, which depends on the resolution with which the curve is examined. It is not very useful when applied to curves like the Koch curve so that *Hausdorff* [483] proposed a modified definition for the length, L, proportional to $L(\Delta x)^{D-1}$, where L is the usual length measured when the resolution is Δx, and D is a number chosen so that L will be dependent of Δx, at least in the limit $\Delta x \to 0$, which implies that D (= $log(4)/log(3) \neq 1$).

Accordingly, we can define the quantum-mechanical path of a particle if we imagine measuring the position of a free particle with a spatial resolution Δx at times separated by an interval Δt. The path is then defined as the curve determined by drawing straight lines between the points where the particle was located at sequential times (at the classical level it will be just a straight line). The localization of a particle within a region of the size Δx results, according to the Heisenberg uncertainty principle, in an uncertainty in the momentum of order $h/(2\pi\Delta x)$. As the particle is more and more precisely located in space its path is thus being increasingly erratic. There is another property of the fractal Koch curve to be mentioned in this respect of quantum-mechanical path and which is above-mentioned self-similarity. If we view a Koch curve with a resolution $\Delta x' = (1/3)\Delta x$ then the curve we see is, up to repetitious and translations, just a scaled down version of the curve we saw when the distances being resolved were of the size Δx. The path of quantum-mechanical particle (mass m) will be self-similar if $\Delta L \propto \Delta x$, i.e., to get a self-similar path we must scale the time between position measurements of the particle in proportions to the square of Δx. That is, if $\Delta x \propto 2\pi m(\Delta x)^2/h$, then the resulting path become self-similar. Thus just as the fractal nature of the quantum-mechanical path reflects the Heisenberg uncertainty principle, the conditions of self-similarity is a reflection of the underlying dynamics, i.e., $E = p^2/2m$. When the particle possesses some nonzero average momentum, p (i.e., in the classical limit the particle is moving) since then the translation from the classical result $D = 1$ to the quantum mode $D = 2$ can be seen. It worth noting that the path of a particle undergoing Brownian motion is also a fractal with $D = 2$ [487]. Further interrelation between the classical and quantum path was given in the preceding paragraph 6.7 dealing with quantum diffusion.

For a given patch of the plane, there is another case of the space-filling curve, which now meets every point in the patch and which provides the fundamental building blocks of most living beings. It is called *Peano-Hilbert* [488] curve and can be obtained by another version of the Koch-like construction. In each step, one line segment (initiator) is replaced by nine line segments, which are scaled down by a factor of three. Apparently, the generator has two points of self-intersection, better saying the curve touches itself at two points, and the generator curve fits nicely into a square. Each subsquare can be tiled into nine sub-subsquares (reduced by 1/9 when compared to the whole one) and the space-filling curve traces out all tiles of a subsquare before it enters the next one. It is worth noting that any finite stage of such construction is rather awkward to draw by hand and is even difficult to manage by a plotter under computer control.

We should bear in mind that the coast of a Koch island has the same character as a circle, (which calculated circumference is as inaccurate as the implicated constant, π) but which length approaches infinity having an important consequence in models build upon the spheres (circle projection). In contrast, the area of the Koch island is a finite number. Alternatively, a typical coastline or a morphological image may curiously not exhibit the 'meaningful' length, as it is dependent to the adopted compasses set at a certain distance. By varying the size of our compass setting we can arrive to the above mentioned *log*-vs.-*log* diagram where horizontal axis comprises the logarithm of the inverse compass setting, t, interpreted as the precision of measurements. The vertical axis is reserved for the logarithm of measured length, h. The plot essentially falls onto a straight line exhibiting the power law again and enabling to read off the exponent, d. It regularly approaches the relation $t = 0.45\ h^{0.5}$ which surprisingly is in agreement with the classical *Newtonian law of motion*, which implies that the distance fallen, h, is proportional to the square of the drop time, t, i.e., $h = g\ t^2/2$. Inserting the gravitational acceleration of 9.81 [m/s^2] it yields the similar power relation, $t = 0.452\ h^{0.5}$.

Correspondingly, we can measure the length of the Koch curve with different compass settings are smaller that the value obtained for the real coast measurement $d \approx 0.36$. This approach becomes important even for some physical measurements such that of porosity, where we employ different scale setting by using different size of molecules, typically from mercury to nitrogen. For a basal metabolic rate it ensues a power function of body mass with exponent $d = 0.5$. Therefore we may generalize the dimension found in the alternative also to shapes that are not strictly self-similar curves such as seashore, interfaces and the like. The *compass dimension* (sometimes called 'divider' or 'ruler dimension') is thus defined as $D = 1 + d$, remaining that d is the true slope in the *log*-vs.-*log* diagram of the measured length versus reciprocal precision. For the measured coast, for example, we can say that the coast has a fractal (compass) dimension of about 1.36. Of course, the fractal dimension of a straight line remains one because $1 = 1 + 0$.

In conclusion, the curves, surfaces and volumes are in nature often so complex that normally ordinary measurements may become meaningless. The way of measuring the degree of complexity by evaluating how fast length (surface or volume) increases, if we measure it with respect to smaller and smaller scales, is rewarding to provide new properties – often called dimension – in sense of fractal, self-similarity, capacity or information – all are special forms of the well-known *Mandelbrot's* fractal dimension [489].

Worthy of extra noting is the *box-counting dimension*, which is useful for structures that are not at all self-similar, thus appearing wild and disorganized. For example a cloud (of lines, dots or shadows, or even morphological images) can be put onto a regular mesh with the mesh size, s, and simply counted regarding the number of grid boxes which contain some trace of the structure involved. Certainly the accounted number, *N(s)*, is dependent on our choice of s, and would change with their progressively smaller sizes. *N(s)* plotted against *1/s* gives a line, which exhibits the slope close to 1.45 and which confirms no difference if the boxes are superimposed over a coastline to account the number of intersects, which historical roots point toward *Hausdorff's* work from the year 1918 [483].

10.6. Fractal dimensions, non-random and natural fractals, the construction of Sierpiñski gaskets

One of the most instructive examples and a much-studied phenomenon is the case *of non-random (geometric) fractals,* best represented by the so-called *Sierpiñski gasket* [465], cf. Fig. 66. Such a gasket is assembled in the same growth rule as a child brings together the castle from building blocks and is also enlightening the network models convenient in structural chemistry. First, we join three tiles together to create a large triangle, with one missing tile inside, which produces an object of the mass $M=3$ and the edge $L=2$. The effect of stage one is to produce a unit with the lower density, $\rho(L) = M(L)/L^2$. We can repeat this rule over, and over again, until we run out of tiles (physics) or until the structure is infinite (mathematics).

There arise two important consequences. Firstly, the density, $\rho(L)$, decreases monotonically with L, without boundaries, so that by iterating sufficiently we can achieve an object with as low density as we wish. Secondly, $\rho(L)$ decreases with L in a predictable fashion following a simple power law with fractal dimension of the inherent allometric plot equal to above cited value of 1.58. Such a kind of 'multiple-reduction in a copy-machine' is based on the collection of contractions, which uses the transformation providing the reduction by different factors and in different directions. We should note that such a similarity transformation maintains angles unchanged, while a more general transformations may not, which give the basis for defining the so-called *admissible transformations,* which permit scaling, shearing, rotation and translation, as well as reflection (mirroring) – all well known and standardized in the field of chemical crystallography (*Bravais*).

Configurations that are illustrated in Fig. 66 can be of help in the elucidation of structures of percolation clusters. We can see that an infinite cluster contains holes of all sizes, similar to the *Sierpinski* chain of Gasket Island. The fractal dimension, *d*, describes again how the average mass, *M,* of the cluster (within a sphere of radii, *r*) scales, i.e., following the simple power law, $M(r) \sim r^D$. The correlation length gives the mean size of the finite cluster, ξ, existing below the percolation transition, p_c, at which a new phase is to occur exhibiting an infinite cluster (typical, e.g., for a magnetic phase transition). We can interpret $\xi(p)$ as a typical length up to which the cluster is self-similar and can be regarded as a fractal. For the length scales larger than ξ, the structure is not self-similar and can be considered as homogeneous. The crossover from the fractal behavior at small length scales to homogeneous behavior at larger scales can again be portrayed on the basis of Fig. 66 again. The fractal dimension of the infinite cluster (at p_c) is not independent, but relies on the critical exponents (such as the order parameter in transformation and the correlation length in magnetism and superconductivity).

On the other hand, our nature exhibits numerous examples of the so-called *natural fractals*, i.e., objects that themselves are not true fractals but that have the remarkable feature, in the statistical average of some of their property, to obey the rules given by the standard allometric plots. It closely touch the mathematical research in the field of *chaos*, which can be traced back at least to 1890 when *Poincaré* became curious if planets would continue on indefinitely in roughly their present orbits or might wander off into eternal darkness or crush into the sun. He thus discovered chaotic behavior in the orbital motion of three bodies, which mutually exert gravitational forces on each other. Extended by *Kolmogorov's* view to irregular features of dynamics and within *Smale's* classification of disordered phenomena, chaos found its place as a natural feature taking shape completely on a par with regular behavior of periodic cycles. While a truly general definition for chaos to most spheres of interest is still lacking, mathematicians agree that for the special case of iteration there are four common characteristics of chaos: sensitive dependence on initial conditions, mixing behavior, dense periodic points and period-doubling (bifurcations). Let as briefly remind some of their bases [13,459,460].

It can be shown that many dynamical systems can *regularly* produce a chaotic behavior. One set of associated problems for us is the investigative concern to a difference equation called *logistic mapping* obviously the quadratic transformation, which comes in different forms, such as $x \rightarrow a\,x(1-x)$. This name sounds a little peculiar today as its origin are in economics from which it gives us the term logistic to describe any type of planning process. It derives from the consideration of a whole class of problems in which there are two factors controlling the size of a changing population, x, which varies between 0 and 1. This population passes through a succession of generations, labeled by the suffix n, so we denote the population in the n-th generation by x_n. There is a birth process in which the number of populated species (nuclei, insects, people) would deplete resources, and prevent survival of them all.

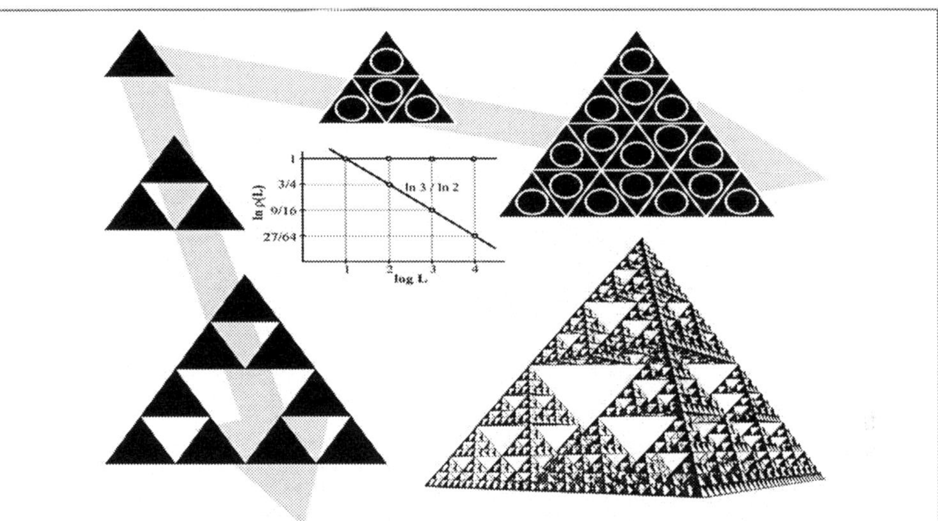

Fig. 66. – On a regular basis, we are traditionally used to build a dense construction of periodical (isogonal) sequences, of, e.g., triangles with the side, L, and the mass (area), M, which is known to provide, for example, the traditional hexagonal crystal structure with the strict 3-dimensional periodicity (upper) [9]. On the other hand, we can build another triangle's aggregation in a different periodic sequence when we purposefully and repeatedly remove the middle part of the remaining triangle creating thus a *porous* composition. The final set of such a construction is a subset of all preceding stages and can be even assembled in three dimension aggregates (bottom). Though the structure would appear somehow bizarre, its porosity, $\rho(L)$, would monotonically decrease in a predictable fashion such as the following series given for the length, L, and the mass, M:

$$L \quad = \quad 1\,(2^0) \quad 2\,(2^1) \quad 4\,(2^2) \quad 8\,(2^3) \quad ....$$
$$M \quad = \quad 1\,(3^0) \quad 3\,(3^1) \quad 9\,(3^2) \quad 27\,(3^3) \quad ....$$
$$\rho(L) = \ M/L^2 \ = \ (3/4)^0 \quad (3/4)^1 \quad (3/4)^2 \quad (3/4)^3 \quad .... \qquad \text{(see the inset plot).}$$

It continues without reaching any limits and is obeying the simple power law $y = a\,x^d$, where a is the slope of plot ln $\rho(L)$ against L. In particular, it demonstrates the relation, $\rho(L) = a\,L^{d-2}$, giving in our case the fractal dimension equal to $\log3/\log2$, which is about 1.58.

There is a negative depletion term proportional to the square of the population. Putting these together we have the non-linear difference equation. By defining the iteration is $x_{n+1} = a\,x_n(1 - x_n)$ and we can illustrate the process graphically upon the superimposed parabola (x^2) and straight line (x) in the interval $0 \le x \le 1$. We can arrive to the two types of iterations by adjusting both the initial point, x_o, and the multiplying coefficient, a. It can either exhibit a sensitive-irregular pattern or non-sensitively stable behavior. Very important phenomenon is thus sensitivity, which either can magnify even the smallest error or dump the larger errors, if the system is finally localized in the stable state. This behavior is called *sensitive dependence on initial conditions* and is central to the problematic of chaos [59,460], though it does not automatically lead to chaos.

Let us recall that the error, E, propagating in the simple linear system as $x \rightarrow a\,x$, increases by the factor a, i.e., through the power law, $(E_n/E_o) = a^n$. This, however, cannot be expected for the quadratic iterator, $x \rightarrow 4x(1-x)$, which is often called as a *generic parabola* (a graph, which precisely fits into a square and which has one of its diagonals on the coordinates bisector). We, nevertheless, can pretend that it is possible so that we can proceed by taking the logarithm to derive that small errors will roughly double in each iteration because the proportionality is approximately $e^{0.7} = 2$. This is true when the parabola is rather flat and errors are compressed but, on the contrary, it is enlarged near the end points. Such reasoning leads to the important concept of *Ljapunov exponents*, λ, which well quantifies the average growth of errors within the amplification ($\lambda=0.42$ and $e^{\lambda} = 1.52$) or the contraction ($\lambda=-1.62$ and $e^{\lambda} = 0.197$). This behavior becomes very important for the limited extent of decimal places often ministered by small pocket calculators, but falls beyond our too short and thus somehow effortless explanation.

10.7. Deterministic chaos, periodic points and logistic functions

Another phenomenon, worth mentioning, is the *mixing behavior,* the intuitive interpretation of which lies in the subdivision of the unit interval into subintervals and requires, at the same time, that by iteration we can get away from any starting interval and are capable to reach any other target interval. If such a requirement is fulfilled for all finite subdivisions, then the iteration of an initial value from any launch interval is spread all over the unit interval. Such orbits are often called *ergodic* and their more detailed analysis become significant in the process of satisfactory smooth command of controlled parameters. Thus, it is of further interest for all those regarding the theory of temperature regulation.

Yet another appreciable occurrence worth noting is the *kneading of dough* [459,460], which provides an intuitive access to many mathematical properties of chaos. Curiously, there is almost nothing random about such a kneading process itself: just stretch, fold, stretch, fold, and so on, up to the idealized situation when the dough is homogeneously stretched to twice the length. Folding the layers does not change the thickness and we can represent the dough by a line segment of layers while neighborhood relations are destroyed. This process becomes imperative in the description of all physical processes of homogenization. Such a stretch-and-fold process is closely related to the discussed quadratic iterator because one application of the transformation, *4x(1-x)*, to the interval [0,1] can be interpreted as a combination of a stretching and folding operation. The joint representation of this quadratic parabola with the *saw-tooth transformation* is also important to mention as it can establish a nonlinear change of coordinates through the function of $h(x) = sin^2 (\pi x/2)$, where, *h,* transforms the unit interval to itself in a one-to-one fashion. This again is of significance in analysis of various profiles of noise or in mathematical treating the different modes of oscillatory heating.

Let us see the famous *Feigenbaum diagram* that has become the most important icon of chaos theory as already. We start from investigating the behavior of the quadratic iterator again looking for all values of the parameter, *a,* between 1 and 4. After a transient phase of a few iterations the orbits settle down to a fixed point for all low values of $a < 3$. For the value of $a > 3$, the final state is not a mere point but a collection of 2, 4, 8 or more points. The resulting fork-tree portrays the qualitative changes in the dynamical behavior of the iterator $x \rightarrow a\,x(1-x)$ witnessing two branches bifurcating. Out of these branches we can see two branches bifurcating again, and so on. This is the so-called *period-doubling regime* and the behavior is called *periodic* with two trajectory states (period of two, cf. Fig. 10.). The length of branches, l_k, relative to each other decreases according to some law, conceivably following a sequence, which is asymptotically geometric, leads to several consequences. It would constitute a ceiling parameter of a, beyond which the branches of the tree could never occur marking the end of the period-doubling regime. It is a very important threshold called the *Feigenbaum point,* equal to 3.5699..., at which the final-state diagram splits into two very distinct parts of order and disorder, the latter being not simply a region of utter chaos as it hides a variety of beautiful structures. When it attains the value of $(1 + \sqrt{8}) \cong 3.8284...$, called the tangent of bifurcation, the stable (laminar) phases alternate with erratic and chaotic behavior showing the interlay between the burst of chaos and order. The more important *universality* threshold, however, is the sequence of magnification called the *Feigenbaum constant,* δ, equal to 4.6692...[490], which obviously arises from the associated iterations, i.e., from the ratio of measures of two succeeding branches, l_k and l_{k+1}, i.e., $\delta = lim_{k \rightarrow \infty} (l_k/l_{k+1})$. Its universality is known for a whole class of iterations generated by functions similar to quadratic function, such as $sin(\pi x)$, and thus is comprehended as general constant of chaos comparable to the fundamental importance of numbers like π and $\sqrt{2}$. The Feigenbaum constant was found verifiable through the real physical experiments, such as sophisticated measurements in electronics (where $\delta = 4.7$) hydrodynamics (4.4), laser feedback (4.3), acoustics (4.8), etc.

Further instructive geometric construction of chaos in continuous system was introduced by *Roessler* [491] in 1976 by solving the system of three differential equations $x' = -(y + z)$, $y' = x + ay$ and $z' = b + xz - cz$, with three adjustable parameters, *a, b* and *c,* which in fact hid the non-linear stretch-and-fold operation. This simple autonomous system has a single non-linear term, the product of x and z in the third equation. Each term in these equations serves its purposes in generating the desired global structure of trajectories. Assuming z to be negligible the two remaining equations can be transformed to the second-order linear oscillator, $x'' - xx' + x = 0$, and with positive parameter a, it exhibits negative damping. It follows that the trajectories of the full system of three first-order equations spiral outwards from the origin. When the orbit has attained some critical distance from the origin, it is first lifted away from the x-y plane and later reinserted into the twisting mode through the feedback of z to x equation and trajectories' fold-back. This procedure of spreading adjacent trajectories is the first sign in the mixing action of chaos as well as the self-regulation of the system as whole, which is typical for modeling the weather and also touches the behavior of fluxes in complex processes of dendritic growth [6,9].

Not only for chemical reactions have differential equations the language in which the modern science encoded the laws of nature. Although the topic of differential equations is, at least, three centuries old and the results have filled libraries, nobody would have thought that it is possible that they could behave disorderly, as found by *Lorenz* in his 1963 experiments [492]. So that we cannot avoid a glimpse trace on the previous logistic equation assuming the population change per unit time, i.e., $\{x(t + \Delta t) - x(t)\}/\Delta t$, which can help to compute the population size as $x(t + \Delta t) = x(t) + \Delta t\, a\, x(t)\, \{1 - x(t)\}$. As $\Delta t \to 0$, the iteration makes a transition into the world of differential equations (famous *Euler's* method) so that $x'(t) = a\, x(t)\, \{1 - x(t)\}$, where the right-hand side is non-linear. We can linearize it by substituting $p(t) = 1/x(t)$, which makes possible to obtain solution in the form $x_o e^{at}/\{\, x_o e^{at} - p_o + 1\}$ showing that the population will eventually go into saturation. On the other hand, if $\Delta t \to 1$, it coincides with the original logistic formula, or, interpreted differently, the growth law for arbitrary time step Δt reduces to the discrete logistic model replacing the expression $(\Delta t a)$ by mere a. In other words, if we have $a > 3$, then chaotic orbits for $\Delta t = 1$, while we orbit converging to 1 for $0 < \Delta t < 2/3$, which provides the so-called stability condition for any numerical approximation, i.e., $(\Delta t\, a) < 2$.

The crucial difference between the discrete logistic system and its continuous derivative-like counterpart is the fact that it is plainly impossible for the dynamics of the differential equation to be chaotic. The reason is that in the one-dimensional system no two trajectories, for the limit $\Delta t \to 0$, can cross each other, thus typically converging to a point or escaping to infinity, which however is not a general consensus in three dimensional system often displaying chaos. These types of differential equations are the most important tools for modeling straightforward ('vector') processes in physics and chemistry, though, no particular analytical solution is regularly available. The associated relationship can be transposably written in the form of a common kinetic equation:

$$d\alpha/dt \cong \alpha(1 - \alpha),$$

where α is the normalized extent of chemical conversion. Factually, this is a well-known form of *Prout-Tompkins* self-catalyzed kinetic model [493] naturally related to various aspects of advanced chemical reactivity.

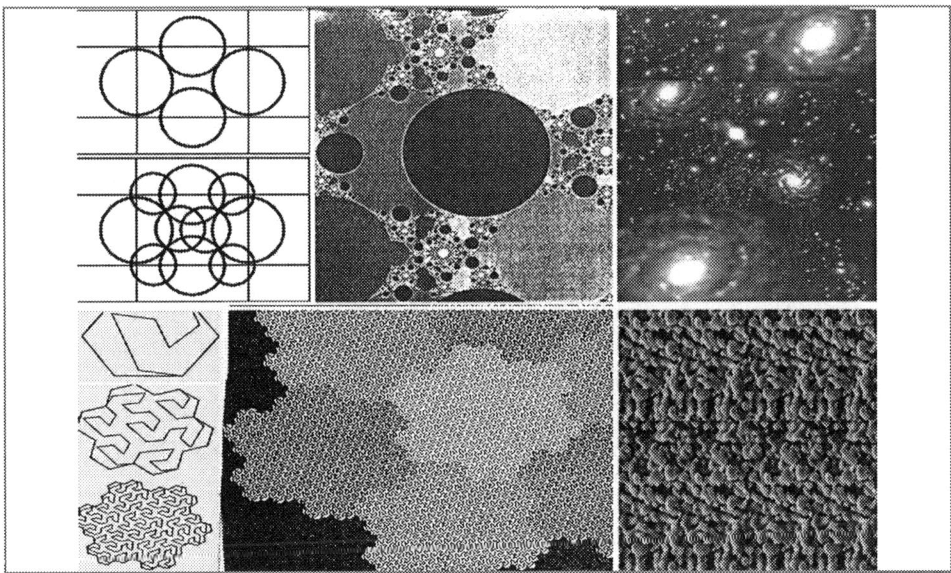

Fig. 67. – Planar pattern of variously positioned circles (upper), which repeated motif (left) gives the image of agglomerates of globular macromolecules spaced in the micro-world (middle) or the macroscopically visual view of the Universe (right). Below the repetitive (ornamental) motif showing its certain resemblance with a microstructure of crystallites.

However, life in the real science is not simple and almost in any physical-chemical system the state cannot be described by a single variable or equation characterized by an integral power exponent (=1), as shown in detail in the previous paragraph. Therefore it was obvious that for generalized purposes of chemical kinetics this logistic-like equation have got to be completed by the non-integral exponents, m and n, as completed in the advanced form of the *Šesták-Berggrenn* equation [480]

$$d\alpha/dt \cong \alpha^m (1 - \alpha)^n.$$

The involvedness of reactants (α) and products ($1 - \alpha$) is attuned to their actual chemical transience (m) and fertility (n), which is timely adjusted by this type of power-law. This uncomplicated but universal equation brought to kinetics additional sphere of more widespread applicability within the model-free domain of kinetic evaluations, see paragraph 11.4.

11. NON-ISOTHERMAL KINETIC BY THERMAL ANALYSIS

11.1. Fundamental aspects of kinetic evaluations

Thermoanalytical (TA) data are of a macroscopic nature as the measured (overall) occurrence (in the sense of observation) is averaged over the whole sample assemblage under study. In spite of this fact, the experimentally resolved shape of TA curves has been widely used as a potential source for the kinetic appreciation of even elementary solid-state reactions taking place within the sample (bulk or its interfaces) [1,3,409,422,425,437, 494-498]. The shape of the TA curve is then taken as the characteristic feature for the reaction dynamics and is then mathematically linked with the analytical form of model functions, $f(\alpha)$, depicted in the previous Chapter 10. We, however, should remind that these diagnostic-like formulas are often derived on the basis of the *simplified* physical-geometrical assumptions of the internal behavior (movement) of reaction interfaces (see previous Figs. 58. and 59.) displaying thus a norm character only. The *basic kinetic relationship* or fundamental equation

$$(d\alpha/dt) = \alpha' = k(T)\, f(\alpha)$$

provides the root for the most traditional evaluation procedures [3], i.e., for all *differential methods* of kinetic data determination when the logarithmic fashion is applied [499] together with the Arrhenius expression for $k(T)$ equal to the traditional form of *Aexp(-E/RT)*. It provides *ln* $\alpha' = $ *ln [Af(α)]* $-$ *E/RT*, which thought linearity (while plotting *ln* α' vs. *1/T*) is an important test of the validity (and suitability) of the introductory basic relationship under the experimental conditions applied. It permits a simple determination of E as a function of α, which experimental uncertainty should lay within about 10% and if superfluous it may reveal further complexity often inherent in multi-step solid-state reactions displaying separate values of E or dissimilar $f(\alpha)$.

More common form is $\{-$ *ln [α'/f(α)]$\}$* $=$ *const.* $-$ *(E/R)* and associated kinetic analysis is carried out by two possible ways, either to find directly a linear dependence between the functions in {brackets or after further differentiation (Δ) [500]. The latter is possible only if the function $f(\alpha)$ is suitably substituted in a simplified manner frequently done by applying a straightforward relation, *(1-α)^n* , representing the so called *reaction order* model, see the previously listed previously in Table 10. I. As a result the a simple plot of *{(Δln α'/Δ ln (1-α)}* vs. *{Δ 1/T/Δ ln (1-α)}* is derived, which can be used for the direct determination of the most preferred kinetic constants: the activation energy, E (as a slope) and the reaction order, n (as an intercept).

Because of the most simplified form of the function $f(\alpha)=(1-\alpha)^n$, the plot provides only *apparent* kinetic parameters, E_{app} and n_{app}, related to the true values of E by the ratio, $E_{app}/E = - [f(\alpha_{max})/f'(\alpha_{max})]\, n_{app}(1 - \alpha_{max})$, where f is the appropriate (relevant) function and f' is its derivative [501,502].

In general, such differential methods of the kinetic data determination are very sensitive to the quality of experimental data, especially to the determination of instantaneous rates, α', and the related differences, Δ. For this reason, and in view of the inaccuracy of this determination in the initial and final reaction stages, they are not so extensively used. However, on rearrangement and upon assuming constant values of α and/or α', the following ratio is found useful for another direct estimation of E from several parallel measurements by the so-called *iso-conversional* and *iso-rate* cross sections, as introduced (and often labeled) by *Flynn* and *Ozawa* [503,504] ($\alpha = const \Rightarrow \Delta\ln\alpha/\Delta(1/T) = E/R = \Delta\ln Z f(\alpha)/\Delta \ln (1/T)$ ($\Leftarrow \alpha' = const$).

Another, rather popular method of kinetic data analysis is based on expressing the maximum value (index *max*) on the dependence of α' *vs.* T, for which it holds, $\alpha'' = 0 = \alpha'_{max} [E/RT_{max} /Z/\phi\, \exp(-E/RT_{max})\, df(\alpha)/d\alpha]$. The result provides simple but rather useful dependence, often called *Kissinger plot* and known since 1959 [505], which in various modifications and reproves shows the basic proportionality $\ln (\phi/T_m) = (E/R) (1/T_m)$.

More mathematical manipulations, such as a direct guess of E from a single point, i, using equation, $E = T_i^2 R\, \alpha' /[f(\alpha)\, g(\alpha)]$ [506], are available, which detailed analysis is the subject of specialized papers on kinetics [1,3,409,422 425,437,494-498,507-510,521], especially observant to the review articles [507] and books [3,425,508-510, 521].

The differential mode of evaluation can even match up with the true experimental conditions when the actual non-uniform heating and factual temperatures are considered. In such a case, we ought to introduce the second derivatives, T'' and α'', which lead to a more complex equation in the form [511]: $\{\alpha''T^2/\alpha'T'\} = df(\alpha)/d\alpha/f(\alpha) \{T^2\alpha'/T'\}+ E/R$. However, it occurs almost impossible to compute this relation with a satisfactory precision because of its extreme sensitivity to noise in the second derivatives, which thus requires enormously high demands on the quality of input data, so that it did not find its way to a wider practical application.

So-called *integral methods* of evaluation became recently more widespread. They are based on a modified, integrated form of the function $f(\alpha)$, which is determined by the following relation, $g(\alpha) = {}_\alpha\!\int d\alpha/f((\alpha) - {}_{T_0}\!\int k(T)\, dT/\phi = (AE)/(\phi R)\, \exp(-E/RT)\, \pi(x)/x = (AE)/(\phi R)\, p(x)$, where ϕ is the constant heating rate applied, $\pi(x)$ is an approximation of the temperature integral (often used in the intact representation as the $p(x)$ function [3,510,512], where $x = E/RT$). There are many approximations of this temperature integral [1,3,510] arising

320

from the simplest $\pi(x) = 1/x$ and $(1-2/x)/x$ (based on the expansion) to more accurate $\pi(x) = 1/(x+2)$ up to the most sufficient $\pi(x) = (x+4)/(x^2 + 6x + 6)$ (based on the rational approximations). It should be emphasized, however, that the determi-nation and sufficiently precise calculation of the $p(x)$ function should be seen as a marginal problem though it has led to exceptionally wide publication activity, which is not cited herewith (e.g. papers by *Segal* [510]). Despite the specific accounts on $p(x)$ accuracy it can be shown that the mathematical nature tends to simple solutions, because in most kinetic calculations the effect of the actual value (of $\pi(x)$ and/or $p(x)$) is often neglected or, at least, diminished. This factually implies that $k(T)$ is (often unintentionally) considered as a constant and, instead of complex integration, is simply withdrawn in front of the integral, namely: $\int k(T)\, dt \approx k(T) \int dt$, which handling is habitually obscured within complicated mathematics (involved and associated inside the $p(x)$ deciphering).

The frequent practice of evaluation by the individually derived integral methods is then done by plotting *ln g(α)* *vs.* the range of functions of temperature, arising from its logarithm *ln T* [513] over *T* (or better ϑ as a given temperature) [514] to its reciprocal *1/T* [515-517], which are respectively employed in the relation to the sort of $\pi(x)$ approximations individually applied. In result, they differ in the multiplying constants, which, by itself, is a definite evidence for a certain extent of the inherent inexactness of such integral way of modeling.

It is worth highlighting that already a plain application of a logarithmic function brings not only the required data smoothing but also introduces a higher degree of insensitiveness. For that reason, the popular double-logarithmic operation effectively paralyses the methods' discriminability towards individual kinetic models as well as our attached desire to attain as linear proportionality as possible. We may humorously make a note of curious but sometimes actual role of double-logarithmic plotting in sense of a drawing on 'rubber-stretch' paper.

To get a better insight, let us divide these integral methods into three groups according to the method of plotting *ln g(α)* against either of the three above-mentioned temperature functions (*ln T* , *T* or *1/T*) [118,519] giving thus the respective slopes of *tan ϖ* . Consequent estimation of E depends on the respective values, i.e., RT_m *(tan ϖ – 1)*, RT_m^2 *tan ϖ* and/or *(Rtan ϖ – 2RT)*. The involved approximation can be tested [520] using the asymptotic expansion of a series with a dimensionless parameter replacing E. It can be seen that the latter dependence is twice as good as that of the former two dependencies. In this way, we can confirm the numerical results of several authors who found that the plain dependence of *ln g(α)* vs. *T* yields E with an error of about 15% at best, whereas the *ln g(α) vs. 1/T* plot can decrease the error by half. This is relative to the entire series given by set of *T's* and *1/T's*.

Another important characteristic of integral methods is the problem of *overlapping* that endows with the same course of the *ln g(α)*-functions for two or more different reactions mechanism. As best example, the models describing diffusion (*Jander*) and phase boundary reaction [3,422,497] can be mentioned.

The above evaluation methods can be approximately systemized as follows (note that apostrophe ' means the derivative $\alpha' = d\alpha/dt$):

Variables	Model function	Rate constant	Name of method	Comprehension
$\alpha,\ T$	$g(\alpha)$	$p(x)$	integral	low discriminability
$\alpha,\ \alpha',\ T$	$f(\alpha)$	$k(T)$	differential	high sensitivity
$\Delta\alpha,\ \Delta\alpha',\ \Delta T$	$\Delta f(\alpha) \Rightarrow m,n$	$K(T) \Rightarrow E$	difference-differential	greater sensitivity
$\alpha,\ \alpha',\ \alpha'',\ T,\ T'$	$df(\alpha)/[d\alpha f(\alpha)$	$k(T),\ dk(T)/dT$	double-differential	too high sensitivity

It is worth noting that the incommutably obvious passage between integral and differential representation *must be reversible* under all circumstances, though it may recollect the dilemma of early-argued problem of the equality of isothermal and non-isothermal rates [404,405] (as already mentioned in Chapter 9.). This brainteaser, which was recently opened to renewed discussion [407] somehow readdress the query, what is the true meaning of partial derivatives of time-temperature dependent degree of conversion $\alpha = \alpha(t,T)$ mathematically leading to obvious but curiously interpreted $\alpha' = (\partial\alpha/dt)_T + (\partial\alpha/dt)_t\ T'$, which is the lesson of mathematics applied to the mutual derivation and integration of the basic kinetic equation.

Notwithstanding, some authors thus keep still convinced that apparently 'isothermal' term $(\partial\alpha/dt)_T$ has the significance of the isothermal reaction rate and the additional term $(\partial\alpha/dt)_t\ T'$ is either zero or has a specific 'non-isothermal' meaning. Recent treatments [407] recall this problem by applying the re-derivation of the integral form of kinetic equation. It is clear that this integral equation was derived upon the integration of basic kinetic equation in its primary derivative form, so that any reversing mathematical procedure must be compatible with the original form of primary differential equation because α depends on t through the upper limit of integral only. Hence, any extra-derived multiplying term *[1 + E/RT]* [370] is somewhat curious and confusing, see previous Fig. 56., and misleads to incorrect results. We have to keep in mind that the starting point of the standard isokinetic hypothesis is the independency of the transformation rate on its thermal history. Therefore the transformed fraction calculated upon the integration must be clearly dependent on the *whole T(t) path* and any further assumption of the coexistence of an $\alpha(T,t)$ function, dependent on the actual values of time and temperature, is fully incompatible with this formulation.

In fact, this can serve as a good example of an inappropriately posed question where an incorrect formulation of constitutional equations [282,405] is based purely on a mathematical treatment (which can appear physically problematic and rationally sightless) instead of the required application of logical analysis first. This approach was once unsuccessfully applied in the kinetic practice of glass crystallization [522] and thus it is not necessary to repeat that no doubts can be raised concerning the mathematical rules for the differentiation of implicit functions.

11.2. Formal kinetic models and the role of an accommodation function

It is clear that the experimental curves, measured for solid-state reactions under thermoanalytical study, cannot be perfectly tied with the conventionally derived kinetic model functions (cf. previous table 10.I.), thus making impossible the full specification of any real process due to the complexity involved. The resultant description based on the so-called apparent kinetic parameters, deviates from the true portrayal and the associated true kinetic values, which is also a trivial mathematical consequence of the straight application of basic kinetic equation. Therefore, it was found useful to introduce a kind of pervasive des-cription by means of a simple empirical function, $h(\alpha)$, containing the smallest possible number of constant. It provides some flexibility, sufficient to match mathematically the real course of a process as closely as possible. In such case, the kinetic model of a heterogeneous reaction is assumed as a *distorted* case of a simpler (ideal) instance of homogeneous kinetic prototype $\{f(\alpha) \approx (1-\alpha)^n\}$ [3,523,524]. It is mathematically treated by the introduction of a multiplying function $\underline{a}(\alpha)$, i.e., $h(\alpha) = f(\alpha)\,\underline{a}(\alpha)$, for which we coined the term [523] '*accommodation function*' and which is accountable for a certain 'defect state' (imperfection, nonideality, error in the same sense as was treated the role of interface, e.g., during the new phase formation).

It is worth mentioning that $\underline{a}(\alpha)$ cannot be simply replaced by *any* time-dependent function, such as $f(t) = t^{(p-1)}$, because in this case the meaning of basic kinetic equation would alter yielding a contentious form, $\alpha' = k_t(T)\, t^{p-1}\, f(\alpha)$. This mode was once popular and serviced in metallurgy, where it was applied in the form of the so-called *Austin-Rickett* equation [525]. From the viewpoint of kinetic evaluation it, however, is inconsistent as this equation contains on its right-hand side two variables of the same nature (α and t) but in different connotation so that the kinetic constant $k_t(T)$ is not a true kinetic constant. As a result, the parallel use of these both variables, provides incompatible values of kinetic data, which can be prevented by simple manipulation and re-substitution. Practically, the *Austin-Rickett* equation can be straightforwardly transferred back to the standard kinetic form [3] to contain either variable α or t on its own by, e.g., a simplified assumption that $\alpha' \cong t^{(p-1)}$ and $\alpha \cong t^p/p$ and $\alpha' \cong \alpha^{(1-1/p)} \equiv \alpha^m$.

Another case of mathematical inter-convertibility is the JMAYK model function, $(1-\alpha)[-\ln (1-\alpha)]^p$, which can be transferred to another but related two parameter form of SB equation [480], as both functions $(1-\alpha)$ and $[-\ln (1-\alpha)]^p$ can be expanded in infinite series, recombined and converted back to the multiple of functions α^m and $(1-\alpha)^n$.

It follows that the so-called *empirical kinetic model function* can be generally described by all-purpose, three-exponent relation, first introduced by (and often named after the authors as) *Šesták* and *Berggren* (SB) equation [480], $h(\alpha) = \alpha^m (1 - \alpha)^n [-\ln (1 - \alpha)]^p$. It is practically applicable as either form, *SB* equation, $\alpha^m (1 - \alpha)^n$, and/or *modified Johnson, Mehl, Avrami, Yerofeev* and *Kolmogorov* (JMAYK) equation, $(1 - \alpha)^n [-\ln (1 - \alpha)]^p$ (related to its original form, $- \ln(1 - \alpha) = (k_T t)^r$, through the exponents p and r, i.e., $p = \{1 - 1/r\}$).

It is broadly believed that an equation, which would contain as many as all three variable exponents (such as m, n and p), would be capable of describing any shape of a TA curve. In reality, however, such model equability is *over-determined* by using all three exponents at once and thus the model is factually competent to describe almost 'anything' [526]. Therefore only the single exponents, such as, m, (classical reaction order) or, p, (traditional *JMAYK* equation with $n=1$) or the two appropriately paired exponents are suitable for meaningful evaluation such as the classical model of self-catalyzes ($m=1$ and $n=1$), which is thus a special case of the general *SB* equation. The justified and proper complementary couples are only $(m+n)$ or $(m+p)$, whilst the pair $(n+p)$ exhibits an inconsistent role of two alike-functions [397], which are in an inadmissible (self-supporting) state.

It is worth noting that the most common pair, $(n+p)$, further abbreviated as the SB-equation, involves a special but time-honored case of the autocatalytic equation, where $n=1$ and $m=1$. The SB-equation $\alpha^m(1-\alpha)^n$ subsists the generalized use of a basic logistic function, $x(1 - x)$, which is customarily exploited to depict the case of population growth. It consists of the two essential but counteracting parts, the first responsible for *mortality*, $x \cong \alpha^m$, (i.e., reactant disappearance and the product formation) and the other for *fertility*, $(1-x) \cong (1 - \alpha)^n$ (i.e., a kind of products' hindrance generally accepted as an 'autocatalytic' effect). The non-integral exponents, m and n, play thus the role similar to all circumstances, which act for the true fractal dimensions [9,478-482,523].

As in the case of a phase-boundary-controlled reaction, the conventional kinetic model describing the diffusion-controlled reactions is ideally based on the expression of geometrical constraints of the movement of reaction interface. Extension of the reaction geometry to non-integral values is again a common and plausible way to formalize the empirical kinetic model functions [422,423,481,482]. As already mentioned in Chapter 10., these models, derived for the phase-boundary-controlled reactions (abbreviated R_n) and for the random

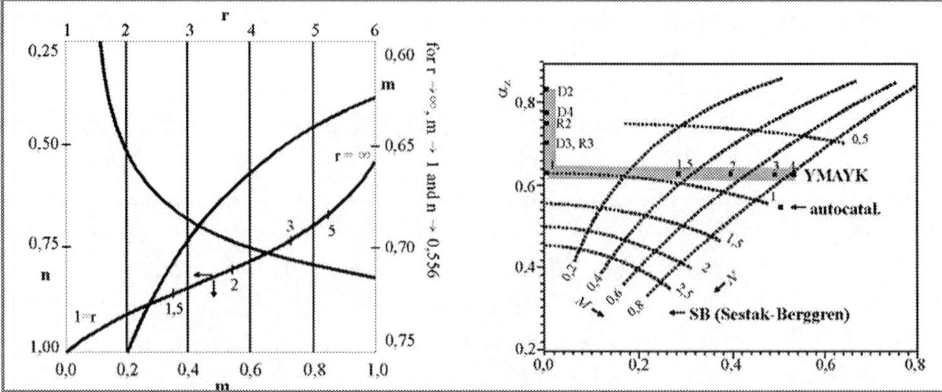

Fig. 68. – Intercorrelation of kinetic parameters characterizing reaction mechanism. Left, the relationship between the single exponent, r (as a characteristic exponent of JMAYK equation, $-ln(1 - \alpha) = (k_{(T)} t^r)$) and the twin exponents, m and n (as characteristic exponents of SB equation $(1 - \alpha)^n \alpha^m$) demonstrating mutual relation between the two most frequent models, JMAYK and SB. Right, the plot of characteristic values of α_Y^* and α_Z^* derived on the basis of accredited functions $y(\alpha) = A f(\alpha)$ and $z(\alpha) = f(\alpha) g(\alpha)$, which stay invariant on temperature and heating. The dotted lines show the idealized effect of kinetic exponents, m and n, of the SB kinetic model though such discrete values cannot be expected while studying any real solid-state process. It remains to determine the condition for the functions maximum, which can be derived as $y'(\alpha) = A f'(\alpha_Y^*) = 0$ and $z'(\alpha) = f'(\alpha_Z^*) g(\alpha_Z^*) + 1 = 0$, where the characteristic values of α_Y^* and α_Z^* correspond to the maximum of the $y(\alpha)$ and $z(\alpha)$ functions for different kinetic models, i.e., *Jander* and R2: $\alpha_Y^* = 0$ and $\alpha_Z^* = 0.75$, R3: $\alpha_Y^* = 0$ and $\alpha_Z^* = 0.704$, 2D-diffusion: $\alpha_Y^* = 0$ and $\alpha_Z^* = 0.834$, *Ginstling-Brounstein*: $\alpha_Y^* = 0$ and $\alpha_Z^* = 0.776$, JMAYK: $\alpha_Y^* = \{1 - exp(1/m-1)\}$ and $\alpha_Z^* = 0.632$, RO: $\alpha_Y^* = 0$ and $\alpha_Z^* = \{1 - n^{1/(1-n)}\}$, SB: $\alpha_Y^* = \{m/(m+n)\}$ and $\alpha_Z^* > \alpha_Y^*$. The shadowed area marks certain theoretical threshold for kinetic models applicability.

nucleation and growth (JMAYK abbreviated as A_m), can easily be formalized with non-integral (fractal) dimensions, such as $R_n = 1 \leq n \leq 3$ and A_m , with $n=1$, $0.5 \leq m \leq 4$, see Fig. 68. Employing the non-integral values of the dimension $1 \leq n \leq 3$, we obtain the relation $r/r_o = (1 - \alpha)^{1/n}$, where r and r_o are the radii of the reactant particles at $t = 0$ and $t = t$. The non-integral values are described by reactions on a fractal domain, the hallmarks of which are anomalous order and tome-dependent reaction rate constants. These anomalies stem either from incompatibility of ideal geometrical models with real reaction morphology and/or from the non-randomness of the reactant distribution (cf. Fig. 62.). For the diffusion-controlled reactions the non-integral kinetic exponents can be assumed from the proportionality of the rate of volume shrinkage to the amount of diffused substances at certain time and under the constant concentration gradient along the direction of diffusion. i.e., $\alpha' = k / [\pm x - (1 - \alpha)^{2/n-1}]$ where k is the apparent rate constant and exponent n varying $1 \leq n \leq 2$ ($x=1$), $n=2$ ($x=0$) and $2 \leq n \leq 3$ ($x = -1$). This relation is taken as the kinetic rate equation for the diffusion-controlled reaction with geometrical fractals [527,528] corresponding to the conventional parabolic law.

It should be noticed that SB-equation, with both non-integral exponents, $(n+m)$, is capable to mathematically fit almost any type of thermoanalytical curves, supporting thus indistinctive representation of a reaction mechanism, which would well correlate to the fractal image of true reaction morphology (cf. previous Fig. 58.). It provides the best mathematical fit, but gives little information about the modeled mechanism. On the other hand, the goal of most studies is more pragmatic, i.e., the reportable determination of a specific reaction mechanism, which would correlate to one of the confident physical-geometric models even though there is no concurrence of observable morphology between the true reaction image observed and the theoretical 'view' symbolized by the model assumed, cf. Chapter 10. It may provide certain, even idealized, information but, factually, it does not keep in fully distinguishing touch with the reaction reality.

The decision what tactic is the best to select is entirely in the hands of experimenters (and theoreticians) and their individual attitude how they want to present and interpret their results. It certainly depends of what would be the result's further destiny (application); citing *Garn* [508] *"materials and reactions themselves do not know any of our mathematical simulations"*. This chapter is obviously incapable of including all of the exceptionally extensive literature so far published on non-isothermal kinetics [3,425,496,508-510] and thus it is powerless to incorporate all peculiarities doubtlessly involved. It is worth noting that only a small fraction of the adequate literature is cited herewith and the thermoanalytical public is still waiting for a comprehensive book to be published on this motivating topic [437].

11.3. Practicality and peculiarity of non-isothermal approach
Apparent values of activation energies

For an appropriate derivation of kinetic parameters the previously shown procedures should always be reconsidered for and fitted to a *real* thermoanalytical application in order to become authentically relevant to the actual state of non-isothermal conditions and the type of process under study. In many cases we have ready-to-use commercial programs, available as accessories of most marketed instruments so that we have to take care about the *apparent* values thus received as outward response to our trialing. Individually specific but often rather sophisticated programs involve their own peculiarities, which we do not want to comment or deal with more extensively but all such individualities must be carefully judged by the users themselves.

First, we often forget to pay adequate attention to the more detailed temperature-dependence of the integration of basic nucleation-growth equations under non-isothermal conditions. It has been already proved [529] that the standard, and so far widely applied procedures, yield very similar dependences in comparison with that, which is found with the application of more complicated derivations involving all variables and functions [3,437]. The results effectively differ only by a constant in the pre-exponential terms depending to the numerical approximations employed in the integration. This means that the standard form of JMYAK equation can be used throughout all non-isothermal treatments [3,529].

A simple preliminary test of the JMAYK applicability to each studied case is worth of mentioning. The simple multiple of temperature, T, and the maximum reaction rate, $d\alpha/dt$, which should be confined to the value of 0.63 ± 0.02, can be used to check its appropriateness. Another handy test is the value of shape index, i.e., the ratio of intersections, b_1 and b_2, of the in inflection slopes of the observed peak with the linearly interpolated peak baseline, which should show a linear relationship of the kind $b_1/b_2 = 0.52 + 0.916\ \{(T_{i1}/T_{i2})\text{-}1\}$, where T_i's are the respective inflection-point temperatures [397,530].

Conventional analysis of the basic JMAYK equation shows that the overall values of activation energies, E_{app}, usually determined on the basis of DTA/DSC measurements, can be roughly correlated on the basis of the partial energies of nucleation, E_{nucl}, growth, E_{growth} and diffusion, E_{diff}. It follows that $E_{app} = (a\ E_{nucl} + b\ d\ E_{growth})/(a + b\ d)$, where the denominators $(a + b\ d)$ equal the power exponent, r, of the integral form of JMAYK equation and the coefficients b and d indicate the nucleation velocity and the growth dimension. The value of b corresponds to 1 or ½ for the movement of the growth front controlled by chemical boundary (chemical) reaction or diffusion, respectively. For example, the apparent values can be read as follows [3,397,433]:

		Chemical, b=1, and diffusional, b=½	
Instantaneous	1-D growth	r=1 ; E_{growth}	r=0.5 ; $E_{diff}/2$
nucleation	2-D growth	r=2 ; 2 E_{growth}	r=1 ; E_{diff}
(*saturation*)	3-D growth	r=3 ; 3 E_{growth}	r=1.5 ; 3 $E_{diff}/2$
Constant rate	1-D growth	r=2 ; $(E_{growth} + E_{nucl})$	r=1.5 ; $(E_{diff}/2 + E_{nucl})$
(*homogeneous*	2-D growth	r=3 ; $(2\ E_{growth} + E_{nucl})$	r=2 ; $(E_{diff}/2 + E_{nucl})$
nucleation)	3-D growth	r=4 ; $(3\ E_{growth} + E_{nucl})$	r=2.5 ; $(3\ E_{diff}/2 + E_{nucl})$

The value of E_{app} can be easily determined from the classical *Kissinger* plot [3,505] for a series of peak apexes at different heating rates irrespective of the value of exponent, r.

Adopting the heat-dependent concentration of nuclei [383], we employ a modified plot in the form of $ln\ (\phi^{(e+d)}/T_{apex})$ versus $(\text{-}\ d\ E_{growth}/RT_{apex})$ applicable to the growth of bulk nuclei where the number of nuclei is inversely proportional to the heating rate ($a \approx 1$). This is limited, however, to such

crystallization where the nucleation and growth temperature regions are not overlapping. If $E_{nucl} = 0$ then E_{app} simplifies to the following ratio:
$(d\,E_{growth} + 2\{a + d - 1\}E\,T_{apex})/(d + a)$ where $E_{growth} << 2\{r + d - 1\}R\,T_{apex}.$

With the decreasing particle size, the number of surface nuclei gradually increases, becoming responsible for the peak shape. The particle-size-independent part of E_{app} can be explained by the nucleus formation energy, which is a function of the curved surface. Concave and convex curvature decreases or increases, respectively, the work of nucleus formation. Extrapolation of E_{app} to the flat surface can be correlated with E_{growth} for $a=1$ and $d=3$, being typical for silica glasses.

It is, however, often complicated by secondary nucleation at the reaction front exhibited by decreasing E_{app} with rising temperature for the various particle zones. The invariant part of E_{app} then falls between the microscopically observed values of E_{growth} and E_{nucl} , the latter being frequently a characteristic value for yet possible bulk nucleation for the boundary composition of the given glass. Particular effect due to the addition of various surface-active dopants can be detected for the nano-sized crystallization of metallic glasses, which is different from the conventional nucleation-growth and is often characterized by the higher values of the exponent, r (<4), see next.

Assuming that the as-quenched glass has a number of nucleation sites equal to the sum of a constant number of quenched-in nuclei, T^{o}_{apex} , and that, depending on the consequent time-temperature treatment, T_{apex} , the difference between the apex peak temperatures for the as-quenched and nucleated glass becomes proportional to the number of nuclei formed during thermal treatment. The characteristic nucleation curve can then be obtained by plotting $(T_{apex} - T^{o}_{apex})$ against temperature, $T,$ or even simply using only DTA/DSC peak width at its half maximum. The situation becomes more complicated for less-stable metallic glasses because the simultaneous processes associated with primary, secondary and eutectic crystallization often taking place not far from each other's characteristic composition. However, another complementary technique is required to elucidate the associated kinetic mechanism.

However, there arise acute problems when the reaction mechanism undertake a more complex, multi-step pathway [531-535]. It is worth noting that such a composite process might not be detected be mere variation in the apparent values of E but a more multifaceted kinetics is necessarily applied or the other, often non-parametric (model-free methods) ought to be considered as more convenient to such application [533].

Non-parametric approach:
An alternative view on the kinetics of solid-state reactions
Such an unconventional treatment called non-parametric kinetics was introduced by *Serra, Nomen* and *Sempre* [533] and is literally based on our

previously accepted assumption that the reaction rate can be expressed as the product of two mutually independent functions $k(T)$ and $f(\alpha)$. However, the temperature dependence of $k(T)$ *need not be of the Arrhenius type* and can thus assume an arbitrary function and the 'model' function $f(\alpha)$ may not be straightforwardly connected with a specific reaction mechanism and can again assume any accidental-like relationship. These both functions are just temperature and conversion components of the kinetic 'hypersurface'. If the reaction rate is measured from several experiments at different heating rates and organized in a matrix system whose rows correspond to different (but constant) degrees of conversion and whose columns correspond (by interpolation) to different (but constant) temperatures than a suitable matrix decomposition algorithm can provide two characteristic vectors. These vectors can be further analyzed by examining the resulting plots of rates against conversions possibly capable to check the validity of Arrhenius –type behavior or the fittingness of another (more simple) functions. It, however, requires rather extensive number of experimentally determined points and wider working ranges of temperatures but such a model-free method allows the isolation of temperature dependences involved without making any assumption about the reaction model.

As conferred by *Šimon* [531], the parameters occurring in both functions are only certain perceived quantities (similarly to the above portrayal of 'apparentness'), which, in general, possess no mechanistic interpretation. Such values are composite and their representation in terms of individual sub-processes and associative subordinated values can be found obscure. Since $k(T)$ is not the standard rate constant, there is no reason to be restricted by an exponential relationship and thus other functions can be used leading to a closed form of the temperature integral [531].

This single-step kinetics approximation involves the imperative condition of separability of both temperature and conversion functions and any couple of such autonomous functions lead to a acceptable description of accommodating kinetics. However, it has been reasoned that if a couple of separable functions cannot be found, it indicates that the single-step kinetics approximation is too crude and the description of the kinetic hypersurface may become inappropriate [531]. The temperature and conversion functions contain enough adjustable parameters so that their values are attuned in the procedure of fitting in order to reach the best fit between the experimental and calculated data. The separability of temperature and conversion functions must thus imply that the values of adjustable parameters are supposed to be unvarying in the whole range of conversions and temperatures.

It is generally recognized that the isoconversional methods lead to the dependence of adjustable parameters in the temperature function on conversion. This fact has provided the concept of variable activation energy (see e.g. [496,534] and the references cited therein). As shown in [531], dependence of activation energy on conversion violates the conditions reparability and thus, in the case of variable activation energy, the basic assumption of the single-step kinetics approximation is incompetent for the associated description and calculated data are inadequate. This factually represents the "logical trap" of the concept of variable activation energy since it becomes mathematically incorrect and inherently self-inconsistent. Deductions drawn from the dependence of activation energy on conversion can hardly be considered trustworthy and should be judged very critically and carefully.

We should accentuate again that the main contribution of the above concept of single-step kinetics approximation is to elucidate the non-physical meaning of equations involved so that it is just a mathematical tool enabling a description of the kinetics of solid-state reactions without any deeper insight into their mechanism. The correct mathematical description should recover the values of conversion and the rate of the reaction under study for given values of time and temperature. Since the adjustable parameters represent just apparent quantities, *no conclusions* should be drawn from their values and so it is close to the use of fractal power exponents discussed beforehand but some useful conclusions can be drawn from the values of reaction rates or isoconversional temperatures and times such as it has been done in the study of induction periods [535]..

11.4. Applied non-isothermal studies under special experimental conditions
Special growth-mode of crystallization of metallic glasses

Generally, the specific case of crystallization of glassy state may comply with the nucleation-growth mechanism of JMAYK kinetic law even though this equation does not entirely correspond to the experimentally obtained data for all rapidly quenched ribbons of metallic glasses. Thus the JMAYK equation has been traditionally accepted for kinetic fitting of so called conventional specimens such as the compositions resembling the cases of $Fe_{80}B_{20}$ [536], $Pd_{82}Si_{18}$ [537] or $Fe_{75}Si_{15}B_{10}$ [538] . The shape of continuous-heating DSC/DTA peak, which is supposed to be compatible the JMYAK equation, is routinely asymmetrical irrespectively to the power exponent, m, but the peak always exhibits a slower rise on its low-temperature (onset) side [390]. Such asymmetry is a characteristic feature for the majority of simple interface and/or diffusion processes explored in the previous Chapter 10 [539,3]. Any pre-annealing of such a class of glassy samples increases its initially transformed fraction, which shifts the peak to lower temperatures and makes it broaden. If the power exponent is greater than one, the JMAYK like isothermal signal shows a peak for a minimum degree of conversion $(\alpha_{min} \cong 1 - exp [(1 - m)/m])$ or nonzero time $(t_{min} \cong \tau[(m - 1)m]^{1/m})$.

On the other hand, the specific micro-spherulitic-like morphology of nano-crystalline phases in the novel class of so called 'finemetals' (mesoscopic

330

scale of 'finemets' or 'nanoperm' fabricates based on the Fe-Mo-Nb-Cu-Si-B alloys) often rules out the above-mentioned standard evaluation because their reactivity may eventually perturb the regular signal, which results from a JMAYK-model of n-dimensional growth from the fixed number of pre-existing nucleation sites,, see previous Table 10.I. The nano-crystal enlargement within finemetals is usually assumed to be the variety of primary crystallization in the initially nano-crystalline matrix. Therefore, several proposals appeared trying to involve the crucial effect of the as-believed effect of long-range diffusion, which is apparently the rate-controlling process of microcrystalline grain growth exhibiting specific peculiarities (anomalous density, thermal stability, relatively low crystalline content, etc.). Among others let us mention the model of rapid diffusion field impingement [540] or the gradual reduction of the actual growth rate when accounting for the switch of the reaction mechanism from the surface-controlled to long-range-diffusion mechanism [541]. Another suitable alternative to the JMAYK crystallization is the mechanism of so called *normal-grain-growth-like mode*, abbreviated as the *Atkinson's* NGG model [454], which has been effectively applied by *Illekova* [542]. It factually reflects the process of coarsening of the microcrystalline phase, justifiable in most cases when the mean grain size falls below 10 nm. In this instance the DSC/DTA exotermic peaks become different, as shown in Fig. 69. The exoperimentally determined DSC peaks possess atypical symmetry, which is the result of rationale arising from the differences in JMAYK and NGG mechanisms [539,543]. The explicatory outcome worth of highlight is the process of pre-annealing, see Fig. 69.

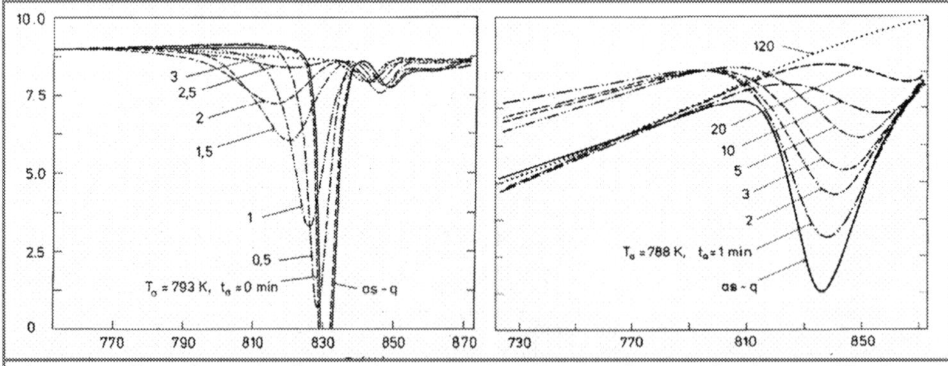

Fig. 69. – Effect of pre-annealing on the course of DSC traces (first run) for the crystallization of glassy ribbons (sample inweight of about 15mg, composed of 10-20µm thick and 10mm wide ribbons cut in several pieces and pressed in a Pt-cell, heating 40 K/min under nitrogen inert atmosphere under the progressively increased annealing time t_0). The annealing temperature was fixed at 793 K for the quenched alloy composed of $Fe_{75}Si_{15}B_{10}$ (left) and at T= 788 K for the quenched finemetal of a comparable composition $Fe_{74}Cu_1Nb_3Si_{13}B_9$ (right). Courtesy of *Emilie Illeková*, Bratislava,   Slovakia

which increases the initial micro-grain radius shifting thus the onset of NGG peak to higher temperatures leading to a narrower transformation range. Atkinson's original idea [454] was that larger grains in the already fine-crystalline samples increase their size at the expense of smaller ones which, however, does not fully correlate with the morphological observations of these heterogeneous systems of finemetals. The associated complex process of devitrification and nano-growth is a kind of multistage progression consisting of primary crystallization and coarsening within nano-crystalline structure, which display the lower power exponent m, often to fall below 2 (down to 0.15). Consequentially, it become important is to investigate the particularity of all effects attributable to the addition of various surface activators and dopants, which happen to be effective for the development of nano-composites and their properties, which end result may differs from that often recognized for conventional nucleation.

Effect of the environment:
nonisothermal studies carried out close to equilibrium

As mentioned above, for all generally assumed transformations, the most frequently applied isothermally derived theories are directly applicable to non-isothermal kinetic evaluations [3,529]. A more detailed analysis imposes, however, that such a non-isothermal generalization is only possible along the lines of the so called 'isokinetic' hypothesis, i.e., the invariance of the rate equation under any thermal conditions. It was shown [402,529,544,545] that the non-isothermal derivation includes the operation of integrated Arrhenius exponential, i.e., the above mentioned $p(x)$ function [545]. Application of suitable approximations lead to the change of the pre-exponential factor only maintaining thus the residual model function unchanged. Thanks to such coincidence, all functions mentioned in Table 10.I. are applicable for a straightforward submission to kinetic evaluations under non-isothermal conditions [3].

Another obstruction may be seen when employing a more general temperature program, which is proportional to the reciprocal temperature, i.e., $dt = \phi\, d\, (1/T)$. Here the function $p(x)$ simplifies in the form of a difference dT, which can be replaced by $d(1/T)$ so that $p(x) \sim (ZE/R\phi)\ [(-E/R)(1/T - 1/T_o)]$, which instead of being a complication, would on the contrary have allowed a more straightforward kinetic appraisal. Although such temperature regulation is rare, it would essentially facilitate the kinetic data calculation, requiring, however, another systematic. Instrument manufacturers' should consider the unproblematic extension of their commercially available digital programs to become applicable for another, newly specified case of kinetic investigations.

Another special case of non-uniform heating is the cyclic heating introduced by *Ordway* in the fifties [546] and actually applied by *Reich* [547] to determine kinetic parameters under isoconversional conditions. If the temperature changes stepwise from T_i to T_{i+1} then it holds that $E/R = (ln$ $\alpha_i/\alpha_{i+1})/(1/T_i)/(1/T_{i+1})$. If the temperature is varied in harmonic cycles, then for the two mentioned temperatures the rate of the change can be obtained by constructing a straight line through the apices of sine waves. This method is analogous to work with several heating rates, but does not require the sample replacement. It also links with the modulated methods of thermoanalytical calorimetry, discussed in the following Chapter 12.

Often neglected but one of the most important cases is the assumption of differences arising from the effect of *changing the equilibrium background* of the process (see chapter 9, Fig. 45.). It is illustrated on the example of various processes taking place for the tetragonal-cubic phase transformation of spinels, $Mn_xFe_{3-x}O_4$ (t) $\rightarrow$ $Mn_xFe_{3-x}O_4$ (c), cf. Fig. 46. The simple kinetic analysis of plotting the model relation against reciprocal temperature revealed that the change of the slope, often ascribed to the change of reaction mechanism, is just caused by switching between the part of the process proceeding under changing equilibrium background and the remainder invariant part controlled by the process itself [548].

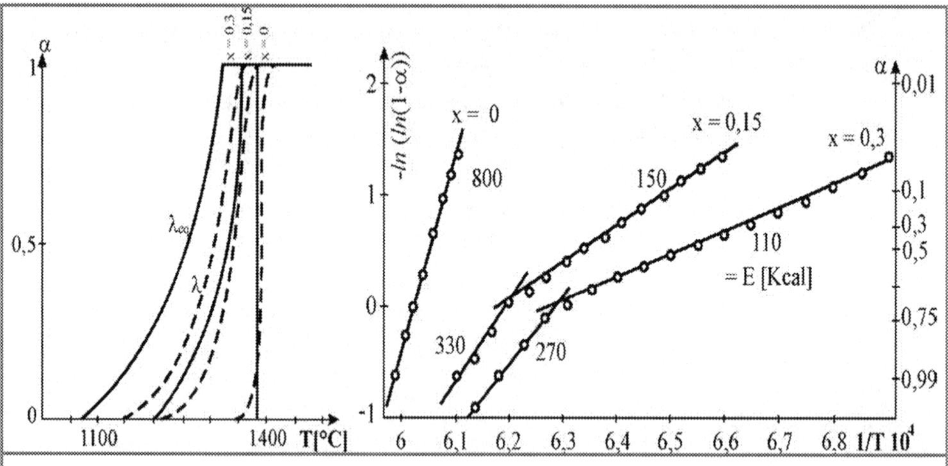

Fig. 70. – Formal kinetic account of the complex dissociation process of solid-state solutions of hematite-type (h) to spinels (sp), $Mn_xFe_{3-x}O_{9/2}$ (h) $\rightarrow$ $Mn_xFe_{3-x}O_4$ (sp) + ¼ O_2 [548]. Left, the profiles of the equilibrium backgrounds (solid) and the corresponding kinetic advancements (dashed). Right, formal kinetic evaluation using the logarithmic (JMAYK) model equation with the marked Mn-stoichiometry (*x*) and values of apparent activation energies (E). The breaks on the linear interpolation (at $x \cong 0.15$ and 0.3) indicate the instant where the equilibrium background reaches its final value ($\lambda_{eq} = 1$) terminating thus its interference with the kinetics ($\alpha = \lambda/\lambda_{eq} = \lambda$).

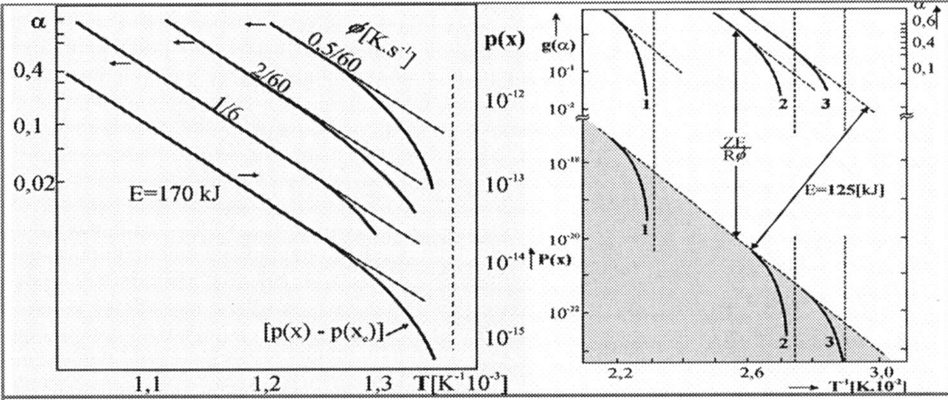

Fig. 71. – Left: The effect of heating rate, ϕ, on the integral kinetic evaluation of a process, which proceeds in the vicinity of equilibrium temperature of decomposition, T_{eq} , effecting thus the value of the function p(x) used for approximation. It is illustrated on the case of one of the most problematic study – the decomposition of $CaCO_3$. If the kinetic plot is markedly curved for small values of α (proximity to equilibrium), the extent of linear region used for the determination of E may significantly shrink as a result of decreasing capability of the sample state to equilibrate when departing too fast ($>\phi$) from its proximity conditions near equilibrium. At a certain stage (ϕ_{crit}, effective $\{p(x)-p(x_{eq})\}$) all curves become bandy making possible to find an almost arbitrary fit, which often provides a wide spectrum of non-characteristic constants (e.g., typical expansion of E-values). Right: The example of kinetic integral evaluation by simple plotting $ln\ g(\alpha)$ versus $1/T$ using experimental data for the dehydration of sulphate hemihydrate, $CaSO_4.1/2\ H_2O$, which dehydrates close to the equilibrium. The following vapor pressures, (1) ~ 10^2, (20 ~ 2,5 and (3) ~ 0.6 kPa were applied during the experiment. At the bottom figure the real shape of function $p(x)$ is marked as solid lines, whose linear extrapolation (dashed lines) yields the real value of activation energy, which simplified approximation is thus problematical and often misleading in search for the adequate characteristic values.

Proximity to equilibrium often plays an important but underestimated role, which is effective in the beginning of a reaction where the driving force is minimal. In such a case some integration terms of $p(x)$-function cannot be neglected as conventionally accomplished in most kinetic studies [3]. The resulting function $[p(x) - p(x_o)]$ is more complex and the associated general plot of $g(\alpha)$ vs. $1/T$ develops into non-linearity. This consequence can be illustrated on the decomposition of $CaCO_3$ and the dehydration of sulphate hemihydrate ($CaSO_4\ 1/2\ H_2O$), see Fig. 71. [549], where too low heating results in the effect that most of the dehydration process is taking place close to equilibrium and the resulting non-linear dependence can be made linear only with large errors. The apparent E value that is determined is usually larger than the actual E value; its correct value must be obtained by extrapolation of curves plotted either for different pressures, illustrated later, or for different heating rates, to get away from the access of overlapping equilibrium.

A further complication is brought about by considering the effect of the reverse reaction, e.g., in the well-known *Bradley* relationship [3,550], i.e., replacing in eq. 12.6 the standard rate constant, $k(T)$, by the *term $k(T)[1 - exp(\Delta G/RT)]$*. The ΔG value then actually acts as a driving force for the reaction and, thermodynamically, has the significance of the Gibbs energy, i.e., for $T \rightarrow T_{eq}$, $\Delta G \rightarrow 0$ and $\alpha' \sim f(\alpha)\, k(\Delta G/RT)$ whereas $T >> T_{eq}$, $\Delta G \rightarrow \infty$. Integration yields a complex relation of $[p(y - x) - p(y_{eq} - x_{eq})]$ where additional variable, y, depends on $\Delta G/RT$, the detailed solution of which we revealed elsewhere [550].

Further complication can be expected when the process under study has a more complex nature, involving several reaction steps each characterized by its own different mechanism and activation energy [426,435,532,534,552,553]. The work of *Blažejowski* and *Mianowski* [554] showed that thermogravimetric curves can be modeled by an equation relating to logarithm of the conversion degree as a function of temperature. It follows from the *van't Hoff's* isobar, which determines the dissociation enthalpy, ΔH_{dis}, that this quantity is proportional to the ratio of *ln* $\alpha'/\{(1/T) - (1/T_P)\}$. Providing that the equilibrium state is reached at constant ΔH_{dis} the relation *ln* $\alpha = a_o - a_1/T - a_2\, lnT$ becomes applicable using three a's as approximate constants.

In some special cases, where we must account for auxiliary and autonomous variables (already discussed in the preceding Chapter 10, the thermal decomposition can proceed in a mixed-control kinetic status. In such a case, both diffusional and chemical processes simultaneously control the overall reaction course with the same strength. Moreover, we have to account, at the same time, for the effect of partial pressure, p_r, of a reactant gas.

Malecki [555,560] derived a more complex equation than that listed in previous Table 10.I, which, however, has a more multifaceted outward appearance: $1 - (2/3)\, \alpha - (1 - \alpha)^{2/3} + s\, [1 - (1 - \alpha)^{1/3}] = k_r\, t$. It involves certain complex parameters, $k_r = 2\, v_m\, \Delta p\, D/ R^2_o$ and $s = 2\, D/(R_o\, k)$, which are composed of standard quantities of D, k, v_m, R_o, and Δp expressing (besides the classical degree of decomposition, α) respectively the diffusion coefficient, the reaction rate constant for a chemical reaction, the molar volume of the substrate, the radius of reacting sphere-shaped grains and the difference between the equilibrium partial pressure of volatile reactant and its external ambient pressure $(p_r - p_o)$. The apparent activation energy, E_{app} , is then given by more complicated relation, such as $E_{app} = [R\, T^2\, D\, (dlnk/dT) + R_o\, k\, f\, R\, T^2\, (dlnD/dT)]/(k\, R_o\, f + D) + R\, T^2\, (dln\, \Delta p/dT)$ where $f = \{1 - (1 - \alpha)^{1/3}\}$. It can be simplified for a negligible pressure ($p_r \cong 0$ of external reactant gas) to the equanimity related to the activation energies of chemical reaction, E_k , and diffusion, E_D, through $E_{app} = (s\, E_k + 2\, f\, E_D)/(2\, f + s) + \Delta H$. This model has been approved when studying the thermal decomposition of Co_3O_4 [555] (within the 2 – 20 [kPa] range of P_{O2}).

11.5. Optimal kinetic evaluation procedures of experimental data
Ruling the most probable group of kinetic models

Due to the inseparable assortment of different $f(\alpha)$'s [or $h(\alpha)$'s] the kinetic data evaluation based on a single-run method exhibits certain degree of ambiguity [422,556]. This insufficiency can be eliminated if the characteristic value of activation energy, E, is a priori applied, which requests its beforehand knowledge or better determination, e.g., through the application of its separate assessment. This execution was proposed by *Málek* [557] for more universal diagnostic purposes, for which it was found useful to define two special functions, $y(\alpha)$ and $z(\alpha)$. These functions can be easily obtained by a simple transformation of the experimental data [422,556-558], i.e., using $y(\alpha) = (d\alpha/d\theta) = \alpha' \exp(x) = A h(\alpha)$ and $z(\alpha) = (d\alpha/d\theta) \theta = h(\alpha) g(\alpha) = \pi(x)\alpha' T/\phi$ where $x=E/RT$ and $\theta = {}_o\!\int exp(-E/RT)dt$. The product of the function $h(\alpha)$ and $g(\alpha)$ is a useful diagnostic tool for determining the most appropriate kinetic model function, for which all these functions reach a maximum at a characteristic value of α_p . By plotting the $y(\alpha)$ function vs. α, both normalized within the interval <0,1>, the shape of the function $h(\alpha)$ is revealed. The function $y(\alpha)$ is, therefore, characteristic of a given kinetic model, as shown in Fig. 72. We should call attention to the important aspect that the shapes of functions $y(\alpha)$ are strongly affected by the value of E. Hence, the value of previously determined and assumingly constant E is decisive for a reliable determination of kinetic models because of the mutual correlation of kinetic parameters involved. Similarly we can discuss the mathematical properties of the function $z(\alpha)$. Its maximum is at α_p^{∞} for all kinetic models and, interestingly, it does not practically depend on the value of E used for its calculation.

In order to see this behavior we have to modify the basic kinetic equation to the form $\alpha' = h(\alpha) g(\alpha) \phi / \{T \pi(x)\}$, which can be differentiated $\alpha'' = [\phi / \{T \pi(x)\}]^2 h(\alpha) g(\alpha) [h'(\alpha) g(\alpha) + x \pi(x)]$ and set equal to zero, thus deriving once more the mathematical conditions of the peak as, $- h'(\alpha_p) g(\alpha_p) = x_p \pi (x_p)$. When x_p is infinite, we can put the limit $\{-h'(\alpha_p^{\infty}) g(\alpha_p^{\infty})\}$ equal to unity, which can be evaluated numerically providing the dependence of α_p^{∞} on the non-integral value of exponent n, illustrated for example for diffusion-controlled models in Fig. 72. It is apparent that the empirical diffusion model D_n is characterized as concave, having the maximum at $\alpha_{max} = 0$, using the $y(\alpha)$ function, and $- 0.774 \leq \alpha_p^{\infty} \leq$ (in the $z(\alpha)$ function). A schematic view of the empirical kinetic model determination by means of $y(\alpha)$ and $z(\alpha)$ functions can be derived from tabled equations and completed by the empirical model functions based on geometrical fractals. The RO_n and $JMAYK_m$ functions can be related to the conventional kinetic model functions $f(\alpha)$ within the integral kinetic exponents. However, the $SB_{n,m}$ function has a true empirical (and fractal) character, which in turn makes it possible to describe all various types of shapes of thermoanalytical curves determined for solid-state reactions (just fulfilling $\alpha_{max} \neq 0$). The kinetic parameter m/n is then given by $\alpha_{max}/(1 - \alpha_{max})$.

It is surprising that the mutual correlation of kinetic parameters as well as apparent kinetic models is often ignored even in the commercially available kinetic soft wares. Therefore we recommend a package which first includes the calculation of E from several sets of kinetic data at various heating This value of E is used as input quantity to another program calculation the above defined functions of $y(\alpha)$ and $z(\alpha)$. The shapes of these functions then enable the proposal of the most probable kinetic model rates [557]. Then the pre-exponential factor and kinetic exponents are calculated. The procedure can be repeated for all heating rates yielding several sets of kinetic data corresponding to various heating rates. If the model mechanism of the process does not change during whole thermoanalytical experiment, it approves its reliability and the consistency of the kinetic model thus determined can be assessed by comparing experimental and theoretically calculated curves. Consequently, the isothermal-like $\alpha - t$ diagrams can be derived in order to predict the behavior of the studied system under isothermal conditions.

Accounting for the evaluation best reliability, we can see that the models of reaction mechanisms can be sorted into three groups, R, A and D, see Fig. 72. andprobably it also gives the extent of determinability and model differentiability by integral calculation methods. *Ozao* and *Ochiai* introduced the model uncertainty due to the fractal nature of pulverized solid particles [481,572]. The area within the limits of side curves of each the given bunch of curves in Fig.72. became extended (fuzzy) and the individual curves with integral exponents loose their dominance in favor of multiple curves (lying within each shaded area) with various non-integral exponents. This, again, supports the viewpoint that through the kinetic evaluation of a single non-isothermal curve we can only obtain information about the kind of process, i.e., assortment within the range of models R, A and D.

The recent literature is still revealing yet novel mathematical treatments [438,496,558-560] the analysis of which would definitely require the publication of a specialized book dealing only with the expert domain of nonisothermal kinetics [437]. There is also multiple computer programs exhibiting a rather sophisticated level of mathematical evaluation of kinetic data, which we do not want to repeat nor examine. Among others, our simple numerical method was proposed using an arbitrary temperature-time relationship. The algorithm searches for E and rate constants by means of minimization of the average square deviation between the theoretically computed and experimentally traced curves on a scale of the logarithm of reduced time, which is expressed as the integral of the Arrhenius exponential.

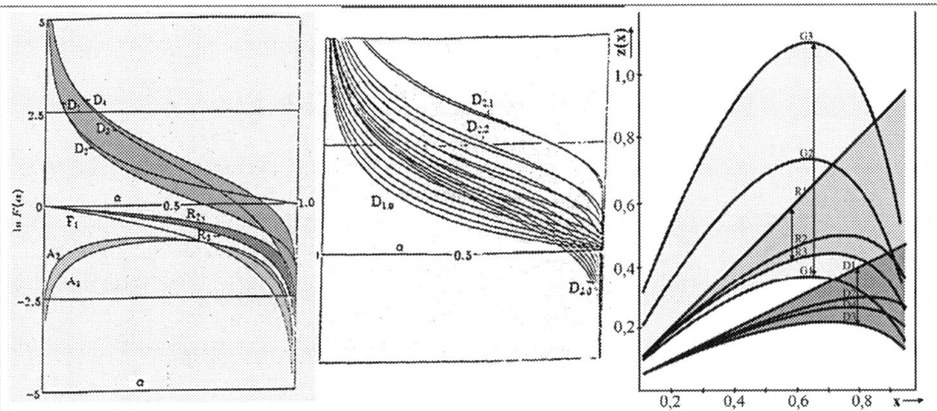

Fig. 72. – Distinctiveness graph of the functions of *ln f(α)* vs. α showing three characteristic areas of self-similar models abbreviated as nucleation (A), phase-boundary (R) and diffusion (D) controlled processes corresponding to equations listed in the table 10.I.. The middle plot displays the wider variety of possible diffusion controlled processes with a non-integral (fractal *{1-D)/D}*) geometry when compared with that on left (upper shaded area) as modeled on basis of the strict Euclidean geometry on left. The models involve such a modified course of characterization, which allows intermediate cases to exist (for $2<D_D \leq 3$ it holds $d\alpha/dt = \{(1- \alpha)^{(1-D)/D}-1\}^{-1}$ etc. [481]). Such approach seemingly enables to cover a larger area with the array of kinetic curves making possible their increased packing, e.g., from $D=2$, which coincides with the classical D_2 curve (shown in the left graph), we can see the curves' shift to the upper side of the plot with increasing D's as far as to 1.9 and than their moving downwards from $D=2.1$ to 3.0 to crowd the normally unfilled area (defined between the conventional D_2 and D_4 curves). Thee same curve swelling applies to the other groups (A's and R's, see left) and it shows well the almost meaningless attempt to define precisely the solitary particularity of individual reaction (courtesy by *Riko Ozao*, Kanagawa, Japan). Right is the characteristic graph of $z(\alpha) = f(\alpha)\,g(\alpha)$, which is plotted against α showing again the individual types of such 'master plot curves' typical for the following three kind of processes: nucleation-growth (G1↔G3, light shadow range), phase-boundary (chemical, R1↔R3, shadow range) and diffusional (D1↔D4, dark shadow range). This three-stage separation and computer evaluation yield just three group of most probable process, which is an acceptable answer in solving reaction kinetics (for the model illustration and mathematical expression see the Table, page 285 [422]. Courtesy by *Jří Málek*, Pardubice, Czechia.

The standard eighteen models listed previously are included for the sequential analysis and arrangements of the roots of mean square deviations provide a good approximation of the original kinetic curves [422,557].

Evaluation based on polynomial regression

The simple (single-step) processes are infrequent in nature. For the multi step reactions, commonly revealed in macromolecular blends of polymers, the number of used parameters becomes copious and thus it is necessary to undergo

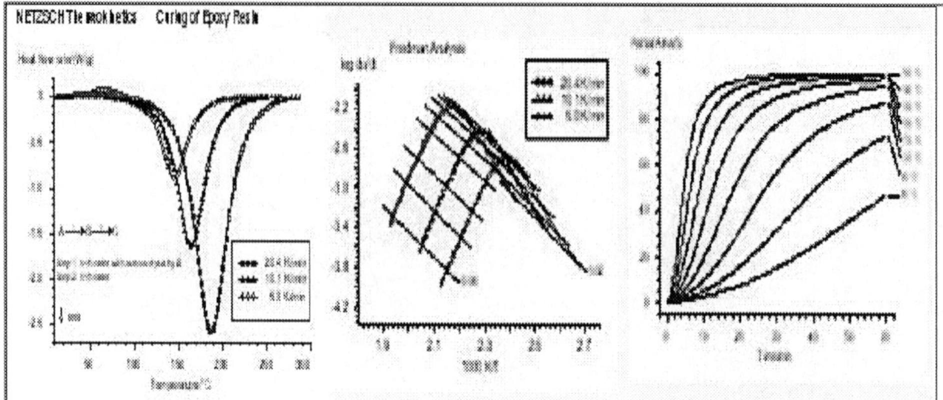

Fig.73. – Actualized graphical recording available for the DSC signal besides the standard digital output. The example of curing of epoxy resin was chosen as such a kind of materials can be best fitted by the Netzsch kinetic software. Left: DSC measurements for different heating rates with corresponding simulated signal curves calculated using the model of 2 consecutive steps of n-th order with autocatalysis and n-order reactions. Middle: the so called Friedman kinetic evaluation with iso-conversion lines. At the reaction start (from right) the curves slope is higher then iso-conversion lines slope, which provides deduction that a reaction starts with an accelerating step. Right: the simulated value of partial peak area (percentage of the total peak area) for the isothermal curing with different temperatures

several measurements under the altered conditions and have the results consequently evaluated by software suitable to reveal adequate kinetic model. One of the several commercially available kinetic assessments is the Netzsch thermokinetic evaluation package [561,562], which is a good example for using polynomial regression to ascertain a particular attitude to the kinetic evaluation based on the model-free and model-rooted approach. The model-free method can calculate reaction parameters without any assumption about the kinetic model or reaction types employing traditional methods by *Ozawa-Flynn-Wall* and *Friedman* [499,503,504] and *ASTM-Kissinger* [505]. The results bear information about the process acceleration or hindrance (e.g., of the first reaction step) and its further consequences. For the multi-step process the rate of each step involved is described as a function depending on the concentration of reactant, step parameters and the position of the step in the kinetic model of the whole process. The assumption about the kinetic model is used in the variant of model-based method where as many as three-step reaction with simultaneous steps A→B→C→D are analyzed using the concentration data of a, b, c and d of the corresponding reactants and their changes towards the overall rates. By using additional balance equation a+b+c+d = 1 the reactant concentration is found and the differential TA signal (DSC/DTA) is integrated. The traditional parameters (*E* and *A*) as well as the specific parameters (reaction order, character of autocatalysis, dimension of nucleation, etc.) are familiarized through the condition of minimizing by least squares analysis. It finally provides the answer of how many steps are involved, what the connection of each step is to the other in parallel, consecutive and independent mode, what the contribution of each step is to the overall signal observed, the formal reaction type of each step, the unknown parameters involved and how we can predict the system behavior under the other (arbitrary) conditions determinable by investigator. The application and optimization of the process under study is predetermined to achieving the high quality of product by the retrospective selection of the most suitable temperature program and is especially relevant for the kinetic analysis of various macromolecular systems (such as inorganic clay minerals, silicates, gels or organic plastics, polymers, biomaterials and cells.

11.6. Controlled rate and temperature oscillation modes

We should also account for the resolution power of controlled rate thermal analysis (CRTA) enabling to control the transformation at such a low rate that the remaining temperature and pressure gradients through the sample are themselves low enough to avoid overlapping, often successive transformation steps. Therefore the term $A\ exp(-E/RT)f(\alpha)$ is constant and its differentiating with respect to α, respecting the conditions of both minima, α_{min} and α_{inf} provides $dT/d\alpha = -T^2/ER\ f'(\alpha_{min})/f(\alpha_{min}) = 0$. The theoretical CRTA curves can be divided into three groups characterized by general shapes and, thus, the common rules regarding the shape of an experimental curve α vs. T can be formulated as follows [563]:

(i) The mechanism of nucleation and subsequent growth of nuclei, i.e., JMAYK model $(\alpha_{min}= 1- exp\{1-n/n)$ as well as empirical SB model $(\alpha_{min}=m/\{m-n\})$ lead to the curves with temperature minimum.

(ii) The boundary-controlled mechanisms as well as the reaction-order models give the shape without minima or inflexion points.

(iii) The diffusion processes provide the curves with an inflection point, the location of which is affected by the value of reduced activation energy, $x=E/RT$ (e.g., for classical *Ginstling-Brounshtein* model as $1-\{(2 + 3x)/4x\}^3)$.

The principle of activation energy calculation is to bring the rate of reaction to switch between two preset values – whose ratio, r, is conventionally chosen to lie within 1 to 4. About tens of jumps are usually performed to analyze one individual step. Each temperature jump allows determining a separate value of activation energy, since it provides a couple of temperatures obtained by extrapolation for exactly the same degree of reaction. Since the degree of reaction remains virtually unchanged during the rate-jump and furnishes that $f(\alpha)$ is not changed, we can assume that $E =(ln\ r\ RT_1T_2)/(T_1-T_2)$, where T_1 and T_2 are the sample temperatures just before and after a rate jump, unnecessary to know the entire reaction mechanism $f(\alpha)$. After evaluating the value of E by the jump method, a single CRTA curve is subjected to the determination of the α

340

dependence as characterized by $f(\alpha)$ and the Arrhenius pre-exponential factor by following the kinetic equation in different forms.

An agreed approach for evaluating E is known as the temperature jump [564,565], only different in the controlling parameter, i.e., the transformation rate or temperature. From the viewpoint of kinetic analysis, the modulated temperature control introduced in the thermoanalytical technique [566,567] can be recognized as a sophisticated temperature control based on the temperature jump method, avoiding over- and undershooting during the jump. However, such a sudden change in the reaction rate or temperature may sometimes be a plausible source of further complications due to the influences of heat and mass transfer phenomena. It should always be borne in mind that the reliability of the kinetic results is determined not only by the methodology of the kinetic measurements and the consequent mathematical treatment, but also by the kinetic characteristics of the reaction under investigation.

11.7. Kinetic compensation effect

A wide variety of the physical, chemical and biological phenomena are traditionally described by means of the exponential law, $G_o exp(-E_d/kT)$, where

the activation energy of a kinetic process, E_a, depends on the input parameter, a, distinguishing the material system under study (for example, related to the history of the system investigated, such as quenching rate, variously imposed external fields of pressure, electric/magnetic/gravity etc). Here, T is the absolute temperature, k is the *Boltzmann* constant and G_o is the pre-exponential factor (pre-factor). For this, so-called *Arrhenius* relation [568], there exists in many cases (especially within limited ranges of the input parameter $\underline{a}$ and/or T) extra correlation between E_a and G_o which is observed and called *thermodynamic compensation law* [568-570], or more often as the kinetic compensation effect [3,570-571] – commonly abbreviated as KCE. It holds that $ln\ (\ G_o/G_{oo}) = E_a/kT_o$, where G_{oo} and T_o (so-called iso-kinetic temperature) are assumed constants). The geometrical meaning of such a KCE correlation is quite obvious: the empirical dependences, fulfilling exponential relation and plotted in the so-called Arrhenius diagram (*ln G vs 1/T*) are arranged in a bundle of straight lines with a single and common intersection point with the coordinates $1/T_o$ and lnG_{oo}. If the above formulae are simultaneously valid, the transformation to an equivalent form is admissible, $G = G_{oo}\ exp\ (-F/kT)$ and $F = E_a\ (\ 1 - T/T_o)$. It is worth noting that it is similar to the original relation but G_{oo} is a constant here and F is a linear function of temperature. The direct comparison with the Gibbs statistical canonical distribution leads to the conclusion that F should have the meaning of isothermal potential (isothermal free energy) which is adequate for the description of isothermal kinetic processes. From this point of view the validity of KCE is mathematically equivalent to the linear dependence of the free energy on the temperature and bears no essentially new physical content – *it is a pure mathematical consequence* [3,570] of the approximations used. It is evident that in many cases such a linear approximation can serve as a satisfactory description of experimental data, but is often used to describe their material diversity.

Appearance of two new quantities, G_{oo} and T_o, may, however, be used for an alternative description and parameterization of a system involving kinetic processes. Taking in account the *Helmholtz* relation between the free energy coefficient and the entropy [my book], $-(\partial F/\partial T)_a = S\,(a)$, we can immediately obtain the entropy changes related to the changes of input parameters $\underline{a}$ in the form $S(a_1) - S(a_2) = (E(a_1) - E(a_2))/T_o$. The right side of this equation can be directly evaluated from the observed data without difficulties. The term on the left-hand side can be interpreted as a "frozen entropy" related to the parameter $\underline{a}$. It is an additional term contributing to the total entropy of a given material system, and may be expressed in a generalized form $S\,(a) = E_d/T_o$. Using the Einstein fluctuation formula [572], we can obtain for the relative change of the thermodynamic probability, W, of the system with respect to the input parameter $\underline{a}$ the following relation: $\ln(W_1/W_2) = \{E(a_1) - E(a_2)\}/kT_o$. The right-hand side may thus serve as a measure of the rigidity and/or stability of the system with respect of the changes in $\underline{a}$. As such, it can be employed to draw up certain consequences to technological conditioning of the system preparation (chemical composition, cooling rate, annealing, mechanical history, external pressure, etc.).

The change of so called "frozen entropy", E_d/T_o, characterizes, on the one hand, the internal configuration of the material system under study and, on the other hand, has a simultaneous meaning of the coefficient in the linear term of the free energy. Consequently, such a change of E_d/T_o must be inevitably associated with the change of the reaction mechanism. We can exemplify it by the correlation between the activation energy of conductance and the corresponding pre-factor in the field effect transistor [573]. The sudden change of the plot for large gate biases (small E_a) corresponding to the electron transport just in the vicinity of the interface between the insulator (SiO_2 or SiN_x) and amorphous (glassy) semiconductor (a-Si) shows the switch in the conductivity mechanism, see Fig. 74.

In the field of chemical kinetics the KCE was first pointed out by *Zawadski* and *Bretsznayder* [571] while studying the thermal decomposition of $CaCO_3$ under various pressure of CO_2 ($F_a \approx P_{CO_2}$) but in the field of solid state physics it was often called the thermodynamic compensation rule [569], originally derived upon conductivity studies on various oxides [575]. In the field of non-isothermal studies it was first noticed through the mathematical interplay between the kinetic parameters A and E by *Šesták* [574].. Early studies pointed that non-linearity in the *Arrhenius* plots gives the evidence of a complex process [3, 566, 570], that the mathematical correlation of the exponential pre-factor and

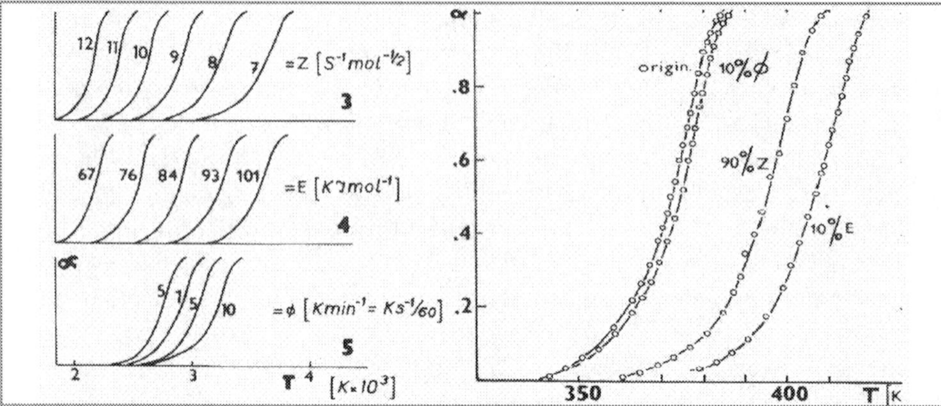

Fig. 74. – Upper, left, the explicatory shift of the position of a series of theoretically calculated TA curves {at a simplest f(α)=(1-a)} due to the changes of the inherent parameters, i.e., pre-exponential factor Z and activation energy E showing the latent effect of Z↔E overlap [574], which is particularly evident on the right graph, where the curves can move along the T-coordinate with the either percent change (Z or E). The change of a heating rate ϕ, however, causes the change of shape. Below, middle, the illustration graph of an Arrhenius plot displaying a mathematically natural link between *ln A* and *E* due to the so called kinetic compensation effect. Left. Apparent (app) values of *E* versus that of pre-exponential factor *A* for the non-isothermal dehydration of crushed crystals of lithium sulfate hydrate [570], which roughly follows the relation $E_{app} = R\, T_I\, T_{II}\, ln\, x\, /\, \Delta T$ where the variable x is dependent to the group of evaluation method and difference $\Delta T = T_I - T_{II}$ is the reaction interval. Right: Actual interpretation of the dependence of the pre-exponential factor, σ_o, on the activation energy, E_a, for different gate biases measured for two field-effect transistors. This explanation of kinetic compensation effect makes it possible to take this field-effect *E* as a measure of the band bending at the solid interface, which is common in solid-state physics [573]. It witnesses about the presence of the surface reservoir of states, which are pinning the free move of the Fermi level when the bend-bending is formed in the electron structure under the gate voltage.

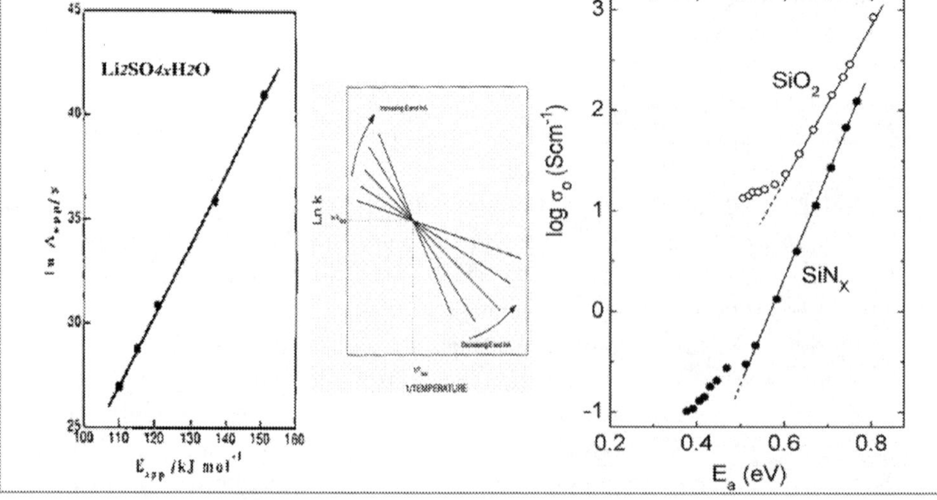

activation energy is inseparable [576] or the determination of the KCE is incorrect by mere linear regression of E_a vs. G_o ,because these quantities are mutually dependent being just derived from the original kinetic data. There were even attempts to relate the KCE to the geometrical shape of the χ-square function [577]. However, the KCE was mostly dealt with in the field of chemical kinetics being a frequent target of critique, but more often under attempt to be explained and solved both from various experimental [578-582] and theoretical [574,580-586] points of view, as reviewed by *Koga* [570]. Among others, it was pointed out that the KCE can be affected by changes in the defect concentrations, is sensitive to the sample mass or may even results from a mere computational error [587]. Even separation of the false and true KCE by two sets of iso-kinetic temperatures was investigated distinguishing physicochemical factors and experimental artifact [588]. The most important, however, were mathematical treatments showing that the KCE is a *factual mathematical consequence of the formulae employed* – exponential rate constant [568,570,589] including the effect of distortion caused by mere application of incorrect or inappropriate kinetic model functions [584].

12. THERMOMETRY AND CALORIMETRY

12.1. Heat determination by calorimetry

Various thermometric assessments have been in the center of retailored techniques used to detect a wide variety of heat effects and properties. The traditional operation aims to measure, for example, heat capacities, total enthalpy changes, transitions and phase change heats, heats of adsorption, solution, mixing, and chemical reactions. The measured data can be used in a variety of clever ways to determine other quantities. Special role was executed by methods associated with enough adequate temperature measurements, which reveals an extensive history coming back to the first use of the word 'calorimeter' introduced by the work *Wilcke* and later used by *Laplace, Lavoisier* as already discussed in the previous Chapter 4.

Calorimetry is a direct and often the only way of obtaining the data in the area of thermal properties of materials, today especially aimed to higher temperatures. Detailed descriptions are available in various books [3,9,590-596] and reviews [597-599]. Although the measurements of heat changes is common to all calorimeters, they defer in how heat exchanges are actually detected, how the temperature changes during the process of making a measurement are determined, how the changes that cause heat effects to occur are initiated, what materials of construction are used, what temperature and pressure ranges of operation are employed, and so on. We are not going to describe herewith the individual peculiarities of instrumentation as we merely focus our attention to a brief methodical classification.

If the calorimeter is viewed as a certain 'black box' [3], whose input information are thermal processes and the output figures are the changes in temperature or functions derived from these changes. The result of the measurement is an integral change whose temperature dependence is complicated by the specific character of the given calorimeter and of the measuring conditions. The dependence between the studied and measured quantity is given by a set of differential equations, which are difficult to solve in the general case. For this reason, most calorimetric measurements are based on calibration. A purely mathematical solution is the domain of a few theoretical schools [3,594,596] and will not be specified here in more detail.

If the heat, Q, is liberated in the sample, a part of this heat accumulates in the calorimetric sample-block system (see Fig. 75) and causes an increase in the temperature. The remaining heat is conducted through the surrounding jacket into the thermostat. The two parts of the thermal energy are closely related. A mathematical description is given by the basic calorimetric equation, often

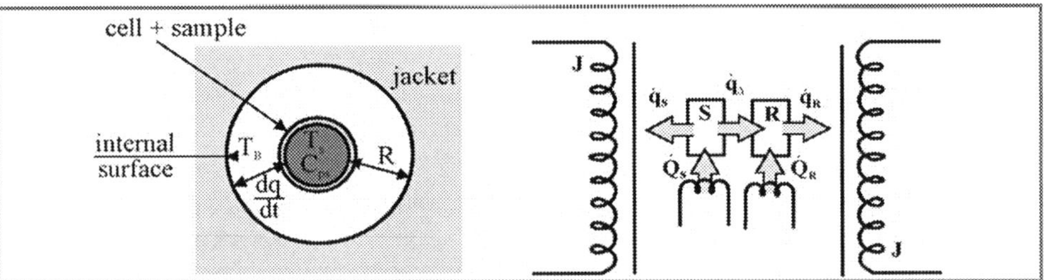

Fig. 75. – Calorimetric arrangements [3]. Left, the designation of characteristic temperatures, T, and spontaneous heat fluxes, q, for the geometrically concentric arrangement of a model calorimeter inside the thermostat/furnace, J. Right, the scheme of the accountable thermal fluxes participating in a two-crucible (twin-cells) arrangement of a generally assumed thermoanalytical measurement. Besides the spontaneous thermal fluxes q, a compensating flux, Q, is accounted as generated from the attached micro-heaters). The subscripts indicate the sample, S, and reference, R, as well as the created differences, Δ.

called the *Tian* equation [600] $Q' = Cp\, T'_B + \Lambda\, (T_B - T_J)$, where the prime marks the time derivative. The principal characteristics of a calorimeter are the calorimeter capacity, Cp, its heat transfer (effective thermal conductivity), Λ, and the inherent heat flux, q', occurring at the interface between the sample-block, B, and surrounding jacket, J. The temperature difference, $T_B - T_J$, is used to classify calorimeters, i.e., diathermal $(T_B \neq T_J)$, isodiathermal $(T_B - T_J) =$ *constant* and $d(T_B - T_J) \to 0$, adiathermal $(T_B \neq T_J)$ and $\Lambda \to 0$, adiabatic $(T_B = T_J)$, isothermal $(T_B = T_J = const.)$ and isoberobolic $(T_B - T_J) \to 0$. The most common version of the instrument is the diathermal calorimeter where the thermal changes in the sample are determined from the temperature difference between the sample-block and jacket. The chief condition is, however, the precise determination of temperatures. With an isodiathermal calorimeter, a constant difference of the block and jacket temperatures is maintained during the measurement, thus also ensuring a constant heat loss by introducing extra heat flux to the sample from an internal source (often called microhetaer). The energy changes are then determined from the energy supplied to the source. If the block temperature is maintained constant, the instrument is often called a calorimeter with constant heat flux. For low Q' values, the heat loss can be decreased to minimum by a suitable instrument construction and this version is the adiathermal calorimeter. Adiabatic calorimeter suppresses heat losses by maintaining the block and jacket temperatures at the same temperature. Adiabatic conditions are more difficult to assure at both higher temperatures and faster heat exchanges so that it is preferably employed at low temperatures, traditionally using *Dewar* vessel as insulating unit.

Eliminating the thermal gradients between the block and the jacket using electronic regulation, which, however, requires sophisticated circuits, can attain the lowest heat transfer and a more complex setup. For this reason, the calorimeters have become experimentally very multifaceted instruments. With compensation "quasiadiabatic" calorimeter, the block and jacket temperatures are kept identical and constant during the measurement as the thermal changes in the sample are suitably compensated, so that the block temperature remains the same. If the heat is compensated by phase transitions in the bath in which the calorimetric block is immersed, the instrument are often termed transformation calorimeter. Quasi-isothermal calorimeters are again instruments with thermal compensation provided by electric microheating and heat removal is accomplished by forced flow of a liquid, or by the well-established conduction through a system of thermocouple wires or even supplemented by *Peltier* effect. The method in which the heat is transferred through a thermocouple system is often called *Tian-Calvet* calorimetry. A specific group consists of isoperibolic calorimeters, which essentially operate adiabatically with an isothermal jacket. Temperature gradients do not exceed a few degrees and the separation of the block from jacket (a vacuum or low-pressure air gap) should ensure that the heat transfer obeys the Newton's cooling law. There are other version such as the throw-in calorimeter where the sample is preheated to high temperatures and then dropped into a calorimetric block, and combustion calorimeter where the sample is burned in the calorimetric block (often under pure oxygen atmosphere). A separate field forms enthalpiometric analysis, which includes liquid flow-through calorimeters and thermometric titrations lying beyond this brief introduction.

12.2. Origins of modern thermal analysis

Special territory of calorimetry-like measurements, more fittingly characterized within the boundaries of thermometric examinations [3], have molded a special field, which is often kept detached from 'true calorimetry' depicted above. In traditional terminology, these measurements are generally called *thermal analysis*, which [1,3,15,44,138,601,602] did not, as frequently inferred, suddenly appear in their modern version in the year 1887. Their roots extended back to the eighteenth century where temperature became better understood as an observable and experimentally decisive quantity, which thus became a monitorable parameter. Indeed, its development was gradual and somewhat international so that it is difficult to ascribe an exact date. First accepted definition of thermal analysis permits, however, identification of the earliest documented experiment to meet current criteria. In Uppsala 1829, *Rudberg* [44] recorded inverse cooling-rate data for lead, tin, zinc and various alloys. Although this contribution was recognized even in Russia (*Menshutkin)* it was overlooked in the interim and it is, therefore, worthwhile to give a brief account here.

The equipment used consisted of an iron crucible suspended by thin platinum wire at the center of a large double-walled iron vessel provided with a

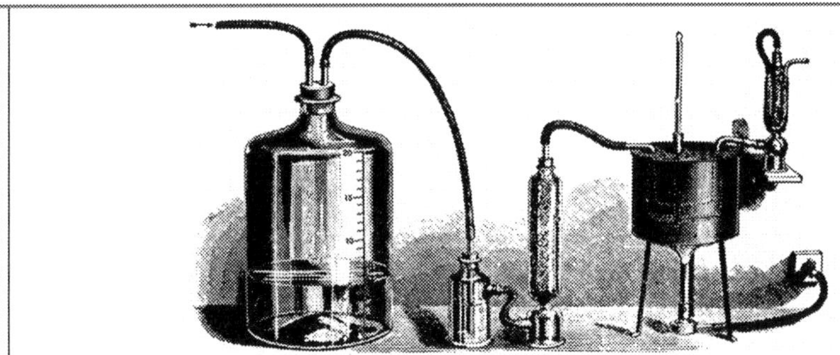

Fig. 76. – Apparatus used by *Hannay* in the year1877 in order to obtain isothermal mass-change curves.

tight-fitting, dished with iron lid, through which passed a thermometer with its bulb in the sample. The inner surface of the outer container and the outer surface of the crucible were blackened to permit the maximum achievement of heat transfer. The spaces between two walls of the outer large vessel, as well as the top lid, were filled with snow to ensure that the inner walls were always kept at zero temperature. In this way a controlled temperature program was ensured once the crucible with molten metal or alloy had been positioned inside and the lid closed. Once the experiment was set up *Rudberg* noted and tabulated the times taken by the mercury in thermometer to fall through each 10 degrees interval. The longest interval then included the freezing point.

The experimental conditions were, if anything else, superior to those used by careful experimentalist, *Robert-Austin,* some 60 years later. The next experiment that falls into the category of thermal analysis was done in 1837 when *Frankeheim* described a method of determining cooling curves on temperature vs. time. This method was often called by his name as well as by another, e.g., the so-called *"Hannay's* time method", when temperature is increased every time so that the plot resembled what we would now call 'isothermal mass-change curves' (see Fig.76.). *Le Chatelier* adopted a somehow more fruitful approach in 1883 that immersed the bulb of thermometer in the sample in an oil bath maintaining a constant temperature difference of 20° between this thermometer and another one placed in the bath. He plotted time temperature curve easily convertible to the sample vs. environmental temperatures, factually introducing the 'constant-rate' or 'quasi-isothermal' program. At that time, thermocouples were liable to give varying outputs and *Le Chatelier* was first to attribute an arrest at about red heat in the output of the platinum-iridium alloy to a possible phase transition. He deduced that thermocouple varying output could result from contamination of one wire by diffusion from the other one possibly arising also from the non-uniformity of wires themselves. The better homogeneity of platinum-rhodium alloy led him to the standard platinum – platinum/rhodium couple. Almost 70 years after the observation of thermoelectricity, its use in thermometry was finally vindicated.

The development of thermocouple, as an accurate temperature measuring device, was rapidly followed by *Osmond* (1886) who investigated the heating and cooling behavior of iron and steel with a view to elucidating the effects of carbon so that he introduced thermal analysis to then most important field of metallurgy. *Roberts-Austen* (1891), however, is known to construct a device to give a continuous record of the output from thermocouple ands he termed it as 'Thermoelectric Pyrometer', see Fig. 77.

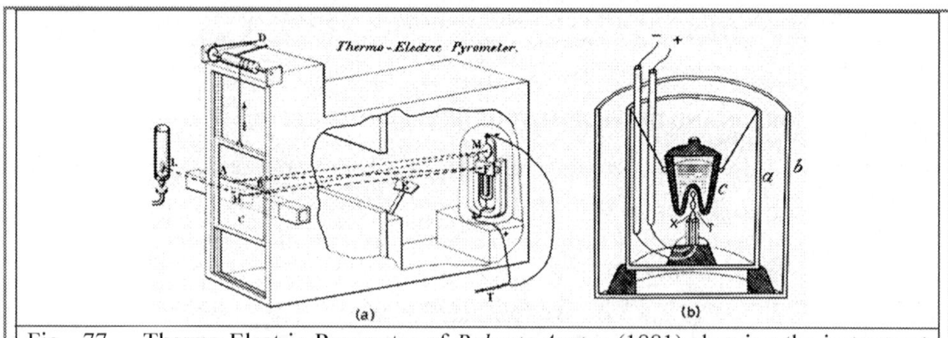

Fig. 77. – Thermo-Electric Pyrometer of *Roberts-Austen* (1881) showing the instrument (left) and its cooling arrangement (right) with particularity sample holder.

Though the sample holder was a design reminiscent of modern equipment, its capacity was extremely large decreasing thus the sensitivity but giving a rather good measure for reproducibility. It was quickly realized that a galvanometer was rather insensitive to pick up small thermal effects. This disadvantage was improved by coupling two galvanometers concurrently and later the reflected light beam was directed to the light-tight box together with the slit system enabling exposition of the repositioned photographic plate. *Stanfield* (1899) published heating curves for gold and almost stumbled upon the idea of DTA (*Differential Thermal Analysis*) when maintaining the 'cold' junction at a constant elevated temperature measuring thus the differences between two high temperatures. *Roberts-Austen* consequently devised the system of measuring the temperature difference between the sample and a suitable reference material placed side-by-side in the same thermal environment, thus initiating development of DTA instruments. Among other well-known inventors, Russian *Kurnakov* [601] should be noticed as he improved registration building his pyrometer on the photographic, continuously recording drum, which, however, restricted his recording time to mere 10 minutes.

The term thermal analysis was introduced by *Tamman* (1903) [603] who demonstrated theoretically the value of cooling curves in phase-equilibrium studies of binary systems. By 1908, the heating or cooling curves, along with their rate derivatives and inverse curves, assumed enough sufficient importance to warrant a first review and more detailed theoretical inspection, *Burgess* [604].

Not less important was the development of heat sources where coal and gas were almost completely replaced by electricity as the only source of controllable heat. In 1895, *Charpy* described in detail the construction of wire-wound, electrical resistance, tube furnaces that virtually revolutionized heating and temperature regulation [605]. Control of heating rate had to be active to avoid possibility of irregularities; however, little attention was paid to it as long as the heat source delivered a smooth temperature-time curve. All early users mention temperature control by altering the current and many descriptions indicate that this was done by manual or clockwork based operation of a rheostat in series with the furnace winding, the system still in practical use up to late fifties. The first automatic control was published by *Friedrich* in 1912, which used a resisttance box with a specially shaped, clock-driven stepped cam on top. As the cam rotated it displaced a pawl outwards at each step and this in turn caused the brush to move on to the next contact, thus reducing the resistance of furnace winding. Suitable choice of resistance and profiling of the cam achieved the desired heating profile. There came also the reduction of sample size from 25 g down to 2.5 g , which reduced the uncertainty in melting point determination from about 2 °C to 0.5 °C. Rates of about 20 K/min were fairly common during the early period later decreased to about quarter. It was *Burgess* [604] who considered significance of various curves in detail and concluded that the area of the inverse-rate curve is proportional to the quantity of heat generated dived by the rate of cooling.

The few papers published in the period up to 1920 gave little experimental details so that *White* was first to show theoretically in 1909 the desirability of smaller samples. He described an exhaustive study of the effect of experimental variables on the shape of heating curves as well as the influence of temperature gradients and heat fluxes taking place within both the furnace and the sample [606]. It is obvious that DTA was initially more an empirical technique, although the experimentalists were generally aware of its quantitative potentialities. The early quantitative studies were treated semi-empirically and based more on instinctive reasoning and *Andrews* (1925) was first to use *Newton's* law while *Berg* (1942) gave the early bases of DTA theory [601] (independently simplified by *Speil*). In 1939 *Norton* published his classical paper on techniques where he made rather excessive claims for its value both in the identification and quantitative analysis exemplifying clay mixtures [607]. *Vold* (1948) [608] and *Smyth* (1951) [609] proposed a more advanced DTA theory, but the first detailed theories, absent from restrictions, became accessible [3] by *Keer, Kulp, Evans, Blumberg, Erikson, Soule, Boersma, Deeg, Nagasawa, Tsuzuki, Barshad, etc.*, during the 1950s. For example, *Erikson* (1953) reported that the temperature at the center of a cylindrical sample is approximately the same as that for an infinitely long cylinder if its length is at least twice its diameter. Similarly, the diameter of a disk-shaped sample must be at least four times its thickness.

Most commercial DTA instruments can be classified as a double non-stationary calorimeter in which the thermal behaviors of sample are compared with a correspondingly mounted, inert reference. It implies control of heat flux from surroundings and heat itself is understood to be a kind of physico-chemical reagent, which, however, could not be directly measured but calculated on the basis of the measurable temperature gradients. We should remark that heat flow is intermediate by mass-less phonons so that the inherent flux does not exhibit inertia, as is the case for the flow of electrons. The thermal inertia of apparatus (as observed in DTA experiments) is thus caused by heating a real body and is affected by the properties of materials, which structure the sample under study.

Theoretical analysis of DTA is based on the calculation of heat flux balances introduced by *Factor* and *Hanks* [610], detailed in 1975 by *Grey* [611], which premises were completed in 1982 by the consistent theory of *Šesták* [1,3,612,613,616]. It was embedded within a 'caloric-like' framework (cf. Chapter 4) as based on macroscopic heat flows between large bodies (cells, thermostats). The need of a more quantitative calibration brought about the committed work of ICTAC [1,3,15,613-616,620] and the consequently published recommendations providing a set of the suitable calibration compounds. It was extended by use of pulse heating by defined rectangular or triangular electric strokes (see Chapter 13).

Calorimetric 'pure' (i.e. heat inertia absent) became the method of DSC (*Differential Scanning Calorimetry*), which is monitoring the difference between the compensating heat fluxes while the samples are maintained in the pre-selected temperature program (*Eyraud* 1954) [617]. This is possible providing two extra micro-heaters are respectively attached to both the sample and the reference in order to maintain their temperature difference as minimal as experimentally possible. Such a measuring regime is thus attained only by this alteration of experimental set up where the temperature difference is not used for the measurement itself but is exclusively employed for regulation. It became a favored way for attaining the most precise measurements of heat capacity, which is close to the condition of adiabatic calorimetry. It is technically restricted, however, to the temperature range up to about 700 °C, where heat radiation becomes decisive and makes regulation complicated.

Another modification was found necessary for high-resolution temperature derivatives to match to the 'noise' in the heat flow signal. Instead of the standard way of eliminating such 'noise/fluctuations' by more appropriate tuning of an instrument, or by intermediary measurements of the signal in distinct windows, the fluctuations were incorporated in a controlled and regulated way. The temperature oscillations (often sinusoidal) were superimposed on the heating curve and thus incorporated in the entire experimentation – the method known as *temperature-modulated* DTA/DSC (*Reading* 1993 [618]). This was preceded by the method of so-called *periodic thermal analysis* (*Proks* 1969), which was aimed at removing the kinetic problem of undercooling by cycling temperature [619]. Practically the temperature was alternated over its narrow range and the ample investigated was placed directly onto a thermocouple junction) until the equilibrium temperature for the coexistence of two phases was attained.

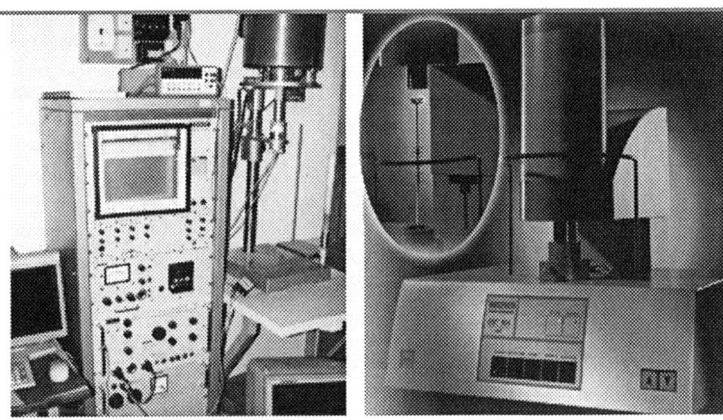

Fig. 78. – Photo of the once very popular and widespread high-temperature DTA instrument (produced by the Netzsch Gerätebau GmbH in the 1960s), that survives in operation in the author's laboratory while subjected to computerization and modern data processing. Right is shown the new, third-generation version of the high-temperature DTA (DSC 404 C Pegasus) produced recently by Netzsch Gerätebau GmbH. The experimental data shown within this book were measured using either of these instruments.

In the 1960, various thermoananytical instruments became available on the market [3,15,620], see Fig. 78., and since then the experienced and technically sophisticated development has matured the instruments to a very advanced level, which certainly includes a comprehensive computer control and data processing. Their description is the subject of numerous manufacturers' booklets and manuals [562], addresses on websites, etc., so that it falls beyond the scope of this more theoretically aimed book.

12.3. Measurements of thermal diffusivity

Modern technologies are always looking for such measuring methods [621] that provide reliable data on cooperative thermophysical parameters in a short enough time. Whatsoever, investigated materials often involve non-equilibrated states where the inherent heat treatment may affect the minute material properties. Most wanted procedure would be the joint determination of threefold data such as specific heat, c_p, thermal diffusivity, a, and thermal conductivity, λ but mostly a single parameter is resolved and then the crucial problem of consequent data consistency from different sources (λ = a c_p ρ, where ρ is density) must be solved. A thermometric procedure, which is quite similar to the convenient relaxation calorimetric method for measuring heat capacities is the pulse-heating technique detailed elsewhere [591,622,623]. The principle of operation is to heat the sample using a precise amount of power for different length of time. Between each throb of power, the sample is allowed to return to its initial state (of temperature). The longer the duration of pulse energy, the higher is the response temperature. By monitoring the time of pulse and the final temperature after each pulse, we can experimentally establish the temperature derivative, dT/dt, with respect to time and find the heat capacity Cp by using the relation $Cp = Q_{pulse}/(dT/dt)$ where Q_{pulse} is the power added to the sample corrected for the heat loss. This scheme is conventionally used to measure the heat capacity of electrically conducting materials at higher temperatures by passing an electrical current through the material. The unique feature of this heat-pulse calorimeters was skillfully used to measure even radioactive materials because the heat leak path from the sample to a heat reservoir can be adjusted in such a way that the leak compensates only for the heat generated by the radioactive decay [624].

The so called pulse and/or stepwise transient methods [622,623] are found beneficial where the thermophysical parameters can be jointly found from the temperature function and the temperature response upon the thermal disturbance applied to the measured sample. The specimen consists of three parts, see Fig. 79, where the planar heat source is clamped between the first and the second parts. The heat pulse is produced by the Joule heat effect in a planar electrical resistor and the temperature response is scanned by thermocouples to distinguish the transient temperature, $T(h,t)$ where h is the sample separating distance and t is the observation time. The standard one-point evaluation procedure considers a maximum of the temperature response for calculation of the thermophysical parameters involved, namely, the specific heat, $c_p = Q/\sqrt{(2\pi e \rho h T_m)}$, thermal diffusivity, $q = h^2/2t_m$ and thermal conductivity, $\lambda = Qh/(2\sqrt{\{2\pi e t_m T_m\}}$ where T_m is the maximum of temperature response at the allied time t_m and e denotes the Euler number and ρ density. A simplified but useful industrial modification is called the hot ball method [623] and is based on the combined generation of the heat flux (within the ball skeleton attached to the measured specimen) from an interior point-heat-source upon a simultaneous sensing the temperature by a thermometer placed parallel in the center of the heat source.

Due to the ever-increasing number of various materials that are used in high-temperature applications, knowledge of their instantaneous thermophysical properties is of a predominant importance for commercial producers. In this relation the most widely used method remains thermal diffusivity, which then provides various possibility to compute thermal conductivity from the product

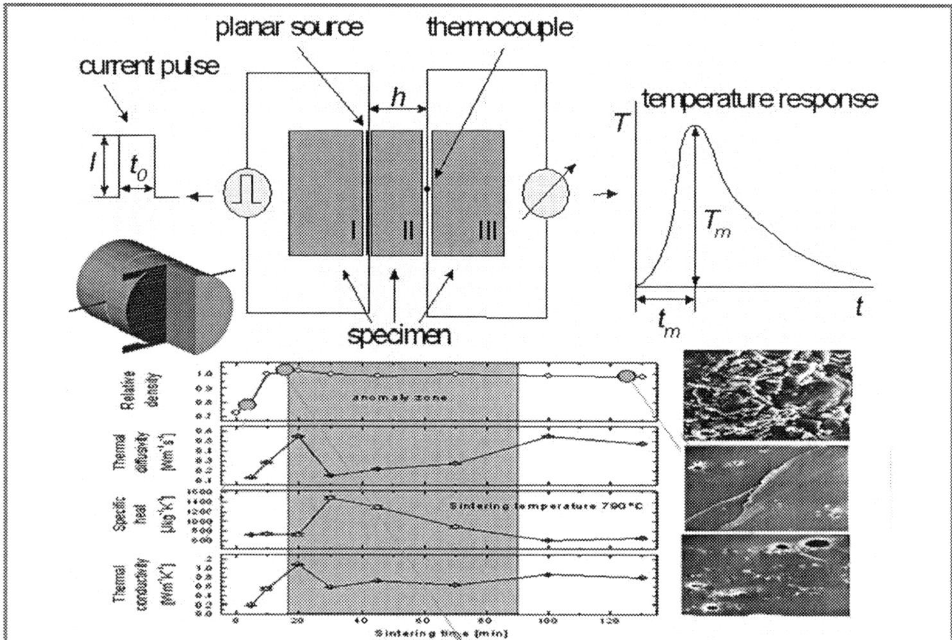

Fig. 79. – Principle of the pulse transient method viewing the current pulse (left) and the temperature response (right). Left below, the concrete specimen set up is displayed showing the heat source alongside thermocouples. Bottom, the actual application of the above method to the investigation of a true non-equilibrium sintering process of powdered oxide glass. A set of specimens was heat treated within various time periods at 790°C. Then the shared measurements of thermophysical parameters were performed at room temperature. The deviation from equilibrium values of the specific heat is noticeable and is connected with anharmonic structures that are created in the boundaries of particle regions. The three photos of microstructures are associated with the circled states marked on the curve of density [622]. By courtesy of *Ludovit Kubičár*, Bratislava, Slovakia.

with the material density and specific heat. In this relation, the laser flash method is a most versatile technique, initially developed for solid samples [625,626]). The laser pulse irradiates the front surface of a disk shape sample and the resultant temperature response of the opposite (back) surface is measured as a function of time. The response time-temperature curve is then used to determine thermal diffusivity. This method can be extended to melts using two- or three- layers cells consisting of an upper cover, in-between layers of liquid and supporting metallic plate in various experimental modifications [627-629]. The theoretical temperature decay is determined under following simplifying conditions: (i) the system is maintained at fashion of thermal equillbrium before and after laser irradiation, (ii) the heat flow is as one-dimensional, (iii) layers are semi-infinite, (iv) the contact thermal resistance of the interfaces involved is negligible and (v) the temperature distribution in the supporting (metallic) layer is isothermal (because of its thinness larger than that

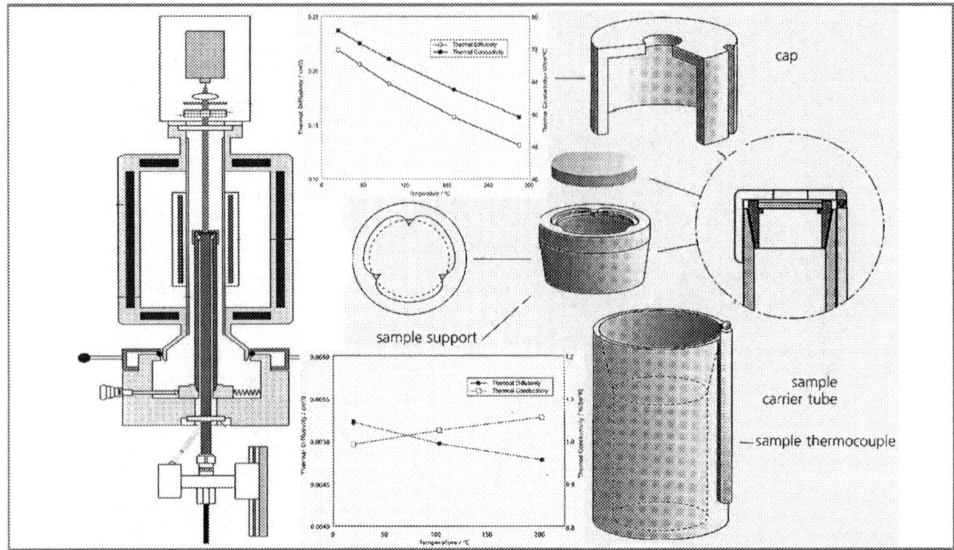

Laser flush method (scheme left) for the determination of diffusivity, commercially produced by the Netzsch instruments at different modes (LFA 427, 437 or 447) using various measuring set ups and sample arrangements (right). Insets show two MicroFlash applications, upper the thermal diffusivity and parallel calculated conductivity (by the aid of Netzsch DSC 404, cf. Fig.85. for electrolytic iron and bottom that for oxide glass.

of measured nonmetallic layers).

One favorite instrument is the laser flash apparatus commercially produced by the NETZSCH corporation [661], see the above Figure, which is used in about 2/3 of all applied cases. This contact-less and non-destructive method is of a simple geometry, with easy sample preparation and wide range of temperature applicability employing small sample size. The completely closed system is a unique instrument showing excellent accuracy and reproducibility with the need for little time to accomplish a single measurement. The vertical alignment of laser output with the IR detector allows simple, horizontal insertion of the sample. The short distance involves minimal loss of laser energy and energy impinging upon the IR detector. Measurements can be carried out in a static or dynamic oxidizing or inert atmosphere or vacuum. The data can be computer-corrected for heat loss or finite pulse time using any number of mathematical models. In addition, the evaluation of two- or three- layer samples and of the varying contact resistance between layers is possible. For adiabatic-like conditions, thermal diffusivity is determined by the simple equation $a = 0.1388\ L^2/t_{0.5}$ where L is the sample thickness and $t_{0.5}$ is time at 50% of the maximum temperature increase. The elegance of this approach lies in the fact that the troublesome measurements of the absolute quantity of laser energy absorbed by the sample and the resulting measurement of the absolute temperature increase is replaced with a more accurate and direct measurement of time and relative temperature changes.

12.4. Classification of thermoanalytical methods – heat balance

The instrumental set up of DTA/DSC has been published at various levels of sophistication and subjected to complex analyses, usually geared to the details of particular apparatuses so that they bring too specific knowledge for a generally oriented reader. The two most common lines of attacks are:

(i) Homogeneous temperature distribution in the calorimetric samples [610] and

(ii) Complete incorporation of the effects of temperature gradients and heat fluxes [631] cf. Figs. 75. and 80. Our approach herewith is to give a more transparent treatment of DTA/DSC [3,613] based on the presentation of *Gray*, [611,632] and *Šesták* and his coworkers [1,612,616].

The popular technique of differential thermal analysis (DTA) belongs among the indirect dynamic thermal techniques in which the change of the sample state is indicated by the temperature difference between the sample and geometrically similar inert reference held under identical experimental conditions. The method of heating and cooling curves, long popular in metallurgy, paved the way to its development by inserting the comparing site. Further sophistication was achieved by further incorporation of a sample-attached microheater. The advantage of this widely used method is a relatively easy verification of differences in thermal regimes of both specimens and the determination of zero trace during the test measurements, currently replaced by in-built computer programming.

Fig. 75. shows a schematic of the DTA/DSC standard arrangement used in the customary sphere of dynamic calorimetry. The sample and reference compartments can be in contact with the common block, B, which is at temperature T_B , through the respective thermal resistances. The actual sample (S) and reference (R) are in thermal contact with the two respective holders/cells through the relevant thermal resistances again, which, however, are often assumed to be effectively zero, i.e., the temperatures of the sample and reference are equal to those of the respective holders. The methods of analysis can be characterized by envisaging the basic flux, q'_S , to take place between the sample investigated and its environmental jacket, J, that is $q'_S = \Lambda_S (T_S - T_J)$, where Λ is the effective thermal conductance (which is the inverse of the respective thermal resistance). In some high-temperature systems, the heat transfer may occur through radiative as well as conductive paths. In the twin arrangement the standard sample heat exchange is also accompanied by additional fluxes between the reference sample and its environment, $q'_R = \Lambda_R (T_R - T_J)$, and between the sample and the reference, $q'_\Delta = \Lambda_\Delta (T_S - T_R)$. It then holds for the change in the sample empathy that $\Delta H'_S = q'_S + q'_\Delta + Q'_S$ and for

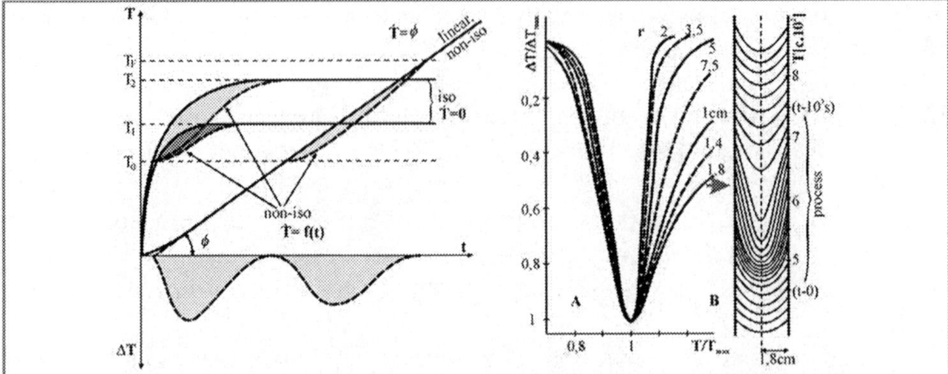

Fig. 80. – Some aspects of DTA-based effects. Left, the sketch showing the development of a DTA peak under idealized and authentic temperature profiles of the sample in isothermal and nonisothermal regimes. The DTA peak is determined as a response of the sample temperature being different from the programmed temperature of the sample surroundings due to the heat sink. The temperature differences actually resemble the traces of both the isothermal and normal DTA record. The dashed lines mark the real temperature change in the sample due to both the externally applied heating and the internally absorbed (or generated) heat of reaction (shaded). The solid lines express the idealized temperature changes habitually assumed to be representative enough even for the sample temperature and identical to the thermally controlled surroundings. Right, the set of normalized DTA peaks, calculated for a real material placed as variously diameter samples, illustrate the actual profiles of temperature gradients in a cylindrical sample (with a limiting diameter of 18 mm) due to the thermal gradient developed during an ongoing process.

the reference sample $\Delta H'_R = q'_R + q'_A + Q'_R$, where Q' is the compensation heat flux which is to be supp-lied by the excess microheater (often built into the measuring head). The scheme of all participating thermal fluxes in the twin crucible arrangement is shown in Fig.75. In the result, we can derive basic equations for the individual setups:

i) For the determination of heating and cooling curves the solution leads to the single-crucible method, where the temperature of the sample, T_S , reads as $T_S = \Delta H_S/\Lambda_S\ \alpha' + Cp_S/\Lambda_S\ T'_S + T_J$

ii) For the classical differential thermal analysis (DTA), where the sample and reference are placed in the same block and the radiation transfer of heat, is neglected, we obtain $\Delta T = [- \Delta HS\ \alpha' + (CpS - CpR)\phi + CpS\ \Delta T' - \Delta\Lambda\ (TR - TJ)] / (\Lambda S + 2\Delta\Lambda)$, where $\Delta\Lambda = \Lambda S - \Lambda R$ and $T'J = \phi = T'S - \Delta T'$. This equation can be simplified by introducing the assumption that $\Delta\Lambda = 0$ (i.e., the both sites in the crucible block are identical). If $\Delta\Lambda$ also equals zero, then the DTA arrangement involves the common case of two separate crucibles providing the most widespread instrumentation following $\Delta T = [- \Delta HS\ \alpha' + (CpS - CpR)\phi + CpS\ \Delta T'] / \Lambda S$ where the overall coefficient of heat transfer ΛS is called the DTA apparatus constant (= KDTA) which is conveniently determined by the calibration of the instrument by the compounds of known properties or by direct use of electric heat pulses. We can proceed to specify similarly the equations for for analogous measurement of both the spontaneous thermal fluxes by the single crucible method, the differences of the external thermal fluxes or even the heat-flux differential scanning calorimetry (often called differential *Calvet* calorimetry).

iii) However, the method of power-compensating differential scanning calorimetry (DSC) bears a special importance calling for a separate notice as it exhibits a quite different measuring principle. In the thermal balance the compensating thermal flux, Q', appears to maintain both crucibles at the same temperatures $(T_S \cong T_R)$ so that the traditional difference, ΔT_{DTA} , is kept at minimum (close to zero) and thus serves for the regulation purposes only. Then we have $\Delta Q' = - \Delta H_S \, \alpha' + (Cp_S - Cp_R)\phi + \Delta \Lambda \, (T - T_J)$. All the equations derived above can be summarized within the following schema using the general summation of subsequent terms, each being responsible for a separate function:

$$\mathit{enthalpy} \; + \; \mathit{heat} \; + \; \mathit{inertia} \; + \; \mathit{transient} \; = \; \mathit{measured} \\ \mathit{quantity}$$

It ensues that the effects of enthalpy change, heating rate and heat transfer appear in the value of the measured quantity in all of the methods described. However, the inertia term is specific for DTA, as well as for heat-flux DSC, and expresses a correction due to the sample mass thermal inertia owing its heat capacity of real material. It can be visualized as the sample hindrance against immediate 'pouring' heat into its heat capacity 'reservoir'. It is similar to the definite time-period necessary for filling a bottle by liquid (easy to compare with 'caloricum' as an outdated fluid theory discussed in Chapter 4, the philosophy of which is, however, inherent in the herewith mathematics). This somehow strange effect is often neglected and should be accentuated because the existence of thermal inertial shows its face as the definite distinction between the two most popular and commercially available methods: *(heat-flux) power compensating DSC* and *heat-flux (measuring) DSC*. Though the latter is a qualitatively higher stage of calorimetry as matured from DTA instrumentation, it remains only a more sophisticated DTA. It factually ensures a more exact heat transfer and its accurate determination, but it still must involve the heat inertia term in its evaluation, in the same way as the classical DTA does.

If the heat transfers, between the sample/reference and its holders are nonzero and unbalanced the area under the peak remains proportional to the enthalpy, ΔH, but the proportionality constants includes a term arising from the definite value of Λ_{SR} so that $\Delta H \cong \Lambda_S [1 + 2(\Lambda_{SR}/\Lambda_S)]$. The coupling between the

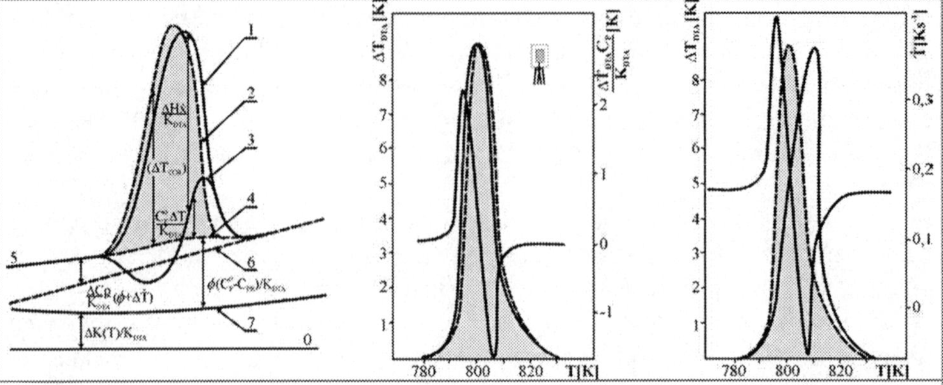

Fig. 81. – Analysis of DTA responses. Left, the graphical representation of the individual contributions from the DTA equation (see text), abbreviated as ΔT in its dependence on the individual terms differences: $+(0\leftrightarrow7)$ $-(0\leftrightarrow7)$ $-(7\leftrightarrow4)$ $-(6\leftrightarrow5)$ $-(4\leftrightarrow3)$ $+(3\leftrightarrow1)$. The composition of an experimentally measured DTA peak then reads: 0 is the recorded baseline, 1 the as-received DTA peak (as a matter of the apparatus response), 2 the corrected DTA peak (actually corresponding to the thermal behavior of the process under observation), 3 the real (s-shaped) background (understand as the actual DTA baseline), 4 the approximated DTA peak baseline (payoff of our customary practice), 5 the recording baseline as the actual DTA trace, 6 the shift in the baseline due to the unequal heat capacities of samples, and 7 the difference in the heat transfer coefficients between the sample and reference. Calorimetrically pure (DSC-like) curve (shaded) can be only obtained on application of all correcting procedures involved. Middle, the actual effect of heat inertia (solid line, shaded peak) with respect to the originally detected DTA peak (dashed) showing the extent of actual corrections involved when accounting for the concrete inclusion of s-shaped DTA background. Right, the tangible consequence of true temperatures indicated inside the sample (solid line, shaded) against the programmed outer temperature, here redrawn to the scale of real temperatures (such as $T_{real} = T_{linear} + \Delta T_{DTA}$). It is noteworthy that the s-shape curve (middle) points up the course of the instantaneous values of the heating rate, if the temperature deviations due to the DTA peak are accounted for an ideally constant heating applied from the surroundings.

sample and reference cells in the disk-type-measuring heads can be significant and such links should not be neglected as done above. Similarly, the central mode of balanced temperature measurements gives another modified solution. A useful model [633] was suggested on basis of analogy between thermal and electric resistances and capacities.

12.5. DSC and DTA as quantitative instruments

The idea of DSC arrangement was first applied by *Eyraud* [617] and further developed by *O'Neill* [634]. Whereas they initially realized the internal resistivity microheater by simple passing the electric current directly through the mass of the sample composed of the test material mixed with graphite powder as an electric conductor, *Speros* and *Woodhouse* [635] already made use of a classical platinum spiral immersed in the sample.

Modern commercial instruments employs microheaters attached to the surface of the sample cells. One of the first more detailed descriptions of DSC theory and its applicability was given by *Flynn* [636], who introduced the notion of time constant of temperature program cropping up from the difference in the microheater power and the interface sample conductivity. It is assumed that DSC operates in a stationary program, with a constant increase in the jacket temperature and with differences in the heat drop, the heat capacities and the masses of the two DSC cells being constant and reproducible functions of temperature.

In [613] the DSC equation is usefully modified by assuming *that $Q'' \cong (T'_B - T'_S)\Lambda$,* to encompass the form $\Delta H_S = -Q' - Cp_S Q''/\Lambda + (Cp_S - Cp_R) T'_B$ and assuming that the electronic control system maintains the temperature of both the sample and reference cells at the temperature of the block, T_B , by controlling the power applied to their respective microheater. This portrays well that the power developed within the sample cell is the sum contribution of the terms placed on the right-hand side: (i) the first term is related to the directly measured signal, (ii) the second term to the first derivative of the measured signal and (iii) the third term to the heat capacity difference between the sample and reference cells. The thermal resistance *(1/Λ)* appears as a coefficient in front of the second DSC term while in the DTA equation it serves as the gross proportionality constant for the measured signal. Therefore, it ensues the capability of DSC instruments to reduce thermal resistance to bottommost, thereby allowing decreasing the time constant without simultaneous decreasing DSC sensitivity. On the other hand, in DTA equipment there is a trade-off between the need for a large resistance for higher sensitivity but a low time-constant *($\tau = Cp_S/\Lambda$)* for high resolution. A small resistance ensures that the lag of the sample temperature behind that of the sample holder is minimized.

Several studies comparing the advantages and disadvantages of the heat-flux and power-compensating DSC methods were conducted, showing not essential differences. The lower values of enthalpies sometimes obtained from the heat-flux measurements were found to be due to the base line correction as a result of algorithm used by computer. However, parameter-space-induced errors, which are caused by changes in the time constant of the sample arising from changes either in the heat exchange coefficient or in the heat capacity of the sample, are less significant in the power-compensating DSC.

The temperature modulated DSC (often abbreviated as TMDSC) has aroused attention as a rather new technique that was conceived in the early 1990s and immediately patented and commercialized. The temperature signal is here modulated as a sinusoidal wave and is superimposed over the traditional monotonous curve of heating. This technique offers the possibility of deconvoluting the signal into an in-phase and an out-of-phase response, which can be used to calculate heat flows or heat capacities. It was preceded by experiments performed in the early 1980s [637,638] and is actually a consequential development of the so-called ac-calorimetry [602].

It would be difficult to survey all possible and existing applications of DSC, which is one the most commonly, used calorimetric methods. Among these, we can mention the determination of impurities, evaluation of phase transformations of the first and second orders, determination of heat capacities and heat conductivities, reaction kinetics, study of metastable states and glasses, determination of water clusters, thermal energy storage, etc..

The traditional method of DTA has deeper roots and it has been described elsewhere [3,15,602,640]. DTA applications were first based on experience and semi-quantitative evaluations so that the early articles, e.g., [641], should be noticed for using a computer in the verification of the effects influencing the shape of DTA peak. The DTA peak area was verified to be proportional to the heat of reaction (in the ΔT vs. t coordinates) and found dependent on the heating rate and distorted by increasing crucible diameter. It is, however, almost independent of the positioning of the thermocouple junction within the sample (it affects the apex temperature).

Through the commercial perfection and instrument sophistication the most DTA instruments reached the state of a double non-stationary calorimeter in which the thermal behavior of the sample (T_S) and of the reference material (T_R), placed in the thermally insulated crucibles, is compared. The relatively complicated relationship between the measured quantity, $\Delta T = T_S - T_R$, and the required enthalpy changes of the sample, ΔH, has been the main drawback in the calorimetric use of DTA.

Doubts about the quality of the enthalpy measurement by DTA stem from the fact that the changes in ΔT are assumedly affected by many freely selectable parameters of the experimental arrangement. So the suitable formulation of DTA equation, neither too complicated nor too simple, was the main objective for the progress of methodology. In our opinion the most important study was that done by *Faktor* and *Hanks* in 1959 [610], which was used as a basis for our further refinement in defining thus the fundamental DTA equation [1,612,616],

$$\Delta T = [\Delta H_S\, \alpha' - (Cp_{oS} - Cp_R)\phi - \Delta Cp_S\, (\phi + \Delta T') - Cp_{oS}\, \Delta T' + \Delta K(T))] / K_{DTA}$$

where the term Cp_{oS} is a modification of Cp_S as $(Cp_o + \Delta Cp\, \alpha)$, the collective thermal transfer term $\Delta K(T)$ is the sum of the individual heat transfer contributions on all levels (conduction as well as radiation) and the so-called DTA instrument constant, K_{DTA}, is the temperature dependent parameter, characteristic for each instrument and showing how the overall contributions of heat transfers change with temperature. The detailed graphical representation of individual terms is illustrated in Fig. 81.

There are four main requirements for the correct DTA measurement [3]:

i) Attainment of a monotonous increase in the temperature (preferably $T' = \phi = const.$). If included in the apparatus software, there is danger that the perfectly linearized baseline could become an artifact of computer smoothing so that it is advisable to see the untreated baseline too.

ii) Correct determination of the peak background, i.e., the method of the peak bottom line linearization (either joining the onset and outset temperatures or extrapolating the inwards and outwards baseline against the normal of the peak apex, cf. Fig. 81). Actual s-shaped background is symmetrical (both deviations are equal) and for $\Delta Cp \rightarrow 0$ does not affect the determination of the peak area.

iii) Appropriate determination of characteristic points, as the onset, outset and tip of peaks, based on a suitable approximation of the base line and extrapolation of peak branches.

iv) Experimental resolution of the temperature dependence of the DTA instrument constant, which is often inherent in the software of the instruments commercially available.

It should be re-emphasized that during any DTA measurements the heating rate is changed due to the DTA deflection itself, see Fig. 81b. Actually, as soon as completely controlled thermal conditions of the sample are achieved, the entire DTA peak will disappear. Fig. 82 well demonstrates the disagreement between non-stationary DTA traces for a real sample and its equilibrium-adjacent rectification when accounting for corrections due to the heat inertia discussed above and those caused by the inherent changes in heating rate.

12.6. DTA calibration and the use of defined electrical pulses

Calibration of a DTA involves adjustment of instrumental electronics, handling and manipulation of the data in order to ensure the accuracy of the measured quantities: temperature, heat capacity and enthalpy [614,615,621]. Temperature sensors such as thermocouples, resistivity thermometers or thermistors may experience drifts that affect the mathematical relationship between the voltage or resistance and the absolute temperature. Also, significant differences between the true internal temperature of a sample with poor thermal conductivity and the temperature recorded by a probe in contact with the sample cup can develop when the sample is subjected to faster temperature scans. The important quantity measured in DTA experiments is the ΔT output from which enthalpy or heat capacity information is extracted. The proportionality constant must thus be determined using a known enthalpy or heat capacity – the power-compensated DSC requires lower attentiveness as it works already in units of power. The factors such as mass of the specimen, its form and placement, interfaces and surface within the sample and at its contact to holder, atmosphere link and exchange, temperature sink, and other experimental effects influence the magnitude of the power levels recorded in DTA (and DSC) instruments, recently often hidden in computerized control.

One of the most straightforward pathways to formulating a relationship between the measured and actual temperatures is the use of standard substances with defined temperatures of certain transformations connected with a suitable thermal effect (peak). This procedure was extensively studied during the 1970s by the Standardization Committee of ICTA and chiefly directed towards explaining discrepancies in the published data and towards selecting and testing standards suitable for DTA. In cooperation with the US national Bureau of Standards there were published and made available four certified sets of the DTA temperature standards covering the temperature region of about 1000 degrees.

Most common calibration substances involve transitions and/or melting (in ^{o}C) and they can include: O_2 (*boiling point, – 182,96*), CO_2 (*sublimation point, – 78,48*), cyklohexene (-87 & 6.5), Hg (*solidification point, – 38,86*), dichlorethane (- 35), H_2O (*melting point, 0.0*), diphenylether (27), o-terphenyl (56), biphenyl (69), H_2O (*boiling point, 100*), benzoic acid (*solidification point, 122, 37*), KNO_3 (128), In (*solidification, 156.63*), $RbNO_3$ (164 & 285), Sn (232), Bi (*solidification, 271,44*), $KClO_4$ (301), Cd (*solidification, 321,11*), Zn (*solidification, 419,58*), Ag_2SO_4 (426), CsCl (476), SiO_2-quartz (573), K_2SO_4 (583), Sb (*solidification, 630,74*), Al (*solidification, 660,37*), K_2CrO_4 (668), $BaCO_3$ (808), $SrCO_3$ (925), Ag (*solidification, 961,93*), ZnS (1020), Au (*solidification, 1064,43*), Mn_3O_4 (1172), Li_2TiO_3 (1212) or Si (1412) (italic figures note the fixed definition points of the *International Practical Temperature Scale*). For more extended figures, including associated enthalpy data, see Tables in the chapter's appendix as transferred from ref. [3]. The deviation of a well-done determination in ordinary DTA-based laboratory can usually vary between 2 to 6 ^{o}C (not accounting the higher temperature measurements usually executed above 900 ^{o}C). Nonetheless, calibration procedures became a standard accessory of most commercial software associated with marketable apparatuses and thus given instrumental sets of calibration substances. The both involved processes of transition and melting, however, limit the range of application in various fields because both processes (as well as materials involving salts, oxides and metals) have a different character fitting to specific calibration modes only (material of sample holders, atmosphere, sample size and shape, etc.). Further, other inconsistency arising from different heat absorption of the radiation-exposed surface (sample and its holder), quality of adherence between the sample and inner walls of the holder, etc., must be taken in account. Often a distinctive adaptation such as sample compression or even blending with grains of well-conducting inert admixtures (noble metals Ag, Au, Pt, Ir or inert ceramics Al_2O_3, ZrO_2), special cutting (or

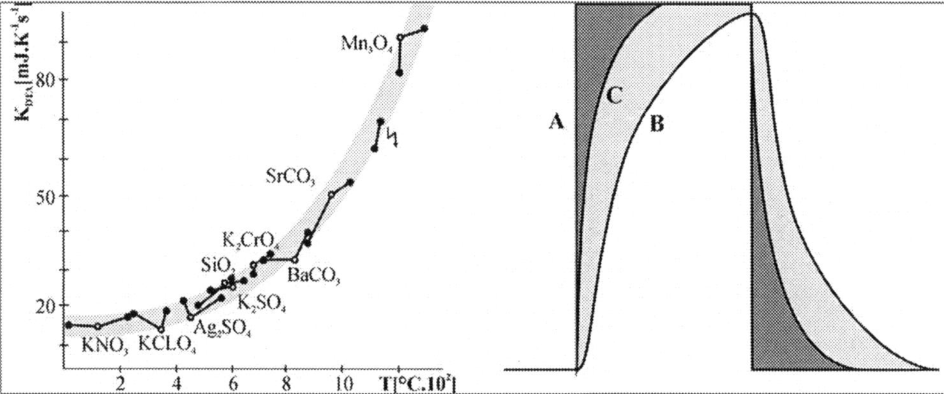

Fig. 82. – Use of the generator of defined heat pulses. Left, an example of a combined calibration plot for a set of DTA measurements using the heats of transformation (of the ICTAC recommended substances), supplemented with the heat of pulses obtained by electric resistive flush-heating applied before and after each transformation (the related calibration points are connected by a line and the substance specified by formulae). Right, the analysis of a response peak showing the assessment of a rectangular thermal pulse (A) (launched by electric flush-heating into the sample bulk using the micro-heater placed inside the material under study) in comparison with the actual response by DTA instrument (B). Upon the numerical application of DTA equation, a corrected DTA peak is then obtained (C). The shaded areas illustrate the effect of the sample thermal inertia and the remaining dark shaded areas reveal the probable effect of thermal gradients remaining within the sample body due the excess unequilibrated heat.

even holder-enwrapping by) layers and ribbons, or favored blacking of the exposed surfaces by fine-powder platinum can improve the quality of calibration (as well as the exactness of consequent measurement, which, certainly, must be done in the exactly same manner as the calibration).

Even more complicated is the guarantee of quantitative determinations by calibrating peak areas to represent a defined amount of enthalpy change. The same test compound did not exhibit sufficiently correct results, so that no ICTA recommendation was issued. In the laboratory scale, we can use certain compounds and their tabulated data, but the result is questionable due to the various levels of the data accuracy. It is recommendable to use the sets of solid solution because they are likely to exhibit comparable degree of uncertainty, such as Na_2CO_3-$CaCO_3$ or $BaCO_3$-$SrCO_3$ systems [642] or various sesquioxides (manganese spinels) [643]. The temperature of the cubic-to-tetragonal transformation of $MnCrO_3$-Mn_3O_4 increases linearly with increasing manganese content, from about 200 °C up to 1172 °C for pure Mn_3O_4 , which was successfully employed as a high-temperature standard [644], see the appendix.

The use of the Joule heat effect from a resistance element on passage of electric charge is the preferable method for an 'absolute' calorimetric calibration. It certainly requires a special setup of the measuring head enabling the attachment of the micro-heater either on the crucible surface (similarly to DSC) or by direct burying it into the mass of (often-powdered) sample. A combination of both experimental lines (substances and pulses) may provide a successful result as shown in Fig. 81. In this case, we use a series of rectangular pulses generated by an electronic circuit [645] assuring the same amount of delivered heat (0.24 J) by automatic alteration of paired inputs of voltage and current[1].

We also used a rectangular heat stroke for experimental verification of the DTA equation. The resulting curves of the input and output signals are also shown in Fig. 81.b. The responding signal is washed out due to thermal phenomena involved but the effect of thermal inertia is clearly seen and can be separated by the application of the DTA equation. This mathematical correction factually decreases the degree of rectangularity distortion. The remaining misfit is due to the thermal gradients (cf. Fig. 79.) and inherent impediment in the heat transfer.

In addition, for a relatively longer frontal part of the response peak we can directly determine the overall data of the mean heat capacity involved, C_p and the global heat transfer coefficient, Λ. It is in fact a traditional technique known as a *ballistic method* often applied in calorimetry when measuring amplitude ΔT_{max} and the peak area A_{max} for prolonged time ($t \rightarrow t\infty$), which can easily be

---

[1*] Back at the beginning of the seventies, we put in use a generator of defined heating pulses. A commercial but slightly modified DTA measuring head was equipped by electric heating element consisting of a platinum wire (diameter 0.1 mm) coil with the total electric resistance of 1.5 Ohm, which was immersed in the material studied (sample). The pre-selected amount of heat was ¼ Joule, easy adjustable by simple selection of input voltage and current pairs, which provided rectangular heat pulses with reproducibility better than 3% and independent of temperature effects. Connecting wires (0.5 mm) were assorted in different ways (e.g., through a capillary lead aligned or even joined with the temperature measurement), which was necessary to respect both the given setup of the measuring head and the guaranty to satisfy the least drop on the incoming voltage. Alternatively, we also used a DTA measuring head modified to accommodate three independent cells enabling the measurement of mutual temperature differences. Beside some other advantages it showed that application of electric pulses into the reference material (thus actually simulating an endothermic process with regard to the sample) had a minimal effect on accuracy. In order to match a more authentic course of a real process we modified our electric circuit to produce the triangular heating pulses, which was found suitable for a better resolution of some kinetic problems. Resulting calibration was very reliable up to about 700 C, which was in agreement with the experience of DSC builders, where the temperature limit is roughly the same. The attachment of externally controlled heat pulsater verified the suitability for such, at that time, advanced solution for improved electric calibration, which also helped to better analyze the peak response necessary to authorize importance of the partial heat contributions, cf. Fig. 10b. It was only a pity that no commercial producer, neither ICTAC, became active in their wider application conservative with the application of the ICTAC calibration substances. It would be much easier if a computerized unit were used and built into commercial instruments.

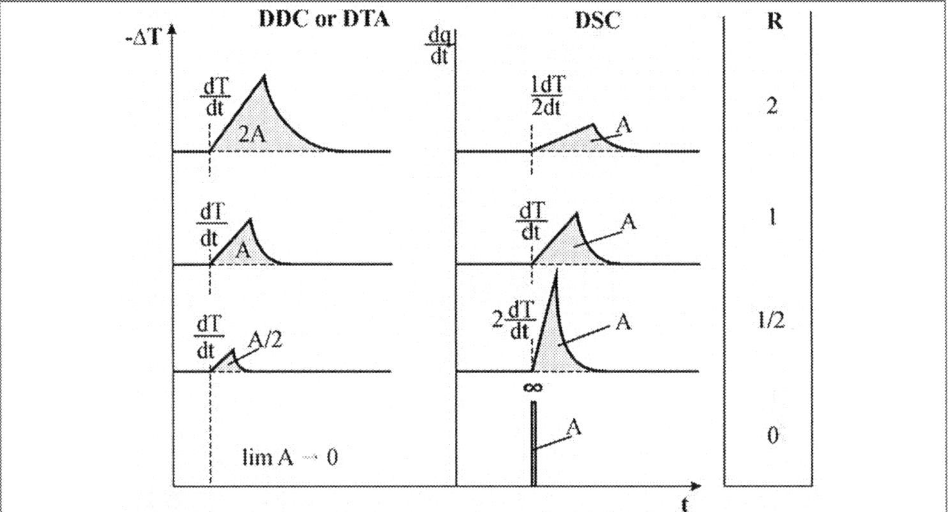

Fig. 83. – Schematic comparison of the reciprocal sensitivity of DSC and DTA methods indicating an analogous thermal process under comparable experimental conditions when the heat transfer coefficient is the only varied property. It points up the level of acceptability and possible degree of replacement of DSC and DTA.

modified for DTA measurement. Then we can determine the time constant of the apparatus, $\tau = A_{max}/\Delta T_{max}$, and knowing the rate of heat input, Q', (by, e.g., feeding a constant current) the instrumental heat transfer coefficient can be determined as $\Lambda = Q'/\Delta T_{max}$. The time constant can also be found from the decaying curve associated with the sample cooling (the tail of DTA peak) after the termination of heating. By linearization of the standard *Tian* equation it yields, $\tau = (t - t_o)/(ln\ T_o - ln\ T) = cotan\ \omega$, where ω is the angle of the straight line obtained by plotting $ln\ T_t$ against time t.

Another type of a triangular pulse has negligible frontal part but the tail is similarly valuable for determining the time constant of apparatus. However, this 'saw-tooth' pulse may become a welcome tool for possible calibration of reaction dynamics, which would be a desirable instrument regarding a better determinability and characterization of the kinetics of processes.

For experimental practice, it is interesting to compare the sensitivity of DTA and DSC for a given experimental arrangement from which a certain value of the heat transfer coefficient follows. It can be seen from Fig. 82. that the magnitude (and also the area) of a DTA peak is inversely proportional to the magnitude of heat transfer [1,636]. With a good thermal conductivity, the DTA peak actually disappears, whereas it becomes more pronounced at low thermal conductivity. On the other hand, a change in heat transfer should not affect the magnitude of a DSWC peak, provided that the work is not carried out under extreme conditions.

It is obvious that a substantial change of heat transfer can change the character of DSC measurement, which is encountered in practice in temperature regions where the temperature dependence ceases to be linear (conduction). Therefore most DSC instruments does not exceed working limit of about 800 C (the approximate onset of radiation as the cubic type of heat transfer).

12.7. Practical cases of applications

There is a wide sphere of applicability of DTA/DSC technique, which is regularly described in satisfactory details in the individual apparatus manuals or other books [1,15,602,613,640,646]. The usage can be sorted in to two classes: The methods based on the (1) modified instrumentation such as (i) high-pressure studies [647] or (ii) differential hydrothermal analysis [648] and the measurements applicable under the (2) ordinary apparatus' set up such (iii) determination of phase boundaries [646,649], (iv) impurity measurements [650,651] and (ivi) reaction kinetics (see previous Chapter [3,425,508-510,521]. Specifically constructed instrumentations [1,15,602] falls beyond the scope of this book so that we shall concentrate on the second type of applications.

Phase diagrams

A successive DTA/DSC investigation of the samples of different compositions provides us with a series of traces useful for exploratory determination of a phase boundary [646,649], see Fig. 83. It, however, requires certain expertise in a conscientious process of a gradual refinement of the positioning of characteristic points, such as eutectic, peritectic or distectic, which are most important for the final determination of the shape of the phase diagram. Their position can be determined as the intercept of suitable extrapolation preferably fitted from both sides, usually by plotting the areas of successive peaks of fusion or solidification versus composition [646]. If such a plot has axes calibrated to the peak areas, it can reveal the enthalpy change associated with the extrapolated maximum peak area. However, it should be noticed that in the region between the solid and liquid phases the less is the solid phase melted per time unit, the steeper is the liquid phase curve. Therefore the DTA/DSC peaks of samples whose composition differs little from that of the eutectics are strongly concave, showing thus less sensitivity than those of the compositions far apart. Although it seems easy to construct a phase diagram based on DTA/DSC curves, it is rarely a case of very simple or ideally behaving systems. Monitoring the reactions of solids mixtures and solutions (particularly powders in ceramic oxide systems), a complex pattern is often obtained exhibiting an assortment of the desired (equilibrium-like phenomena) together with delayed or partially occurring reactions (hysteresis due to nucleation, transport, etc.). One of the most misleading phenomena is the plausible interference of metastable phases, i.e., side effects due to the partial or full

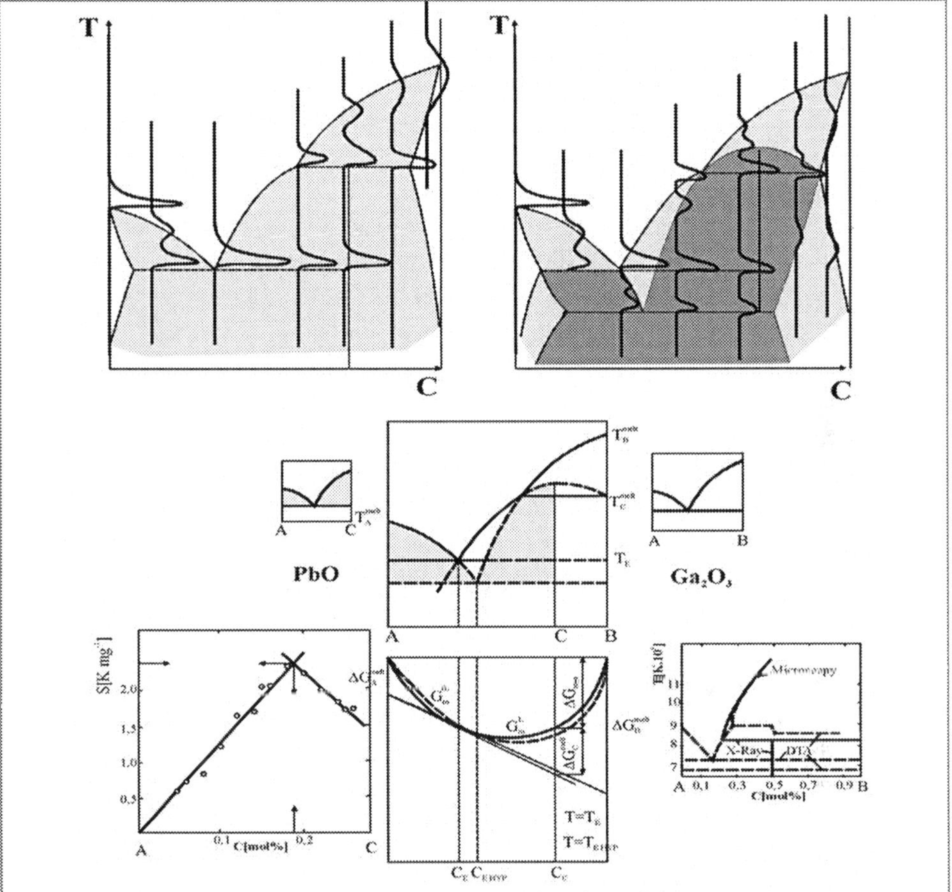

Fig. 84. – Practical case of DTA application for the determination of a eutectic-peritectic phase diagram. Upper left, the schematic picture of experimental DTA responses obtained under the ideal passage through the phase boundaries of the two-component system on heating. Upper right: The same as before but assuming the surplus effects due to the possible interference of changes taking place on the extrapolated lines down to the metastable regions (shadowed). Below, in the middle row, it is explicated more exhaustively as the superposition of two simpler eutectic phase diagram. Bottom, there is an actual case of deciphering the phase diagram of pseudobinary BaO-Ga_2O_3. The most important experimentally accessible figure is the exact position the eutectic point, which would be difficult to find experimentally on the mere basis of a pre-mixed compositional fit. Near eutectic points, it seems more easy to use a wider compositional range and extrapolate both branches of the heat determined by straight lines to intersect them in the desired eutectic point (left). It must be certainly completed by the experimentally determined DTA traces, necessary to indicate the characteristic temperatures responsible for the positioning of horizontal lines, as well as by RTG data fixing the area compositions. This, however, is not always a clear-cut solution so that a complementary mathematical evaluation based on the established models (e.g. Fig. 37.) must be completed in order to gain an overall realistic image [9, 646].

detection of extrapolated phase boundaries (cf. Figs. 48.-51.) down into lower-temperature, often metastable regions. Therefore it is always necessary to combine practical continuous heating/cooling TA measurements with complementary data on high-temperature equilibrated samples investigated in-situ (high-temperature XRD) or after effective quenching or by other techniques of magnetic, electric or structural investigation.

DTA/DSC data can be of assistance in glass viscosity prediction and the associated evaluation of glass stability through ready-available non-isothermal data. It is based on the determination of the temperature range of glass transformation interval, Tg, and its relation to the so-called *Hrubý* coefficient [359,652], cf. Fig. 54. A more sophisticated correlation provides interrelation between the experimental activation energy, E_{DTA}, and those for shear viscosity, $E\eta$, on the basis of the relative constant width of T_g, (i.e., difference between the onset and outset temperatures). It reveals a rough temperature dependence of the logarithm of shear viscosity on the measured temperature, T, using the simple relation $log \ \eta \ = 11.3 + \{4.8/2.3 \ \Delta(1/Tg)\}(1/T - 1/Tg)$ [653], etc., its detailed discussion is beyond the scope of this text.

Certainly, we can anticipate progress in the derived methods. Among many others we can mention the development of a laser flush apparatus for simultaneous measurements of thermal diffusivity and specific heat capacity by *Shinzato* [654], or localized micro-thermal analysis (*Price* [655]. Practical applications in materials science qualified, for example, searching for non-stoichiometry in high-T_c superconductors (*Ozawa* [656]), defining glass transitions in amorphous materials (*Hutchinson* [657]) or fitting heat capacity for the spin crossover systems (*Šimon et al* [658]) which enumeration can be found elsewhere [15,620]).

Heat capacity

A classical example of specific heat determination [659] was carried out emphasizing the ratio method [660]. The first measurement is done with empty crucibles (the baseline run), in the second measurement a standard material of known specific heat (synthetic sapphire) is employed in the same crucible, see Figs. 85. and 86. In the third run the sample being analyzed is put into the same crucible, providing the necessary data for the determination of the sample heat capacity, $Cp_{sam}(T)$, through the ratio of $m_{cal} \{V_{sam}(T) - V_B(T)\}$ vs. $m_{sam} \{V_{cal}(T) - V_B(T)\}$ multiplied by $Cp_{cal}(T)$, where respectively m_{cal} and m_{sam} are the mass of calibrating substance (*cal*) and sample (*sam*), and $V_B(T)$, $V_{cal}(T)$ and $V_{sam}(T)$ are the measured differences in thermocouple voltages between the sample and reference (B), and the differences observed during calibration (cal) and sample measurement (sam). Concrete outputs are shown in Fig. 85 and 86. Left, the measurement of the apparent specific heat of water is depicted, being specific for its positioning within the sensitive temperature region ($- 35 < T < 80 \ °C$),

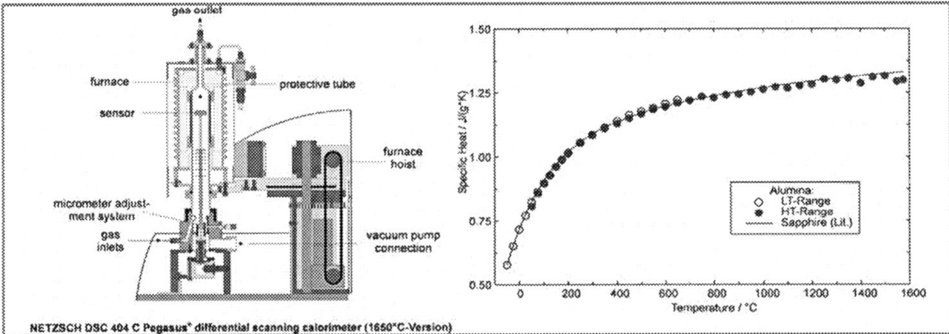

Fig. 85. – Schema of the high temperature instrument (Netzsch DSC 404 C Pegasus, version up to 1650 °C). Right, the calibration curve of specific heats obtained on α-alumina samples between -50 and 1600 °C using platinum crucibles with lids under an inert atmosphere.

that is normally difficult to ascertain. The melting point is clearly detected at 0 °C and the heat of fusion over-laps the specific heat between 0 and 30 °C (its value is close to the literature data). The accuracy of the measured specific heat between 30 and 60 °C is below 2%.

Another extremely sensitive case is the specific heat determination of partially amorphous titanium- chromium alloy due to its extreme sensitivity to oxidation requiring a particularly purified atmosphere of argon. The both heating (solid) and cooling (dashed) curves were recorded exhibiting that on heating the measured specific heat was slightly lower compared to the cooling run. Outside the transition ranges, a good agreement was achieved between heating and cooling runs.

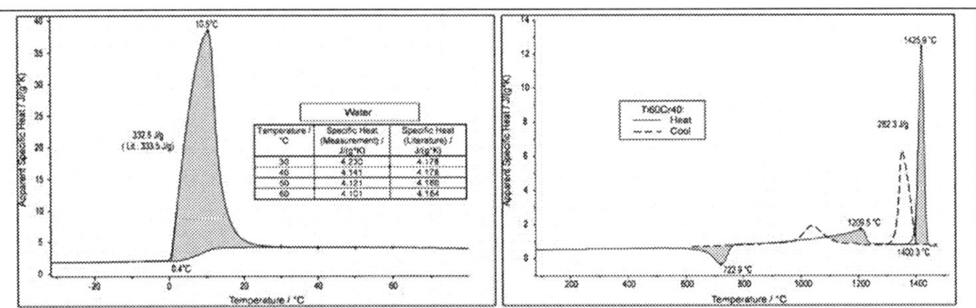

Fig. 86. – Left, apparent specific heat of water below and above its melting point is shown using the alumina crucibles with lid and carried out under helium atmosphere. It includes the inset with a comparison of the measured specific heat with the data tabulated. Right, the apparent specific heat of glassy $Ti_{60}Cr_{40}$ samples are publicized between the room and high temperatures (up to 1500 °C) employing platinum crucibles with alumina liners and lids with the inner crucible surface protected by yttria coating (to avoid reaction between the titanium alloy and the alumina liner). Courtesy of Netzsch application laboratory, Selb, Germany.

Impurity measurements

Let us mention here only one of the earliest applications of DTA/DSC: the rapid determination of the impurity (of mostly organic compounds) without the requirement for using a reference material [650,651,662,663]. The procedure for quantitative evaluation of purity is based on the thermodynamic relationship for the depression of the melting point by the presence of usually unknown contamination. For the derivation of this procedure, it is assumed that (i) the solute forms an ideal solution upon melting, (ii) the solvent and solute are immisible in the solid phase, (iii) the concentration of impurities can be expressed as a mole fraction of solute and (iv) the enthalpy of fusion of the sample is unchanging over the range of temperatures investigated.

Based on the standard relations for the melting point depression $(T_{im}-T_o)$ due to the impurity possession, im, valid for the ideal solution and the constant enthalpy of fusion, ΔH, it provides that $x_{im} = \Delta H/R(1/T_{im} - 1/T_o)$ where x_{im} and T_{im} are the impurity mole fraction and the melting temperature of impure material and T_o is the melting temperature for pure material. Because the approximation of $T_{im} \cong T_o$ it simplifies as $x_{im} = \Delta H/(RT_o^2)\ (T_o - T_{im})$ yielding for the fraction, $F = x_{im}/x_2$, of the sample melted at temperature T (given by the ratio of the composition of the sample x_{im} to the composition of the equilibrated liquid x_2) the relationship: $F = RT_o^2\ x_{im}/\{\Delta H(T_o - T)\}$. This relation is then used in the convenient form of $T = T_o - (x_{im}RT_o^2)/\Delta H\ (1/F)$ furnishing the plot of the sample temperature (T) versus the reciprocal of the fraction melted $(1/F)$ which straight line has a slope equal to $(-x_{im}RT_o^2)/\Delta H$ and intercept T_o. Values of the fraction melted at various temperatures are obtained from a series of partial integration of the DTA/DSC peak, where $F(T) = \Delta H(T)/\Delta H_F$. This sort of the *van't Hoff* plot shows often a nonlinearity, which is attributed to the underestimation of the fraction melted in the early part of the melting process. The movement of *dH/dt* curve away from the base line is customarily so gradual that it is sometimes difficult to estimate the exact beginning of the melting. Preselecting few reference points in order to determine a suitable correction factor [664] can eliminate this arbitrariness. Another limitation is the factual formation of solid solutions [665] as well as the precarious applicability of the method to determine ultra small impurity levels. Because the technique is measuring small differences in pure materials, appropriate care during the preparation of the specimen is critical.

Optimization of DSC resolution

Each of the DSC apparatuses has a certain attribute and quite a number of given specifications that cannot be altered and that have to be accepted though the operational conditions may be adjusted within a certain frame conformant to the given instrument employed and the type of sample tested [666]. The most

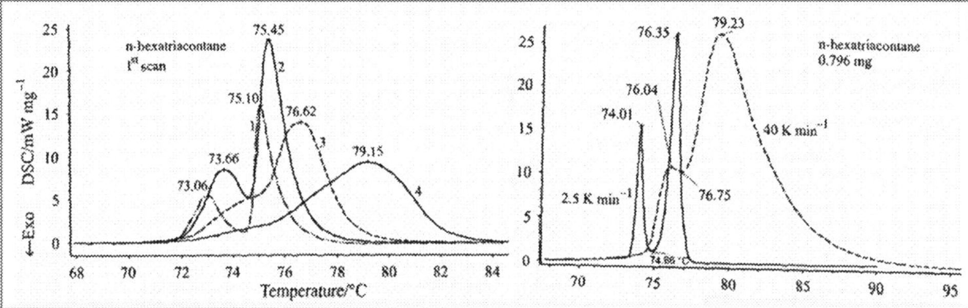

Fig. 87 – Left: dependence of the resolution sensitivity on sample mass for the recommended test substance of n-hexatriacontane [see table] – first scan of DSC curves (Netzsch apparatus 304 Phoenix equipped with a special μ-sensor) treated under the heating rate of 10 K/min. Right: with the same test sample the DSC resolution is improved showing uncorrected curves (dashed, relative scaling at y-axis) by means of reduction of heating rate. Courtesy of Netzsch application laboratory, Selb, Germany.

convenient approach and best results may be eventually obtained by a procedure based on experience and the experimenter's sense of the instrument's finest performance but so far there are no clear rules for this. Additionally, a new yet unknown behavior or a newly generated chemical entity or a supplementary phase could also lead to a desired invention and a certain fine-tuning in the laboratory is therewith the 'state-of-the-art level' [667]. The substance n-haxatriacontane was selected and proposed as the new test substance for a quantification of resolution similarly to the previously introduced substance $4,4'$-azoxyanisole [668]. Experimental evidence showed two most important parameters influencing the DSC resolution, namely the thermal resistance of the instrument and, under certain conditions, also the thermal resistance of the sample material including the sample pan in use. As the resolution of a DSC instrument we can choose the ability to identify overlapping caloric events by the existence of a minimum between the uncorrected experimentally observed peaks. The previously conferred sensitivity [668] was given as the quotient of the peak height for the second transition of $4,4'$-azoxyanisole and the peak-to-peak noise read right before and after the transition. Several candidates for resolution testing [650,669] are listed in the Table in the appendix and for the evaluation procedures applied for the elucidation of scanned data the commercial accessories were employed as follows: (i) Netzsch-Proteus software, (ii) correction for thermal resistance and (iii) peak separation software. The best resolution factor R_λ was defined in respect to both the broadened λ-transition of n-haxatriacontane and its melting as $R_\lambda - h_\lambda/h_{min}$ where h_λ is the peak height for the transition/melting observed and h_{min} is the height of the minimum of the DSC curve in between the two peaks measured in respect to the baseline. Mathematical representation can then be applied in the form of logarithmic dependence $R_\lambda = a_\lambda \log m_a$ where a is the coefficient for the dependence of the resolution factor with mass m_a.

12.8. Temperature modulated mode

Rather high-speed measurements of the sample heat capacity are a characteristic feature of DSC methodology. It is based on a linear response of the heat-flow rate, q', to the rate of change in temperature, T', which is proportional to the temperature difference, ΔT. For a steady state with negligible temperature gradients it follows that the product of sample mass, m, and its specific heat capacity, Cp, is ideally equivalent to the ratio of q'/T'. It can be practically expressed according to *Wunderlich* [551,671] as $K_{DSC}\ \Delta T/T' + Cs\ \Delta T'$, where Cs stands for the specific heat capacity of whole set of the sample with its holder and the first term serves as a correction for the larger changes of Cs with temperature relative to that of reference, C_R. As long as the change in the difference between reference and sample temperatures becomes negligible $(\Delta T' \cong 0)$, that the heat capacity is directly proportional to flux rate. The sample temperature is usually calibrated with the onsets of the melting peak of two or more melting-point standards and the value of K_{DSC} must be established as a function of temperature by performing a calibration run, typically with sapphire as the sample. The same is true in the case of saw-tooth modulation where the heating and cooling segments are long enough to reach the steady state in-between switching. It is also possible to insert occasional isothermal segments between the modulations to monitor the baseline drift of the calorimeter.

In the true, temperature-modulated DSC, the experiment is carried out with an underlying change of temperature, which is represented by sliding average over one modulation period. The sample temperature and heat-flow is changing with time, as dictated by the underlying heating rate and the modulation, and are written as $T_S(t)$ and $q'(t)$. The measurement of Cp remains simple as long as the conditions of steady state and negligible gradients can be maintained. If the sample response is linear, the sliding averages over one modulation period of T_S and q', abbreviated as $<T_S>$ and $<q'>$, yield identical curves as for the standard DSC. A simple subtraction of these averages from their instantaneous values yields the 'pseudo-isothermal (reversing) signal'. For the sinusoidal modulation, the quasi-isothermal analysis is known as it provides the following expression for the heat capacity, $(C_S - C_R) = A_\omega/A_\phi\ \sqrt{(1+\{C_R\ \omega/K_{DSC}\}^2)}$ where A_ω and A_ϕ are the modulation amplitude of the heat-flow and that of the heating rate $(A_{Ts}\ \omega)$, respectively. The frequency ω is given in [rad^{-1}] $(\omega=2\pi/p$, where p is the modulation period in seconds). It can be seen that if the reference calorimetric holder is an empty pan/cell, then $(C_S - C_R) = m\ Cp$ and so A_ϕ represents the maximum amplitude of the rate of temperature change without

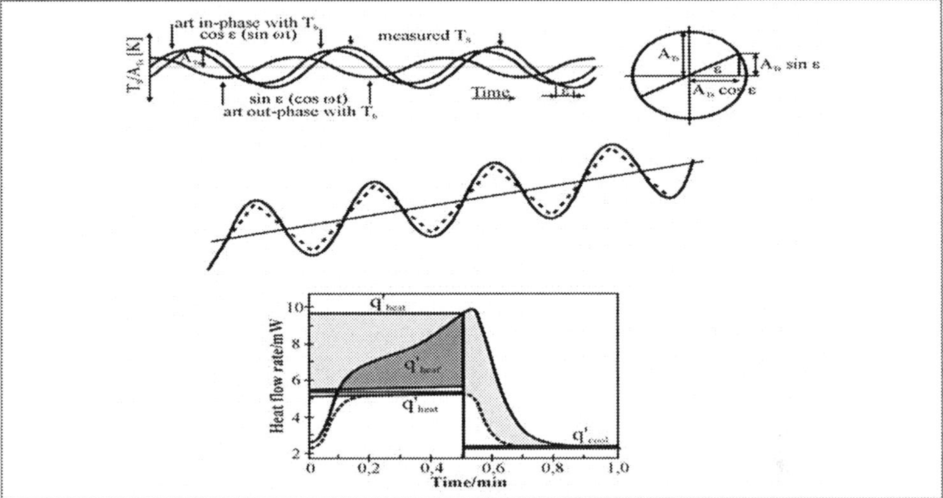

Fig. 88. -Schematic drawings of temperature modulations [9]. Upper, characteristics of the sinusoidal signal modulations and its underlying heating rate with the measured part including the in- and out-phase parts. Middle, the comparison of saw-tooth (dashed) and sinusoidal modulation superimposed on the linearly increased temperature. Bottom, a detailed heat flow picture for the saw-tooth signal showing the baseline-subtracted heat-flow rates at two temperatures (solid and dashed lines) separated by 60 K. Experimentally observed modulation phase for the temperature 451 K (dashed – before the major onset of PET melting) and at 512 K (solid   within the major PET melting peak), where the solid line exhibits a typical departure from reversible behavior. Comparing the heat-flow rate to the data taken at 451 K allows for an approximate separation of the melting, as marked by the vertical shading in the heating segment and diagonal shading in the cooling segment of the modulation cycle. Comparing the diagonally shaded areas in the heating and cooling segments suggests a rough equivalence, so that the marked q'$_{heat}$ accounts practically for all irreversible melting. The cooling segment, however, when represented by q'$_{cool}$ contains no latent heat, and similarly the term q'$_{heat'}$ is a measure of the heat-flow rate on heating without latent heat.

the effect of underlying constant heating. If the calibration and measurements are performed at frequency ω using the same reference site, the square root becomes constant and can be made part of a calibration constant.

It is clear that the conditions of steady state and negligible temperature gradients within the calorimetric sites are more stringent for the dynamic version (TMDSC) than for the standard DSC. In fact, with a small temperature gradient set up within the sample during the modulation, each modulation cycle has smaller amplitude of heat-flow rate, which depends on the unknown thermal conductivities and resistances. A negligible temperature gradient within the sample requires thus the sample calorimeter to oscillate in its entirety. It also requires almost negligible thermal resistance between all parts of calorimeter, such as thermometer, sample holder and the sample itself. The phase lag between the heater and the sample must be entirely controlled by the thermal diffusivity of the path to the sample. As soon as any shift and a change in maximum amplitude develop during the modulation, it is impossible to appropriately handle the analysis under the above-mentioned equations and the relationships becomes more complicated. We know well that even at the standard DSC setup we are not completely aware of the 'truly entire' sample temperature conditions and superimposed "complication" of temperature further modulation makes our restricted knowledge yet less complete. It even stirs up some concern whether the dynamic techniques can provide an adequately realistic picture and not only a virtual, mathematical or artifact-like image convincingly evaluated at our best will.

A definite progress was made recently by studying TMDSC with the simpler saw-tooth modulations [671]. Providing that the *Fourier* equation of heat flows holds, the solution for the different events in the DSC attitude are additive. Steady state, however, is lost each time when a sharp temperature change occurs, i.e., T'' attains a real value. An empirical solution to this problem was to modify the description by introducing a kind of 'time constant, τ, dependent not only on the heat capacity of samples and cooling constants but also on the mass, thermal conductivities, interface hindrances, etc. So that $(C_S - C_R) = A_\omega/A_\phi \ \sqrt{(1+\{\tau \ \omega\}^2)}$ where τ depends on the calorimeter type and all possible cross-flows between the samples and holders. Its real manifestation can be assessed by the inverse plot of the uncorrected heat capacity against the square of frequency. It is true for longer frequencies only, say > 200 s, while for the shorter frequencies, reaching down to 5 s, τ and remains a continuous function $of \omega$. Naturally, to conduct a series of measurements at many frequencies for each temperature represents a considerable increase in the effort to establish the required data, but it can be eased by application of different frequencies at a single run.

For the analysis with the standard DSC method, the initial data after each change in the heating must be discarded until a steady state is reached. Hence, if the heating/cooling segments are too short to attain the equilibrium-like state and thus minimize temperature gradients within the sample, the evaluation does not give useable results. It may be possible, however, to extrapolate the immediate conditions to a certain state of steadiness or, of course, to modify the modulation mode as to find the steady-state situation experimentally.

In the current TMDSC approach, the non-reversible contribution of the latent heat can only be assessed indirectly by subtracting the irreversible heat capacity from the total one or by analyzing the time domain. Any error in reversing as well as total heat capacity will then be transferred to the non-reversing heat capacity. The analysis using the standard DSC method allows a direct measurement of the non-reversing heat capacity by determining the difference in the heat capacities at their steady state. We can call this quantity as the imbalance in heat capacities so that $(m \ Cp_{imbal}) = (q'/T')_{heat} - (q'/T')_{cool}$. The advantage of the imbalance measure is that it is a directly measurable quantity of the non-reversible contribution and avoids, under the requested conditions, the contribution of potential errors. The irreversible heat-flow rate can be calculated by separating q'_{heat} into its reversible, underlying and irreversible contributions $(q'_{rev}$, q'_{under} and $q'_{irrev})$ which can provide the relation $q'_{irrev} = m\, Cp_{inal}\, /\, (1/q'_{heat} - 1/q'_{cool})$ if the irreversible parts are equal.

TABLE – appendix A

Characterization of selected substances suggested as candidates for a DSC resolution test[666]

Substance	$T_{trans,,I}$ [°C]	$T_{tran,,II}$ [°C]	$\Delta T_{tran,(I\text{-}II)}$ [°C]	$\Delta H_{trans,I}$ kJ/mol	$\Delta H_{trans,II}$ kJ/mol	$\Delta H_{trans,(I+II)}$ kJ/mol
4,4'-azoxyanisole	117 [668]	134 [668]	17	0.52	31	31.5
	118 [*]	136 [*]	18			
cimetidine [#]	$T_{melt,A}$	$T_{melt,D}$				
	140.30±0.	140.65±0.	0.35+0.06	39.7±0.	41.0±0.5	40.4 [1]
n-tricosane [669]	40.5	47.5	7.0	21.8	54.0	75.8
n-tricosane (1 K/min) [+]	40.3	47.9	7.6	23.9	52.4	76.3
n-hexatriacontane [669]	T_A 74.0	T_{melt} 76.0	2.0	31	89	120
n-hexatriacontane [2]	73.0	75.4	2.4	40	82	122

[*] Sigma-Aldrich (www.sigmaalsdrich.com);
[#] A.Bauer-Brandl, E. Marti, etal, J.Thermal.Anal.Calor. 57(1999)7;
[&] A.A. Schaerer, C.J. Bosso, etal, J.Amer.Chem.Soc. 77(1955)2017;
[+] as applied in [666]; [1] as a mixture of 50 to 50% in weight;
[2] data given for the sample of 41 µg, heating rate 10 K/min, second heating and the curve treatment by the Netzsch software "Peak Separation"

TABLE – apendix B

Enthalpy Changes during Fusion of Some Compounds not Containing Oxygen

(G) Gschneidner K. A.: Solid State Physics, Vol. 16, Academic Press, New York 1964.

(K) Kubashewski O. et al: Metal Thermo-chemistry, Pergamon, London 1967.

(B) Barin L, Knacke O.: Thermochem. Propert. Inorg. Subst., Springer, Berlin 1973, suppl. 1977.

(+ +) Mataspina L. et al: Rev. Int. Hautes Temp. Refract. 8 (1971) 211; High temper. — High press. 8 (1971) 211.

(H) Hrubý A., Urbánek P.: in "Methods of calibration and standardization in TA" (Holba P. ed.), DT ČSVTS, Ústi n. L. 1977, p. 138.   (L) Garbato L., Ledda F. : Thermoch. Acta 19 (1977) 267

[+] Wade K., Branister A. J.: Compreh. Inorg. Chemistry, Vol. 1, Pergamon, New York 1973.

[x] Mills K. ., Thermodyn. Data Inorg. Sulphides, Selenides and Tellurides, Butterworth, London 1974.

[°] Holm B. J.: Acta Chem. Scand. 27 (1973) 2043; in "Thermal Analysis (Proc. Ist ESTA)", Heyden, London 1976, p. 112.

Cont. – Enthalpy Changes during Fusion of Some Compounds not Containing Oxygen

Compound	Melting point [°C]	C_P and ΔC_P of solid phase at T fusion [J mol⁻¹ K⁻¹]		ΔH fusion in [KJ mol⁻¹] tabulated in			ΔH procedural values in [KJ mol⁻¹] determined by DTA	
				(G)	(K)	(B)	(H)	(L)
Hg	-39	28.55	—	—	—	2.33[++]	—	—
Ga	30	26.64	0.1	—	—	5.59	—	—
In	157	30.35	-0.8	3.27	3.27	3.29[++]	3.3	3.3
Sn	232	30.76	1.0	7.02[++]	7.16	7.03	7.5	7.2
Bi	272	29.84	-6.6	10.87	10.88[++]	11.3	10.1	10.9
Cd	321	29.55	-1.8	6.2	6.41[++]	6.2	5.8	6.3
Pb	327	29.42	-1.2	4.77	4.83[++]	4.78	4.0	4.7
Zn	420	29.60	-1.8	—	7.29	7.33	—	6.9
CuCl	430	62.9	-4.1	10.24	10.26	—	—	10.9
Te	450°	35.11	-2.6	17.5	17.5	17.5	18.8	17.6
InSb	525	56.41	-5.6	49.4	—	50.66[++]	48.1	48.6
AgI	558	56.52	-2.1	9.42[+]	9.42	9.42	—	8.8
(147 trans.)		(66.75)	(10.2)	(—)	(—)	(6.15)	(—)	(—)
Sb	631	30.90	-0.4	19.85	19.81[++]	19.89	17.4	20.2
Al	660	33.76	2.0	10.72	10.80[++]	10.72	10.5	10.9
GaSb	707			61.13[+]	—	—	—	58.2
KBr	734	68.35	-1.6	—	25.54	25.54	—	25.7
NaCl	801	63.51	-6.2	28.18[+]	28.05	28.18	—	27.7
Ag₂S	830	82.77	-10.4	—	—	7.87	—	—
(176 trans.)		(82.6)	(-1.1)	(—)	(3.97[x])	(3.94)	(—)	(—)
(676 trans.)		(83.91)	(1.1)	(—)	(—)	(0.5)	(—)	(—)
InAs	942	54.79	-5.1	61.1[+]	—	77.04	—	61.9
Ge	940	28.78	1.2	36.84[++]	36.84	36.97	38.6	34.3
Ag	961	31.95	-1.6	11.64	11.1-	11.3	11.3	11.4
Na₂AlF₆	1012	355.8	40.6	113.4°	111.8	107.3	116.7°	—
(565 trans.)		(305.2)	(-23.1)	(—)	(9.3)	(8.5)	(—)	(—)
(880 trans.)		(280.5)	(75.4)	(—)	(—)	(3.8)	(—)	(—)
AlSb	1060	56.38	-2.7	—	—	82.06	—	—
Au	1064	33.38	2.4	—	12.77	12.56	—	12.9
Cu	1083	29.94	-1.5	13.02[++]	12.98	13.27	13.3	13.0
FeS	1195	65.71	-5.5	—	—	32.36	—	—
(138 trans.)		(69.33)	(-3.5)	(—)	(4.4[x])	(2.39)	(—)	(—)
(325 trans.)		(72.85)	(15.8)	(—)	(—)	(0.5)	(—)	(—)
Si	1412	29.22	-2.0	—	—	50.24		
CaF₂	1418	125.8	25.7	—	—	29.73		
(1151 trans.)		(103.4)	(-19.6)	(—)	(—)	4.77		
Ni	1459	36.21	6.9	—	—	17.48		
Co	1495	37.79	2.75	—	—	16.2		
(427 trans.)		(31.1)	(-4.85)	(—)	(—)	(4.52)		

TABLE – appendix C
Enthalpy Changes for Phase Transformations of Selected Oxide Compounds

| Compound | Phase trans-forma-tion[°C] | C_P and ΔC_P in [J mol^{-1} K^{-1}] at transf. T | | ΔH tabulated values [in kJ mol^{-1}] | | | ΔH procedural values determined by DTA, [in kJ mol^{-1}] | | | | DSC meas. |
				[B]	[R]	[G]	Old ICTA test (variance)	[M]	[N]	DDR test[x]	[GR]
KNO_3 ($s_2{\to}s_1$)	128	108.59	-12.0	5.11	5.44	—	13.6 (± 86%)	—	5.60	6.11	5.0
($s_3{\to}s_1$)	(128)	(—)	(—)	(—)	(2.34)	(—)	(—)	(2.93)	(—)	(—)	(2.52)
Na_2SO_4	249	162.96	-10.69	10.82	—	—	—	—	—	—	—
	(707)	(198.69)	(—)	(0.34)							
$KClO_4$	299.5	162.36	—	13.77	13.77	—	32.7 (±85%)	13.82	18.55	17.04	1434
Na_2CrO_4	421	198.3	12.5	9.59	—	—	—	—	—	—	—
Ag_2SO_4	430	178.83	—	15.7	7.95	—	43.1 (±110%)	17.58	18.59	—	16.12
Na_2CO_3	450	192.27	48.8	0.69	—	—	—	—	—	—	—
SiO_2 (quartz)	573	75.49	1.9	0.73	0.63	0.63	0.9 (±99%)	0.81	0.59	—	0.39
K_2SO_4	583	203.32	14.62	8.96	8.12	—	—	8.79	8.71	6.15	5.63
Li_2SO_4	586	195.3	6.5	27.21	28.47	—	—	—	—	—	—
K_2CrO_4	666	194.18	-1.6	13.73	10.26	—	8.2 (±22%)	7.45	9.42	7.85	6.92
Bi_2O_3 ($s_1{\to}s_2$)	727	136.34	-10.2	57.12	—	—	—	36.84[+]	39.8[+]	—	29.56[+]
$BaCO_3$	810	138.79	-16.1	18.84	19.68	—	25.7 (±59%)	18.0	22.15	19.57	18.71
	(968)	(154.91)	(-8.4)	(2.93)							
$ZnSO_4$	754	154.66	—	19.68	—	—	—	—	—	—	—
$PbSO_4$	866	195.08	10.9	17.17	17.0	—	—	—	—	—	—
$SrCO_3$	925	131.59	10.8	19.68	—	—	15.9 (±65%)	17.17	18.50	—	—
ZnS	1020	57.28	-1.3	13.4	—	—	—	—	—	—	—
Mn_3O_4	1 172	210.05	—	20.93	18.84	18.84	—	—	16.74	—	—
$CaSiO_3$	1 190	191.51	1.1	7.12	—	—	—	—	—	—	—
Li_2TiO_3	1212	161.61	-3.48	11.61	—	—	—	—	—	—	—
Ca_2SiO_4	1420	214.12	9.78	4.44	1.47	—	—	—	—	—	—
	(675)	(191.51)	(1.3)	(3.27)	(3.22)	—	—	—	—	—	—

[B] Barin L, Knacke O.: Thermochem. Propert. Ingorg. Substances, Springer, Berlin 1973, suppl. 1977.

[R] Rossini F. D. et al: Selected Values Chem. Thermodyn. Properties, US Gavern. Print. Office, NBS Circular 500, Washington 1952.

[G] Glushko V. P. et al: Termickeskiye Konstanty Veshchestv, Akademija Nauk, Moscow 1966 (Vol. I.)- 1978 (Vol. VIII).

[M] Mackenzie R. C., Ritchie D. F. S.: in "Thermal Analysis (3. ICTA)", Vol. I, Birkhauser, Basel 1972, p. 441.

[N] Nevřiva M., Holba P., Šesták J.: in "Thermal Analysis (4. ICTA)", Vol. 3., Akademiai Kiadó, Budapest 1975, p. 725.

[GR] Gray A. P.: ibid. p. 888;

[x] unpublished results (FIA Frai

13. THERMOPHYSICAL EXAMINATION AND TEMPERATURE CONTROL

13.1. Measurements and modes of assessment

In concluding the book we cannot avoid to point out some particulars about thermophysical measurements, which are generally understood to be the conversion of a *selected physical quantity,* worthy of attention due to its intended interpretability with respect to the *basic quantity*, i.e., the quantity that can be identified and revealed with the available instrumentation [1,3,15,672,673]. Such basic quantities are usually electric voltage and current, or sometimes frequency. If the originally recorded physical quantities are not electrical, various *transducers* are employed to convert them duly (e.g., pressure, temperature, etc.). The measurements thus yields data on the *sample state* on the basis of the response of physical quantities investigated under certain measuring conditions applied, i.e., *experimental conditions* to which the measured object (that is called *sample*) is subjected. The initial conditions are understood as a certain set of these physical parameters adjusted and then applied to the sample.

In order to carry out a measurement on the sample, A, (see Fig. 89.) it is necessary that a suitable set of generators be in the block, B, acting on the sample whose response is followed in the measuring system with inputs connected to the sample. Service block, C, controls the proper setting of the generators and evaluation of the response of the measuring instrument, according to the requirements for the measurement set by block, D. The service block coordinates the overall procedure and development of the measurement, sometimes on the basis of an interpretation of the results obtained. Thus, from the point of view of the system, the experiments represent a certain flux of information through three communication level as the contacts: $\underline{a}$, $\underline{b}$ and $\underline{c}$. Contact $\underline{a}$ represents the actual measuring apparatus which is an extremely variable and constructional heterogeneous element in each experimental arrangement. Block B generally contains the necessary converters on the input side (heating coils, electromagnets, etc.), as well as on the output side (thermoelectric cells, tensometers, etc.). If the individual parts of block B have a certain local autonomy in their functional control the block is termed an *automated measuring system.*

Block D contains a set of unambiguously chosen requirements defined so as to obtain the measurement results in an acceptable time interval and reliable format. It is actually a prescription for the activity of previous service blocks, C, and contains the formulated algorithm of the measuring problem. The usual requirements for measurements actually bear typical algorithmic properties,

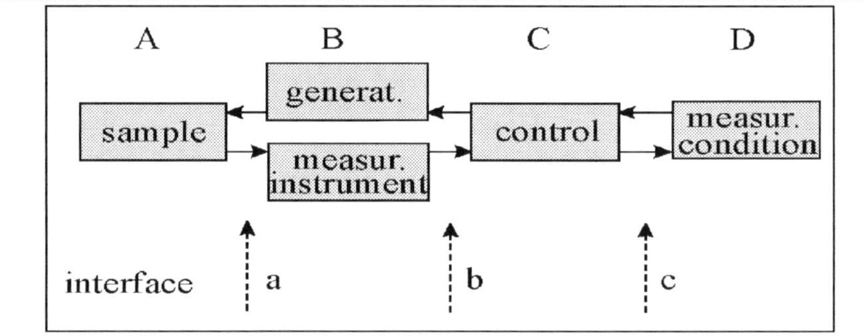

Fig. 89. – Block schema illustrating the basic arrangement of a physical experiment, where the individual blocks A, B and C represent the essential parts of setup (e.g., sample, measuring and controlling assembly, services and established experimental conditioning). Very important role is played by interfaces (depicted as a, b and c showing the mutual contacts taking place during the measurements), namely: the sample versus detector, detector versus service and service versus surroundings.

such as unambiguous attainment of certain data at a final time, accuracy qualifying factors, etc. The service block must react to both the recording of the algorithm of the investigated problem and the correct control of setting the generators and measuring instruments. i.e., interpret the information at contacts b and c in both directions. Auxiliary service functions, such as the sample and algorithm exchange, calibration, etc., also belong here. The service block, C, must have the greatest flexibility in predicting possible effects requiring certain inbuilt smartness of algorithm applied. The manual servicing often brings certain subjective effects into the process of measurement, which can be suppressed by various inbuilt checks. On the other hand, the good quality service can compensate for poor quality in the other parts of the scheme depicted in Fig. 89, which can be abused so that the imperfect measurements are improved by intelligence, diligence and perseverance in the service and use. Programmable interpreters – computers have effectively improved the performance of measurements replacing communication between service and algorithm by formal language in the strictly defined machine code.

A problem is the tendency of recent instruments to become over-sophisticated, in the form of impersonal "black boxes" (or less provoking "gray boxes") that are easy to operate providing computed or, at least, pre-analyzed results and nicely smoothed and methodically granted data. Attached computers, equipped with powerful programs, allow instantaneous analysis and solution of most mathematical and interpretative tasks. Nonetheless, it occurs persuasive but effortless to pass scientist's responsibility for the results evaluation and interpretation straightforwardly in the hands of computers. If a user agrees with such a policy, that means he accepts the ready-calculated data without taking notice of the error analysis and interpretative ambitions, he becomes inattentive to the instrument calibration and performance. In extreme, he does not even aspire to learn what physical-chemical processes have been really measured, which it is not the manufacture's, instrument's or computer's fault, but it is sluggishness of the scientific operator.

In this respect, progress in thermophysical and thermoanalytical instrumentation is apparently proceeding in two opposite directions:
(i) towards "push-and-button" instruments with more and more functions being delegated to the computer and, in somewhat opposite direction,
(ii) towards more "transparent" (flexible and easy to observe) instruments with well allocated all functions, that still require the researcher's involvement, experience and thus his explicit 'know-how' and perceptive scientific 'sense'.

While the first type have their place in routine analysis, mostly required in the industry, the transparent-type piece of equipment (substituting once popular, laboratory self-made instruments, where each piece was intimately known in its operative functioning) should become more appreciated in the scientific and industrial research.

One of the most frequently controlled experimental parameter, defining the sample and its surroundings, is *temperature* [672,673] leading thus to the general class of *thermophysical measurements* [1,3,15,394,602,613]. Derived methods of *thermal analysis* include all such measurements that follow the change in the state of the sample, occurring at stepwise or constantly changing temperature implemented around the sample (see next section). Most systematically, however, thermal analysis is understood to include mostly the dynamic procedures of temperature changes (heating, oscillations). In this sense, thermal analysis is really an *analysis* (cf. previous Fig. 4.) as it enables identification of the chemical phases and their mixtures on the basis of the order and character of the observed changes for a given type of temperature change. Just as in classical chemical analysis, where chemical substances act on the sample and data on the reagents consumption and the laws of conservation of mass yield information on the composition of the sample, the 'reagent' effect of heat (better saying the heat exchange between the sample and its surroundings) is found responsive to the shape of thermoanalytical curve obtained and is interpreted using the laws of thermodynamics, see schema in Fig. 90.

The *basic thermoanalytical methods* reveal changes in the state of the sample by direct determination of the sample single property. In addition to such most widespread measurements of the pre-selected physical property of the sample [3], *indirect measurements* can also be carried out to follow the properties of the sample adjacent environment as an implied response to changes in the sample properties. If the sample property serves, at the same time, as a measured quantity to indicate the requested property of the test material, we deal with *single arrangement,* such as temperature in the case of heating and cooling curves. *Differential methods* are those in which the measured quantity is the difference between that for the test sample and that for a reference site. The behavior of the reference specimen (a specific site) can be simulated by computer, which factually analyses the course of the temperature increase if the sample were to behave as reference material. Consequent superposition of the actual (including changes caused by the reactions involved) and the idealized (reference) curve makes it possible to calculate differences in any desired form (such as the common DTA-like trace). This method, moreover, does not require a specific temperature program as it works with any type of a smooth temperature change, e.g., self-exponential achieved by the regular sample insertion into a preheated furnace.

Simultaneous methods are those that permit determination of two or more physical parameters using a single sample at the same time. However, the determination of a parameter and its difference with a reference site is not considered as a simultaneous method. *Complementary methods* are those where the same sample is not used during simultaneous or consecutive measurements. *Coupled simultaneous techniques* cover the application of two or more techniques to the same sample when several instruments are connected through an interface. *Discontinuous methods* generally include the application of two or more simultaneous techniques on the same sample, where the sampling for the second technique is discontinuous. *Oscillatory and jumps methods* cover dynamic procedures where the controlling parameter is time-temperature modulated.

Modes of thermal analysis can be conveniently classified in the following five general groups [3] so that each measuring method is connected with a certain change in the following properties of the sample:

(i) the *content of volatile component* (change in mass – Thermogravimetry, evolution of gas and its composition – Evolved gas analysis);

(ii) *thermal* (such as temperature – Heating and cooling curves, DTA, Heat flux – DSC, Power compensation – DSC; heat and pressure – Hydrothermal analysis);

(iii) other *thermodynamic properties* (electric polarization – Dielectric TA, magnetization – Magnetometry, dimension – Dilatometry, deformation – Mechanical measurements, dynamic modulus – Dynamic thermomechanometry);

(iv) *flux properties* (electric current – Electrometry; gass flow – Permeability; temperature gradient – Thermal conductivity; concentration gradients – Diffusion measurements; release of radioactive gas – Emanation analysis);

(v) *structural characteristics*, where the instantaneous state of the sample is given by the shape of the functional dependence (i.e., spectrum recorded for X-ray analysis or spectroscopy provided by oscillatory scanning). They are sometimes called special methods for structural measurements. They can cover optical properties (e.g., photometry, refractometry, microscopy, luminescence

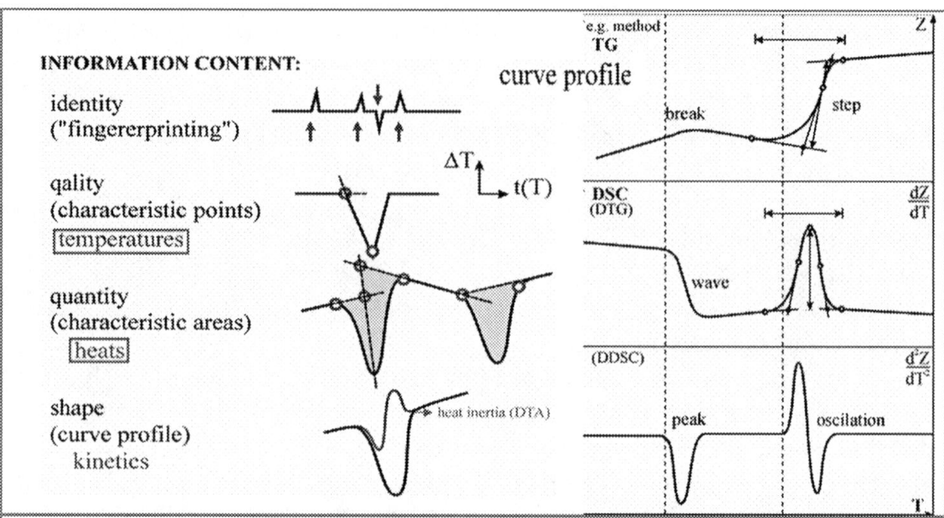

Fig. 90 – Left: chart of feasible thermoanalytical effects (exemplified by the most common crest-like curves) and the inherent degree of information contents. Right: typical shapes of the thermoanalytical curves recorded by individual instruments in dependence to the monitored physical property, Z, as a function of temperature, T. Simpler shapes can be gradually converted into more complicated ones by dual differentiation (break to wave through mathematical manipulation, dZ/dT) or by simple geometrical combination (wave is composed of two subsequent breaks, etc.). Characteristic points marked by circles are described in the text.

and spectroscopy) or diffraction (X-rays) and acoustic (sonometry, acoustometry) characteristics.

Detailed table, providing a more complete survey of thermoanalytical methods comprehended more universally as thermophysical measurements, was published in the appendix of my previous book [3]. Though conceptually at little variance from the classification published by the Nomenclature Committee of ICTAC [15,594,674] it upholds its 'thermoanalytical' philosophy. Linguistic and methodological aspects of the ICTAC accepted nomenclature, and the associated classifications, were the long-lasting subject of ICTAC activity, which is accessible elsewhere, being also the subject of internal analysis for the qualified formulation of the Czech-Slovak nomenclature [675]..

The concept of experimental arrangement (often referred to and specified as the 'measuring head') includes the positioning of the sample(s) and measuring sensor(s), the type of contact between the sample and its environment, etc. [3,15,508,551,620,676,677]. These data are necessary for describing the experiment, especially in relation to its possible reproducibility.

Most important is the *measuring system isolation*, which can assure no exchange of heat and mass (*isolated*), enabling only heat exchange (*closed*) and making possible that all components are exchangeable (*open*). Most usually, only one particular volatile component is exchanged with the surroundings, to give a system that is termed *partially open* [278,286]. Last two types of isolation are most difficult to handle, although most common, so that we shall deal with them in more details (thermogravimetry) later on. In the actual arrangement, two types of experimental arrangement can be differentiated, the *simple arrangement* common for basic instrumentation and the *double arrangement* used in differential measurements having respectively *single* or *twin holders* to contain or to support the test specimen (sample containing cells). The *sample* is understood to be a part of the actually studied material (including the material used for dilution, if necessary) which is located in the sample holder. The reference specimen is that part of the reference (inert) material whose behavior is known (usually not responding to changes in the state of surrounding, particularly thermally inactive over the temperature range of interest).

We can distinguish the *single point measurements* of the so-called *centered temperature* where the thermal gradient existing within the sample bulk is averaged by high enough thermal conductivity of the sample and/or its holder. *Multipoint measurements* (thermopile) allow us either to scan temperature gradients or localize the reaction interface or read the mean surface temperature of the bulk. When seeking a gradientless setup, the sample size has to be miniaturized, which consequently brings problems with the appropriate determination of the sample starting weight/volume/shape and with a threshold dilemma where is the limit when we start observing surface properties of very small samples, which thus prevails characteristics provided by bulk. More faithful measurements can be assisted by specially shaped sample holders, e.g., multistory crucibles where the sample is thin-layered on each ribbon, as was found suitable in thermogravimetry [678-682] but which can conflict with common requirements of the sufficiently sensitive, simultaneous and location confined measurements of heat and temperature. Hence the development of apparatuses designed to indicate either intensive (e.g., *temperature*) or extensive (e.g., *heat*) properties *may not follow the same path* in further development.

The classification and utilization of thermal methods are matters of tradition and of specialization for individual scientific workers in the given field of science. Among traditional methods, thermoanalysts would customarily include those methods, which are associated with the detection of thermal properties (DTA, DSC) complemented by the detection of sample mass (TG) and size (dilatometry). Determination of other non-thermal properties is thought to belong more to the field of thermal physics so that magnetic, conductivity, structural measurements, etc. can thus be thermoanalytically judged as complementary. It is evident that non-traditional methods used in one branch of learning are superior in other fields of studies, e.g., solid-state chemistry would not exist without XRD or SEM, nor would physics without electric and magnetic measurements. In the latter cases, DTA and DSC would then serve as a supplementary source of information, often considered of exploratory value, only. Simultaneously, there arises another problem: how to compare correctly the results from different kinds of static (XRD, SEM, magnetic) and dynamic (DTA, TG) measurements. Individual methods can also reveal a different picture of the process and/or material investigated depending on the nature of physical property and the manner of measurement, which is related on the degree of observability or detectability allied to the elementary subject concerned.

13.2 Temperature control

Knowledge of the *sample temperature* [1,3,23,672,673] is a unifying and necessary requirement for all successful temperature treatments and/or analysis. Moreover temperature is the basic quantity that is involved in <u>all</u> types of physical-chemical measurements regardless of the physical quantity chosen to represent the sample state and whehter temperature is accounted for in observation or not. As mentioned above, in thermal analysis it is traditional to investigate the state changes exhibited by sample exposed to steady conditions of a constant temperature environment subjected to certain temperature program traditionally operational as a result of external heating or cooling. Consequently, the changes observed during the measurement possess characteristic temperature-dependent shapes responsible for individual processes under investigation, which are mutually connected either geometrically or mathematically. This transformation, however, is formal and does not give rise to other thermoanalytical methods.

Most methods of experimental physics, however, study equilibrated states of the sample under (preferably) stationary temperature conditions (usually chosen as isothermal). Apart from the direct low- and high-temperature investigations, most thermo-physical measurements are carried out at room temperatures, which, however, require a suitable method of quenching so as to preserve the equilibrated (mostly high-temperature) state of the sample down to the laboratory temperature. Such a thermal pretreatment is usually necessary to carry out outside the standard measuring head.

An important aspect of recent development in thermal analysis is the diversification of temperature control modes, which besides classical isothermal runs (including stepwise heating and cooling) and constant-rate heating and cooling account at present for the *in-situ* temperature control of sample history. These are, in particular, the temperature jump and rate jump methods, the repeated temperature scanning and finally innovative and the most progressively escalating method of temperature modulation. The last two methods are rather different in data processing. In the temperature modulated methods, the oscillating change of a physical property is observed and analyzed by *Fourier* analysis. The amplitude of oscillating rate of transformation is compared with that of temperature modulation imposed by controlling program. On the other

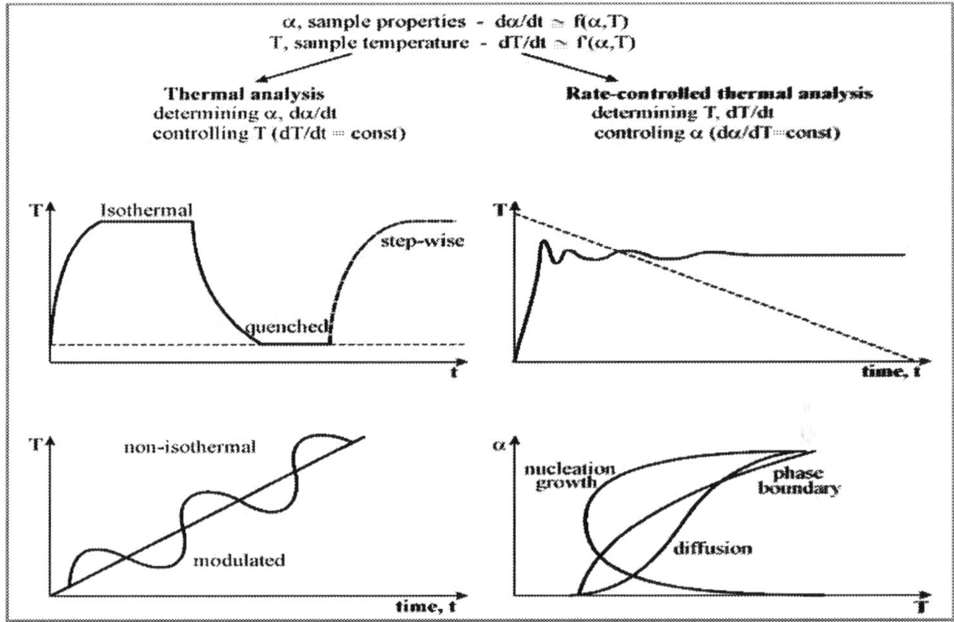

Fig. 91 – Two principal modes of the sample thermal treatment by classical thermal analysis (Controlling Sample Temperature) and modified method of the so-called 'Rate Controlled Thermal Analysis', controlling, on the contrary, the instant rate of the reaction under study.

hand, the repeated heating and cooling of the sample often employs triangular saw-teeth pulses and, on assumption that the rate of conversion is the same at a given temperature/conversion, equivalent isothermal curves of the transformation against its rate are extracted. The method controlling the sample history tries to elucidate the nature of an observed physical transition by using a thermal instrument as a tool for both the sample heat treatment (to provide the desired thermal record) and the entire observation consequently made during a single measurement. There is yet another mode of temperature control called the sample controlled-rate thermal analysis, which is different in the nature of regulation because we control the temperature as a response to the required course of reaction rate (obligatorily taken as constant), see Fig 91. It is evident that the wide range of controlling modes put an increased demand on the improvement of regulators and computer command.

Most thermal regulators in commercial production traditionally control heating and/or cooling of a furnace upon a solitary response signal from an often single-point temperature sensor placed in a certain thermal contact with the heater (which customarily surrounds the measuring head, see Fig. 92.). However, such a conservative system often lacks a feedback connection to provide interrelation between the required temperature of the sample and the actual temperature program controlling the temperature of a heater. It is now approved due to the multi-point detection, heater segmentation, heat shielding, etc., which are the most appropriate setups providing an advanced capability for computerized control. The feedback response to the entire sample temperature was ordinarily available only in fine calorimetric apparatuses where, moreover, heat generators and/or sinks are incorporated to the measuring head to help precisely maintain the pre-selected temperature conditions. However, new spheres of oscillatory programming put the stage of controllers a step forward.

For a successful temperature control, the sample location is very important as it can be placed within a homogeneous or heterogeneous gradient of temperature distribution, which is a result of the perfectionism in the design of the surrounding heater. The sample is normally exposed to a conservative treatment, which means the application of continuous, constant or even oscillatory temperature program conveniently employing a static arrangement of the measuring head (sample) and the furnace (heater). So far, such a furnace temperature control is a common arrangement for all commercially produced thermoanalytical instruments and other apparatuses used in physical technology (e.g., single crystal growth). Slow temperature changes are thus relatively easily attainable by any commercially available thermoanalytical equipment, while rapid changes require a special set up. For fast cooling the application of high velocity flowing gas is usually applied (practically carried out by using helium because of its high thermal conductivity) and the stream is directed against a sample holder, which should preferably exhibit larger surface (ribbons, filaments). Fast heating can be achieved with the aid of concentrated external heat source (e.g., focused microheater, laser or other high-energy beam). Quenching procedures play very important role in the material tailoring and thus is a natural part of generalized thermal treatment [683-688] which was dealt in details in my previous book [9].

Under standard instrumentation, however, the temperature jumps are difficult to enforce accurately due to the thermal inertia of surrounding furnaces, which are the controlled heat source. Therefore, sharp stepwise changes are imposed more conveniently using a modified assembly where the sample is inserted into and shifted along the preheated furnace with the known temperature profile, cf. Fig. 92. This design complies with a less conventional experimental arrangement of the mutually movable sample and the furnace, which requires a definite, continuous and preferably linear distribution of temperatures (steady gradient) along the furnace. Continual temperature control can be then carried out mechanically with sufficiently high precision by sensitively moving either the sample against stationary furnace or, *vice versa*, the furnace against stationary sample. In contrast with traditional (temperature controlling) systems any type of programmed changes can be realized, even relatively fast jumps from one temperature to another and back, the exactitude of which depends on the inertia of sample holder only. In addition, it also allows

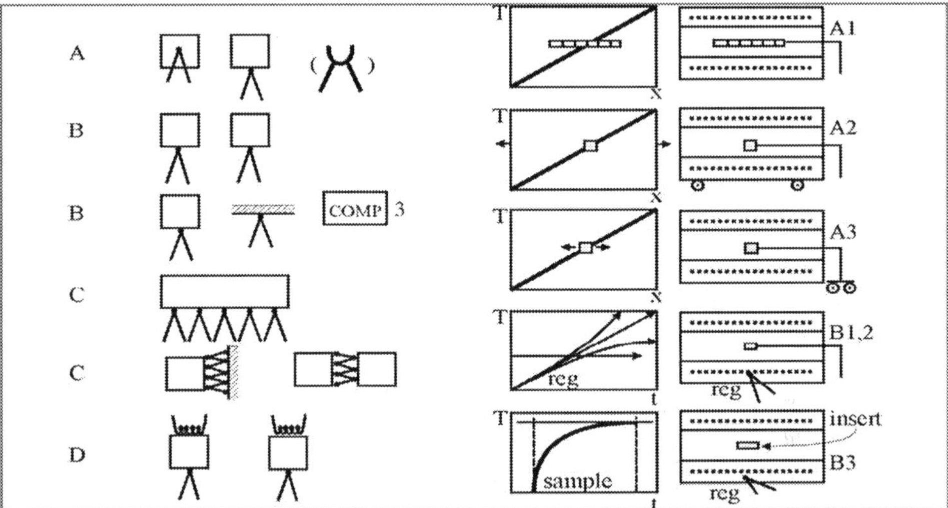

Fig. 92 – Left: Temperature sensors positioning [6]. A – direct, single (heating and cooling curves determination), sensor inside the sample, in the contact with the sample holder, extreme case of "sample inside the sensor". B – direct, twin, two geometrically similar specimens. Sample versus reference, sample versus standard conditions (reference site) and sample versus standard conditions created by computer. C – Multiple, gradient (*Calvet*-type microcalorimetry), several sensors (thermopile) placed along the temperature gradient between the sample and furnace or between the samples. D – Compensation, nongradient, microheater generation of heat to compensate the temperature difference between the samples. Right: Types of the temperature control using the alternative furnace design. A1– Multiple (stationary) samples placed in (and along) the furnace gradient. A2 – Single stationary sample placed inside the movable furnace furnished with constant temperature gradient (typical for "zone melting" instruments). A3 – Variant, using the movable sample in stationary furnace (typical for *"Bridgman"* type of crystal growth). B12 – Traditional set up for controlling the furnace temperature with the sample fixed inside (isothermal, linear or even hyperbolical temperature increase). B3 – Spontaneous (often non-regulated, nearly exponential) heating of a suddenly inserted sample to the pre-tempered furnace (typical also for a "drop-in calorimetry"). In my vision, it may well serve in near future for a promising, rather novel DTA-like method with the inherent (exponential) heating (naturally reached on the sample insertion), where the reference site would be simulated by computer on the calculated basis of "zero-run" with the reference sample or, better, just mathematically, when subtracting the behavior of thermal inertia – possibly a task of fully computerized set up.

inflicting relatively very low rates of cooling often needed for specific time-consuming procedures (such as crystal growth experiments).

We predict possible future trends in the instrumentation entirely based on computer handling out the thermal data without any need of temperature regulation (cf. Fig. 92.). The low-inertia measuring head is inserted into a suitable preheated furnace and the heating curve is recorded for both the sample and reference. The computer program then makes it possible to evaluate any desired type of outcome (DTA) just by appropriate subtraction and combination of active (sample) and passive (reference) curves computationally suited to fictitious conditions of the requested temperature program. In the age of computers, such a relatively simple arrangement would substantially reduce the need of expensive instrumentation based on temperature control.

The temperature controller (applicable for any ordinary type of a heater), together with the controlled system consisting of a sample, its holder and temperature sensor, form a closed dynamic system in which the temperature is both measured and compared with the preset value given by the temperature program [3]. Depending on the sign and magnitude of the detected difference, a regulation action is taken aiming to decrease this difference down to the required and thus permitted level. Upon introducing heating bursts, which can be normally represented by periodical heat pulses (or harmonic oscillation) at the acting heater, the consequent response of the detecting system at the sensor output is also periodic of the same frequency but with lower (generally different) amplitude. Sometimes, this response has a distorted shape and often a phase-sift caused by the transport delay due to the heat inertia of the furnace structure. The regulation function is to prevent (uncontrolled) fluctuation in the system by damping the output signal of the sensor, so that the output signal of the regulator (and thus new input signal for heating) is appropriately reduced. The precision of regulation is thus given by the error permitted in maintaining the regulated quantity within the given limits (often specified in percentage of the original value of the regulated quantity in a stationary state). The related dynamic precision of regulation then expresses the error with which the regulated quantity follows a change in the controlled quantity. The area due to regulation under the curve of the transition process, which should be minimal, characterizes the quality of regulation. The time of regulation is the duration of the transition process and must be minimal particularly for oscillatory regimes.

Because of high quality and fast progression of digital regulators available on the market, the more detailed description of individual parts of a controlling system falls beyond the scope of this text and the reader is referred to special (but older) literature or my previous book [1-3]. We just want to repeat some fundamental terms used in basic theoretical concepts that encounter the regulated quantity (X) that is the output value of the system and also the characteristic quantity of the regulated system, which has to be maintained by the regulator at a required level corresponding to the present value of the controlling quantity (W). Their difference $(X - W = \Delta X)$ is the error fed to the input of the regulator, whose output is the action quantity (Y). The latter quantity acts on the regulated system and ensures that the regulated quantity is kept at the required level. The perturbation quantity (Z) acts on the regulated system and affects the regulated quantity, being thus actually an input quantity of the system. Regulating circuits have two input quantities and thus two transfers have to be considered, i.e., the control transfer, X/W, and the perturbation transfer, X/Z,. Basically, regulated systems are divided into static and dynamic, according to the kind of response: the first tend to stabilize spontaneously (having a characteristic constant of auto-regulation) while the second never stabilize just on its own accord. Differential equations are used to specify the relationship between the output and the input signals with respect to time. The type of differential equation distinguishes linear regulation (with constant coefficients) from a more complex but broad-spectrum regulation with non-linear characteristics. The transfer function and the transition and frequency characteristics add the dynamic properties. In both simple and complex regulation circuits the temperature regulators perform certain mathematical operations, such as multiplication, division, integration and differentiation and the regulators are often named after the operations they perform. In order to find the dynamic properties of a selected regulator, a unit pulse is usually fed to the input, the response being regarded as the transition characteristic. The selection of an appropriate regulator is then determined by the functional properties of the regulated system in question.

Worth mentioning is also the use of power elements to control heating. They extend from an obsolete continuous mechanical powering of voltage by means of an autotransformer [1,15,602], along a discontinuous automaton with a relay system of variation of the powering voltage, either for a constant magnitude or constant time of heating pulses, to a thyristor opening or frequency control of the power half-waves. The interference from the line voltage higher harmonics is minimal with the whole-wave regulation disregarding thus the phase regulation. Altogether, it is clear that the entire process of temperature regulation is a delicate procedure because it involves both the knowledge of intact temperature and the nature of temperature transfer. Any heat input, introduced by the heater or due to the sample enthalpy changes, pushes the system out of the previous steady state, which is exaggerated by the peculiarities of transfer characteristics associated with the particular construction (the measuring head and heater and the rapidity of their mutual response).

For temperature modulation methods the focal point of the system's sufficiently fast reaction became a very insightful stance to see whether the measured temperature of imposed oscillation is known accurately or is merely predicted or even just believed. Correspondingly a question might be enhanced whether the consequent mathematical analysis of modulation is trustworthy or is only a desired but fictitious mathematical outcome. Despite the very advanced regulation and measuring technique and smoothly looking results, we cannot be completely without doubts whether some thermal artifacts do not distort the traced data. Moreover, it has not been fully open to discussion as yet, whether the thermally uneven interfaces for miniature samples can become a source of unwanted fluctuations providing a periodic solution on its own, which consequence was discussed in Chapter 5. It was shown that the spot-localized interfacial heat transfer across holders' junctures can even be in competition with the temperature oscillations purposely furnished by external regulation.

13.3 Temperature detection

Temperature sensors are constructional elements for measuring temperature and employ the functional dependence of a certain physical property of the sensor material on temperature, which is customarily recognized and well defined [672,673]. Practically, resistance or thermoelectric thermometers are most often used, whereas thermistors, ion thermometers, optical pyrometers and low-temperature gas, magnetic or acoustic thermometers are employed less frequently.

Resistance thermometers are the passive elements in electric circuits, i.e., a current powers them and the dependence of their resistance on temperature is measured. Remember that the measuring current heats the thermometer, so that it must be as low as possible. Metals whose electric resistance increases roughly quadratically with increasing temperature are usually used. The actual sensor is a platinum, nickel, copper, palladium or indium wire or, at low temperatures, germanium wire. The most common Pt-thermometers employ wires 0.05 to 0.1 mm thick, wound on mica holder or inside a ceramic capillary or otherwise specially packed or self-supported. The range of standard applicability extends up to about 700 K (special designs exceeding even 1200 K), with time constant depending on the sensor type being in the order of seconds.

For the international practical temperature scale the platinum thermometers serve as standard (interpolation) instruments with characteristics values such the reduced resistance, the temperature coefficient of the resistance and the platinum temperature. Interpolation polynomials of the third to fifth degrees, using as the principal reference points 0, 100 and 419.58 °C (fusion of Zn) are used most frequently (as a standard accessory of commercial products). Another form of resistance thermometers are films or otherwise deposited layers (0.01 to 0.1 μm thick). They can be suitably covered to protect against corrosive media and platinum deposited on ceramics can withstand temperatures up to 1850 K. Non-metals and semiconductor elements can be found useful at low temperatures.

Thermoelectric thermometers (often called *thermocouples*) are contact thermometers, whose sensor is a thermoelectric couple. In spite of the number of apparent drawbacks, they have retained their importance for most thermoanalytical instruments. These are active measuring elements, whose thermoelectric voltage depends on many factors, such as chemical and structural homogeneity, quality of junction, material aging (grain coarsening, component evaporation, external diffusion) and adjustable calibration (connection to etalon, reference junction). The indisputable advantages of thermoelectric couples include the very small dimensions of the sensing element (junction), a small time constant, rather high thermoelectric voltage, and wide temperature range (exceptionally up to 3000 K, depending on the type [3,673]).

Thermoelectric couples are tested by using secondary standards of the first or second order and are often accompanied by a certificate expressing the relationship between the signal and the actual value. The measuring uncertainty is then considered as the width of a range within which the accurate value lies with 99% probability – which is, for the most common PtRh10-Pt thermocouple from 0.5 to 2 K from temperatures up to 1200 °C. The constant of interpolation quadratic equation are approximately $a_o = 2.982x10^2$, $a_1 = 8.238$ and $a_2 = 1.645x10^{-3}$ (based on melting data of Sb, Ag and Au).

The thermoelectric voltage developed in a thermoelectric couple is a function of the temperature difference between the measured and reference sites. The reference ends are connected with further conductors in an electrically and thermally insulated space to avoid parasitic voltages (compensation leads, shielded conductors, shortest path to encircle smallest possible, holder induced leaks). Temperature measurements by thermoanalytical instruments are conventionally calibrated by a set of ICTAC-approved standards, convenient for both the DTA/DSC [594,614] and TG [615] techniques.

13.4. Treatment of the output signal

The manner, in which the value of the studied physical property is obtained, as a particular signal at the output of the measuring instrument, is determined by a number of experimental factors which can be localized in three basic parts of the measuring apparatus.

The test sample

The sample response to changes in its surroundings, particularly assuming temperature, depends on its size and shape, thermal conductivity and capacity and other kinetic factors affecting the rate of production and/or sink of heat during the process. The instantaneous value of the measured quantity then depends on the character of this quantity and on the responding (consonant) physical properties of the sample investigated. Experimental arrangement is also associated with two contentious viewpoints of data interpretation, such as [682]:
(i) scientific search for reaction kinetics and
(ii) practical need for engineering applications.
The latter often requires:
(a) determination of the optimum temperature region (from the economic viewpoint) for carrying out the industrial process, (b) prediction of the yield after a specific time of a thermal process, and (c) estimation of the time necessary to achieve a specific yield.

On the other hand, the kinetics seeks the determination of characteristic parameters (whatever one wishes to call them) and of the dependence of their

392

values on the process progression as well as the definition of the character of the process mechanism (macro/micro-nature of the rate controlling process). Engineers aspire to use large samples to approach practical conditions of pilot processing while scientists try to diminish disturbing factors of gradients by sample miniaturization.

The measuring head

It provides the thermal and other (flow) controlled resistance (lag [670]) at the contacting surfaces between the sample, sample holder and the sensor, or the chamber. *Weighing* is one of the oldest and the most useful analytical methods. When it became associated with temperature changes, it matured into the method called *thermogravimetry* (TG) [676-682]. Traditionally, thermobalances have been used mainly to study thermal decompositions of substances and materials; broadly speaking, in "materials characterization". However, increasingly research is being done on controlled chemical reactions (including synthesis) at elevated temperatures, explored gravimetrically. The sphere of increasing importance of vapor thermogravimetry includes not only traditional sorption research, but also the study of equilibrium by conducting TG measurements at controlled partial vapor pressures and the desired partial pressure can be achieved either by controlled vacuum, or by dilution of the vapor with an inert gas. Modern TG instruments are capable of producing astonishingly precise data (e.g. some can record microgram weight changes of large samples, weighing up to 100 g)[1]. At the same time, if less-than optimum conditions are applied, the balance readings can be heavily distorted, because balances do not distinguish between the *forces of interest* and *disturbing forces*. The full extent of the balance performance is achievable and spurious phenomena can be controlled, provided that they are understood.

The most common sources of TG disturbances and errors in the sequence of the frequency of their occurrence [682] are: unstable buoyancy forces, convection forces, electrostatic forces, and condensation of volatile products on sample suspension, thermal expansion of the balance beam (as a severe problem in horizontal TG's) or turbulent drag forces from gas flow. Note that the first item is the *static* buoyancy (related to the gas density), and the last one is the influence of dynamic forces of flow and of thermal convection. It is a common

---

[1*] A special beam-balance was even used to measure the gravitational constant. In this experimental set up the field masses are two vertically movable 7-tonne cylindrical tanks of mercury that modify the apparent weights of two identical 1,1 kg test masses hanging from wires in the axial high-vacuum chamber. The beam balance, which can attach to either of the test masses, measures the difference between their apparent weights when the field masses are between them, and then again when the field masses sit outside them. A cycle of measurements with different working offsets (copper, tantalum) and repeated rearrangement of the field masses is long-lasting delicate experiment, which yielded the gravity constant with only 33-ppm uncertainty.

mistake to use the term "buoyancy" when meaning drag. Both drag and buoyancy depend on the size and the shape of the pan plus sample, on their distance from the baffle tube, as well as on the velocity and the density of the gas. Any instability of those factors will cause erroneous weight readings. TG experiments often require a controlled environment around the sample. Operating the instrument filled with static gas would be expected to be better for the stability of the balance readings than maintaining a gas flow, but the reality is the opposite. A TG run with no forced gas flow produces an uneven and irreproducible temperature build-up in the system, resulting in irreproducible weight baseline deviations. In most instruments, it is almost impossible to have a static gas during a TG experiment. Systems containing any gas other than air must be tightly sealed or air will diffuse into the system. If the system is sealed, the pressure will build up and an unstable pressure will result. In both cases the weight readings will drift due to the buoyancy changes which result from heating.

Factors other than buoyancy also make flow beneficial for TG. Gas flowing downwards helps keep the furnace heat away from the balance mechanism. Although some TG balance mechanisms are protected against corrosion, exposure to corrosive gases should be restricted to the reaction zone (hang-down tube). Thermal decomposition often produces volatile and condensable matter, which settles on everything in its path. Whether the balance mechanism is corrosion resistant or not, there can be no protection against contamination – if the contaminants reach the balance mechanism. Some TG instruments can be used to study very large samples of plastics, rubber, coal etc., without any contamination of the moving parts of the system. These instruments require a constant flow of carrier gas. So, a flow of gas, both through the balance chamber and through the reactor tube, prevents the temperature and pressure building up and the baseline is stable. Flow also protects against corrosion and contamination. So, if "to flow – or not to flow?" is the question, then the answer is supporting the procedure of dynamic atmospheres. Another general recommendation is to ensure that there is no gradual change in the composition of the gases inside the TG system (both locally and generally) due to incomplete replacement of the previous gas with the new one when the gases are switched. The balance chamber takes longest to switch gases completely. It may be possible with some TG models to evacuate the previous gas from the system and fill it with the new gas. Obstruction of the outlet, changing of the flow rates or the gas pressure, and shutting the purge gas off, even for a short while, can result in an unnoticed surge of the reactor gas into the balance chamber. The start of a run may have to be postponed until the intruding gas has had time to escape.

There are some golden rules of gas handling worth mentioning for any TG measurement: (i) It is necessary to establish (often experimentally) the optimum flows of the gases before the run and (ii) It is not recommendable to change any settings during the entire series of tests; if gases are switched or mixed, the total flow-rate should be kept constant.

In this view, the capability of changing the flow rate (not switching the gases) by computer would be a feature of little practical value. TG instruments are also able to operate with samples exposed to corrosive gases such as ammonia, water vapor, solvent vapors, HCl, SO_2, fumes and smoke of burning rubber or plastics, coal, etc. In most of these extreme environments, the balance chamber is isolated from the reaction chamber by purging and/or by the procedure known as "Gas-Flow Separation". An additional and severe problem is encountered in TG studies in vapors of liquids. Unless proper techniques are used, the data may become erroneous. Attack by aggressive environments, condensation and contamination, as well as unstable buoyancy forces, are best addressed by two kinds of flow pattern: (i) Downward, concurrent flow of the gas which purges the balance chamber ("purge gas") and the reaction gas or (ii) Counter-directional flow ("gas-flow separation") of the purge gas and the reaction gas. The first flow pattern can be generally recommended, due to its simplicity and effectiveness, whereas the latter is invaluable in those difficult cases, where no purge gas can be allowed in the environment of the sample. Gas-flow separation is possible only if: (i) the purge gas used is much lighter than the reaction gas (helium is the gas of choice), (ii) both gases are flowing continuously or (iii) a suitable type of baffling is used.

The measuring instrument

As a whole it accounts for the delays between the instantaneous values of the measured quantity provided by the sensor, i.e., the effect of electronic treatment of the basic quantities using converters. The value of beginning of the recorded signal, Z_{record} , is then related to the actual value of beginning of the studied process, Z_{proc} , by the relation ship, $Z_{record} = Z_{proc} + \phi\ (\tau_1 + \tau_2 + \tau_3\ )$, where ϕ is the rate of change of the controlled parameter in the vicinity of the sample and τ are the *time constants* corresponding to the processes given under points 1 to 3. Some experimentations have *detectors located outside* the chamber with the sample such as is typical for analysis of gaseous products in a stream of carrier gas or emanation TA. The description of the process, which takes place due to the interface-connecting tube is then coded into gas flow as changes in the concentration of selected products. To this the detector responds and the values read at the detector output are apparently shifted in time with respect to the resultant process. It is caused by slowness of transport delay and because of distortion of the concentration profile as a result of variability of the flow of the carrier gas in the connecting tube. The detector responds to the process occurring in the measuring head only after time, $\tau = l/2\ v$, where l and r is the tube length and radius and v is the mean flow-rate of the gas at a given cross-section $\pi\ r^2$. An originally sharp concentration edge is washed out and transformed to a wedge-like profile on passage through the connecting tube and the overall volume of gas, v_{transp} entering the detector, or $V_{transp} = \pi r^2 v \left[(\tau - t) + (l/2 \, v)^2 \, (1/\tau - 1/t) \right]$.

Kinetic aspects

We should note again that simple outcome by means of thermoanalytical kinetics (cf. previous Chapter 11) is generally not employed in practical engineering studies. A few engineering researchers employ the literature models of their preference, simply, using the algorithms they have purchased with their instruments, with sometimes misleading and ungrounded claims, where yet another data reporting (publication) can be the only benefit. The discrepancies and inconsistencies reported in the literature are sometimes large. Some calculated values are obviously nonsensical, e.g., negative activation energies, and there is often virtually no correlation of the values obtained from thermal analysis experiments with those typical for larger-scale industrial processes. Therefore the chances of practical utilization of the kinetic data to predict real-life processes are greatly diminished.

To minimize diffusion effects, established kinetic practice requires that samples should be as small as possible (thin layers spread on multistory crucible [574,689]). However, the smaller the sample, the greater is the ratio of its surface to its bulk and this may overemphasize surface reactions and make correlation with large-scale processes poorer. Experience shows, however, that even very small samples (less than 1 mg) are far from being small enough to be free of diffusion inhibition. To justify the obvious errors, the adjectives: "apparent", "formal" and "procedural" are used in conjunction with the otherwise strictly-defined terms of "activation energy" and "reaction order" as established in homogeneous chemical kinetics.

The value of the activation energy calculated for a reversible decomposition of a solid may thus be just a procedural value of no real physical meaning other than a number characterizing the individual thermoanalytical experiment. Many sophisticated kinetic methods neglect the distinction between micro-kinetics (molecular level, true chemical kinetics) and macro-kinetics (overall processes in the whole sample bulk). The conventional process of the type, $A_{solid} = B_{solid} + C_{gas}$, as applied to the entire solid sample, consists of many elementary processes, some of them of purely physical in nature (e.g. heat transfer, diffusion). The experimentally obtained kinetic data may then refer to only one of these processes, which is the slowest one, often difficult to link up with a particular reaction mechanism.

13.5. Characterization of experimental curves

The graphical or digital recording of sequences of experimental data, called the thermoanalytical curve, is the basic form of thermal analysis output and thus represents the primary source of information on the behavior of the sample [1,3,15,551,680]. Although most such curves are digitized the interpretation ability and experience of the scientist and his subjective "at first glance" evaluation remains important. A further advantage of graphical recording is its lucidity, which is advantageous for illustrative comparison of the actual course with theoretical or otherwise obtained one. The recording of the dependence of the measured quantity on time or temperature can be divided into two regions: (a) *Baselines*, i.e., regions with monotonous, longitudinal and even linear curves, corresponding to the steady changes in the state of the sample, (b) *Effects* (singularities) on the measured curves, i.e., regions in which at least the slope of the tangent to the curve changes (so-called stepwise changes).

The part of the baseline preceding the changes is the *initial baseline* and the part after the completion of the changes is the *terminal baseline*. While a horizontal baseline is typical (and desirable) for most TA methods (TG, DTA, DSC) it may be a generally curved line (dilatometry), an exponential line (emanation TA) or even a hyperbolical line (thermomagnetometry). In most combuter-controlled instruments the baseline is self-corrected by inbuilt programming, providing a nice but sometimes misleading result (straight line). Therefore both types of baselines should be available, native and corrected.

The simplest pattern revealed on the experimental curve is a *break* (or bend). It is characterized by the *beginning* and *end of the break* (and/or by the extrapolated onset point), see Fig. 90. Higher class of singularity is a *step* (or half-wave) which is the most typical shape to appear on experimental curves. It joins baselines of different levels and is characterized by the *inflection point* in addition to previous beginning and end points. It is composed of two consequent breaks. The difference between the value of the measured quantity at the beginning and end is the *height* of the step and the difference in its time or temperature coordinates is the *length* of the step (also called *reaction interval*).

A *peak* is that part of the curve departing from, and subsequently returning to the baseline and is composed of two consecutive steps or four breaks. The peak is characterized by its *maximum point* besides the (*frontal and terminal*) inflection points and beginning and end points. The extrapolated *onset* is the intersection of the extrapolated initial baseline and the tangent to the frontal inflection point. The extrapolated end point (*outset*) is obtained analogously. The linearly interpolated baseline is then found by suitable joining these points. The actual peak background may, however, be different as is found for DTA peak in the form of its S-shaped base-boundary, see Chapter 12. The *tip* of the peak is the point through which the tangent constructed at the maximum deflection parallel to the peak background passes. The peak *height* is thus vertical distance to the time or temperature axis between the value of the measured quantity at the peak tip and the extrapolated baseline. The peak *area* is that enclosed between the peak and its interpolated baseline.

Generally, simpler shapes can be converted into more complicated ones by dual differentiation and/or combination. A step is the derivative of a break and similarly a peak is the derivative of a step. Differentiation of a peak yields an oscillation with three inflection points and two extremes. This shape is most sensitive to the determination of characteristic points (the initial and end points of the peak appear roughly as the inflection points, while the inflection points of the peak appear as extremes which are most noticeable). The oscillation, however, no longer contains the original quantitative information, such as the peak height or area. The gradual succession of derivatives bears the same sequence as the progressive extraction of thermodynamic quantities, e.g. from heat/entropy to heat capacity.

The information coded in a thermoanalytical curves can be found gradually richer, along the increasing reliance on the type of experiment carried out, as illustrated in Fig 90.

It can be divided into:

(a) qualitative information about the *identity* (fingerprinting) of the test sample
$\Rightarrow$ spectrum effects,
(b) location in the temperature axis, i.e., the *temperature region* in which a state change of the sample occurs $\rightarrow$ position of the effect,
(c) quantitative data on the *magnitude* of the change $\Rightarrow$ size of the effect and
(d) *course* and rate of the change (kinetics) $\Rightarrow$ shape of the effect.

From the point of view of *information gain* (the determination of effectiveness of our efforts) the TA experiment must be considered as means for decreasing the inexactness of our knowledge. If the position of a TA effect is found directly from the tables without measurements, then irrelevant (zero) information is obtained. If, however, one of the four possible phases of a material is found on the basis of identification of a TA effect between the other singularities on the TA curve, then the information gain equals two

Objective evaluation for furnishing suitable equipment for a TA laboratory, automated data treatment and the choice of most relevant information from possibly parallel figures and successful solution of a given problem, lead to a more responsible direction of the progressive enhancement of research work. The main problem of such quasi-stationary measurements is the intricacy of deciphering all the complex information hidden in a single curve.

13.6. Purpose of the measurement – exemplifying thermogravimetry

Most experiments are carried out in order to evaluate the course of a certain process and can be carried out in the following manner [682]: (i) By studying one or a few partial, isolated processes, using suitable arrangement of the experimental set up. The results can be interpreted within a suitable physical framework; (ii) By studying the process as a whole, possibly simulating industrial conditions, and interpreting the result either empirically or within engineering convenience. One of the most discussed cases is the decomposition of limestone in vacuum and/or under different atmospheres using its real piece or thick layered sample [690]. Reaction interface can thus move from the outer surface (crucible walls) towards the center of the sample as a macroscopic boundary whose movement is controlled by physical processes related to heat and mass flows. Similar flows can also be located along and across the reaction interface [691] to determine consequently the fractal roots of interface geometry.

For thermogravimetry the decomposition rate of the same sample, in the same vessel, but at various values of the isothermal temperature should preferably be recorded. If the process is *diffusion-controlled,* measuring the magnitude of the influence of diffusion-related factors, e.g., the size and shape of the sample, or can confirm this by varying the partial pressure of the volatile decomposition product, up and down from the basic conditions of the experiment. The magnitude of the resulting variation in the reaction rate is then compared to the magnitude of the change calculated for a diffusion-controlled process. If, for example, the decomposition of a hydrate is studied in an environment of 10 mbar of water vapor, the run should be repeated twice: say under 9 mbar and under 11 mbar (such instruments are now commercially available). If the magnitude of the change in the reaction rate by diffusion-related factors is negligible, but the temperature dependence is strong, the process may follow the *Arrhenius* equation and kinetic models based on it. If the magnitude of the measured temperature-related changes in the reaction rate is in the range typical of diffusion-controlled processes, the process probably is diffusion-controlled. A simplified way of determining whether a given reaction is mass-transport controlled would be to compare (in a non-quantitative way) the TG curves obtained when the degree of the transport hindrance of removing the volatile products of the decomposition is changed. This could be diminished by spreading the sample into a thinner layer placed onto the large enough surface such as on the shaped sample holder having the plateau (multistore or ribbed crucible), first introduced to TG practice by *Šesták* at the turn of sixties [689], see Fig.93., later commercially exploited [680,681]. On the contrary, the diffusion can be slowed down when the powder sample is well packed into capillaries [574]. Alternatively the same sample can be situated in a more voluminous crucible, which is covered by a more or less tightened lid to restrain volatile products from evaporation. As a result, the various labyrinth-type and also multistore crucibles were conceived by the brother *Pauliks* and put in a very popular practice [680,681], which is known under the name of *derivatography,* see Fig. 94, initiating a various domain of extended applicability [692].

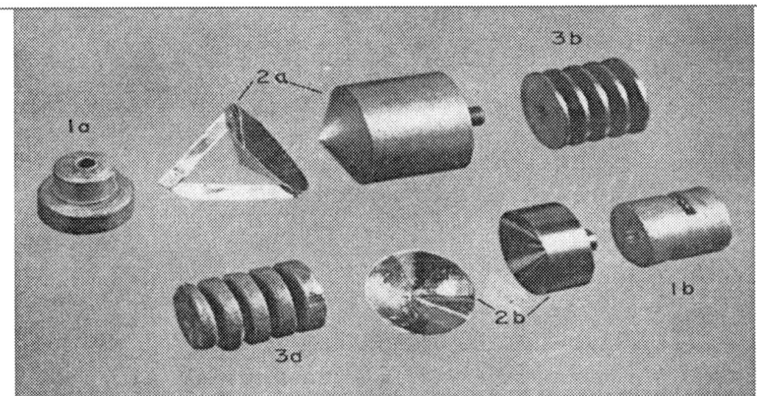

Fig. 93 – Practically employed self-constructed crucible-holders suitable to enhance the study of the decomposition of solid substances (designed by the author and self-made in the laboratory of the Institute of Physics at the turn of the 1960s [574,689]): 1a) Perforated silver block with a multiple of shallow hollows (containing the investigated sample in the outer-wall) while the viewed hollow in the inner wall serves for the setting in of the corundum double-capillary with a termocouple. The version with the narrow hollows (about 1mm diameter capillaries) and with different depth, see 1b, was used to investigate diffusion progress through the given sample depth. 2a) A conical silver block with a very light fitting lid (left) on which the investigated sample is spread out in a thin layer and which is set to fit on the tapering top of the metal block (located in a furnace with the stabilized temperature). This arrangement was found convenient for isothermal decompositions because it helped to attained suitably rapid temperature equilibration between the inserted lid with the sample and the preheated block held at selected temperature. 2b) this is the similar arrangement as before but made of platinum with the sample-cone adjustment in reverse position. 3) The ribbed crucibles intended to carry out non-isothermal decompositions for a thin layer of sample spread on the ribs of such a multistory holder in order to diminish the temperature gradient within the sample using two construction materials: platinum (b) and silver (a).

Relatively minor changes in the reactant diffusion paths should not significantly affect the rate of bond-breaking steps. Heat exchange, however, may be affected more, but that is likely to be a negligible factor. If heat transfer controls the process (hence applying only to programmed temperature experiments, or the initial stage of heating to the isothermal temperature), this can be confirmed by slightly increasing the ramping rate. If heat transfer is the controlling process [691], the rate of the weight loss will not increase, or the approximately straight segment of the TG curve (plateau on the DTG curve) will become straighter and longer. The temperature range of the decomposition process will not shift towards higher temperatures. The magnitudes of the influences of temperature-related factors, e.g., changing the thickness of the walls of a highly-conductive sample holder; adding or removing thermal insulation; comparing the decomposition in shiny and in black sample holders, or by making the temperature oscillate (square wave) very slightly around the set isothermal level, can all be investigated. Where heat flow is rate controlling,

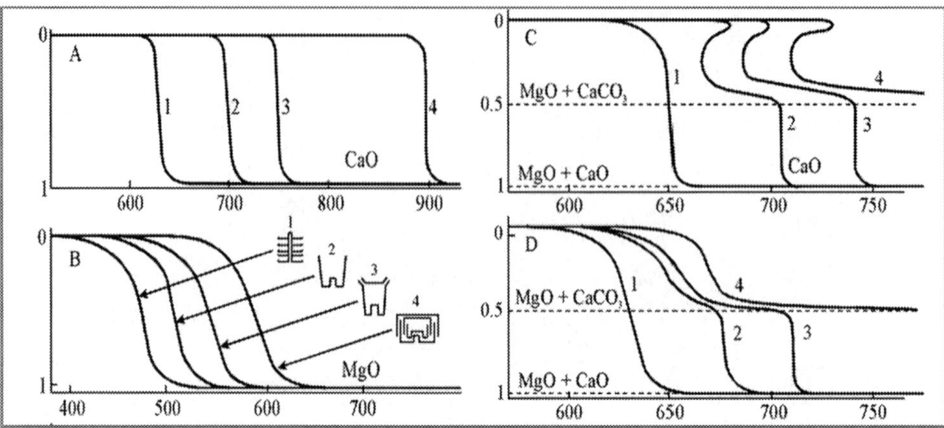

Fig. 94 – Examples of the so called 'quasi-isothermal' curves determined by once very popular thermoanalytical instrument 'Derivatograph' [Q-TG [680,681] (produced commercially by the brothers *Pauliks* in Hungary from the beginning of sixties). A special case of monitoring the sample temperature during thermogravimetric measurements (for preferably decomposition-type processes) employed different types of gas-tightened crucibles capable of enriching the disconnection of individual reaction steps and hence providing a more selective differentiation of the consecutive responses. It literally falls into the more general so called 'constant decomposition rate thermal analysis' [681,692,696]. The individual curves correspond to the crucibles employed, i.e., (1) multistory (ribbed) sample holder enabling free gaseous products exchange from the sample thin layer, (2) a deep crucible without any cover self-hindering easy path of gaseous products through the sample mass, (3) yet more constricted passage by well adjusted lid and (4) stepwise escape of gaseous products through a multiple-lid arrangement, the composite crucible with labyrinth. The crucibles were applied to investigate following samples: (A) Decomposition of calcite ($CaCO_3$), which is strongly dependent on the partial pressure of CO_2 any increase of which above the sample surface is necessarily accompanied by a shift in the decomposition temperature above its normal equilibrium value. (B) For magnesite, $MgCO_3$, the decomposition already proceeds above its equilibrium temperature so that even the decomposition temperature stays roughly unchanged with increasing CO_2 partial pressure but the curve shapes are affected. (C) Decomposition of dolomite, $MgCa(CO_3)_2$, is the symbiosis of the two previous cases. After a delay caused by equilibrium set-up of decomposition due to the accumulation of CO_2 , the measured curves have a tendency to return to their original decomposition course. (D) Addition of a suitable component (nucleator such as 2% of NaCl) can lead to disappearence of the incubation period and the early decomposition can proceed at lower temperatures while the second-bottom parts remain approximately unaffected. More detailed analysis of such environmental effects due to the gaseous products buildup become very important in distinguishing the macroscopic and microscopic effects responsible for the determination of overall kinetics. Courtesy of *Ferenc Paulik*, Budapest, Hungary.

the response should be directly proportional to the magnitude of the temperature jumps.

To illustrate the problems discussed, we refer to some published TG results for the hydrates of copper sulfate and magnesium sulfate [693], which were obtained over a broad range of conditions of hindrance of water vapor escape. What causes these hydrates, commonly believed to be stable at room temperature, to decompose rapidly and release their water much below the boiling point of water? Let us visualize the processes involved, and discuss in detail the concept of decomposition temperature.

The observed behavior of the hydrates can be explained if the kinetics of these decompositions is limited by the rate of escape of the gaseous product, C (water). Even at room temperature, the decomposition is fast whenever the volatile product is being rapidly removed. When C is not removed, the decomposition almost stops and the reactant seems (erroneously) to be stable. It would be unreasonable to interpret this behavior in any micro-kinetic terms. This means that the decompositions of these, and many other hydrates, which behave similarly, are diffusion-limited (but the controlling factor is effectively the concentration of C). At a given temperature, either solid, A or B, is the energetically stable form, but not both. When the temperature of the substance rises, at the point marked [682,693] as "PIDT", i.e., *Procedure-Independent (equilibrium-like) Decomposition Temperature,* the decomposition and the recombination have equal likelihoods, see Fig 95.

Above the PIDT point, the decomposition into B + C is energetically favored, but the decomposition would be suppressed if molecules of C are present in the immediate vicinity of the surface of A. In that case the equilibrating temperature is increased to a new value; let us call it the "ADT" for the "*Augmented (pseudo-equilibrium) Decomposition Temperature*".

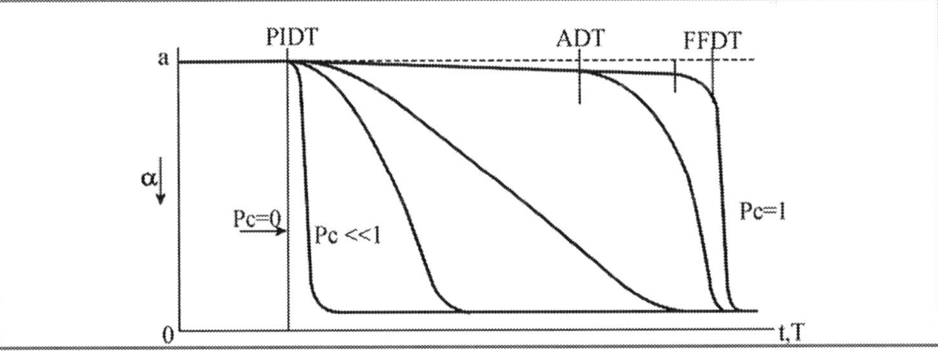

Fig. 95. – Graphical representation of thermal decomposition at various constant values of concentration of the gaseous products [682] – from the left vertical line, at $p_c \cong 0$, to the far right curve, at $p_c \cong 1$. Explanation of proposed definitions for TG curves obtained at various degree of hindrance of escape of the gaseous products of the decomposition: PIDT – procedure independent decomposition temperature ('pivoting point'), ADT – augmented decomposition temperature (often called procedural decomposition temperature) and FFDT – forced flow decomposition temperature. The dashed line is extension of the initial, perfectly horizontal section of the TG curve. Courtesy of *Jerzy Czarnecki*, LaHabra, USA

If the pressure of C is constant with time and uniform in the whole sample, and there is no air or other gas in the vessel, the ADT is still a sharp point. If it is not a sharp point, this indicates a coexistence region of solid mixtures of A and B. Such coexistence means the existence of a reaction front, or possibly enforced metastability, of A beyond its true stability range. Such metastability can occur when the partial pressure, p_c, is not a constant value, that is if p_c varies with time, location, or both. The section of the TG curve associated with the ADT is the temperature range where both A and B forms co-exist in a negative feedback of a dynamic pseudo-equilibrium (see Figs. 93 and 94 again). Labyrinth crucibles [3,574,680,689] are found as useful qualitative tools for making available saturation conditions of the volatile decomposition products, and improving the reproducibility of the data.

Let us return to the concept of the PIDT point that separates two sections of a TG curve: the *perfectly* horizontal section (if it exists), and the inclined one. It is usually possible to see the PIDT if the TG curve was obtained without diffusion hindrance (samples usually not larger than 1 mg, under vacuum). Even if these two slopes differ almost insignificantly, that elusive point remains the borderline between the stability ranges of the substances A and B. By hindering theescape of gas we can decrease the slope of the inclined section even to zero, and then the point PIDT would *seemingly* disappear. The reason why the distinction between the perfectly horizontal and the inclined section of TG curves is important is as follows. Below the PIDT, curves are horizontal because decomposition cannot occur due to thermodynamic factors. Above the PIDT, curves can still be horizontal because decomposition cannot *proceed* due to accumulation of the product. Another important difference between the two sections is that only in the inclined section is the slope diffusion-dependent, because in the true-horizontal region molecules of the gaseous product C do not form. It may not be possible to see the PIDT point on a TG curve and only the ADT instead, but this does not make the ADT the same as the PIDT. Even if the PIDT is hidden, its value is fixed. Extrapolation of such a TG curve produces the ADT, not the PIDT.

If the escape of gas C is *completely* free, the ADT will disappear and the true pseudo-equilibrium PIDT emerges. Such elimination should be ensured and verified, not just assumed. A suggested procedure, which satisfies such requirements, would be as follows: (i) place not more than 1 mg of the powdered sample in a TG sample holder, at ambient pressure and temperature, and record the mass for at least several hours, (ii) continue the recording for several more hours while evacuating the TG system, (iii) keep evacuating and apply very slow linear heating, (iv) the lowest temperature that will start a measurable slope, is the PIDT. There may be cases when the PIDT cannot be obtained because the substance can decompose under vacuum even at room temperature (so no true horizontal baseline is obtained under vacuum). The PIDT then lies in a lower temperature range than that being considered.

Does a "decomposition temperature" actually exist? By definition, as long as the reaction is reversible, there is a never-ending competition between decomposition and recombination. At equilibrium, molecules of C are constantly being produced and consumed, but if the liberated C molecules are removed, then we have a non-equilibrium situation and substance A decomposes irreversibly until it disappears completely. These processes take place at any temperature within the reversibility range. Most such decompositions are endothermic and the entropy of decomposition is also positive, so the Gibbs energy change for decomposition will only be negative at high T and positive at low T (ΔG at temperature T will be given by $RT\ ln\ (Q_p/K_p)$ or $RT\ ln(p/p_{eq})$, where Q_p is the reaction quotient, p is the actual pressure of C, and p_{eq} is the equilibrium pressure of C at T). When p is kept small, ΔG will be negative at all temperatures T, but the rate of decomposition will be determined by the Arrhenius equation. Therefore there is no unique decomposition temperature. The term "decomposition temperature" has a thermodynamic meaning of the switching point at which the decomposition reaction replaces the recombination reaction as the thermodynamically favored process (with other conditions such as pressures of any gases held constant).

Not only the physical meaning, but also the earlier metrological definitions of "Decomposition Temperature" do not seem to be consistent. The terms "Decomposition Temperature", "Procedural Decomposition Temperature" ("PDT") or "Initial Temperature" ('T_i') are often defined as the lowest temperature at which the rate or the cumulative weight change of the decomposition is measurable. The symbol T_1 is also used for that, sometimes marked as '$T_{0.01}$' (for the transformation ratio, $d\alpha/dt$, as low as about 0.01). A "Temperature of Thermodynamic Stability" ('T_{th}') is defined as the "temperature at which the partial pressure of the gaseous product equals the ambient pressure." It is believed that for the decomposition to take place, the temperature must be "the lowest temperature at which the decomposition is measurable" (T_1, $T_{0.01}$, T_i, PDT). The physical meaning of lowest temperature, at which the rate or the cumulative decomposition is *measurable*, is inevitably narrowed to the particular conditions and sensitivity of the thermoanalytical experiment. In the first approximation it has no absolute value, in the same way, as the beginning of a TG-recorded evaporation of a liquid below its boiling point would be a merely procedural value, lacking any absolute physical meaning. It has also very limited practical significance. The literature abounds in examples of the fact that appreciable decomposition is often recorded at temperatures much below the values generally accepted or expected. For example copper sulfate pentahydrate is capable of producing TG curves, for which the T_1, $T_{0.01}$ or T_i can be sometimes as high as 100 °C, and as low as 40 °C. One can justifiably interpret these facts as follows: copper sulfate pentahydrate decomposes below room temperature, but it appears stable because the decomposition in a closed container at room temperature is hindered. Therefore neither 40°C nor 100 °C are values of any absolute meaning. If the TG examination of a small sample of that hydrate had started, not at 20 °C but below room temperature, the apparent decomposition temperature would be even lower than 40°C. For very carefully planned TG experiments, "practically horizontal" curves can be obtained and the PIDT, measured as the "first detectable deviation", then becomes an absolute value. Relying on the ADT as an indication of the limits of thermal stability of a substance can be a dangerous mistake. The only reliable value representing thermal stability is the PIDT.

The future role of TG instruments is anticipated to become gradually reduced to verification of the validity of highly computerized modeling that would include all the possible processes discussed above. Such an extensive theoretical treatment, comprehensively covering all the various aspects of a detailed scientific and engineering description, would become the principal method for the prediction of the behavior of materials. It would essentially lower the number of instrumental measurements and save both time and labor. Experimentation would, however, need to be perfected to such an extent that each measurement is reliable and meaningful. Not all measurements are yet adequately reliable. Artifacts are easily produced if we allow a computer to plan an experiment for us. This is frequently the case when "rate-controlled TG" is used. Many thermal processes are characterized by TG curves consisting of more than one weight-loss step. This may result from a sample being a mixture, or from the reactant undergoing a multi-step thermal transformation. These TG steps may overlap partly or completely. When the stoichiometry of a multi-step TG process is determined, the weight value at the horizontal section of the TG curve is used as the indicator of the stoichiometry. This is the basis for "compositional analysis". If no horizontal section between the steps exists, the inflection point is used for that purpose. The inflection point on a TG curve can be precisely determined from the DTG curve, but it is *not* a direct indication of the stoichiometry in the common situation of incompletely separated TG steps. The mathematical addition of partly overlapping individual TG steps (or of any other curves of similar shape) produces TG curves whose inflection points consistently fall below the stoichiometric values. This fact (belonging to analytical geometry) is seldom realized in thermal analysis. It is frequently observed that the location of the inflection point varies for TG curves obtained under different conditions. These shifts are attributed to experimental parameters, such as sample geometry. In the case of partly-overlapping TG steps, the common practice of using the inflection points to obtain stoichiometric values, may lead to systematic errors of up to 20%. Moreover, the common practice of attributing those "shifts in the stoichiometric values" to changes in

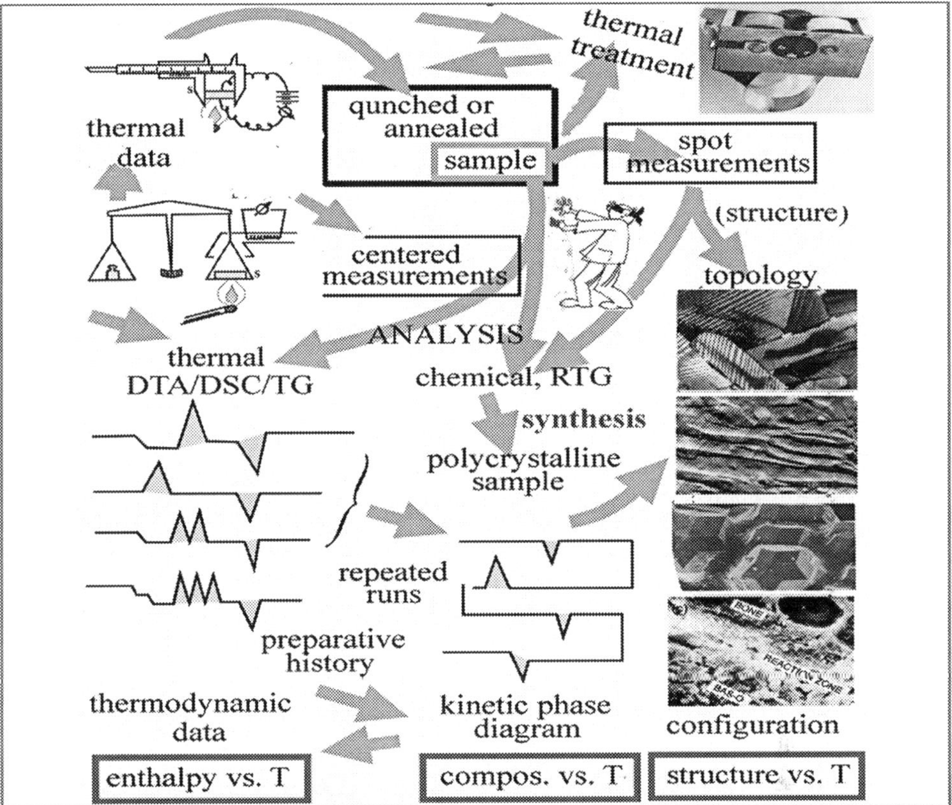

Fig. 96. – Practice chart showing the routes rationally applied to the study of materials when one is careful to distinguish the contradictory characters of bulk (centered thermoanalytical) and spot (localized microscopic) measurements applied on the variously heat-treated samples [9]. Right column shows examples of microscopic pictures (downwards): Direct observation in polarized light of the fractionally solidified Cu-rich liquid of $YBa_2Cu_3O_x$ composition, cf. Fig. 35., (peritectically embedding the non-superconducting (211) in the superconducting (123) matrix and segregating unsolved yttrium oxide and yttrium deficient liquid on grain boundaries). Interface boundary of the orthogonal and tetragonal superconducting phases shows the specific ordering strips of CuO-chains and their twinning due to the sudden commencement of perpendicular growth caused by cumulated elastic stress. Below photo shows the creep-induced anisotropy of the $Fe_{80}Cr_2B_{14}Si_4$ metallic glass (with T_{curie}=347 C and T_{cryst}=470 C) demonstrated by special SEM observations of magnetic anisotropy (strips). Beside the in-plane magnetization we can see particularly visualized magnetic domains whose magnetization is oriented perpendicularly to the continuous as-quenched ribbon obtained by casting melt onto the outside of rotating drum. Yet below, a SEM photograph of the extraordinary triacontahedral faceting that occurs in some glassy materials obtained upon rapid freezing – herewith is the typical crystallite grown within the quenched melts of Al_6Li_3Cu alloy. Bottom, a SEM view of the interface of the bone tissue with the bio-active Na_2O-CaO-SiO_2-P_2O_5-based glass-ceramic implant observed after healing. The morphological images help to decipher the structural process and thus should be a complementary part of all centered thermal studies but cannot just be handled for a sound interpretation alone.

the reaction mechanism seems to be disputable, unless supported by other facts. A practical way of finding the inflection point is to use an appropriate deconvolution program on the DTG curve. The largest errors can result when the single steps that make a combined two-step TG curve are symmetrical. Asymmetrical steps produce smaller errors, because the weight changes are much slower at the beginning of the second step, than at the end of the first one. Therefore, the error caused by including the initial weight changes of the second step into those of the first step is less significant.

13.7. Controversial character of bulk and spot observations

In closing stages let us mention a very important but often neglected problem, which is the mutually observed but somehow controversial existence of two basic kinds of observation on localized and averaged levels thus difficult to correlate. One is based on the indirect average measurement of the mean property of a whole sample, which is executed by the standard thermophysical and thermoanalytical measurements (e.g., DTA, DSC, XRD, thermogravimetry, dilatometry or magnetometry). The other is the appearance method consisting of visualized local-measurements directed at a precise spot on the sample surface or its intentionally made fracture, which is a traditional morphology measurement (such as light or electron microscopy). Both approaches have their loyal defenders and energetic rejecters often depending to what are their subject instrument and the direction of study. Though measurements on larger (more voluminous) samples are more sensitive to a mean property determination, particularly when represented by thermal (DTA, DSC), mass (TG) or magnetization measurements, they always involve undefined temperature (and mass gradients) in its bulk [694] (Fig. 96). Such gradients are often unnoticed or later observed as disagreeable changes of the result fine structure because the total sample size is simplistically defined by single values of temperature (in the same way as weight or magnetization). Even miniaturization of such a sample is not always a recommendable elucidation due to the undesirably increasing effect of the property of its surface over that of bulk although a specially shaped sample holders were devised to help the problem, cf. preceding Fig. 92. The averaged type of measurements are often sharply criticized when employed as the input data for the verification of kinetic models, which are based on specific physico-geometrical premises related to the idealized morphology pictures.

On the other hand, we can recognize the advantage of the spot measurements, which actually reveal a real picture of the inner structure, but we have to admit that the way how we see it, that is the entire selection of insight windows and their multiplicity, is often in the hands of the observer and his experience as to what is more and less important to bring to his view and then observe in more details. Therefore, such a morphological observation, on the contrary, can be equally questioned as being based on a very localized and almost negligible figure, which provides the exposed-surface pattern representing thus almost insignificant part of the sample. Moreover, it is observed under conditions much different from those occurring at the entire experimentation. Furthermore, considerable changes of the observed surface can occur not only as a result of the actual structure of the sample (and its due transformation) but also owning to the mechanical and chemical treatments. The preparation of the microscopically ready-to-observe sample, as well as the other effects attributable to experimental condition of the entire observations is often neglected. There is no any appropriate selection or recommendation; we have to restore our understanding and final attitude accounting on both aspects – limiting viewpoints having little networking capacity. There are not many reliable reports showing the direct impact of the measuring finger ('gadget'), i.e., the way that we actually 'see' the sample (by, e.g., light or electron beam). The scanning electron microprobe analysis (SEM) [695] is one of the oldest methods using the focused electron beam having energy typically from 1 to 30 kV and a diameter from 5 nm to some micrometers. There can be seen a striking similarity with the previously discussed laser, which can equally be used as both, the measuring or the energy-source tool. SEM just disposes with the different energy source so that it is obvious that the interaction of the charged particles involved (electrons) can reveal the changes in both the structure and the chemical composition of the sample under investigation. [694], cf. Fig 96.

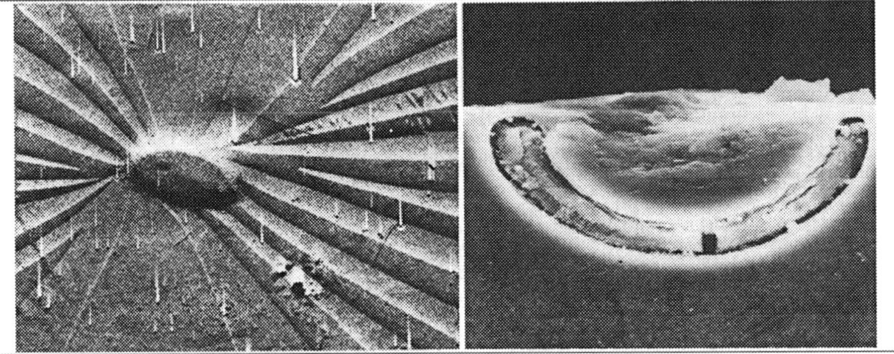

Fig. 97 – Two illustrative examples of the surface damage caused by the action of electron beam during a SEM measurements (under the chosen intensity of 10^{-6} A). Left: photograph of the metallic glass (composition $Fe_{40}Ni_{40}B_{20}$ with $T_{curie} \approx 350$ C and $T_{cryst} \approx 400$ C), showing the re-crystallized area (middle dark spot) as a result of electron beam impact (diameter 1 μ) which is surrounded by yet remaining magnetic domains striation (characteristic for annealed as-cast ribbons, cf. previous Fig. 96.). Right: oxide glass (composition $60SiO_2$-$10Al_2O_3$-$10MgO$-$10Na_2O$, with $T_{cryst} \approx 780$ C and $T_{melt} \approx 1250$ C) revealing the crystallized region along the concave interface as well as the re-solidified area in the inner part as a result of the electron beam impact (diameter 10 μ). The occurrence of melting is proved by disappearance of liquation spots characteristic for the originally phase separated glass [694]. Courtesy of *Václav Hulínský* and *Karel Jurek* (Prague, Czechia).

It is assumed that about 40–80 % of the power delivered into the excited volume is transformed into heat. Heat is generated in the excited pear-shaped volume below the surface and its quantity is determined by the irradiation parameters (such as accelerating voltage), the primary current, the beam diameter and by the material parameter abbreviated as *'f'*, which is defined as the fraction of the primary electron energy absorbed in the sample. The quantity *(1 – f)* represents back scattered or secondary electrons. For the evaluation of temperature increase the same equation for thermal conductivity can be employed as shown in the previous sections. The temperature rise in the center of the probe was calculated for the case of a hemispherical homogeneous source in a semi-infinite homogeneous sample with its outer surface kept at room temperature [686] yielding relation $\Delta T \sim U\,I\,/\,\Lambda\,r_o$ *where ΔT, U, I, Λ and r_o* are the maximum temperature rise in the center of the probe, electric voltage, current, thermal conductivity of the sample and the probe radius, respectively.

It is evident that a direct temperature measurement in the excited volume is possible only at very high currents of the electron beam. Moreover, it is impossible to study the temperature field created inside and outside the bombarded volume. The masterminded idea was to use a material that can change it structure with the temperature assuring that this change is instant, irreversible and observable with a sufficient resolution. Glass, glass-ceramics, or other magnetic materials with visualized domains can serve for that purpose. The experimental result approved the calculated temperature field together with the SEM picture at the sample scale for glass-ceramics exposed to a stationary beam of 200 μm diameter, accelerating voltage 50 kV and the absorbed current 7560 nA. Coming from the pot edge of the SEM image we can see the following areas, dark area of melted diopside crystals, this area continuously grades into the larger area containing large diopside crystals that become denser and smaller with the increasing depth. The next area contains individual rods of pyroxene that convert to spherolites near the lowest part of the crystallized volume.

The *a priori* fear that any SMA investigation may result in the sample over-heating is not tangible as the small exposed region permits very strong cooling by thermal conduction to the bulk. Nevertheless, the experimenter must be aware of the potential risk associated with the electron bombardment of the viewed sample surface, which, moreover, is exposed to vacuum providing little heat transfer to the surroundings. The maintenance of yet acceptable surface temperature by good heat sink to neighboring parts is the only securing parameter for realistic measurements, certainly not avoiding possible creation of certain (even mild) temperature gradients (which we ceaselessly face in all other type of measurements).

13.8 Particularities of temperature modulation

For superimposed temperature transitions, two fundamental concepts are accessible: either temperature modulation or repeated temperature scanning [696-698].. In temperature modulation, the amplitude of the temperature oscillation is generally small and a response of the sample to this perturbation by temperature oscillation is observed. The experimental response can be the measure of heat flow, sample mass, etc. The responses are usually analyzed by *Fourier* analysis, and are separated into 'in-phase' and 'out-of-phase' responses, though their physical meanings are discussed from various points of view (see Chapter 12). The repeated temperature scanning, on the other hand, allows the sample heating and/or cooling in a large temperature range necessary to observe the sample behavior by this temperature fashion. Such amplitude of temperature oscillation lies, however, in a wider range, say from 50°C and 200°C. Until now, the sample mass changes are under such a mode of investigation. By extracting data of the mass and the mass change rate at a given temperature, we can get a relation between the mass and the mass change rate; the resulting curve is equivalent to that obtained by a commensurate isothermal run. Thus we can get equivalent isothermal curves at multiple different temperatures. Another application of this method is observation of non-stoichiometry, in which we can obtain so called *Lissajous* figures by plotting the mass versus the temperature.

However, the essential point in both these methods is the same. When the property, under observation by thermoanalytical methods, depends only on the temperature, we get a hold of a single curve in the conventional plot of the property versus the temperature (*Lissajous* figure) and we do not observe 'out-of-phase' component in the response. This single curve obtained is independent on temperature control modes, such as constant-rate heating, sample-controlled thermal analysis or other temperature modulation varieties. The sample behavior becomes dissimilar, when the property become dependent on the temperature and if the time gets involved in the kinetic effects, such as in relaxation, etc.. In this case, it takes a certain time for the sample to reach the stable (or equilibrium-like) state, and due to this kinetic time-obstruction, we find out a new, ellipsoid-like cyclic curve and we also observe the out-of-phase response. It can be said that the out-of-phase response and the ellipsoid-like cyclic curve (in the characteristic *Lissajous* figure) are observed when the sample property is dependent on its previous thermal history (namely known as hysteresis). One example is the glass transition often observed for organic and inorganic polymers by means of tm DSC and another example is the case of non-stoichiometry detected by repeated temperature scanning. In the latter case, the large amplitude of temperature oscillation is preferable for detecting the effect by enlarging the change.

Observing these phenomena at very low frequencies (or at very low rates of heating and/or cooling), we do not perceive the out-of-phase response (nor

410

the ellipsoid-like cyclic curve but a single curve) because enough time is given to the sample to relax reaching its stable state. In contrast with this, at very high frequencies or at very high rates of heating and/or cooling the change in the observed property cannot be detected, because no time to change is allowed to the sample adjustment. At a frequency in which the rate necessary to reach to the stable state is comparable to the frequency or the rate of heating and/or cooling, we observe maximum effect or the maximum out-of-phase response. Thus, the appropriate fine-tuning of frequency dependence is very important to get fitting insight to the sample behavior. For instance, it helped to elucidate that in the $YBa_2Cu_3O_{7-\delta}$ system three types of oxygen at three different crystalline sites are involved, which distinguishes its non-stoichiometry [9,697]. This was found as a consequence of associated kinetics because the frequency dependencies differ from each other.

To identify chemical reactions by the temperature oscillation, a similar but somewhat different viewpoint is needed. For simple chemical reactions, in which a single elementary process is involved, the rate of chemical reaction depends only on the temperature and the state of conversion. Thus, we can obtain a single isothermal-equivalent curve for the given temperature. For the other type of chemical reactions, which involve multiple elementary processes of different temperature dependence, the rate of chemical reaction is also restricted by the sample previous thermal history. Therefore, we cannot obtain a single equivalent to isothermal curve, as the heating mode is different from that under cooling. Interesting illustrations can be found for various depolymerization (or unzipping) of macromolecules (polymers). In these reactions, radicals are often formed by some elementary processes at weak links, at chain ends or at random sites in the main chain and the radical are annihilated by radical recombination or disproportionation. The rate of mass loss caused by unzipping is proportional to the radical amount, and the radical concentration reaches to a constant in steady state, where the rate of radical annihilation becomes equal to the rate of radical formation; the constant radical concentration in the steady state is dependent on the temperature. Thus this reaction proceeds by a first order reaction because the radical amount is proportional to the sample mass in the steady state. Supposing that time to reach to the steady state is comparable to the frequency or the rate of heating and cooling, the out of phase response would be observed and we would observe that equivalent isothermal curve. Evidently, on cooling it would be different, which stems from the time delay to reach the steady state.

The concept and theoretical considerations of the sample behavior, which responds to the temperature oscillation is, thus, quite similar to the sample behavior responding to other oscillation, such as polarization of dielectrics to alternating electric field and deformation of viscoelastic substances to dynamic mechanical stress, and these would also be applicable to other thermoanalytical methods, such as thermomechanical analysis, etc.. The essential point is the existence of stable (or equilibrium) state and the disposition to reach stability. For example, let us mention cyclic voltammetry, in which electric current by electrode reaction is plotted against applied voltage, while the voltage is repeatedly scanned over a wide range. When we observe dehydration and hydration process under constant water partial pressure by wide-range temperature oscillation, we would get a similar cyclic curve of the mass or the mass change rate versus the temperature, and the curve would become frequency dependent.

In order to describe as well as consider the oscillating behavior, the *Euler* equation and other methods of the determination of unknown parameters is necessary assuming the steady state of oscillatory behavior (*Fourier* analysis and *Lissajous* figures). However, it should be noted that these mathematical methods are regularly based on the linearity of processes so that any decrease in the amplitude and shift of the phase angle (inevitably occurring during propagation of the temperature wave along the instrument and into the sample), may cause errors. Therefore moderately thin specimens are preferably attentive to the sample low thermal diffusivity (polymers and powders) and the quality of thermal contact between the specimen and the instrument.

AFTERWORD AND ACKNOWLEDGEMENT

In respect to more demanding thermodynamic, thermophysical and thermoanalytical treatments, particularly for the material processing under real experimental conditions (near- and non-equilibrium) and also at specially temperature-modulated modes of investigations, warnings were sounded [699] about the lack of attention paid by educational institutions to thermodynamics (and associated theory of thermal analysis) and its connected physical and analytical sciences, forecasting a somewhat decreasing competence level of customers in using the related 'as-cast' techniques. The anticipated development of a yet more specialized application of generalized aspects of thermal analysis [700] needs also its deeper level of edification in different themes where the 'thermo-physical' measuring attitude and 'thermo-dynamic' philosophy can be extendedly applied and operated [701]. Much of the potentiality of thermal analysis lays in the future development and operation of associated techniques and its sophisticated state has been the result of creativity, skill and theoretical worth of various scientists in developing the presently matured state-of-art.

We can eventually anticipate certain trends for future instrumentation, which would be entirely based on computer handling out the thermal data without any need of temperature regulation. A low-inertia measuring head can be inserted into a suitable preheated furnace and the heating curve is recorded for both the sample and the reference and the computer program makes it possible to evaluate any desired type of outcome (like DTA/DSC) just by an appropriate subtraction and combination of active (sample) and passive (reference) heating curves. In the age of computer handling, such a relatively simple arrangement would substantially reduce the need of yet expensive instrumentation based on a traditional temperature control.

We, however, should become adequately attentive to unseen limits of sophistication, which inexhaustible erudition may bring hidden side-effects for a multifaceted instrumental design, e.g., intentionally modulated temperature programs (and its unconventional application to analyze complex processes) witnessed in the light of innate source of natural oscillation modes due to the localized surface-to-surface impedance forcing the heat transfer pulsation (usually to appear within miniature sample holders, cf. paragraph 5.4). In such very small and almost infinitesimal thermally controlled arrangements some up until now undefined principles of microscopic uncertainty may be found to restrain the simultaneously reliable determination of temperature and heat flux (similarly like quantum determination of particle position and momentum). It can encompass such a connotation that we cannot determine the temperature precisely enough for large heat flows and vice versa.

Thermal physics, and thermodynamics as its natural component, can be considered to lay in the heart of natural science and science itself is not a discipline contradictory to and incompatible with artistic (and even moral or akin religious) feelings of human beings. Science, as any art does, requires imagination and intuition and thus our scientific concern should be tolerant of new scientific motifs, which sometimes are thought bizarre and thus abandoned or delayed by scientific narrowness. As an immodest example, consider the brilliant works of *Gibbs, Röntgen, Bělousov* or *Feingenbaum*, which were buried and hidden from common knowledge for some time in obscure and insignificant annals. Thus in reviewing new ideas we should be thoughtful but broadminded.

At the same time we should suppliantly perceive and respect the mystic faith of some of the greatest physicists, such as *Heisenberg, Schrödinger, Einstein* or *Planck* who in searching the deepest corners of nature became religious. Such an attitude did not contradict their widely respected scientific credit. Moreover, their scientific attempt to achieve better understanding of the laws of nature actually matured to a very private approach, which manifested their own internal vision of the Great Architecture of the Universe.

Even the modern world with its progressive technologies and exceptional theories and with a more broadly assumed cross-disciplinarity, which cannot exclude philosophical thoughts, is understood as the admiration of the miracle of life, which has retained its influential role in the challenge of striving to maintain a sustainable civilization on both the levels of matter and mind. The struggle of Mankind for a better life, while striking to stay alive in the bosom of Mother Nature and pursuing eternal fire/heat, should be understood as a fight for lower disorder and the utmost level of information, and not a mere seeking for sufficient energy. Its comprehension was a modest objective of my treatise.

I am sincerely hopeful that there will come, sooner or later, some novel scientific discoveries that will possibly arise from curious ideas (such as the zero-point electromagnetic radiation [22]) and which would make the survival of our civilization more feasible. If not, we should keep learning how to be congenial with our modest means to our gracious Earth, which has awarded us the land where we can live; a land that was offered to our forefathers and, hopefully, will remain serviceable to our children, too.

A specific place in this book captures the novel understanding of periodic precipitation of reaction products and oscillatory behavior of certain chemical reactions, which solution has bothered scientists for almost hundred years. Its description involves the *Planck's* universal constant so that there has not been consensus concerning its legitimacy as various scholars regard it either accidental (without any deeper physical meaning) or enigmatic (with something very important behind). We believe [36] that the most consequential criterion resides in how to decide whether a particular physical problem belongs to the domain of classical or quantum physics, i.e., if the quantity of type of action

414

(relevant to a given physical problem) is truly comparable with *Planck's* quantum. It seems plausible that for its approval it can be adequate to clarify the condition under which the numerical values of classical diffusion (*Fick*) constant attains that spheres of the quantum (*Fürth*) motion [269]. This convergence can provides certain resulting conditions for acting species, i.e., for ball-like particles (thus simulating molecules), which exhibit no observable persistency when moving through low viscous media. In such a case *Einstein-Smoluchowski* formula [267] is applicable in the form $kT/(3\pi\eta a) >> h/(4\pi M)$. Beside the standard constants (k, h, π) we can conveniently substitute for dynamic viscosity that of liquid water, $\eta_{H2O} = 10^{-3}$ [kg/(ms)], and for temperature its room value, $T_{room} = 300$ [K], which yields the inequality $8.4x10^{15} >> a/M$, where a [m] is the size and M [kg] is the molecular weight ($x\ 1.67x10^{-27}$). Accounting for particular ions and their radii we can find that the most suitable candidates for such a quantum behavior (Brownian particles) capable of self-organization are sodium, calcium and potassium. It means that these cations exhibit most excellent decoupling from the classical noise (of thermally agitated mechanical impacts) so that they can move more straightforward becoming thus 'quantum' sensitive. It may have a crucial consequence in speculations about the creation of life, which may not be merely associated with the formation of sufficiently complex organic molecules but also, and perhaps foremost, to the salty environment of oceans, which provided essential structure to give rise the necessary communication environment allowing message transfer between molecules.

A similar analysis is yet due in the special case of the most mobile protons (H^+), which are most important in all life processes (e.g., thermo- and electrodynamics of the propagation of nerve signals [258]) together with the above mentioned Na^+, K^+ and Ca^{++}. We have faith in this novel enlightenment, and are keen to see yet more relevant experiments and theoretical proofs focused on the boundary range of nano- and macroscale, which is the intermediate area of mesocopic thermodynamics, which seem us to be desirably unlocked for a fruitful future.

The contents of this book and inherent theories may recall various critical assessments, which are acceptable in the current, rather curious state of our knowledge, extending in this transcript from the factual forte of science to some spheres of interdisciplinarity, touching even the humanities [9], which was actual in the past and would be confidently similar in the future as the adequate signs of an opposite subject evolution.

The sciences, databases and information exchange have all expanded tremendously over the past decades. Important example of the many changes in thoughts and scientific confrontations was the convincing preference of the vibrational theory over the flow caloric theory in thermal physics though the equations even now hold the 'caloric fingerprint'. However, the novel subject matter

Fig. 98. – Group photography of the ICTA Council meeting in the castle Liblice near Prague, which took place at the occasion 8th ICTA Conference in Bratislava 1985 (former Czechoslovakia), celebrating the 20th anniversary of ICTA foundation. From left: Edward.L. Charsley (England), behind Michael E. Brown (South Africa), Bordas S. Alsinas (Spain), late Walter Eysel (Germany), late Vladislav V. Lazarev (Russia), late Paul D. Garn (USA), John O. Hill (Australia), John Crighton (England), Tommy Wadsen (Sweden), Joseph H. Flynn (USA), Patric K. Gallagher (USA), Hans-Joachim Seifert (Germany), Slade St.J. Warne (Australia), behind Klaus Heide (Germany), Vladimír Balek (Czechia), late Viktor Jesenák (Slovakia), Milan Hucl (Slovakia), Jaroslav Šesták (Czechia), late Jaroslav Rosický (Czechia), behind Shmuel Yariv (Izrael), right Erwin Marti (Switzerland) and Giuseppe Della Gatta (Italy). Bottom are exampled the early front pages of below mentioned journals.

of the so-called 'dark' energy and/or matter might somehow endure then provisional position of caloric in explaining the problem of antigravity. Even today the world of physics is not fully cohesive as it somehow divides articles into the authorized (and by reviewers well accepted) papers and those contributions that are not in favor with the standard contentment of regular journals. They, thus, form an appreciable sphere of refugee (often-labeled dissident) physics well known from the Internet and subject of other (someway-expelled) journals [702]. Comparable, but less imperative state appeared even on a much smaller scale within the thermoanalytical family, where those who reach to have control over the society [331] start to put emphasis rather on bureaucracy instead of the scientific worth, creating thus the state of two bulletins-news, one being official ICTAC News (but sometimes censored) [703] and the other autonomous open for any contributions [704].

Special notice should be paid to the lengthy efforts and services of *International Confederation of Thermal Analysis and Calorimetry* (ICTAC) as an important forerunner of the field of thermal analysis. The first international conference on thermal analysis held in the Northern Polytechnic in London, April 1965 (see the following figure), paved the way to the newborn opportunity for a better international environment for thermal analysis [705] assisted by the great pioneers such as *B.R. Currell, D.A. Smith, R.C. Mackenzie, P.D. Garn, R. Barta, M. Harmelin, W.W. Wendlant, J.P. Redfern, L. Erdey, D. Dollimore, C.B. Murphy, H.G. McAdie, L.G. Berg, M.J. Frazer, W. Gerard, G. Lombardi, F. Paulik, C.J. Keattch, G. Berggren, J. Šesták, K. Heide, W.L. Charsley, R. Otsuka, C. Duval, M. Vanis*, etc., in their effort to establish such a constructive scientific forum to be cooperative for all thermoanalysts. An international platform of thermal sciences then began in earnest when ICTA was established in Aberdeen 1965, which has productively kept going until now appreciative the precedents of friendly manners and scientific merit (past ICTA/ICTAC presidents *L.G. Berg, R. Bárta, G. Lombardi, H.G. McAdie, H. Kambe, P.K. Gallagher, J. Oswald, H.J. Seifert, S.St.J. Warne, T. Ozawa, E.L. Charsley*).

It was supported by the foundation of international journal of *Thermochimica Acta* in the year 1970 by Elsevier and, for a long time, edited by *Wesley W. Wendlandt*, with the help of a wide-ranging international board (such as B.R. *Currell, T. Ozawa, L. Reich, J. Šesták, A.P. Gray, R.M. Izatt, M. Harmelin, H.G. McAdie, H.G. Wiedemann, E.M. Barrall, T.R. Ingraham, R.N. Rogers, J. Chiu, H. Dichtl, P.O. Lumme, R.C. Wilhoit*, etc.). It was just one year ahead of the foundation of another specialized *Journal of Thermal Analysis*, which was brought into being by *Judit Simon* (who has been serving as the editor-in-chief even today) and launched under the supervision Hungarian Academy of Sciences (Académia Kiadó) in Budapest (*L. Erdey, E. Buzagh, F. and J. Paulik brothers, G. Liptay, J.P. Redfern, R. Bárta, L.G. Berg, G. Lombardi, R.C. Mackenzie, C. Duval, P.D. Garn, S.K. Bhattacharyya, A.V. Nikolaev, C.B. Murphy, J.F. Johanson*, etc.,) to aid preferably the worthwhile East European science suffering then behind the egregious 'iron curtain'.

Worth noting is also the 60 years anniversary of continuous series of the US Calorimetry Conferences, which evolved from a loosely knit group operating in the 1940s to a highly organised assembly working after the 1990s (among many others let us mention *H.M. Huffman, D.R. Stull, G. Waddington, G.S. Parks, S. Sunner, J.J. Christensen, E.F. Westrum, J.P. McCullough, D.R. Stull, D.W. Osborne, W.D. Good, P.A.G. O'Hare, P.R. Brown, W.N. Hubbart, R. Hultgren, R.M. Izatt, D.J. Eatough, J. Boerio-Goates, J.B. Ott*).

Consequentially, the *Journal of Chemical Thermodynamics* began publication in the year 1969 (edited by *L.M. McGlasham, E.F. Westrum, H.A. Skinner* and followed by others).

417

INTERNATIONAL SYMPOSIUM ON THERMAL ANALYSIS

13th and 14th April, 1965

CHEMISTRY DEPARTMENT
NORTHERN POLYTECHNIC
HOLLOWAY ROAD · LONDON · N.7

WEDNESDAY, 14th APRIL

MORNING SESSION

Chairman: Dr. R. C. MACKENZIE

9.30 a.m.	Theatre A	"The Causes of Error in Thermogravimetry" Dr. M. HARMELIN (C.N.R.S., Paris)
10.45 a.m.	Coffee	
11.15 a.m.	Theatre A	"Miscellaneous Thermal Methods" Prof. W. W. WENDLANDT (Texas Technological College)
12.30 p.m.	Lunch	

AFTERNOON SESSION

2.00 p.m.	Theatre A	Chairman: Dr. J. P. REDFERN "Kinetic Studies by Thermogravimetry" Dr. G. GUIOCHON (Ecole Polytechnique, Paris)
3.00 p.m.	Tea	
3.30 p.m.	Theatre B	"Applications of Thermal Analysis to Inorganic Compounds II" (Papers B/5-9) Chairman: Dr. M. J. FRAZER
	Theatre C	"Applications of Thermal Analysis to the Study of Organic Compounds (including Polymers)" (Papers C/5-9) Chairman: Mr. D. A. SMITH
	Theatre D	"Techniques of Thermal Analysis II" (Papers D/5-9) Chairman: Dr. R. C. MACKENZIE

Close—5.30 p.m.

TUESDAY, 13th APRIL

MORNING SESSION

Chairman: Dr. J. P. REDFERN

9.20 a.m.	Theatre A	Introduction by Dr. W. Gerrard (Head of the Chemistry Dept., Northern Polytechnic)
9.30 a.m.	Theatre A	"Atmosphere Effects in Differential Thermal and Thermogravimetric Analysis" Prof. P. D. GARN (University of Akron, Ohio)
10.45 a.m.	Coffee	
11.15 a.m.	Theatre A	"Derivatography" Prof. L. ERDEY (Technical University of Budapest)
12.30 p.m.	Lunch	

AFTERNOON SESSION

2.00 p.m.	Theatre A	"General Aspects of Thermal Analysis" (Papers A/1-4) Chairman: Prof. P. D. GARN
3.30 p.m.	Tea	
4.00 p.m.	Theatre B	"Applications of Thermal Analysis to Inorganic Compounds I" (Papers B/1-4) Chairman: Prof. W. W. WENDLANDT
	Theatre C	"Applications of Thermal Analysis to Polymer Studies" (Papers C/1-4) Chairman: Mr. D. A. SMITH
	Theatre D	"Techniques of Thermal Analysis I" (Papers D/1-4) Chairman: Dr. B. R. CURRELL

Close—5.30 p.m.

KEY LECTURES "1"

"General Aspects of Thermal Analysis"

Chairman: Professor P.D. Garn

Theatre A   Tuesday: 2.00 - 3.30 p.m.

A1   "The Standardisation of Experimental Conditions in Thermal Analysis"

F. Paulik and J. Paulik (Technical University of Budapest)

A2   "Procedural Variables Influencing the Thermogravimetric Study of Polymer Decomposition"

D.A. Smith (National College of Rubber Technology, London)

A3   "The Use of Thermal Standards in Thermogravimetric Analysis"

C.J. Keattch (John Laing Research and Development Ltd., Borehamwood)

A4   "The Error of Kinetic Data Obtained from Thermogravimetric Traces at a Rising Temperature"

J. Sestak (Institute of Solid State Physics, Czechoslovakia)

In wrapping up I found it fair to include some brief characteristics of some foremost thermoananlysts [706] similarly to the short disposition data of selected remarkable scientists and scholars of the past, which were already published in my previous book [9] (together with their 60 miniature photos). Unfortunately, the lack of sufficient space herewith prevented me to include a adequately extended listing of all noteworthy personalities in this edition. It will be revealed on my websites [704] in more details yet open for further additions. Therefore, I found it necessary to limit the entries in this book appendix just to dead personalities and those who already passed their age of sixty-five.

It is also my great pleasure and content duty to greatly acknowledge the rewarding atmosphere of my home Institute of Physics incorporated with the Academy of Sciences of Czech Republic in Prague, where I have spent a fruitful forty-five years. During the last two decades, I have had the honor to appreciatively work with the group of theoretical physicists and experimenters that are widely educated, par excellence persons with many adroitness and proficiency. It was a good school for improving my background of a solid-state chemists as well as my scientific culture.

In a similar way as with my previous book [9], I owe my deepest thanks to many of my best-respected friends, forward-thinking persons and brilliant coworkers, just mentioning *RNDr. Jiří J. Mareš, PhD,* theoretical and experimental physicists, as well as my scientific and beer-drinking companion *Prof. Zdeněk Strnad, MEng., PhD.,* the director of spin-off laboratory 'LASAK'. Over and above, I have also to acknowledge my former, very talented students and the present supreme thermoanalysts, *Prof. Jiří Málek, MEng., PhD. DSc.,* the rector of Pardubice University and *Dr. Nobuyoshi Koga, PhD.,* professor at Hiroshima University. Certainly, I should not forget my early group leader, *RNDr., JUDr. Arnošt Bergstein, PhD.,* (1914-1973), originally lawyer who survived a nacist's concentration camp half-dead and who after this enforced internment found it incongruous to continue his law career under the growing regime full of bursting extrajudiciality and who thus became a devoted thermoanalyst (virtually "litigating samples"), subsequently oppressed again by communistic totalitarianism.

My deepest and most sincere appreciation, however, is paid to my family, particularly my wife *Věra,* who graduated at both the Czech Institute of Chemical Technology in Prague (1968 – M.Eng.) and the US University of Missouri at Rolla (1970 – M.S.) and who, for a long time, has also coauthored our articles (being the expert on the growth and utilization of single crystals of various semiconductors) and ceaselessly corrected my transcripts. She truly helped me to complete all my books by her friendly support, patience, enthusiasm and encouragement – kindheartedly tolerant to my ceaseless chaotic nature. Similarly my heartfelt thanks are directed to my children, *Alžběta* (28) and *Pavel* (26), who always provided beneficial critique, helped to keep a pleasant family environment, and also aided my photography attempts – to all of whom I have most earnestly dedicated the book.

Last, but not least, I would like to acknowledge technical and financial support of the *Academy of Sciences* of Czech Republic in Prague, *West Bohemian University* in Pilsen (Department of Mechanics, project of MŠMT No. 4977751303 "Defects predication in heterogeneous materials as a function of mechanical and biomechanical systems"), enterprises *NETZSCH Gerätebau* GmbH., Selb (Germany) and *LASAK,* s.r.o., Praha (Czechia) as well as the *Grant Agency* of Czech Republic (project No. 522/04/0384 "Thermodynamic study of the formation of biological glasses and the preservation of biodiversity of plants"). The *Municipal Government* of Prague 5 and their cultural grant 56/0/OSK/2005 is also acknowledged. Some parts of the book were written during the author's autumn 2004 stay at the School of Energy Sciences, as a visiting professor at *Kyoto University* (Japan).

Deep appreciation is paid to the book reviewers and their refining comments: *Prof. Robert Černý, MEng., PhD., DSc.,* of the Czech Technical University in Prague and *Prof. Peter Šimon, MEng., PhD.,* of the Slovak Technical University in Bratislava, and to my publishing contacts at Elsevier, *Derek Coleman* and *Michiel Thijssen.*

Allow me to repeat for the second time that I tried to join so far uncommon spheres of privity within the ordinary domains of science and culture, using the occasion to portray my broader conception of science of heat/energy [9] that obviously affected my 'extempore' trips to some deeper recesses of nature, which vision, however, may likely become the future driving force for our better awareness of any avant-garde science [158, 194,335,707-711]. I have completed this authorship in my best will and with a view to stimulate the interdisciplinarity of knowledge, thus seen in the primary understanding of the Pythagorean term 'philosopher': *"the one who seeks to uncover the secrets of nature".*

In conclusion let me accentuate that in addition to critical commentaries , I am awaiting the natural reaction of those who bear different views on the inherent subjects. I ask all my readers for tolerance when looking upon all the passages of this book, which, due to their dispersed nature, I could not write with equal feeling and powers, and I am not self-righteous enough to be able to have done so. I would just like to quote *"Nobody should assume that what I am saying is true. It is not given to us to know what is true in this sense. But everybody knows that I write this scientific treaty in an implicit unwritten understanding among the scientists that I can be absolutely believed to be what I believe"* [712].

420

LITERATURE

[1] J. Šesták, V. Šatava, W.W.Wendladt "The Study of Heterogeneous Processes by Thermal Analysis", special issue of Thermochim. Acta, Vol. 13, Elsevier, Amsterdam 1973
[2] J. Šesták "Měření termofyzikálních vlastností pevných látek" (Measurements of thermophysical properties of solids), Academia, Praha 1982 (in Czech)
[3] J. Šesták "Thermophysical Properties of Solids: their measurements and theoretical thermal analysis", Elsevier, Amsterdam 1984
[4] J. Šesták "Teoreticheskij termicheskij analyz" (Theoretical thermal analysis), Mir, Moscow 1988 (in Russian)
[5] Z. Chvoj, Z. Kožíšek, J. Šesták "Non-equilibrium Processes of Melt Solidification and Metastable Phases Formation: a review", special issue of Thermochim. Acta, Vol. 153., Elsevier, Amsterdam 1989
[6] Z. Chvoj, J. Šesták, A. Tříska (edts) "Kinetic Phase Diagrams: non-equilibrium phase transitions", Elsevier, Amsterdam 1991
[7] J. Šesták, Z. Strnad, A. Tříska (edts), "Speciální technologie a moderní materialy", (Special Technologies and Modern Materials), Academia, Praha 1993 (in Czech)
[8] J. Šesták (ed), "Vitrification, Transformation and Crystallization of Glasses", special issue of Thermochim. Acta, Vol. 280/281, Elsevier, Amsterdam 1996
[9] J. Šesták "Heat, Thermal Analysis and Society", Nucleus, Hradec Králové 2004
[10] J. Šesták, NATAS/Mettler Award lecture "Rational Approach to the Study of Processes by Thermal Analysis" in "Thermal Analysis", Proc. of the 4th ICTA in Budapest, Vol.1, Academia Kiadó, Budapest 1975, p.3. and ICTAC/TA Award lecture "Thermal Treatment and Analysis: the art of near-equilibrium studies" in J. Thermal Anal. 40(1993) 1293
[11] J. A. Comenius: "Disquisitiones de Caloris et Frigoris Natura", Amsterdam 1659
[12] Č. Strouhal "Thermika" (Thermal science), JČMF, Praha 1908 (in Czech)
[13] I. Prigogine, I. Stangers "Order out of Chaos", Bantam, New York 1984
[14] J.D. Barrow "Impossibility – Limits of Science and Science of Limits", Vintage, New York 1999
[15] from the earliest book by W.W. Wendlandt "Thermal Methods of Analysis", Wiley, New York 1962 to the latest by M.E. Brown "Introduction to Thermal Analysis: techniques and applications" (2nd edition) Kluwer, Dortrecht 2004
[16] M.E. Brown, J. Málek, J. Mimkes and N. Koga (edts) " Thermal Studies beyond 2000", special issue of J. Thermal Analysis dedicated to J. Šesták, Vol. 60., No. 3., Akademia Kiadó/Kluwer, Dortrecht 2000.
[17] I.M. Havel "Scale Dimensions in Nature" Int. J. General Syst. 24(1996)295.
[18] L. Nottale "Fractal Space-time and Microphysics: towards a theory of scale relativity", World Scientific, New York 1993
[19] P. Vopěnka "Úhelný kámen Evropské vzdělanosti a moci: rozpravy o geometrii" (Interviews on Geometrization I, II, III and IV), PRAH, Praha 2000 (in Czech) and "Mathematics in the Alternative Set Theory" Leipzig 1979
[20] M. Holeček "Averaging in Continuum Thermomechanics" (textbook by ZČU Plzeň, in print 2005]
[21] L.A. Zedah "From Computing with Numbers to Computing with Worlds: from manipulation of measurements to manipulation of perception" plenary lecture at the of 2nd IPMM (Intelligent Processing and Manufacturing of Materials) Honolulu, Hawaii 2001 (CD proceedings)
[22] J.J. Mareš "On the Development of the Temperature Concept" J. Thermal Anal. Calor 60(2000)1081

[23] Ya.A. Smorodinskii "Tjemperatura" (Temperature) Nauka, Moscow 1987 (in Russian)
[24] M.-A. Poincare, Essais de physique, Manget 1790
[25] T. Basaram, X. Ilken, Energy and Buildings 27(1998)1
[26] J. Šesták "Information Content and Scientific Profiles" in ICTAC News, 31/1(1998)52
and "Is Scientometry Harmful" in ICTAC News 31/2(1998)66
[27] J. Fiala, Thermochim. Acta 110(1987)11 and Archiwum Nauki o Materialech
12(1991)85
[28] R.E. Maizel "How to find Chemical Information" Wiley, New York 1987
[29] J. Fiala, J. Šesták "Databases in material sciences" J. Thermal Anal. Calor. 60(2000)1101
[30] H.G. Wells "World Brain" Doubleday, New York 1938
[31] E. Garfield "Citation Indexing" Wiley, New York 1979
[32] H. Small, J. Informat. Sci. 11(1985)47 and Scientometrics 26(1993)5
[33] J.D. Barrow "Artful Universe", Oxford University Press, Oxford 1995
[34] J. Meel, Intro-address at the 3rd Int. Conf. on Intelligent Processing and Manufacturing of
Materials, Vancouver 2001, Proceedings available in the CD form
[35] Č. Bárta, *jun.,* Č. Bárta, *sen.,* J. Šesták "Thermophysical Research under Microgravity"
key lecture at the of 3rd IPMM (Intelligent Processing and Manufacturing of Materials)
Vancouver, Canada 2001 (CD proceedings)
[36] J.J. Mareš, .J. Stávek, J. Šesták, "Quantum aspects of self-organized periodic chemical
reactions" J. Chem. Phys., 121(2004)1499
[37] A. Capocci, F. Slanina, Y.C. Zhang "Filtering information in a connected network"
Physica A 317(2003)259
[38] G. Green "The Elegant Universe – Superstrings and Hidden Dimensions", W Norton,
New York 1999
[39] A.P. LaViolette "Subquantum Kinetics", Staslane, New York 1994
[40] A. Einstein., "Ather and Relativitatstheorie" Lecture presented at the University of
Leyden, Springer, Berlin 1920 and book "Einstein and the Ether" by L. Kostro, Apeiron,
Montreal 2000
[41] J. Šesták, R.C. Mackenzie, "Heat/fire concept and its journey from prehistoric time into
the third millennium" key lecture at the 12th ICTAC (Copenhagen) in J. Thermal Anal. Calor.
64(2001)129
[42] J.D. Barrow "Artful Universe" Cambridge Univ. Press 1994; P. Back "How Nature
Works: science of self-organized criticality", Copernicus, New York 1996; J. Ziman "Real
Science: what it is and what it means" Cambridge Univ. Press 2000
[43] S.A. Kauffman "Investigations", Oxford Press, New York 2000.
[44] R.C. Mackenzie "History of Thermal Analysis", special issue of Thermochim. Acta, Vol.
73, Elsevier, Amsterdam 1984
[45] S. Carnot "Reflexions sur la puissance motrice do feu", Paris 1824 and its reprint in
Ostwalds Klassiker, Engelmann, Leipzig 1909
[46] R. Clausius "Mechanische Warmetheorie" Vieweg Sohn, Braunschweig 1876
[47] I. Prigogine, P. Glansdorff "Thermodynamic Theory of Structures, Stability and
Fluctuations", Wiley, New York 1971
[48] I Prigogine "From Being to Becoming – Time and Complexity in Physical Sciences"
Freeman, San Francisco 1980
[49] R.C. Mackenzie, Thermochim. Acta 95 (1982) 3
[50] J.C. Maxwell "Theorie der Warme" Marusche Berendt, Breslau 1877
[51] E. Mach "Die Principen der Wärmelehre", Barth, Leipzig 1896
[52] S. G. Brush "The Kind of Motion we call Heat". Vol. I & II , North Holand, Amsterdam
1976

422

[53] J. Šesták "Heat as Manufacturing Power or the Source of Disorder" key lecture at the TERMANAL conference (Slovakia 2001); J. Thermal Anal. Calor. 69(2002)113

[54] H. Boerhaave "Elementa chimiae", Leyden 1732

[55] L. von Bertalanffy "General System Theory", Brazillerr, New York 1968

[56] J.R. Searle "The Mystery of Consciousness" Granta Books, London 1997

[57] G. Nicholis and I. Prigogine "Self-organization in Non-equilibrium Systems", Wiley-Interscience, New York 1977

[58] R.H. Peters "Ecological Implications of Body Size", Univ. Press, Cambridge 1983

[59] M. Schröde "Fractals, Chaos and Power Laws" Freeman, New York 1991

[60] S.S. Stevens "Handbook of Experimental Psychology", 1951 and rewritten in 1988 as "Stevens' Book of Experimental Psychology"

[61] M.A. Nowak, J.B. Plotkin, V.A. Jansen, Nature 2000, 494

[62] G.K. Zipf "Human Behavior and the Principle of Least Effort" Addison/Wesley, Cambridge 1972

[63] M. Levy and S. Solomon, Physica A 90 (1997) 242

[64] Č. Strouhal "Akustica" (Acoustics), JČM, Praha 1902 (in Czech)

[65] I. Johnson "Measured Tones: the interplay of physics and music", Hilger, New York 1992

[66] P.V. Šimerka, "Síla přesvědčení: pokus v duchovní mechanice" (Strength of conviction: attempt to mental mechanics), Časopis pro pěstování matemat. fyziky, 11(1882)75 (in Czech); A. Pánek "Život a působení Šimerky" (Šimerka´s life and actuation), Časopis pro pěstování matemat. fyziky, 17(1888)253 as well as J. Fiala in "Jubilejni Almanach JCSNF", Praha 1987, p. 97

[67] M. Cross, P. Hohenberg, Rev. Modern Phys. 65(1993)1078

[68] J. Palacious "Dimensional Analysis", McMillan, New York 1964

[69] L. Rayleigh, Phil. Mag. 32(1916)529

[70] J. McPhee "Control of Nature", Farrar & Giroux, New York 1989

[71] J.M. Fowler "Energy and the Environment" McGraw Hill, New York 1984

[72] P. Svoboda (ed.) "Zdroje a výroba elektrické energie" (Sources and production of electric energy), special issue of "Československý časopis pro fyziku" (Czech Journal of Physics) A2 (2002), Vol. 52 (in Czech)

[73] C.J. Clevelan (ed.) "Encyclopedia of Energy" (6 volume set), Elsevier 2004

[74] R.C. Neville "Solar Energy Conversion: solar cells" Elsevier, Amsterdam 1995

[75] J. Larminie, A. Dicks "Fuel Cell Systems Explained" Wiley, Chichester 2001

[76] M. J. Moran, H. N. Shapiro "Fundamentals of Engineering Thermodynamics" Wiley, New York 1992

[77] S.E, Jørgenson, Y.M. Svirezhev "Towards a Thermodynamic Theory for Ecologial Systems" Elsevier, Amsterdam 2004

[78] A. W. Culp "Principles of Energy Conversion", McGraw-Hill, New York 1991

[79] P. Gipe "Wind Power" James&James, London 2004

[80] M. Green "Third Generation Photovoltaics: advanced solar energy conversion" Springer, 2003;

[81] J.A. Curry, P.J. Webster "Thermodynamics of Atmosphere" Academic, New York 1999

[82] D.L. Ray "Trashing the Planet" Regnery Gateway, Washington 1992

[83] J. de Swaan Arons, H.J. van der Kooi, K. Sankaranarayanan "Efficiency and Sustainability in the Energy and Chemical Industries: scientific principles and case studies" Dekker, New York 2003;

[84] M.K. Hill `Understanding Environmental Pollution` Cambridge Univ. Press, Cambridge 2004

[85] J. Šesták, Sklář a Keramik (Prague) 26(1976)427 (in Czech) and Wiss. Zeit. Friedrich-Schiller Univ., Jena, Math.-Naturwiss. Reihe, 32(1983)377

[86] J. Šesták, Glastech. Ber. Glass. Sci. Tech. 70C(1997)43; J. Thermal Anal. Calor. 61(2000)305 and Thermochim. Acta 110(1987)427

[87] Proceedings of the IPMM international conferences ('Intelligent Processing and Manufacturing of Materials') all coedited by J. Meech (Vancouver University), namely: Golden Coast (Australia) 1996, Honolulu (Hawaii) 1999, Vancouver (Canada) 2001 and Sendai (Japan) 2003

[88] B.C. Randal "Nanotechnology: molecular speculation on global abundance" Massachusetts Inst. Tech. 1999

[89] K. Goser (ed.) "Nanoelectronics and Nanosystems" Springer 2004;

[90] J. Šesták, P. Lipavský , J. Thermal Anal Calor. 74(2003)365

[91] Z. Strnad, J. Šesták, J. Strnad, N. Koga "Thermodynamics ofnon-bridging oxygen in silica biocompatible glasses" J. Thermal Anal. Calor. 71(2003)927 and 76(2004)17

[92] Z. Strnad, J. Šesták, "Bio-compatible Ceramics" invited lecture at the 3rd Inter. Conf. Intelligent Processing and Manufacturing of Materials, Proceedings Vancouver Univ., p. 123, Vancouver 2001; Biomaterials, 13(1992)317

[93] L.L. Hench and J. Wilson "Introduction to Bioceramics" World Sci. Publ., London, 1993 and "Handbook of Bioactive Ceramics" Vol.1, CRT Press 1990

[94] L.L. Hench, J. Biomed. Mater. Res. 23 (1989) 685 and Thermochim. Acta 280/281(1996)1

[95] E. Sikes "The Fire Bringers" Prometheus Vinctus, London 1912

[96] W. Hough "Fire as an Agent in Human Culture" Smithsonian Inst, Washington 1926

[97] J. M. Frazer "The Myths of the Origin of Fire" Cambridge 1929

[98] G. Bachelard "La psychoanalyse du feu", Gallinard, Paris 1949 (in French)

[99] Z. Neubauer "O přírodě a přirozenosti věcí" (About the Nature and Naturality of Objects), Malvern, Praha 1998 (in Czech)

[100] B. Jonathan "The Presocratic Philosophers" Routledge & Kegan, London 1989

[101] A.A. Long "The Cambridge Companion to Early Greek Philosophy", Cambridge Press, Cambridge 1999

[102] P. Zamarovský "The Roots of Western Philosophy" ČVUT Publ. House, Praha 2000

[103] H.R.. Maturana and F.J. Varela "The Tree of Knowledge", Shabhala, Boston 1987

[104] J.D. Barrow "Pi in the Sky: counting, thinking and being" Oxford University Press, 1992

[105] J.R. Partington "History of Greek Fire and Gunpowder", Cambridge 1960

[106] Theophrastus: "Peri Pyros" (On fire), about 320 BC

[107] D. Mylius: "Philosophia Reformata" , Frankfurt 1622

[108] Norton: "Ordinall of Alchimy", 1477

[109] G. Agricola: "De re Metallica", Basel 1520 – 1556 (Vol. I – VII)

[110] H. Gebelein : "Alchemy, Die Magie des Stoffichen", Diederichs, München 1991

[111] C. Priesner, K. Figala (edts) "Alchemie", München 1998

[112] G. E. Stahl: "Fundamenta Chymiae Dogmaticae et Experimentalis", Parsi 1746

[113] V. Karpenko, HYLE 4(1998)63

[114] J. Lindsay "Origin of Alchemy in Greek-Roman Egypt", London 1970

[115] Meng-Shen 'Sh-liao pen-cchao' about 670 AD

[116] V. Karpenko "Tajemství magických čtverců" (Mystery of Magic Squares), Academia, Praha 1997 (in Czech)

[117] J. Chun-xual, Apeiron 5 (1998) 21

[118] Norton: "Ordinall of Alchimy" 1477

[119] G. Agricola: "De re Metallica" Basel 1520 – 1556 (Vol. I – VII)

[120] Stolcius: "Viridarium chimicum" 1624

[121] Marcus Marci: "De proportione motu" 1639

[122] J.D. Barrow "The Origin of the Universe" Orion, London 1994

[123] R.P. Feynman "The Character of Physical Laws" MIT Press, Massachusetts 1995

[124] S. Hawkins and R. Penrose "The Nature of Space and Time" Princeton University Press, Cambridge 1997

[125] R. Penrose "Shadow of the Mind: approach to the missing science of consciousness" Oxford University Press, Oxford 1994

[126] M. Gina "Storia della Chimica" Tipografico Torinese 1962 (in Italian), Russian translation, Moscow 1975

[127] R. Descartes: "Principa philosophia" 1644

[128] J. B. Robinet "De la Nature" Amsterdam 1766

[129] J.R. Partington "History of Chemistry" MacMillan, New York 1964

[130] B. Bensande-Vincent, I. Stenger "History of Chemistry" Harvard Press, London 1996

[131] J. Šesták "The Role of Order and Disorder in Thermal Science" Part I. "Heat and Society" and Part II. "Universe, Matter and Society" as invited papers in J. Mining Metal. (Bor, former Yugoslavia) 38(2002)1 and 39(2003)1.

[133] R. Fox "The Caloric Theory of Gases: from Lavoisier to Regnault", Claredon Press, Oxford 1971

[134] J.M. Socquet "Essai sur le calorique", Paris 1801; J.D. Forbes "A Review of the Progress of Mathematical and Physical Science between Years 1775 and 1850" Edinburgh 1858; J.T. Merz "History of European Thoughts in Nineteenth Century" (4 Vols) Edinburgh 1896-1914

[135] J. Black "Lectures on the Elements of Chemistry" (J. Robinson, ed), Edinburgh 1803; J. Dalton "A New System of Chemical Philosophy" Manchester 1808 & 1827 (readition by Phil.Lab. New York 1964); H. Davy "Elements of Chemical Philosophy" London 1812;

[136] J.L. Lagrange "Mechanique analytique", 1788; A.L. Lavoisier "Traite elementaire de chimie". Paris 1789: P.S. Laplace "Traite de mechanique celeste" (5 Vols) Paris 1799-1825; C.L. Berthollet "Essai de statique chimique" Paris 1803; J.B.J. Fourier "Theorie analytique de la chauler", Paris 1822; D. Poisson "Theorie mathematique de la chaleur", Paris 1835; J.R. Mayer "Organische Bewegung in ihrem Zusammenhange mit dem Stffweschel" Heilbronn 1845

[137] R. Clausius "Mechanische Warmetheorie" Viewg Sohn, Braunschweig 1876

[138] I. Proks "Evaluation of the Knowledge of Phase Equilibria" first chapter in the book "Kinetic Phase Diagrams" (Z. Chvoj, J. Šesták, A. Tříska, edts.), Elsevier, Amsterdam 1991

[139] R.B. Lindsay "Energy: historical development of the concept" Dowden, Stroudburg 1975

[140] G. Truesdel "Tragicomedy of Classical Thermodynamics" 1971; "Tragicomical History of Thermodynamics: 1822-1854" Springer, Heidelberg 1980

[141] H. Hollinger, M. Zeusen "The Nature of Irreversibility" Reidel, Dorderecht 1985

[142] W.H. Cropper, Amer. J. Phys. 55 (1987) 120

[143] K.L. Chung and Z. Zhao "From Brownian Motion to Schrödinger Equation" Springer, Berlin 1995

[144] W. Nernst "Teoretische Chemie" 1909

[145] W. Lemery "Cours de chemie" 1675

[146] J.J. Berzelius "Lorbok i Kemien" (Textbook on Chemistry), Stockholm 1819 and "Lehrbuch der Chemie" in German transl., Dresden 1825-1831)

[147] D. Mendĕleev "Principles of Chemistry" (English translation edited by G. Kamnesky) Collier, New York 1905

[148] J.D. Van der Waals, P. Kohnstamm "Lehrbuch der Thermodynamic" Leipzig 1908

[149] G. Tammann "Lehrbuch der heterogenen Gleichgewichte" Vieweg, Braunschwein 1924

[150] G.N. Lewis, Proc. Amer. Acad. 43(1907)259; J. Amer. Chem. Soc. 35(1913)1 and G.N. Lewis, M. Randal "Thermodynamics and Free Energy of Chemical Substances" McGraw Hill, New York 1923

[151] J.W. Gibbs "On the Equilibrium of Heterogeneous Substances" Trans. Connect. Acad. 1875 p.108 and Collected works of J.W. Gibbs, Vol. I: "Thermodynamics" Longmans, New York 1928

[152] G.N. Lewis "A plan for stability process" in Econom. J. 35 (1925) 40

[153] L. Onsager "Reciprocal Relations in Irreversible Processes" Physical Review 37 (1931) 405

[154] H.P. Stapp "Mind, Matter and Quantum Mechanics" Springer Verlag, Berlin 1993

[155] A. Bartoli, Nuovo Cimento 15(1884)193

[156] B. Carazza, H. Kragh "A. Bartoli and the Problem of Radiant Heat: a review" Annals Sci. 46 (1989) 183

[157] J.J. Mareš, V. Špička, J. Krištofík, P. Hubík "Bartoli's Heat Engine Working with Zero-point Electromagnetic Radiation" see next, p. 411

[158] Proc. of 1st IC on Quantum limits to the Second law, San Diego '02, (D. P. Sheehan, ed.) AIP Proceedings, 643 (AIP, Melville, 2002)

[159] L. Boltzmann, Wied. Ann. d. Phys. u. Chem. 22(1884)31 & 291

[160] D.C. Cole, Phys Rev. A 42 (1990) 1847

[161] W. Nernst, Verhandl. d. Deutsch. Phys. Ges. 4(1916)83

[162] P. W. Milonni in "Quantum Vacuum" Academic Press Inc., p.239, Boston 1994

[163] G.J. Jordan, H. Fearn, P.W. Milonni "Some theoretical significance and implications of Casimir Effects" in Europ. J. Phys. (2005, in print)].

[164] S. Sounders, H.R. Brown "Casimir Effect in Critical Systems" World Scientific, Singapure 1994

[165] M. Sparnaay "Historical Background of the Casimir Effect" in "Physics in the Making" (A. Sarlemijn, M. Sparnaay, edts.), Elsevier, Amsterdam 1989

[166] P. Yam, Scientific American, Dec. 1997, p. 82

[167] J. Ambjorn, S. Wolfram "Properties of Vacuum: mechanical and thermodynamic views" Ann. Physics 147 (1983) 1

[168] S.K. Lamoreaux "Casimir Forces: a literature review" A. J. Phys. 67(1999)850

[169] H.G.B. Casimir, Proc. Kon. Ned. Akad. Wetenschap 51(1948)793 and H.B. Casimir, D. Polder, Phys. Rev. 73(1948)360

[170] F. London, Zeit. für Physik 63(1930)245]

[171] T.H. Boyer, Phys. Rev. 174(1968)1764

[172] E.M. Lifshitz "Theory of Molecular Attractive Forces between Solids" Soviet Phys. JETP 2(1956)73 and Zh. Eksper. Teoret. Fiz. 29(1955)94 (both in Russian)

[173] S.K. Lamoreaux, Am. J. Phys. 67(1999)850

[174] D.V. Fenby "Heat, its Measurement from Galileo to Lavoisier" Pure Appl. Chem. 59 (1987) 91

[175] P. Kelland "Theory of Heat", Cambridge 1837

[176] W. Thompson "On an Absolute Thermometric Scale: founded on the Carnot's theory of heat motive power" Phil. Mag. 33(1848)313

[177] S. Carnot "Reflexions sur la puissance motrice du feu" Paris 1824; reprinted in Ostwalds Klassiker, Engelmann, Leipzig 1909

426

[178] J. Dalton "New System of Chemical Philosophy" Ostwalds Klassiker, Engelmann, Leipzig 1902 and also Phil. Library, New York 1964
[179] D.S.L. Cardwell "From Watt to Clausius: the rise of thermodynamics" Cornel University Press, New York and Heinemann, London, 1971
[180] J.A. Ruffner "Interpretation of Genesis of Newton's Law of Cooling" Arch. Hist. Exact. Sci. 2(1963)138
[181] H.S. Carlaw, J.C. Jeager "Conduction in Solids", Cleradon, Oxford 1959
[182] H. Lienhard "Heat Transfer Textbook" Phlogiston, Cambridge 2002
[183] R. Černý, P. Rovaníková "Transport Processes" SponPress, London 2002
[184] D. Hardorp, U. Pinkall, Matematisch-Physikalishe Korespondence 281(2000)16
[185] Z. Kalva and J. Šesták "Transdiciplinary aspects of diffusion and magnetocaloric effect" J. Thermal Anal. Calor. 76(2004)1.
[186] E. Nelson "Derivation of the Schrödinger Equation from Newtonian Mechanics" Phys. Rev. 130 (1966) 150
[187] R.E. Collins "Quantum Mechanics as a Classical Diffusion Process" Found. .Phys. Lett. 5 (1992) 63
[188] M. Jammer "The Philosophy of Quantum Mechanics" Wiley, New York 1974
[189] D. Jou, J. Casas-Vazques and G. Lebon "Extended Irreversible Termodynamics", Springer, Berlin/Heidelberg 1993
[190] L. Hes, M. Araujo, V. Djulay, V., Textile Res. J. 66(1996)245
[191] L. Sertorio "Thermodynamics of Complex Systems" World Sci Publ., London 1991
[192] C.F. Bohren, B.A. Albrecht "Atmosphere Thermodynamics" Oxford Univ. Press, New York 1998
[193] T. Feldemann, R. Kosloff, Phys. Rev. E 68(2003)16101
[194] Proc. of 2nd IC on Frontiers of Mesoscopic and Quantum Thermodynamics, Prague 2004, (V. Špička, P. Hubik, edts.), Phys. E., December 2005 (Elsevier, in print)
[195] K. Meyer-Bjerrum "Die Entwicklung des Temperaturbegriffs im Laufe der Zeiten" Vieweg u. Sohn, Braunschweig, 1913
[196] P. Ehrenfest "Princip le Chateliéra – Brauna" Journ. d. russ. phys. Ges., 41(1909)347 (in Russian).
[197] J.J. Mareš, J. Šesták"An attempt to quantum thermal physics" J.Thermal Anal. Calor. 82/3(2005), in print
[198] C. Rumford "An inquiry concerning the source of heat which is excited by friction " Phil.Trans., 88(1789)80
[199] J. P. Joule "On the Mechanical Equivalent of Heat" Phil.Trans., 140(1850)61
[200] P.W. Bridgman "The Nature of Thermodynamics" Harvard University Press, Cambridge 1941
[201] H. L. Callendar "The Caloric Theory of Heat and Carnot's Principle" Proc. Phys. Soc. of London, 23(1911)153
[202] G. Job "Neudarstellung der Wärmelehre – Die Entropie als Wärme" Akad. Verlagsges., Frankfurt am Main 1972
[203] S. Carnot "Réflexions sur la puissance motrice du feu" Bachelier, Paris 1824
[204] C. Truesdell "Rational Thermodynamics" McGraw-Hill, New York, 1969
[205] A. Wehrl "General properties of entropy" Rev. Mod. Phys. 50 (1978) 221
[206] A. Sommerfeld "Das Planksche Wirkungsquantum und seine allgemeine Bedeutung für die Molekularphysik" Phys. Z., 24 (1911) 1062
[207] J.-M. Lévy-Leblond, F. Balibar "Quantics, Rudiments of Quantum Physics" North-Holland, Amsterdam 1990

[208] L. de la Peña, A. M. Cetto "The Quantum Dice – An Introduction to Stochastic Electrodynamics" Academic Publishers, Dordrecht, 1996 and P. W. Milloni "The Quantum Vacuum – An Introduction to Quantum Electrodynamics" Academic Press Inc., Boston, 1994

[209] J.D. Bernal, R.H. Fowller, J. Chem. Phys. 1(1933)515 or L. Pauling, J.Am.Chem. Soc. 57(1935)268

[210] M. Tribus "Information and Thermodynamics: bridging the two cultures", J. Non-equil. Thermodyn. 11(1986)247 and "Thermostatics and Thermodynamics" Van Nostrand, Princeton 1961

[211] P. Ehrenfest, Ann. der Phys. 36(1911)91.

[212] M. Planck "Vorlesungen über die Theorie der Wärmestrahlung" 5. Aufl., J.A. Barth, Leipzig, 1923

[213] T. H. Boyer, Phys. Rev. 182(1969)1374.

[214] E. Elizalde, A. Romeo, Am. J. Phys. 59(1991)711.

[215] L. Spruch, in "Long-Range Casimir Forces" (F. S. Levin, D. A. Micha, Eds.), Plenum Press, New York 1993, p.1

[216] S. K. Lamoreaux, Phys. Rev. Lett. 78(1997)5.

[217] L.D. Landau, E.M. Lifshitz "Elektrodinamika Sploshnykh Sred" (Electrodynamics of Contiuous Medium), Nauka, Moscow 1982 (in Russian).

[218] M. W. Zemansky, R. H. Dittman "Heat and Thermodynamics" 6th ed., McGraw-Hill, Inc., New York 1981

[219] J.J. Mareš, V. Špička, J. Krištofik, P. Hubík, J. Šesták "Filling of a cavity with zero-point electromagnetic radiation" Phys. E, December 2005, Elsevier in press

[220] L. Szilard, Z. Physik 53(1929)905

[221] R. Landauer, IBM J. Res. Dev. 5(1961)183

[222] P.N. Fahn, Found. Phys. 26(1996)71

[223] C.H. Bennett, Int. J. Theor. Phys. 21(1982 905 and Sci. Amer. 55(1987)106

[224] R. Feistel, W. Ebeling "Self-organization, Entropy and Development" Kluwer, Dordrecht 1989

[225] J.C. Iñigues "Revision of Clausius Work" Entropy (1999) p.111, 118 and 126; Acta Negentropica 1(2000)28

[226] C.E. Shannon, Bell System Tech. J. 27(1948)379 and 623

[227] J.F. Brillouin "Science and Information Theory" 1956

[228] K. Denbigh "How subjective is entropy", Chemistry in Britain 17(1981)168

[229] J.F. Bekenstein "Energy cost of information" Phys. Rev. Lett. 46(1981)623

[230] A.N. Kolmogorov, Problems of Information Transmission 1(1965)3

[231] J. Wicken "Generation of Complexity; a thermodynamic and information discussion" J. Theoret. Biology 77(1979)349; "Entropy and Information: suggestion for a common language" Phil. Sci. 54 (1987) 176

[232] J.G. Roederer "On the Concept of Information and its Role in Nature" Entropy 5(2003)1 and in W.H. Zurek (Rd.) "Complexity, Entropy and the Physics of Information" Addison-Wesley, New York 1990

[233] M. Burgin, Cybernetics 4(1990)21 and Entropy 5(2003)21

[234] A. Harkevich, Problems of Cybernetics 4(1960)2

[235] T. Stonier "Information and the Internal Structure of Universe" Springer, New York 1990

[236] J. Šesták, ICTAC News 31/2 (1998) 166 and J. Mining and Metallurgy (Bor, Yugoslavia) 35 (1999) 367

[237] N.I. Blumenfield "Problemy Biologicheskoj Fyziky" (Problems of Biological Physics), Nauka, Moscow 1977 (in Russian)

[238] E.H. Battley "On Entropy and Absorbed Thermal Energy in Biomass: a biologist's perspective" Thermochim. Acta 331 (1999) 1

[239] H.J. Morowitz, Bull. Math. Biophys. 17 (1955) and H.A. Johnson, Science 168 (1970) 1545

[240] P.T. Landsberg, Studium Generale 23 (1970) 1108

[241] S. Kullback "Information Theory and Statistics" Wiley, New York 1959

[242] J.D. Barrow, F.J. Tipler "Anthropic Cosmological Principle" Cleradon, Oxford 1996

[243] I.M. Havel, A. Markoš (edts) "Is there a purpose in nature", Vesmír, Prague 2002

[244] A. Markoš "Povstání živého tvaru" (Forwardness of living configuration), Vesmír, Praha 2002 and "Berušky, andělé a stroje" (Ladybugs, angels and machines) Dokořán, Praha 2004 (both in Czech)

[245] M.A. Nielsen, I.L. Chuang "Quantum Computation and Information" University Press, Cambridge 2000

[246] T. P. Spiller, Materials Today, January 2003, p. 30

[247] P. Bennet, Phys. Rev. Lett. 1985, p.70

[248] A. Einstein. B. Podolsky, N. Rosen, Phys. Z. phys. Chem. 47(1935) 777

[249] R.E. Liesegang, Naturwiss. Wochenschrift 11(1896)353 and Photogr. Arch. 37(1896)321.

[250] W.M. Ord "On the Influence of Colloids upon Crystalline Form and Cohesion" Stanford, London 1879

[251] F.F. Runge, R.E. Liesegang, B.P. Belousov, A. M. Zhabotinsky *Selbsorganisation chemischer Strukturen"* Akademische Verlag, Leipzig, 1987

[252] J. Stávek, M. Šípek, J. Šesták "Application of the Principle of Least Action to some Self-organized Chemical Reactions" Thermochim. Acta 388(2002)440

[253] A. Lima-de-Faria "Evolution without Selection: form and functions of autoevolution" Elsevier, Amsterdam 1988

[254] S. Leduc "Mechanism of Life", Rebman Ltd., London, 1911.

[255] B. Chance, A.K. Chost, E.K. Pye and B. Hess "Biological and Biochemical Oscillations", Academic Press, New York 1973.

[256] P.L.M. Maupertuis "Oevres de Maupertuis" Alyon, Paris 1768, Vol. IV.

[257] J. Stávek and M. Šípek, Cryst. Res. Technol. 30(1995)1033.

[258] J.J. Mareš, J. Šesták, J. Stávek, H. Ševčíková, P. Hubík "Do the periodical chemical reactions reveal Fürth's quantum diffusion limit" Phys. E., December 2005 (Elsevier, in print)

[259] J. Stávek "Diffusion action of chemical waves" Apeiron 10(2003)183, "Diffusion of individual Brownian particles through Young's double-slits" Apeiron 11(2004)175 and "Diffusion of self-organized Brownian particles in the Michelson-Morley experiment" Apeiron 11(2004)373

[260] E. Brunner, Z. phys. Chem. 47(1904)56 and W. Nernst, Z. phys. Chem. 47(1904)52

[261] P. Michaleff, W. Nikoforoff, F. M. Schemjakin, Kolloid Z. 66(1934)197.

[262] B.P. Bělousov, Collection of Short Papers on Radiation Medicine; Med. Publ., Moscow. 1959; p. 145; and B.P. Belousov in "Oscillations and Traveling Waves in Chemical Systems" by R.J. Field and M. Burger (edts), John Wiley & Sons: New York, 1985; p. 605

[263] A.N. Zaikin and A.M. Zhabotinsky, Nature 225(1970)535.

[264] J. Tyson "The Bělousov-Zhabotinsky Reactions" Springer Verlag, Heidelberg 1976.

[265] W. Schaabs, Kolloid Zeitschrift. 137(1954)121

[266] J. A. Christiansen, I. Wulff, Z. physikal. Chem. B26(1934)187.

[267a] M. v. Smoluchowski "Abhandlungen über die Brownsche Bewegung und verwandte Erscheinungen" (Herausgegeben von R. Fürth), Akademische Verlagsgesellschaft, Leipzig, 1923

[267b] A. Einstein "Investigations on the Theory of the Brownian Movement" (R. Fürth, ed) first published 1926, reediton by Dover Publications, New York 1956.

[267c] P. Lévy "Processus Stochastiques at Mouvement Brownian" Gauthier-Villars, Paris 1948

[268] P. Frank, Physik. Zeitschr. 19(1918)516

[269] R. Fürth, Z. für Physik, 81(1933)143.

[270] E. Nelson, Phys. Rev. 150(1966)1079

[271] L. F. Abbot, M. B. Wise, Am. J. Phys. 49(1981)37.

272] H. Cullen "Thermodynamics: an introduction to thermostatics" Wiley, New York 1960 (re-edition 1985))

[273] H. Cullen, Foundation of Physics 4 (1974 423.

[274] L. Tisza "Generalized Thermodynamics" MIT Press, Cambridge 1966

[275] J. Kestin "Course in Thermodynamics" 1966

[276] E.A. Guggenheim "Thermodynamics" North-Holland, Amsterdam 1959

[276] C. Truesdell, S. Bharatha "Concepts and Logic of Classical Thermodynamics as a Theory of Heat Engines" 1988

[277] J. Šesták "Thermodynamic basis for the theoretical description and correct interpretation of thermoanalytical experiments" Thermochim. Acta 28(1979)197

[278] J. Šesták, Z. Chvoj "Irreversible thermodynamics and true thermal dynamics in view of generalized solid-state reaction kinetics" Thermochim. Acta 388(2002)427

[279] S.R. DeGroot, P. Mazur "Non-equilibrium Thermodynamics" Wiley, New York 1962;

[280] I. Prigogine "Introduction to the Thermodynamics of Irreversible Processes" Wiley/Interscience, New York 1967

[281] D.K. Kondepudi, I. Prigogine "Modern Thermodynamics: from Heat Engines to Dissipative Processes" Wiley, London 1998

[282] G. Grioli "Thermodynamics and Constitutive Equations" 1985

[283] R.A. Swalin "Thermodynamics of Solids" Wiley, New York 1972

[284] J.L. Eriksen "Introduction to the Thermodynamics of Solids" Springer, New York 1998

[285] O. Kubaschevsky, C.B. Alcock, P.J. Spenser "Materials Thermochemistry" Pergamon, Oxford 1993

[286] P.Holba, Czech. J. Phys. 42(1992)549

[287] F. D. Richardson, J.H.E. Jeffers, J. Iron Steel Inst. 160(1948)261

[288] J. Šesták, N. Koga, Thermochim. Acta 203(1992)321

[289] W.B White, S.M Johnson, G.B. Dantzig, J. Chem. Phys. 28(1958)751

[290] W.R. Smith, R.W. Missen, Can. J. Chem. Eng. 46(1968)269

[291] W.R. Smith, R.W. Missen "Chemical Reaction Equilibrium Analysis: Theory and Algorithm" Wiley, New York 1982

[292] F. Van Zeggeren, S.H. Storey "The Computation of Chemical Equilibria" Wiley, New York 1982

[293] G. Eriksson, W.T. Thompson, CALPHAD 13(1989)389

[294] P. Voňka, J. Leitner, CALPHAD 19(1995)25 and 305

[295] J. Leitner, P. Chuchvalec, P. Voňka "MSE-THERMO: Integrated computer system for application of chemical thermodynamics in material science, and MSE-DATA: database for pure substances" High Temperature Mater. Sci. 34 (1995) 265

[296] *Czech sources:* Č. Strouhal "Thermika" (Thermics), JČMF, Praha 1908; F. Záviška "Termodynamika" (Thernodynamics), JČMF, Praha 1943 and 1954; M. Brdička "Mechanika

430

kontinua" (Continuum mechanics), Academia, Praha 1959; V. Šatava "Úvod do fyzikální chemie silikátů" (Introduction to physical chemistry of silicates), SNTL, Praha 1960; E. Hála "Úvod do chemické termodynamiky" (Introduction to Chemical Thermodynamics), Academia, Praha 1975; J. Kvasnica "Termodynamika" (Thermodynamics), SNTL, Praha 1965; A. Regner "Electrochemie" (Electrothermochemistry), Academia, Praha 1967; I. Samohýl "Termodynamika irreversibilních procesů v kapalných směsích" (Thermodynamics of irreversible processes in liquid mixtures) Academia, Praha 1985; F. Maršík "Thermodynamika kontinua" (Continuum Thermodynamics), Academia, Praha 1999 (all in Czech): ; M. Šilhavý "Mechanics and Thermodynamics of Continuous Media" Springer, Berlin 1997

[297] A. Reisman "Phase Equilibria" Academic, New York 1970

[298] M. Hillert "Phase Equilibria: phase diagrams and phase transformations", University Press, Cambridge 1998

[299] N. Saunders, A.P. Miadownik "CALPHAD Calculation of Phase Diagrams", Pergamon/Elsevier, New York 1998

[300] M. Margules, Sitzber. Akad. Wiss. Wien 104(1895)1243

[301] J.D. Esdaile, Met. Trans. 2(1971)2277

[302] J.H. Hildebrand, R.L. Scott "Regular Solutions" Prentice-Hall, New Jersey 1962

[303] M. Tyemkin, Acta Physikokhimika SSSR 20(1945)411 (in Russian)

[304] F.R.N. Nabarro, Proc. Roy.Soc. London 52(1940)90

[305] F.F. Abraham "Homogeneous Nucleation Theory" Wiley, New York 1974,

[306] R.S. Elliot, "Eutectic Solidification Processing: crystalline and glassy alloys" Butterworth, London 1989

[307] C. Eckart : Thermodynamics of Irreversible Processes" Phys. Rev. 58(1940)267

[308] J. Maixner "Thermodynamik der irreversiblen Processen" Z. phys. Chem. 538(1943)235

[309] B.D. Coleman, W. Noll "Thermodynamics of Elastic Materials with Heat Conduction" Arch. Rat. Mech. Anal. 17(1964)1 and B.D. Coleman, D.R. Owen "Mathematical Foundations for Thermodynamics" Arch. Rat. Mech. Anal. 54(1974)1

[310] J. Kratochvíl, M. Šilhavý "On Thermodynamics of Non-equilibrium Processes", J. Non-equilib. Thermodyn. 7(1982)339 and Czech. J. Phys. A31(1981)97

[311] J. Šesták "Thermodynamics and Society", key lecture at the IC of Intelligent Processing and Manufacturing of Materials, Honolulu, Hawaii 2001 (CD proceedings of 2nd IPMM)

[312] E. Burmeister, A.L. Dobell "Mathematical Theories of Economic Growth" Macmilln, London 1970

[313] V. Pareto "Cours d'Economie Politique" Rouge, Lausanne 1897

[314] M. Levy, S. Solomon "Power Laws are Logarithmic Boltzmann Laws" Int. J. Mod. Phys. C7(1996)595,

[315] W.J. Reed, B.D. Hughes, Phys. Rev. E66(2002)067103

[316] F. Slanina, Phys. Rev E69(2004)046102

[317] S.Gallucio, G. Caldarelli, M. Marsili, Y.C. Zhang, Physica 245(1997)423

[318] M.H.R. Stanley, L.A.N. Amaral, H.E. Stanley, Nature 379(1996)804

[319] B.B. Mandlbrot "Fractal and Scaling in Finance" Freeman, San Francisco 1982

[320] H.E. Stanley "Introduction to Phase Transitions and Critical Phenomena" Oxford Univ. Press, Oxford 1971

[321] A. Einstein, Ann. Physik 17(1905)549

[322] B.B. Mandelbrot, J. Business 36(1963) 394

[323] P. Levy "Calcul des Probabilites" Gouthier-Villars, Paris 1925

[324] R.N. Mantegna, Physica A 1791991)232

[325] R.U. Ayers "Eco-Thermodynamics: economics and the second law" Ecological Econom. 26 (1998) 189

[326] J. Stepanic, I. Bertovic, J. Kasac "Mediated Character of Economic Interactions" Entropy 5(2003)61

[327] E. Majorana, Scientia 36(1942)58

[328] E.W. Montrol, W.W. Badger "Introduction to Quantitative Aspects of Social Phenomena" Gordon and Breach, New York 1974

[329] N. Georgescu-Roegen "The Entropy Law and the Economic Progress" Cambridge, London 1971

[330] R.N. Mantegna, H.E. Stanley "Introduction to Econophysics: correlation and complexity in finance" Cambridge Univ. Press, Cambridge 2000

[331] A Canetti "Crowds and Power" New York 1981 (In German, Zurich 1983)

[332] R. Gibbsen "Game Theory for Applied Economics" Princeton Univ. Press, New Jersey 1992

[333] J.W. Weibull "Evolutionary Game Theory" MIT Press, Cambridge 1995

[334] D. Chalet, Y.C. Zhang, Physica A 246(1997)407 and A256(1998)514

[335] T. Svobodny "Mathematical Modelling for Industry, Engineering and Economy", Prentice Hall, New Jersey 1998

[336] J. Mimkes "Concepts of Thermodynamics in Economic Growth" in Proc. WEHIA'04 (Springer, Berlin 2004) and Economic Working Paper Archive, WUSTL 2004

[337] W. Kaplan "Advanced Calculus" Addison-Wesley, New York 2003

[338] H Uzawa, Rev.Econ.Studies 29(1961)40 and 30(1963)105

[339] J. Mimkes "Binary Alloys as a Model for the Multicultural Society" J. Thermal Anal. 43 (1995) 521

[340] J. Šesták, ICTAC News 31/2(1998)166 and J. Mining and Metallurgy (Bor, Yugoslavia) 35(1999)367

[341] D. Helbing "Wonderful World of Active Many-particle System" Adv. Solid State Phys. 41(2001)357

[342] J. Mimkes "Society as Many Particle System" J. Thermal Anal. Calor. 60(2000)1055

[343] R. Axerold "Evolution of Cooperation" Penguin, London 1990,

[344] P. Coveney, R. Highedield "Frontiers of Complexity" Fawcett Columbine, NewYork 1995

[345] R. Dawkins "The Selfish Gene" Oxford Press 1989

[346] J. von Neumann, J. Morgenstern "Theory of Games and Economical Behavior" 1944

[347] F. Schweitzer (ed) "Modeling Complexity in Economic and Societal Systems" World Sci., Singapore 2002

[348] K. Sznajd, J. Sznajd, Int. J. Modern Phys. C11(2000)1157

[349] F. Slanina, H. Lavička, Europ. Phys. J B35(2003)279

[350] M.V. Volkenstein "Biophysics" Nauka, Moscow 1988 (in Russian)

[351] S.E. Jørgensen "Integration of Ecosystems Theories: a pattern" Kluwer, Dordrecht 1997

[352] W. Ebeling, A. Engel, R. Feistel "Physik der Evolutionsprozesse" Akademia, Berlin 1990 (in German)

[353] H.T. Odum, Amer. Sci.43(1955)331

[354] H.T. Odum "Environment, Power and Society", Wiley, New York 1971 and "Environmental Accounting: energy and decision making" Wiley, New York 1996

[355] M. Straskraba, S.E. Jørgensen, S.E. Patten, Ecolog.Model. 96(1997)221, 117(1999)41 and 126(2000)249

[356] S.E. Jørgensen, Ecolog. Model. 41(1988)117 and p. 628 in "Fundamentals of Ecological Modeling", Vol.19: "Environmental Models" Elsevier, Amsterdam 1994

432

[357] A.J. Lotka "Elements of Physical Biology" Willimas&Wilkins, Baltimore 1925 and "Elements of Mathematical Biology" Dover, New York 1956
[358] H.J. Morowitz "Entropy for biologists: an introduction to thermodynamics" Academic, New York 1970 and "Foundation of Bioenergetics" Academic, New York 1978
[359] W. Ostwald "Gedanken der Biosphare" Leipzich 1931 (reprinted 1978)
[360] S.E. Jørgensen, Sci. World 1(2001)71 and S.E. Jørgensen, M. Straskraba, Ecol. Model. 153(2001)269
[361] I. Aoki "Entropy production in living systems: from organism to ecosystems" Thermochim. Acta 250(1995)359
[362] B. Hannon "Total energy cost in ecosystems" J. Theoret. Biol. 80(1979)271
[363] S.E. Jørgensen "Parameter estimation and calibration by use of exergy" Ecol. Model. 146(2001)299
[364] Yu.M. Svirezhev "Exergy of the Biosphere" Ecol. Model. 96(1997)7
[365] N.B. Rolov "Rozširenie fazovije prevraščenija" (Broadened Phase Transformations), Zinantne, Riga 1972 (in Russian)
[366] L.D. Landou, Zhurn. experim. tekh. Fiziky 7(1937)7 and 627 (in Russian)
[367] A. B. Pippard "Elements of Classical Thermodynamics" Univ. Press, Cambridge 1966
[368] J. Šesták, Thermochim.Acta 95(1985)459
[369] I. Gutzow, Wiss. Zeit. Friedrich-Schiller Univ. Jena, Math. Natur. 32(1983)363
[370] P. Holba J. Šesták "Kinetics with regard to the equilibrium of processes studied by non-isothermal techniques" Zeit. phys. Chemie NF 80(1972)1
[371] P. Holba, J. Šesták in "TERMANAL'73", Proc. of Slovak Conf. on Thermal Anal., SVST, Bratislava 1973, p.P-1 (in Czech) and P. Holba, in "Thermal Analysis" Proc. of the 4th ICTA, Vol. 1, Akademiai Kiado, Budapest 1975, p. 60.
[372] J. Šesták, Z. Chvoj, F. Fendrich, J. Thermal Anal, 32(1987)1645 and 43(1995)439
[373] D.R. Uhlmann, J. Non-cryst. Sol. 7 (1972) 337, 25 (1977) 42 and 38 (1980) 693
[374] D.R. McFarlane, J. Non-cryst. Sol. 53 (1982) 61
[375] E.D. Zannoto, Thermochim. Acta 280 (1996) 73
[376] J. Colmenero, J.M. Barrandian, J. Non-cryst. Sol. 30 (1979) 263 and Thermochim. Acta 35 (1980) 381
[377] G.B. Boswell, Mat. Res. Bull. 12(1977)331
[378] Z. Strnad "Glass-ceramic Materials" Elsevier, Amsterdam 1987
[379] M.C. Weinberg, D.R. Uhlmann, E.D. Zanotto, J. Amer. Cer. Soc. 72(1989)2054
[380] M.C. Weinberg, J. Amer. Cer. Soc. 74(1991)1905; Thermochim. Acta 194(1992)93 and J. Mining Metal. (Bor, Yugoslavia) 35(1999)197
[381] Z. Kožíšek, Cryst. Res. Tech. 23(1988)1315
[382] M.T. Clavaguera-Mora, M.D. Baro, S. Surinach, N. Clavaguera, J. Mat. Sci. 5(1990)11
[383] N. Koga, J. Šesták "Thermoananlytical kinetics and physico-geometry of the nonisothermal crystallization of glasses" Bol. Soc. Esp. Ceram. Vidr. 31(1992)185
[384] N. Koga, J. Šesták, J. Thermal Anal. Calor. 56(1999)755 and "Nucleation and growth in lithium diborate glass by thermal analysis" J. Amer. Cer. Soc. 83(2000)1753
[385] R.A. Grange, J.M. Kiefer, Trans. ASM 29(1941)85
[386] I. Gutzow, J. Schmelzer "The Vitreous State: thermodynamics, structure, rheology and crystallography" Springer, Berlin 1995
[387] E. Donth "The Glass Transition: relaxation dynamics in liquids and disordered materials" Springer, Berlin 2001
[388] B. Hlaváček, J. Šesták "Thermal Science of Glassy State: microdynamics and microstructure of isotropic amorphous materials" (book in the course of preparation, 2007) and "Mutual interdependence of partition functions in the vicinity of Tg transition" J. Thermal

Anal. Calor. 67(2002)239 and "Forms of vibration and structural changes in liquid and viscous state" J. Thermal Anal. Calor. 80(2005)271
[389] R.O. Davis, G.O. Jones, "Advances in Physics" (Mott N.F. ed.) 2(1953)370 and Proc.Roy.Soc. 217(1953)26
[390] I. Priggogine, R.Defay "Chemical Thermodynamics" (chapter 9), Longmans Green, New York 1954
[391] A.Q. Tool, J. Amer. Cer.Soc. 14(1931)276 and 29(1946)240; J. Res. 34(1945)199
[392] T.C. Moynihan, K.P. Gupta, J. Non-cryst. Sol. 29(1978)143
[393] G.S. Fulcher, J. Amer. Cer. Soc. 8(1925)1644, H. Vogel, Physik Zeit. 22(1921)645
[394] J.M. Hutchinson, Prog. Polym. Sci. 20(1995)703
[395] J. Málek, J. Non-cryst. Sol. 307/310(2002)778
[396] M. Liška, I. Štubňa, J. Antalík, P. Pericha, M. Klyuev, Ceramics (Prague) 40(1996)15 and 85; J. Greguš, M. Chromčíková, M. Liška, L. Vozír, Glasstech. Ber. Sci. Technol. 77C(2004)372
[397] J. Šesták "Use of phenomenological kinetics and the enthalpy-temperature diagrams (and its derivative DTA) for a better understanding of transition processes in glasses" Thermochim. Acta 280/281(1996)175
[398] J. Šesták, Thermochim. Acta 110(1987)427; J. Thermal Anal. 33(1988)75
[399] A. Hrubý, Czech J. Phys. B22(1972)1187 and B23(1973)1623; D.D. Thoronburg, Mat. Res. Bull. 9(1974)1481
[400] O.S. Narayanaswami, J.Amer.Cer.Soc. 54(1971)491; A.C. Angel, K.L. Ngai, G.B. McKenna, P.F..McMillan, S.W. Martin, J.Appl.Phys. 80(2000)3113
[401] J. Šesták, Z. Strnad in "Reactivity of Solids" (Proc. 8. ICRS), Elsevier, Amsterdam 1977, p. 410 and in Proc. of the XI Congress on Glass (J.Goetz, ed.), DT ČVTS, Prague 1977, Vol. II., p. 249
[402] J. Šesták, key lecture "Integration of nucleation-growth equation when considering non-isothermal regime and shared phase separation" at the. 2nd ESTAC (Europ. Symp. on Thermal Analysis) in Proc. "Thermal Analysis"(D. Dolimore, ed) Heyden. London 1981, p. 115
[403] J. Šesták, key lecture at the 3rd ICTA in Davos, in Proc. "Thermal Analysis" (H.G. Wiedemann, ed.) Vol. 1, p.3, Birghauser, Basel 1972
[404] J.R. MacCallum, J. Tanner, Nature 225(1970)1127
[405] J. Šesták, J. Kratochvil "The role of constitutive equations in chemical kinetics" Thermochim. Acta 7(1973)330 and J. Thermal Anal. 5(1973)193
[406] J. Blažejowski, Thermochim. Acta 76(1984)359
[407] A. Mianowski, J. Thermal Anal. Calor. 63(2001)765
[408] J. Šesták, J. Thermal Anal. 16(1979)503
[409] M.E. Brown, J. Thermal Anal. 49(1997)17
[410] A.K. Galwey, M.E. Brown, Proc. R. Soc. London 450(1995)501 and Thermochim. Acta 300(1997)93
[411] E.T. Whittaker, G. Robinson "Chemical Separations and Measurements" Sounders, Philadelphia 1974
[412] S. Vyazovkin, New J. Chem. 24(2000)913
[413] D.A. Young "Decomposition of Solids", Pergamon, Oxford 1966,
[414] S.F. Hulbert, "Models for solid-state reactions in powdered compacts: a review" J. Brit. Ceram. Soc. 6(1969)11
[415] W. Jander, Z.anorg. Chem. 163(1927)1
[416] C. Kroger, G. Ziegler, Glasstech. Ber. 26(1953)346 and 27(1954)199
[417] V.F. Zhuravlev, I.G. Lesokhin, J.Appl. Chem. USSR 21(1948)887 (English transl.)
[418] A.M. Ginstling, B.I.Brounshtein, USSR 23(1950)1327 (English transl.)

[419] R.E. Carter, J. Chem. Phys. 34(1961)2010

[420] W. Komatsu, in "Reactivity of Solids" (G.M. Schwab, ed.) Elsevier, New York 1050, p. 182

[421] M.E. Fine "Phase Transformation in Condensed Systems" McMillan, New York 1964

[422] J. Šesták, J. Málek "Diagnostic limits of phenomenological models of heterogeneous reactions and thermal analysis kinetics" Solid State Ionics, 63/65(1993)254

[423] N. Koga, J. Málek, J. Šesták, H. Tanaka, Netsu Sokutei (Jap. J. Thermal Anal. Calor.) 20(1993)210 (in English)

[424] N. Koga, H. Tanaka, J. Thermal Anal. 41(1994)455 and Thermochim. Acta 383(2002)41

[425] A.K. Galwey, M.E. Brown "Thermal Decomposition of Ionic Solids" Elsevier, Amsterdam 1999 and M.E. Brown, D. Dollimore, A.K. Galway "Reactions in the Solid-State" (Comprehensive Chemical Kinetics-Vol.22), Elsevier, 1980

[426] V. Mamleev, S. Bourbigot, M. LeBras, S. Duquesne, J. Šesták "Modelling of nonisothermal kinetics in thermogravimetry" Phys.Chem.Chem.Phys. 2(2000)4708 and 4796

[427] W.A. Johnson, R.F. Mehl, Trans. AIME 135(1939)416

[428] M. Avrami, J.Chem.Phys. 7(1939)1103; 8(1940)212; 9(1941)177

[429] B.V. Yerofyeyev, Dokl.Akad.Nauk USSR 52(1946)511

[430] A.N. Kolmogorov, Izv.Akad.Nauk USSR (1937)355

[431] P.W.M. Jacobs, F.C. Tompkins in "Solid State Chemistry" (W.E.Garner, ed.), Butterworth, London 1953, p. 184

[432] K.L.Z. Mampel, Phys.Chem. A187(1940)235

[433] J. Šesták, Phys.Chem.Glasses 15(1974)567

[434] J. Šesták, P. Holba, J. Kratochvil, chapter in the book "Heterogeneous Chemical Reactions and Reaction Capability" (M.M. Pavlyuchenko, ed), Nauka i Technika, Minsk 1975, p. 519 (in Russian)

[435] S. Vyazovkin, J. Comput. Chem. 18(1997) and 22(2001)178

[436] M. C. Brown, J Thermal Anal, 49 (1997) 17

[398] 74] F. Maršík : "Termodynamika kontinua" (Continuum thermodynamics), Academia, Praha 1999 (in Czech); 84] M. Šilhavý, "Mechanics and Thermodynamics of Continuous Media", Springer, Berlin 1997

[437] S. Vyazovkin, N. Koga, J. Šesták (edts.) "Kinetic Evaluation of Thermal Analysis Data" (book in the course of preparation, Kluwer, Dordrecht 2006 or 2007

[438] R.C.O. Sebastiao, J.P. Braga, M.I. Yoshida, Thermochim. Acta 412(2004)105 or G.R. Heal, Thermochim. Acta 430(2005)15 and 23

[439] H.L. Anderson, R.K. Boyd, Can.J.Chem. 49(1971)1001 or J. Batailie, G.B. Edelen, J. Kestin, J.Non-equilib.Thermodyn. 3(1978)153

[440] H.E. Stanly, N. Ostrowski (edts), "Growth and Forms" Mortimes Nijhoff, Dordrecht 1986

[441] D.T. J. Hurle (edt), "Handbook of Crystal Growth", Vol.1- Fundamentals, Vol-1b "Transport and Stability" North-Holland, Amsterdam 1993 (Dendrites – chap. 14, p. 899 and chap.15, p.1075)

[442] E. Glicksman, Mater. Sci. Eng, 65(1984)45 and 57

[443] V. Jesenák, key lecture at the. 8th Czechoslovak Conference on Thermal Analysis, Proc. TERMANAL´82, by SVST Bratislava 1982, p.13 (in Slovak)

[444] I. Avramov, T.Hoche, C. Russel, J. Chem. Phys. 110(1999)8676.

[445] P. Gray and S. Scot "Chemical Oscillations and Instabilities", Oxford Press, Oxford 1990.

[446] C. Beck, F. Schlögl "Thermodynamics of Chaotic Systems" Cambridge Univ. Press, Cambridge 1993

[447] Č. Bárta, F. Fendrych, A. Tříska "Methods of the Characterization of Phases" in Z. Chvoj, J. Šesták, A. Tříska (edts), "Kinetic Phase Diagrams: non-equilibrium phase transitions", Elsevier, Amsterdam 1991, p. 519

[448] A. Tockstein, L. Treindl "Chemické oscilace" (Chemical Oscillations) SNTL, Praha 1986 and in Chem. Listy (Prague) 81(1987)561 (both in Czech)

[449] W.D. Kingery, M. Berg, J. Appl. Phys. 26(1955)1206

[450] B.Wong, J.A. Pask, J.Amer.Cer.Soc. 62(1979)138

[451] R.L. Cable, J.E. Burke, Progres in Ceramic Science Vol. 3., Part 4, Pergamon, New York 1963

[452] J.E. Burke, D. Turnbull, Prog. Metal. Phys. 3(1952)220

[453] G.W. Greenwood, Acta Metall. 4(1956)243

[454] H.V. Atkinson "Theories of normal grain growth in single phase systems; a review" Acta Metall. 36(1988)469

[455] C.S. Smith, Trans.Am.Soc.Metals 45(1953)533 and Metal Rev. 9(1964)1

[456] O. Pfeifer, Chimia 39 (1985) 120

[457] D. Avnir "Fractal Approach to Heterogeneous Chemistry" Wiley, New York 1989 and Nature 308 (1984) 261

[458] J. Stávek, J. Ulrich, Cryst. Res. Technol. 29(1994)763) and J. Cryst. Growth 166(1996)1053

[459] A. Bonde, S. Havlin "Fractals and Disordered Systems" Springer, Berlin 1991 and/or H.O. Peitgen, H. Jurgen, D. Saupe "Chaos and Fractals: new frontiers of science" Springer, New York 1992

[460] B.B. Mandelbrot "Gaussian Self-similarity, Fractals, Globallity and 1/f Noise" Springer, New York 2002 and/or K. Falcone "Fractal Geometry" Wiley, Chichester 2003

[461] D. Staufer "Introduction to Percolation Theory" Taylor & Francis, London 1985

[462] S. Alexander, R. Orbach, J. Phys. Lett 43(1982)L625

[463] D.J. Whitehouse, Wear 249(2001)345

[464] M. Holeček, J Thermal Anal. Cal. 60 (2000) 1093 and Physica A 177(1992)236

[465] W. Sierpinski, C.R. Acad. (Paris) 160(1915)302

[466] C.F. Wenzel, Lehre Verwanddtschaft Koerper 8, Dresden 1777

[477] R. Kopelman, Science 241(1988)1620

[478] R. Kopelman, S.J. Parus in "Fractal Aspects of Materials" (D.W. Schaefer, B.B. Mandelbrot, eds.), Material Res. Soc., Boston 1986; J. Chem Phys. 92(1988)1538

[479] R. Kopelman, Phys.Rev.Lett. 56(1986)1742 and J. Stat. Phys. 185(1986)

[480] J. Šesták, G. Berggrenn "Study of the kinetics of the mechanism of solid-state reactions at increasing temperature" Thermochim. Acta. 3(1971)1

[481] M. Ochiai, R. Ozao "Fractal reactions in solids" Bull. Ceram. Soc. Jpn. 26(1991)1181 and J. Thermal Anal. 38(1992)1901

[482] R. Ozao, M. Ochiai, Thermochim. Acta 198(1992)279 and 288

[483] F. Hausdorff, Math. Ann 79(1918)157

[484] P. Beckmann "History of Pi" Golem, Boulder 1971

[485] H. von Koch, Acta Mathematica 30(1906)145

[486] L.F. Abbot, M.B. Wise, Amer. J. Phys. 49(1981)37

[487] E. Nelson, Phys. Rev. 150(1966)1079

[488] G. Peano, Mathemat.Anal. 36(1980)157 and D. Hilbert, Mathemat. Anal. 38(1981)459

[489] B.B. Madeltbrot, Science 155(1967)636

[490] M.J. Feigenbaum, J. Stat. Phys. 19(1978)25

436

[491] O.E. Rössler, Phys.lett. 57A(1976)397
[492] E.N. Lorenz, J. Atm. Sci. 20 (1963) 268
[493] E.G. Prout, F.C. Tompkins, Trans. Farad. Soc. 40(1944)488
[494] J. Maciejewski, J. Thermal Anal. 33(1988)51 and 38(1992)51; Thermochim. Acta 355(2000)145
[495] J. Šesták, J. Thermal Anal. 16(1979)503; Thermochim. Acta 110(1987)87
[496] S. Vyazovkin, Int. Rew. Phys. Chem. 17(1998)407 and 19(2000)45
[497] J. Málek, T. Mitsuhashi, J.M. Criado "Kinetic analysis of solid-state processes" J. Mater. Res. 16(2001)1662
[498] J.M. Criado, J. Morales, Thermochim. Acta 16(1976)382 and 19(1976)305; J. Thermal Anal. 53(1998)389
[499] H.L. Friedman, J.Polym.Sci. C6(1964)183
[500] E.S. Freeman, B. Carroll, J. Phys. Chem. 62(1958)394
[501] N. Koga, J. Šesták, J. Málek, Thermochim. Acta 188(1991)333
[502] J. Málek, J.M. Criado, Thermochim. Acta 236(1994)187
[503] T. Ozawa, Bull. Chem. Soc. Jpn. 38(1965)1881
[504] J.H. Flynn, L.A. Wall, J. Polym. Sci. B4(1966)323
[505] H.E. Kissinger, J. Nat. Res. Bur. Stand. 579(1956)217 and Anal. Chem. 29(1959)1702
[506] C.D. Doyle, J. Appl. Polym. Sci. 5(1961)285; Nature 207(1965)290
[507] *Early reviews:* J.H. Flynn, L.A. Wall, J. Res. Nat. Bur. Stand. 70A (1966)487; J. Šesták, Silikáty (Prague) 11(1967)153; J.G. Murgulescu, E. Segal, St. Cerc. Chim. Tom. (Bucharest) 15(1967)261
[508] P.D. Garn "Thermoanalytical Methods of Investigation" Academic Press, New York 1964
[509] E. Koch "Non-isothermal Reaction Analysis" Academic Press, New York 1977
[510] E. Segal, D. Fatu "Introduction to Nonisothermal Kinetics" , Publ. House of the Romanian Acad. Sci., Bucharest 1983 (in Romanian)
[511] J. Vachuška, M. Vobořil, Thermochim. Acta 2(1971)315
[512] J. Šesták, Thermochim. Acta 3(1971)150
[513] D.W. Van Krevelen, C. Van Heerden, F.J. Hutjens, Fuel 30(1951)253
[514] H.H. Horowitz, G. Metzger, Anal. Chem. 35(1963)1464
[515] A.W. Coats, J.P. Redfern, Nature 20(1964)88,
[516] J.R. MacCallum, J. Tanner, Europ. Polym. J. 6(1970)1033
[517] V. Šatava, Thermochim. Acta 2(1971)423
[518] K.J. Leider, J. Chem.Educ. 49(1972)343 a 61(1984)494
[519] J. Šesták, Thermochim. Acta 110(1987)427
[520] A. Broido A, F.A. Williams, Thermochim. Acta 6(1970)1033
[521] G.O. Piloyan, I.O. Ryabchikov, O.S. Novikova, Nature 3067(19966)1229 ; G.O. Piloyan «Vyedyenie v tyeoryu termicheskogo analysa» (Introduction to the theory of thermal analysis), Nauka, Moscow 1969 (in Russian)
[522] E.A. Marseglia, J. Non-cryst. Solids 41(1980)31 ; H. Yinnon, D. Uhlman, J. Non-cryst. Solids 54(1987)113
[523] J. Šesták, J Thermal Anal. 36(1990)69
[524] N. Koga, J. Málek, Thermochim. Acta 282/283(1996)69
[525] J. Austin, R.L. Rickett, Trans AIME 135(1939)396
[526] M. Maciejewski "Dysocjacja Termiczna Cial Stalych" Publ. Tech. University in Warsaw 1988 (in Polish); J Thermal Anal. 33(1987)1269 and Thermochim. Acta 355(2000)145
[527] R. Ozao, M. Ochiai, J.Ceram.Soc.Jpn. 101(1993)263

[528] J. Pysiak, B. Pacewska, Thermochim. Acta 200(1992)205, J. Thermal Anal. 29(1984)879 and 43(1995)9

[529] J. Kemeny, J. Šesták "Comparison of crystallization kinetic theories derived by isothermal and nonisothermal methods" Thermochim. Acta 110(1987)113

[530] J. Málek, Thermochim. Acta 267(1995)61

[531] P. Šimon, J. Thermal Anal. Calor. 76(2004)123, 79(2005)703 and 81(2005)

[532] S. Vyazovkin, C.A. Weight, J. Phys. Chem. A101(1997)5653 and 8279

[533] R. Serra, J. Sempre, R. Nomen, J. Thermal Anal. 52(1998)933 and Thermochim. Acta 316(1998)37

[534] S. Vyazovkin, J. Thermal Anal. 49(1997)1493; Thermochim. Acta 355(2000)155 and 397(2003)269

[535] A.K. Galwey, Thermochim. Acta 413(2004)123

[536] A.L. Greer, Acta Metal. 30(1982)171

[537] K.F. Kelto, F. Spaepen, Acta Metal. 33(1985)455

[538] E. Illeková, F.A. Kuhnast, I. Matko, Ch. Naguet, Mater. Sci. Eng. A205(1996)166 and A215(1996)150

[539] L.C. Chen, F. Speapen, J. Appl. Phys. 69(1991)679

[540] D.R. Allen, J.C.Foley, J.H. Peperezko, Acta Metal. 46(1998)431

[541] T. Clavaguera-Mora, N.Clavaguera, N. Crespo, T. Pradell, Prog. Mater. Sci. 47(2002)559

[542] E. Illeková, Thermochim. Acta 280/281(1996)289 and 387(2002)47

[543] E. Illeková, J. Non-cryst. Sol. 287(2001)167 and chapter "Kinetic characteristic of nanocrystal formation in metallic glasses" in "Properties and Application of Nano-crystalline Alloys" (B. Idzikowski, P. Švec, M. Miglierini, eds), Kluwer, Dordrecht 2004

[544] D.W. Henderson, J. Thermal Anal. 15(1979)325

[545] T.J.W. DeBroijn, W.A. DeJong, P.J. van den Berg, Thermochim. Acta 45(1981)315

[546] F. Ordway, J. Res. Nat. Bur. Stand. 48(1952)197

[547] L. Reich, Polym. Lett. 2(1963)1109

[548] J. Šesták, P. Holba, M. Nevřiva, A. Bergstein, "TERMANAL'73" Proc. of Thermal Anal Conf. at High Tatras, by SVST Bratislava 1973, p.S-67 (in Czech)

[549] V. Šatava, J. Šesták, Anal. Chem. 45(1973)154

[550] J. Šesták in "Thermal Analysis", Proc. of the 3rd ICTAC in Davos (H.G. Wiedemann, ed), Vol.1, p.3, Birghauser, Basel 1972

[551] B. Wunderlich "Thermal Analysis" Academic, Boston 1990 and B. Wunderlich "Thermal Analysis of Materials" downloadable from Internet

[552] S. Vyazovkin, N. Sbirrazzouli, Macromolecules 29(1996)1867

[553] V. Mamleev, S. Bourbigot, M. LeBras, S. Duquesne, J. Lefebre, J. Thermal Anal. Calor. 70(2002)565 and 78(2004)1009

[554] J. Blažejowski, Thermochim. Acta 68(1983)233 and A. Mianowski, J. Thermal Anal. Calor. 58(2000)747

[555] A. Malecki, J. Thermal Anal. 36(1990)215 and 41(1994)1109

[556] J. Málek, Thermochim. Acta "Kinetic analysis of crystallization processes in amorphous materials" 355(2000)239

[557] J. Málek, Thermochim. Acta 129(1988)293; 138(1989)337 and 200(1992)257; F.J. Gotor, J.M. Criado, J. Málek, N. Koga, J. Phys. Chem. 104(2001)10777; J. Shánělová, J. Málek, M.D. Alcalá, J.M. Criado, J.Non-cryst.Sol. 351(2005)557

[558] B.V. Lvov, Thermochim. Acta 424(2004)183

[559] K. Hungerbühler, Thermochim. Acta 419(2004)1

[560] B. Malecka, E. Drsdocoeda, A. Malecki, Thermochim. Acta 423(2004)13

438

[561] J. Opferman, F. Giblin, J. Mayer, E. Kaisersbrger, Int. Lab. Jan/Feb. 1995, p.25 and Netzsch softwares – www.ngb.netzsch.com

[562] B. Roduit, Thermochim. Acta 388(2002)377 and J. Thermal Anal. Calor. 80(2005)225

[563] J. Málek, J. Šesták, F. Rouquerol, J. Rouquerol, J.M. Criado, A. Ortega "Possibilities of two nonisothermal procedures for kinetic studies" J. Thermal Anal. 38(1992)71

[564] J.H. Flynn, B. Dickens, Thermochim. Acta 15(1976)1

[565] J.M. Criado in the book O.T. Sørensen, J. Rouquerol (edts.) "Sample Controlled Thermal Analysis" Kluwer, Dordrecht 2004 and plenary lecture "Sample controlled thermal analysis and kinetics" at the ICTAC 13th published in J. Thermal Anal. Calor. 80(2005)27

[566] R.L. Blaine, B.K. Hahn, J. Thermal Anal. 54(1998)695

[567] T. Ozawa, Thermochim. Acta 356(2000)173 and key lecture at the 8th ESTAC in Barcelona in J. Thermal Anal Calor. 73(2003)1013

[568] 153] J. Šesták, J.J. Mareš, J. Krištofik, P. Hubík, Glastech. Ber. Glass. Sci. Technol. 73 C1(2000)104

[569] P. Mialhe, J.P. Charles, A. Khoury, J. Phys. D 21(1988)383

[570] N. Koga, Thermochim. Acta "Kinetic compensation effect, a review" 244(1994)1 and 258(1995)145

[571] J. Zawadski, S. Bretsznayder, Z. Electrochem. 41(1935)215

[572] A. Einstein, Ann. der Phys. 22(1907)569

[573] J. Šmíd, J.J. Mareš, L. Štourač, J. Krištofík, J. Non-cryst. Solids 77/78(1985)311

[574] J. Šesták "Errors, shapes and interconnections of TG curves and kinetic data evaluation" Talanta 13(1966)567

[575] W. Mayer, N. Neldel, Z. Tech. Phys. 18(1937)588

[576] J.R. Hullet, Quarterly Rev. 18(1964)227

[577] N. Eisenreich, J. Thermal Anal. 19(1980)289

[578] P.D. Garn, J. Thermal Anal. 7(1975)475 and 13(1978)581

[579] A.K. Galwey, Advan. Catal. 26(1977)247

[580] A.I. Lesnikovich, S.V. Levchik, J. Thermal Anal. 30(1985)237 and 677

[581] A.V. Logvinenko, J. Thermal Anal. 36(1990)1973

[582] A.K. Galwey, M.E. Brown, Thermochim. Acta 300(1997)107 and 386(2001)91

[583] J. Šesták, J. Thermal Anal. 32(1987)325

[584] N. Koga, J. Šesták, J. Thermal Anal. 37(1991)1103

[585] 167] G. Pokol, G. Varhegyi, CRC Crit. Rev., Anal. Chem. 19(1988)65 and G. Pokol, J. Thermal Anal. 60(2000)879

[586] J.H. Flynn, J. Thermal Anal. 36(1990)1579

[587] R.R. Krug, W.G. Hunter, R.A. Grieger, J. Phys. Chem. 80(1976)2335

[588] R.K. Agrawal, J. Thermal Anal. 32(1987)1277 and 35(1989)909

[589] N. Koga, J. Šesták, Thermochim. Acta 182(1991)201

[590] E. Calvet, H. Prat "Recent Progress in Calorimetry" Pergamon, London 1963 [591] J.M. Stuarevant "Calorimetry" in "Physical methods of Chemistry" (A. Weissenberg, B.W. Rossiter, edts.) Wiley/Interscience, New York 1971

[592] J.L. Oscarson, R.M. Izatt "Calorimetry" in "Determination of Thermodynamic Properties" (B.W. Rossiter, R.C. Beatzold, edts.) Wiley, New York 1992

[593] 48] W. Hemminger, G.W.H. Höhne "Grundlagen der Kalorimetrie", Verlag Chemie, Weinheim 1979 and "Calorimetry: fundamentals and practice" Chemie – Deerfiled Beach, Florida 1984

[594] G. Lombardi, J.O. Hill (etds) "For better Thermal Analysis" ICTAC Publ., 1st edition 1975, 3rd edition 1991 and many other, e.g., G.W.H. Höhne, J. Thermal Anal. 37(1987)1991;

S.M. Sarge, E. Gmelin, C.W.G. Höhne, H.K. Cammenga, W. Hemminger, W. Eysel, Thermochim. Acta 247(1994)129 or E. Gmelin, S.M. Sarge, Pure Appl. Chem. 67(1995)1789
[595] G.W.H. Höhne, H.K. Cammenga "Methoden der termische Analyse", Springer, Berlin 1989 and 2003
[596] W. Zelenkiewicz E. Margas (edts.) "Theory of Calorimetry", Kluwer, Dordrecht 2002
[597] I. Lamprecht, Thermochim. Acta 83(1985)81
[598] W, Zelenkiewicz "How do mathematical models of calorimetry work" Thermochim. Acta 420(2004)23 and W, Zelenkiewicz, E.Margas, J.Hatt, Thermochim. Acta 70(1983)257
[599] I. Wadsö, Thermochim. Acta 88(1985)35
[600] A. Tian, Bull. Soc. Chim. France 33 (1923) 427 and Ac. Soc. France 178(1924)705
[601] L.G. Berg "Skorostnoi Kolichestvenyj Fazovyj Analyz" (Rapid Quantitative Phase Analysis), Acad. Nauk, Moscow 1952 (in Russian)
[602] R.C. Mackenzie (ed.) "Differential Thermal Analysis" Academic, London 1970 and 1972
[603] G. Tammann, Z. Anorg. Chem. 43(1905)303
[604] G.K. Burgess, Bull. Bur. Stand. Washington, 4(1908)199
[605] R.C. Mackenzie, Platinum Met. Rev. 26(1982)175
[606] W.P. White, Am. J. Sci. 28(1909)453 and 474; J. Phys. Chem. 24(1920)393
[607] F.H. Norton, J. Am. Ceram. Soc. 22(1939)54
[608] M.J. Vold, Anal.Chem. 21(1949)683
[609] H.T. Smyth, J.Am.Ceram.Soc. 34(1951)221
[610] M.M. Factor, R. Hanks, Trans. Faraday Soc. 63(1959)1129
[611] A.P. Gray, in proc. of the 4th ICTA, "Thermal Analysis", Akademia Kiado, Budapest 1975
[612] P. Holba, J. Šesták, Silikáty (Prague) 20(1976)83 (in Czech)
[613] J.Boerio-Goates, J.E. Callen "Differential Thermal Methods" in "Determination of Thermodynamic Properties" (B.W. Rossiter, R.C. Beatzold, edts.), Wiley, New York 1992
[614] M.J. Richardson, E.L. Charsley "Calibration" in "Handbook of Thermal Analysis and Calorimetry" (M.E. Brown, ed.) Elsevier, Amsterdam 1998, Vol.1, Ch.13
[615] M.E. Brown, T.T. Bhengu, D.K. Sanyal, Thermochim. Acta 242 (1994) 141 and P.K. Gallagher, R. Blaine, E.L. Charsley, N. Koga, R. Ozao, H. Sato, S. Sauerbrunn, D. Schultze, H. Yoshida, J. Thermal Anal. Calor. 72 (2003) 1109
[616] J. Šesták, P. Holba, G. Lombardi "Quantitative evaluation of thermal effects: theory and practice of dynamic calorimetry" Annali di Chimica (Roma) 67(1977)73
[617] L. Eyraud, C.R. Acad. Sci. 238(1954)1411
[618] M. Reading, Trends Polym. Sci. 1993, 248
[619] T. Ozawa, J. Thermal Anal. Calor. 64(2001)109
[620] V.B.F. Mathot (ed.) "Handbook of Calorimetry and Thermal Analysis of Polymers" Carl Hanser, München 1994; T. Hatakeyma, Liu Zhenhai (edts.) "Handbook of Thermal Analysis" Wiley, Chichester 1998; M.E. Brown (ed.) "Handbook of Thermal Analysis and Calorimetry" Elsevier, Amsterdam 1998; R.B. Kemp (ed.) "Handbook of Thermal Analysis and Calorimetry: from macromolecules to man" Elsevier, Amsterdam 1999 and/or M. Sorai (ed.) "Comprehensive Handbook of Calorimetry and Thermal Analysis" Wiley, New York 2004
[621] H.S. Carslaw, J.C. Jeager "Heat Conduction in Solids" Cleradon, London 1959; J. Krempaský "Měranie termofyzikálnych hodnot" (Measurements of Thermophysical Quantities), VSAV, Bratislava 1969 (in Slovak)
[622] L. Kubičár "Pulse Methods of Measuring Basic Thermophysical Parameters" Elsevier, Amsterdam 1990

[623] L. Kubičár, V. Boháč "A step-wise method for measuring thermophysical parameters of materials" Meas. Sci.Technol. 11(2000)252, and "Thermophysical analysis: methodology for puls transient method" Inter. J. Thermophys., in print 2005

[624] R.J. Trainor, G.J. Pokorny, R.B. Snyder, Rev. Sci. Instrument 46(1975)1275

[625] W.J. Parker, R.J. Jemkins, C.P. Buttler, G.L. Abbott, J. Appl. Phys. 32(1961)197

[626] R.E. Taylor, High Temp.-High Press. 11(1979)43

[627] H.J. Lee, R.E. Taylor "Thermal Conductivity", Plenum, New York 1976

[628] J.T. Schriempf. High Temp.-High Press. 4(1972)411 and Z, Fang, R. Taylor, High Temp.-High Press. 19(1987)29

[629] H. Ohta, H. Shibata, A. Tsuzuki, Y. Waseda, Rev. Sci. Instruments 72(2001)1899

[630] C.P. Camirand, Thermochim. Acta 417(2004)1 or H. Fisher, Thermochim. Acta 425(2005)69

[631] T. Ozawa, Bull. Chem. Soc. Jpn. 39 (1966) 2071

[632] A.P. Gray chapter "Simple Generalized Theory for Analysis of Dynamic Thermal Measurements" " in the book "Analytical Calorimetry" (R.S. Porter, J.F. Johnson, edts.), Vol.1., p.209, Plenum Press, New York 1968

[633] P. Claudy, J.C. Commercon, J.M. Letoffe, Thermochim. Acta 68(1983)305 and 317

[634] M.J. O'Neill, Anal. Chem. 36(1964)1238 and 2073

[635] R.A. Speros, R.L. Woodhouse, J. Phys. Chem. 67(1963) 164

[636] J.H. Flynn, Thermochim. Acta 8 (1974) 69 and J. H. Flynn chapter " Theory of DSC" in the book "Analytical Calorimetry" (Porter R.S., Johnson J.F., edts.), Plenum Press, New York 1974, p.3 and 17

[637] S.G. Dixon, S.G. Black, C.T. Butler, A.K. Jain, Analytical Biochemistry, 121(1982)55

[638] N.O. Birge, S.R. Nagel, Phys. Rev. Let. 54(1985)2674

[539] P.F. Sullivan, G. Seidel, Phys. Review 173(1968)679

[602] R.C. Mackenzie (ed) "Handbook of DTA" Chem. Publ., New York 1966,

[640] P. Heines, C. Keattch, N. Reading chapter "Differential Scanning Calorimetry" in the book "Handbook of Thermal Analysis" (M.E. Brown, ed.), Elsevier, Amsterdam 1998

[641] R. Meilling, F.W. Wilburn, R.M. MacIntosh, Anal. Chem. 41(1969)1275

[642] M.D. Judd, M.I. Pope, Thermochim. Acta 7(1978)247

[643] M. Nevřiva, Thermochim. Acta 22(1978)187

[644] P. Holba, M. Nevřiva, J. Šesták, Neorg. Mat. DAN USSR 10(1974)2097 (in Russian) and Silikáty (Prague) 20(1976)99 (in Czech)

[645] H. Svoboda, J. Šesták, in the Proc. 4th ICTA, 'Thermal Analysis", Vol. 3, p.999, Akademiai Kiado, Budapest 1975

[646] M. Nevřiva, J. Šesták in "Thermal Analysis" (Ž.D. Živkovič, ed.) Collection of Papers by Technical University in Bor, Yugoslavia 1984, p. 3

[647] Y. Sawada, H. Henmi, M. Mizutami, M. Kato, Thermochim. Acta 121 (1987) 21

[648] V. Šatava, J. Amer. Cer. Soc. 58 (1975) 357

[649] H.G. Wiedemann, G. Bayer, J.Thermal Anal. 38(1985)1273 and/or R. Ferro, A. Saccone, Thermochim. Acta 418(2004)23

[650] E.E. Marti, Thermochim. Acta 5 (1972) 173 and Rev. Sci. Instruments 58(1987)639

[651] E.F.Joy, J.D.Bonn, A.J. Bernard, Thermochim. Acta 2(1971)57

[652] F. Branda, A. Marotta, A. Buri, J. Non-cryst. Sol. 134(1991)123]

[653] C.T. Moynihan, J. Amer. Cer. Soc. 76(1993)1081

[654] K. Shinzato, T. Baba, J. Thermal Anal. Calor. 64(2001)413

[655] D.M. Price, M. Reading, R.M. Smoth, H.M. Pollock, A. Hammiche, J. Thermal Anal. Calor. 64(2001)309

[656] T. Ozawa, J. Thermal Anal Calor. 72(2003)337

[657] J.M. Hutchinson, J. Thermal Anal Calor. 72(2003)619

[658] B. Papánková, P. Šimon, R. Boča, Sol. State Chem. V – Sol. State Phenomena 90/91(2003)177

[659] J. Blumm, E. Kaisersberger, J. Thermal Anal. Calor. 64(2001)385

[660] J.B. Henderson, W.-D. Emmerich, E. Wasner, J. Thermal Anal. 33(1988)1067

[661] Netzsch Manuals, e.g., FLA 426, 437, 447; DSC 404; Netzsch softwares: www.ngb.netzsch.com

[662] W.P. Brenann, M.P. DeVito, R.I. Fayans, A.P. Gray "Overview of the Calorimetric Purity Measurements" in "Purity Determinations by Thermal Analysis" (R.L. Blaine, C.K. Shoff, edts.), p.5, Amer. Soc. Test. Mater., Philadelphia 1984

[663] J.P. Elder, Thermochim. Acta 34(1979)11, 52(1982)235 and in "Purity Analysis by Dynamic and Isothermal Step DSC" in "Purity Determinations by Thermal Analysis" (R.L. Blaine, C.K. Shoff, edts.), p. 50, Amer. Soc. Test. Mater., Philadelphia 1984

[664] D.L. Sondack, Anal Chem. 44 (1972) 888

[665] W. Brostow, M.A. Macip, M. Sanches-Rubio, M.A. Valerdi, Mater. Chem. Phys. 10(1984)31

[666] E. Marti, E. Kaisersberger, W.-D. Emmerich, J. Thermal Anal.Calor. 77(2004)905

[667] E. Marti, E. Keisersberger, G. Kaiser, W.Y. Ma "Thermoananlytical Characterization of Pharmaceuticals" Netzsch Annual Publ., Selb 2000

[668] P.J. van Ekeren, C.M. Holl, A.J. Witteveen, J.Thermal Anal. 49(1997)1105

[669] E.S.Domalski, E.D. Hearing, J.Phys.Chem.Ref.Data (1990) 881 and (1996) 1

[670] P. Roura, J. Farjas, Thermochim. Acta 430(2005) in print

[671] B. Wunderlich, Thermochim. Acta 330(1999)55 and 348(2000)181, W.Hu, B. Wunderlich, J Thermal Anal. Cal. 66(2001)677

[672] W.K. Roots "Fundamentals of Temperature Control" Academic, New York 1989

[673] T.D. McGee "Principles and Methods of Temperature Measurements", Wiley, New York 1988

[674] English (ICTAC): J. Thermal Anal. 8(1975)197; 13(1978)387 and 21(1981)173; Pure Appl. Chem. 37(1974)439 and Thermochim. Acta 5(1972)71; 28(1979)1 and 46(1981)333;

[675] Czecho-Slovak: J. Šesták, P. Holba, S. Fajnor, Chem.listy (Prague) 77(1983)1292

[676] K. Heide "Dynamische termische Analysenmethoden" Deutscher Verlag, Dresden 1982

[677] R.F. Speyer "Thermal Analysis of Materials", Marcel Dekker, New York 1994

[678] C. Duval "Inorganic Thermogravimetric Analysis" Elsevier, Amsterdam 1962 and/or P. Vallet "Thermogravimetry", Gauthier-Villars, Paris 1972

[679] C.J. Keatch, D. Dollimore "Introduction to Thermogravimetry" Heyden, London 1975

[680] F. Paulik, J. Paulik "Simultanous Thermoananalytical Examinations by Means of Derivatograph", Elsevier, Amsterdam 1981

[681] F. Paulik "Special Trends in Thermal Analysis" Wiley, Chichester 1995.

[682] J.P. Czarnecki, J. Šesták "Practical thermogravimetry" J. Thermal Anal. Calor. 60 (2000) 759

[683] J. Šesták, Sklář a Keramik (Prague) 26(1976)427 (in Czech) and Wiss. Zeit. Friedrich-Schiller Univ., Jena, Math.-Naturwiss. Reihe, 32(1983)377

[684] J. Šesták, Z. Strnad, chapter "Preparation of Fine crystalline and Glassy Materials by Vitrification and Amorfization" p.176 and J. Šesták, J. Čermák, "Application of Lasers in the Production of Materials" p.244, both in the book "Special Technologies and Materials" (J. Šesták, ed.) Academia, Praha 1993 (in Czech)

[685] M.A. Ottni (ed.) "Elements of Rapid Solidification", Springer, Berlin 1997

[686] C. Suryanarayana (ed.) "Non-equilibrium Processing of Materials", Pergamon, Amsterdam 1999

442

[687] B.L. Mordike (ed.) "Laser Treatment of Materials – I" DGM Inform., Bad Nauheim, Germany 1986 and J. Mazumder, J. Mukherjee, B.L. Mordike (edts.) "Laser Treatment of Materials – IV" TMS, Warrendale 1994

[688] D.M. Ohnaka, D.M. Stefenscu (edts.) "Solidification Science and Processing" TMS, Warrendale 1996

[689] J. Šesták, Silikáty (Prague) 7 (1963) 125 (in Czech)

[690] A.W.D. Hills, Chem. Eng. Sci. 23 (1968) 297

[691] J.P. Czarnecki, N. Koga, V. Šestákova, J. Šesták, J. Thermal Anal. 38 (1992) 575

[692] F. Paulik, E. Bessenyey-Paulik, K. Walther-Paulik, J. Thermal Anal. Cal. 58(1999)725; Thermochim. Acta 402(2003)105 and 424(2004)75

[693] J.P. Czarnecki, Thermochim. Acta 192 (1991) 107 and NATAS Notes, Fall 1991, p. 48.

[694] V. Hulinský, K. Jurek, O. Gedeon, Mikrochim. Acta 13 (1996) 325

[695] R. Castaing "Electron Probe Analysis" Advances Elect. Elect. 13(1960)317]

[696] O.T. Sørensen, J. Rouquerol (edts.) "Sample Controlled Thermal Analysis" Kluwer, Dordrecht 2004

[697] T. Ozawa, Thermochim. Acta 356(2000)173 and J. Thermal Anal. Cal. 73(2003)1013

[698] K. Kanari, T. Ozawa, Thermochim. Acta 399(2002)189

[699] V.B.F. Mathot, Thermochim. Acta 355(2000)1

[700] J. Šesták, J. Thermal Anal Calor. "Trends and boundary aspects of the non-traditionally applied science of thermal analysis" 75(2003)1006

[701] M. Reading, D.J. Houston, M. Song, H.M. Pollock, A. Hammiche, D.M. Price, Amer. Lab. January 1998, p.13 and J Thermal Anal. Calor. 60(2000)723

[702] "Apeiron" or "Galilean Thermodynamics" or websites "www.google.com" under entries "tired light" or "alternative physics"

[703] websites of ICTAC – www.ICTAC.org

[704] author's websites – www.fzu.cz/~sestak

[705] G. Lombardi, J. Šesták "History of ICTA", plenary lecture at the Int. Conf. TERMANAL 2005, Stara Lesná (Slovakia), proc. J. Thermal Anal. Calor (in preparation, 2006)

[706] G. Liptay, J. Simon (edts.) "Who is Who in Thermal Analysis and Calorimetry", Academia Kiado, Budapest 2004

[707] J.M.P. Parrando, J. Dinis. "Brownian motion and gambling: from retches to paradoxical games" Contem.Phys. 45(2004)147

[708] E. Frey, K. Kroy "Brownian motion: a paradigm of soft matter and biological physics" Ann. Phys. (Leipzig) 14(2005)20

[709] P. Hanggi, G.L. Ingold "Fundamental aspects of quantum Brownian motion" Chaos 15(2005)111

[710] M.P. Blancowe "Quantum electromechanical systems" Phys. Rep. 305(2004)159

[711] Focus issue: "Brownian motion and diffusion in the 21st century" New Journal of Physics Vo.7, 2005

[712] J. Bronowski "Origins of Knowledge and Imaginations" Yale University Press, USA 1978

INDEX

448

Intelligence,321,325
Intelligent processing of materials,75
Intensive parameter,50,210,212,213,
251,254,276
Interaction parameter,220,221,223,237
Interface,3,72,74,77,79,127,154,155,166,
202,206,212–215,224,226,237,246,264,
276,280–284,287,289,291,292,295,302,
308,310,318,322,329,341,346,351,359,
379,381,389,394,398,405
Internal energy,6,150,151,152,158,204,
205,206,212,227
Internet,13,15,18,19,21,22,24,236,415
Interpolation,246,332,338,390,391
  – of peak baseline,359,362,383
Invariant,122,138,177,259,264,325,328
  – process,256–258,333
Iso-conversion evaluation method,320,339
Isodiathermal,346
Isokinetic temperature,322
Isolation,21,40,62,68,329,383
Isoperibolic,347
Isothermal,137,143,213,255,266,273,278,
279,331,336,345,347,354,372,384,398,
399,409,410
  – calorimeter,109,117,345
  – degree of conversion,256,318
  – kinetics,281,321
Iteration,305,313,315,316
– iterator,314,315

J
Jander (parabolic law),283–285,321,324
Jefferson monograph,218
JMAYK (equation),323,324,326,329,
330,366,385,387
Journals,
  – ICTAC News, 415
  – Thermochimica Acta/Thermal Analysis
   and Calorimetry,10,415

K
KCE–kinetic compensation effect,340,
342,345
Kelvin,8,34,133,144,146,148,170,194,
214,298
Kinetic phase diagram,216,260,264,265
Kinetics,127,192,198,213,253,262,264,
276,278,280,283,287,290,297,302,317,

319,325,317–219,332,336,347,360,365,
391,395,400,401
  – calculation of data,218,320,
  – compensation effect,340,342
  – equation,231,267,279,316,321,322,339
  – parameters,319,322,324,326,332,335
Kissinger plot,319,326,338
Kneading dough, 314
Knowledge,1,7,10,14,16,24,30,46,53,61,
73,85,90,100,122,126,140,175,204,232,
263,352,384,397,414
Koch curve,303,307–310

L
Lambda phase transition,252,255
Landau theory,134,252,253
Liesegang rings,198,199
Log/log plot,5,365
Logarithm,4,28,60–62,113,132,142,146,
148,157,184,219,237,267,272,302,308,
311,315,319,321,333,335,337
  – human sensation,1,7,44,60,61,62,75,
   91,109,160,161
Logistic,313,316
  – equation/function,314,316,317,323
  – mapping,17,313
Lorenz attractor,316

M
Magnetic,35,74,76,77,82,128,142,151,194,
196,242,251,391,406
  – field,162,188,206,207,210
  – measurements,207,369,384,385
  – susceptibility,207,251
  – transformation,313
Magnetization,206,207,210,251,253,257,
381,405,406
Magnetocaloric,14,210
– coefficient,207
Mandelbrot,3,231,310
Mapping,17,219,313
Maurpetuis least action,29,198
Maxwell,42,50,51,86,116,119,129,133,
135,137,142,147,148,153,179,184
  – demon,86,179
  – Boltzmann equation,280
  – transformations,210
Measurements,5,7,9,11,19,25,34,38,50,55,
103,111,117,119,141,145,148,150,151,

APPENDIX: summary characteristics of some important scholars and scientists of the past and a selection of recent thermoanalysts related to the book contents.

Agricola Georgius (Georg Bauer) (1494–1555) Ger. physic. (working in Bohemian town Jachymov) father of mineralogy (devised a system of classification of geological specimens), inventor of modern concept of mining ('De re metallica libri XII' 1556)

Agrippa (von Nettesheim) Cornelius Heinrich (1486–1535) Ger. phys., urged return to beliefs, supporter of mystical philosophy (3 spheres – elements, stars, spirit)

Amontons Guillaume A. (1663–1705) Fr. phys., concept of absolute zero thermodynamic temperature at which gas pressure vanishes, constructor of thermometers and barometers

Ampére Marie Andre (1775–1836) Fr. phys., founder of electrodynamics (Ampere's Law), inventor of galvanometer (book 'Theorie des phenomenes electro-dynamiques' 1826)

Anaxagoras of Clazomenae (500–428 BC) Gr. phil., existence results from ordering of seeds by infinite mind (concept of atoms), theory of perspectives

Anaximenes of Miletos (588–528 BC) Gr. phil., conceptions of physical rather than moral law governing cosmos

Andrews Thomas (1813–1885) Irish chem., critical temperatures of gases, heat of chem. combustion

Apollonius from *Porgy* (≈ 262 – 200 BC) Greek philosopher known for algebraic and geometrical characterization of ellipses, parabolas and hyperbolas

Archangelski Igor V. (1941-) Rus. chem., expert in thermal analysis of carbonic materials

Aristarchos of Samos (3rd century BC) Gr. astron., fixed stars and the sun remain immobile while the Earth revolves about the sun, distance problem by trigonometry

Aristotle (384–322 BC) Gr. phil., (author: Categories; On Interpretation; Prior Analytics; Metaphysics; Ethics; etc.), accepted 4-element theory (+ ether), rejected void, propelling force inversely proportional to resistance, introduced variables into logic, held heat to be center of heat and life

Arrhenius Svante August (1859–1927) Swed. chem, formulated theory of electric dissociation later extended to rate of chem. reactions (Arrhenius constant), discovered expression for latent heat as a function of raising boiling point through dissolved nonvolatile components, appreciated light pressure in cosmic physics, studied reaction velocities, viscosities ('Teorien der Chemis' 1900), he was first to note global warming

Ataka Tooru (1942-) Jap. chem.., calorimetrist, expert in thermodynamics of materials

Avicena (Abu'ali al-Husain ibn Abdallah, Ibn Sina)(980–1037) Pers. physician, author of Canon of Medicine, studied therapeutic measures, used methods of Aristotle, suggested speed of light must finite quantity, rejected metallic transmutations

Avogadro Lorenzo Romano Amedeo Carlo (1776–1856) It. phys., equal volume of gases at (at the same T and P) contain identical number of particles (Avogadro's Law), studied specific heats, expansion

Babbage Charles (1792–1871) Brit. math., referred as the father of computing in recognition of his design of two machines, the difference engine for calculating tables of logarithms by repeated additions performed by trains of gear wheels, and the analytical engine designed to perform a variety of computations using punch-cards

Bacon Francis, Lord Verulam (1561–1626), Brit. phil., inventor of inductive

method and empiricism, identified gravitation force, held heat is motion, showed that salt lower melting point of ice
Baekeland Leo Hendrik (1863–1944) Belg. chem., synthetic resin, plastics (bakelite), contributed electro-chemistry (book 'Some Aspect of Industrial Chemistry' 1914)
Balek Vladimír (1940-), Czech chem.., inventor of emanation thermal analysis method, author of books of same subject
Barone Guido (1937-) Ital. chem., thermal behavior of biomolecules, atmospheric pollution (book 'volumetric and calorimetric techniques' 2003)
Barrow John D. (1942–) Brit. phys., protagonist of cosmology and gravitation theory as well as aspects of the history and philosophy of science, writer of popularizing books
Bárta Rudolf (1897–1985) Czech. chem. expert in cements, cofounder of ICTA
Bartoli Adolfo (1851–1896) It. phys., radiation pressure of light, specific heat of water and its dissociation, heat engine based on light
Becher Johann Joachim (1635–1681), Ger. chem., working in medicine, mineralogy, economics, developer of theory that burning substances are losing their anima ('phlogiston')
Becher Joseph Pitus, sen. (1769-1840) and *Johan Nepomuck, jun.* (1813-1895), Czech pharmcists., founders of renowned herbaceous liquor "Carslbad Becher"
Becquerel Antoine Cesar (1788–1878) Fr. phys., cofounder of electrochemistry ('Elements de l'electrochimie' 1843), used platinum and palladium to measure high temperatures
Beilstein Friedrich Konrad (1838–1906) Rus. chem. And professor in Petersburg Univ., research in anal. and org. chemistry ('Handbuch de Organischen Chemie, 1880)
Bénard Henri (1885–1973) Fr. sci., who made his famous discovery on convection rolls during his thesis based on early studies of *B. Thompson* (see Strutt)

Berg Lev Germanovič (1896–1974) Rus. chem., originator of thermoananlytical instrumentation and theories, co-founder of ICTA
Bergman Tobern Olaf (1735–1784) Swed. chem., founder of mineral chemistry, its classification and quantitative determination of composition, developed theory of chem. affinity
Bergstein Arnošt (1914-1973), Czech. chem., orig. lawyer, inventor of dielectric thermal analysis
Bernoulli Daniel (1700–1782) Swith math., early formulation of principle of energy conservation, pressure as result of particles impact on the container, differential calculus application in theory of probabilities, acoustics ('Hydrodynamica' 1738)
Bernoulli Johann (1667–1748) Swiss math., developer of differential, integral and exponential calculus, law of quantity of conservation – mv^2 ('vis viva')
Bertalanffy (von) Ludwig (1901–1972) Austr. born, Can. biol., research in ordaining conception in biology, inventor of general 'organismic' system theory and comparative physiology (book 'Modern Theories of Development' 1933)
Berthelot Marcelin (1827–1907) Fr. chem., enunciated the principle of maximum work, known for his sharp criticism ('Essai de mecanique chimique fondee sur la thermochimie' 1879)
Berzelius Jons Jacob Baron (1779–1848) Swed. chem., founder of modern chemistry ('Theory of Chemical Proportions' 1814), oxygen as the standard for atomic weights, pioneered gravimetric analysis
Bessel Friedrich Wilhelm (1784–1864), Ger. astron., Bessel's geoid, theory of errors, Bessel's functions
Biot Jean Baptiste (1774–1862) Fr. phys., who invented polariscope, fundamental laws of heat, magnetism
Black Joseph (1728–1799), Brit. chem.. (Glasgow and Edinburgh), helped to lay foundations for quantitative analysis, recognized that heat quantity is different

form of heat intensity, concept of specific heats

Boerhaave Herman (1668–1738) Dutch chem., introduced modern concept of chemistry ('Chemical Textbook' 1732)

Bogdanov Alexander Alexandrovič (1873–1928) Rus. phil., who made sophisticated reinterpretation of Mach's empiriocriticism, known for calculating automata, who proposed that all phys., biolog., and human sciences be unified by treating them as systems of relationships ('Tectology: universal organization')

Bohr Niels Henrik David (1885–1962) Danish phys., known for Bohr's model of atomic structure, spectroscopic data to explain internal structure, electrons in the outer-most shall determine chem. properties ('Atomic Theory and the Description of Nature' 1934)

Boldyrev Vladmír V. (1927-) Rus. Chem.., originator of the reactivity control of solids, author of books of same subject (book "Reactivity of molecular solids" 1999)

Bolos of Mendes (pseudo-Democritos, circa 200 BC) Gr. nat. sci., probably earliest Greek writer on alchemy ('Psychica at Mystica')

Boltzmann Ludwig Eduard (1844–1906) Austr. phys., cofounder of equipartition theory, kinetic theory of gases leading to theory of statistical thermodynamics, Stefan-Boltzmann Law of radiation

Boole George (1815–1864) Brit. logician, invented the symbolic processes of algebra as tools of numerical calculation, in which symbols are used to represent logical operations. In his book 'An Investigation of the Laws of Thought' (1854) the mathematical theories of logic and probabilities was proposed as well as the calculus (taking one of only two values <0 and 1>, i.e., Boolean algebra)

Böttger Johan Friedrich (1682-1719), Ger. alchem., initiator of European production of chinaware (porcelain) using his own recipes

Boyle Robert (1627–1691) Brit. chem., disaffirming the indoctrination of four elements, cofounder of "British Royal Society" and scientific journal "Philosophical Transactions", designed vacuum pump, chemistry of combustion and respiration

Brandštetr Jiří (1931-), Czech calorim., constructer of ‚entalphiograph'

Bravais Auguste (1811–1863) Fr. phys., upon observing natural crystals he grouped them into seven crystal systems (Bravais lattice is called after him), which is an infinite array of discrete points with an arrangement and orientation that appears exactly the same viewed from any point of the array ('Etudes cristalographiques' 1851)

Brillouin Léon (1889-1969) French born US phys., invented the concept of negentropy, father of information theory (book 'Théorie de l'Information' 1959)

Brouwer Luitzen Egbertus Jan (1881 1966) Dutch. math., founder of modern topology, proved that dimensionality of a Cartesian space is topological invariant, worked in point sets

Brown Michael Ewart (1938-) S. African chem.., designer of decomposition kinetics of solids, author of books of same subject

Brown Robert (1773–1858) Scot. botan., studied plant physiology, known for Brownian movement of microscopic particles

Bruno Giordano (1548–1600) It. phil., drew his cosmology from Copernicus and Lucretius, conceived that Earth revolves around the sun, stars being the center of other planetary systems (burned at stake)

Bunsen Robert Wilhelm (1811–1899) cofounder of spectroscopy, investigated variation of melting point with pressure, galvanic battery, new types of labor. Equipments (Bunsen burner)

Burian Josef (1873-1942), Czech. technol. and early propagator of thermal analysis in ceramics

Byelousov (Bělousov) Boris Pavlich (1893–1970) Rus. pediatr., problems of protein regime, known for unsuccessful

456

persuading unusual oscillatory manners of some chemical reaction

Cai E. Xian (1938-), Chin. chem., thermochemistry and thermodynamics (book „Experimental Physical Chemistry" 1980)

Callen Herbert Bernard (1919–1990) US. phys., research on solid-state physics, thermodynamics and statistical mechanics, fluctuation-dissipation theorem (book 'Thermodynamics: introduction to thermostatics' 1960)

Calvet Edouard (1895–1966) Fr. calorim., known for Calvet-Tian calorimetry

Cantor Georg Ferdinand Ludwig Philip (1845–1918) Ger. math., developed theory of sets, defined real, irrational and transfinite numbers

Carnot Sadi Nicolas Leonard (1796–1832) Fr. phys., founder of thermodynamics (Carnot cycle)

Casimír Hendrik B. G. (1909–1980) Dutch phys., inventor of mathematical formalism of quantum-mechanics, hyper-fine structures, zero-point electromagnetic background, thermodynamics, influence of retardation on Van der Waals forces (Casimir forces)

Cavendish Henry (1731–1810) Brit. chem., revealed composition of water, hydrogen, believed that heat is caused by internal motion, anticipated much of the work of the next half century though he published almost nothing, experimentally defined gravitation constant and work on electricity laws

Cayley Arthur (1821–1895) Brit. math., theory of matrices and groups, invariance algebra of matrices, geometry of n-dimensional space

Celsius Anders (1701–1744), Swed. astron., identified Earth's ecliptic obliquity, temperature scale after him

Cesáro Attilio (1942-), Ital. chem., thermodynamics of biomacromolecules

Charles IV (1316–1378) Rome emperor and famous Bohemian king (1346), supporter of idea of a united 'greater Europe' founder of the first middle European university in Prague

Charsley Edward Leonard (1939-) Brit. thermoanal., standardization, industrial application of thermal analysis (book "Thermal Analysis Techniques" 1992)

Christensen Jim James (1931-1991) US phys., industrial calorimetry, nonideality measurements, known for Christensen Memorial Award of US Calorimetry Conferences

Clapeyron Benoit Pierre Emile (1799–1864) Fr. eng., mathemat. theory of elasticity of solids, found relation between conversion of heat, steam, pressure and volume changes, help construction of locomotives

Clausius Rudolf Julius Emmanuel (1822–1888) Ger. math., reconciled Carnot's theory of heat to equivalence of heat and work (2nd Law of thermodynamics), changes of state (Clausius-Clapeyron equation), contributed theory of electrolysis

Colemann D. Bernard (1930–) US. math., researched hydrodynamics of non-classical fluids, theory of wave propagation in materials with memory, co-founder of rational thermodynamics

Comenius (Komensky) Jan Amos (1592-1670) Czech educator and expatriate, known as the 'teacher of nations' who necessitated spontaneity ('Janna Linguarum reserata', 'Orbis Sensualium Pictus' or 'Didactica ragna')

Copernicus Nicolaus (Kopernik Nikolai) (1473–1543) Pol. astron., worked out details of heliostatic theory of solar system, possibly the greatest astronomer since Ptolemy

Coriolis Gustave Gaspard (1792–1843) Fr. phys., developed theory of relative motion, mechanics, powers and motion, modern definition of kinetic energy ($mv^2/2$)

Criado José Maria (1942-) Spain. chem., heterogeneous catalysis, co-inventor of sample controlled thermal analysis

Czarnecki Jerzy (1937-), polish born US chem.., focused on large TG samples and

reversible decomposition, solved interactions with vapors/gases, developed new TG systems (Cahn Inst., TG Design)
Dalton John (1766–1844) Brit. chem., founder of atomic theory ("New System of Chemical Philosophy"1808), developed (Dalton's) Law of partial pressures, arranged table of relative atomic weights, chemical stoichiometry in simple numerical ratios by weight
Davy (Sir) Humprey (1778–1829) Brit. chem., founder of electrochemistry, showed melting of ice pieces by mutual friction below their freezing temperature – heat a form motion, theory of galvanic decomposition, transmission of thermal radiation through vacuum
De Groot Sybren Ruuds (1916–1990) Dutch phys., research on thermodynamics of irreversible phenomena, relativistic theory of statistic electromagnetic phenomena (book 'Thermodynamic of irreversible processes 1951)
Debye Petrus Josephus Wilhelmus (1884–1966) Dutch phys. chem., dipole moments and molecular structure, structure analysis of powdery crystals by means of X-ray diffraction, study of polymers
DellaGatta Giuseppe (1935–) Ital. calorim., authority in solution thermodynamics
Democritos (460–370) Gr. phil., known for his cosmological and atomic theories, behavior of atoms is governed by unbreakable natural laws and their aggregates are formed by kind of hook and eye mechanism, investigated structure of human body regarding soul as material, compiled ethical concepts
Denbigh George Kenneth (1911–1989) Brit. phys. chem., protagonists of thermodynamics, first class educator (book 'Principles of chemical equilibrium 1961)
Descartes Rene du Perron (Cartesius Renatus) (1596–1650) Fr. math. and physic., known for preservation of motion (≈ mv) as an universal principle, studied geometric forms by algebraic means, gave rules of signs, identified matter with

extension (nonexistence of voids, 'Principia Philosophiae' 1644), study of meteorology, causality ("cogito ergo sum")
Dewar (sir) James (1842–1923), Brit. chemists, obtained liquid hydrogen, worked with low temperatures (Dewar flask)
Dharwadkar Sanjiv Ravalnath (1938-) Ind. phys., high-temperature chemistry and thermodynamics
Diesel Rudolph Christian Karl (1858–1913) Fr. eng., inventor of diesel engine by ignition through compression
Diokles of Karystos (≈ 400–350 BC) Gr. phys., idea of 'pneuma', believed that both sexes contribute to embryo formation
Diviš Prokop (Divish Procopius) (1696–1765) Czech phil. and experim., studied hydrodynamics, electrophysiology and electricity, known as the inventor of lightening rods (1753) proposing their wider use
Dollimore David (1927–2000) Brit. chem.., expert in solid-state kinetics, founder of ESTAC, (book "Reactions in Solid-State" 1980)
Du Bois-Reymond Emil Heinrich (1818–1896) Ger. physiol., showed that electric phenomena occur in muscular activity, physiology of muscles, measurable velocity of nerve impulses
Duhem Pierre Maurice Martin (1861–1916) Fr. phys., attempt to construct a general energetic and abstract thermodynamics using axiomatic-deductive approach, theory of elasticity and hydrodynamics ('Le potentiel thermodynamique' 1886)
Dulong Pierre-Louis (1785–1838) Fr. chem., research on refractive indices and specific heats of gases, co-formulated Dulong-Petit's law, devised empirical formula for the heat
Earnest Charles Mansfield (1941-) US chem., expert in geoscience and minerals (book 'Thermal analysis of clays' 1984)
Einstein Albert (1879–1955) Ger. phys., originator of theories of relativity, laws of motion and rest, simultaneity and

interrelation of mass and energy, quantum theory of photoelectric effect, theory of specific heats, Brownian motion, etc. ('Builders of the Universe' 1932), widely characterized elsewhere

Emmerich Wolf-Dieter (1939–) Ger. phys., co-inventor of coupling TA techniques (designer for Netzsch)

Empedokles of Akragas (492–432 BC) Gr. phil., originated classical doctrine on 4 elements, held the Earth to be spherical and planets moving through space, changes could be understood in terms of motion

Epikouros of Samos (342–271) Gr. phil., adopted atomism as mechanistic explanation of universe, pleasure must be life of prudence, honor and justice equations, mechanic of fluids, hydrodynamics, oscillations

Eukleides of Alexandria (365–300 BC) famous mathematician, founder of 'flat' (Euclidian) gemetry

Euler Leonhard (1707–1783) Swiss math., most prolific mathematician, algebraic series, functional notations (Euler numbers), imaginary numbers (i), topology, differential

Eysel Walter (1935-1999) Ger. chem., expert in mineralogical thermal analysis

Fahrenheit Gabriel Daniel (1686–1736) Pol. phys., working in Amsterdam, inventor of alcohol and mercury thermometers dressed with his temperature scale, discovered undercooling of water

Faraday Michael (1791–1867) Brit. phys., one of the greatest experimentalists, explained electromagnetism, introduced concept of magnetic lines, Faraday's laws of electrolysis, unit of electrostatic capacitance named farad

Favre Pierre-Antoine (1813–1880) Fr. chem., known for series of then rather precise calorimetric determinations of heats involved in various chemical reactions

Fechner Gustav Theodor (1801–1887) Ger. phil., pioneering psycho-physics, measuring sensation indirectly in units corresponding to the just noticeable differences between two sensations, cofounder of the famous W-F physiological law (book 'Elemente der Psychophysik' 1860)

Feigebaum Mitchell Jay (1945–) US. phys., famous for discovering the constant 4.6692… for ith bifurcation ($\lim_{i\to\infty} = di/di+1$) named after him (Feigebaum numbers), his disclosures spanned new field of theoretical and experimental mathematics

Fermat De Pierre (1601–1665) Fr. math., devised principle of least time (action) and Fermat's small and big (last) theorems, gravity reciprocal attraction, father of modern theory of numbers, probabilities

Feynman Richard Phillips (1918–1988) US. phys., quantum electrodynamics, devised Feynman diagrams as means for accounting possible particle transformations ('Theory of Fundamental Processes' 1961)

Fibonacci Leonardo of Pisa (~1170–1230) It. math., best known for his book of Abacus, putting thus end to old Roman system of numerical notations, his series are now called Fibonacci's

Fick Adolf Eugen (1829–1901) Ger. physiol. who made important discoveries in every branch of psychology, well-knownfor the Law of diffusion (Ann. Phys. 94(1855)59) named after him

Flynn Joseph Henry (1922-) US phys., known for Flynn kinetic evaluation method

Fourier Jean Baptiste Joseph (1768–1830) Fr. math., evolved mathematical series known by his name and important in harmonic analysis, providing source of all modern methods in mathematical physics, originated Fourier's theorem on vibratory motions

Frankenheim Moritz Ludwig (1801–1861) Ger. phys., modern structural theory of crystals, introduced cooling curves to study materials (temperature vs. time)

Frankland (Sir) Edward (1825–1899) Brit. chem., organo-metallic compounds, effect of atmospheric pressure on combustion, modern concept of valence, studied flame and luminosity

Friedman Aleksandr Aleksandrovich (1888–1925) Rus. math. and astron., known for nonstationary solution of general theory of relativity, cofounder of dynamic meteorology who anticipated theory of Big Bang

Fürth Reinhold Heinrich (1893–1979), Czech born, Ger. and Brit. phys., authority in statistical mechanics and Brown movement who gave stochastic approach to quantum mechanics

Galen Klaudios (Galenos of Pergamum) (129-199 AD), Gr. physiol., systemized and unifies Greek anatomic and medical knowledge, possibly a founder of modern science, believed that mind is located in brain

Galilei Galileo (1564–1642) It. astron., showing that velocity of a falling object is proportional to (g t) but not to its weight, invented hydrostatic balance, discovered numerous stars and planets, analyzed projectile motion, gave apparatus for temperature measurements

Gallagher Kent Patrick (1931-) US thermoanl., authority in solid-state chemistry of materials, electronics, editor of various compendia

Galwey Andrew Knox (1933-) Ir. phys., expert in kinetics of solid-state reactions, author of books on same subject

Gamow George (1904–1968) Rus. born US phys., applied nuclear physics to problems of stellar evolution, proposed theory of origin of chemical elements by successive neutron capture and creation of the universe from a singularity ('Thirty Years that Shock Physics' 1965) and even went to analyze coding for triplet-system of proteins

Garn Paul D. (1920-1999) US chem.., expert in non-isothermal kinetics, early TA promoter (book 'Thermal methods of investigation' 1964)

Gassendi Pierre (Gassendius) (1592–1655) Fr. phil and math., held that atoms differed in size, weight and shape, gaseous

pressure is due to collisions, measured velocity of sound

Gauss Karl Friedrich (1777–1855) Ger. math., fundamental theorems of algebra and ontribution to modern number theory (Gaussian integers), vectorial representation of complex numbers, method of least squares and observational of errors, unit of magnetic field gauss named in his honor

Gay-Lussac Joseph Louis (1778–1850) Fr. chem., discovered law (stating that all gases expand equally for equal increment of temperature), verified law of capillary action, investigated temp. solubility of salts, invented hydrometer, barometer, thermometer

Gibbs Josiah Willard (1839–1903) US. phys., father of thermodynamics of heterogeneous substances where he established theoretical basis of physical chemistry (Gibbs phase rule), vector analysis in crystallography, statistical mechanics, even patented railroad brake

Glauber Johann Rudolph (1604–1670) Ger. chem.., invented various synthetic and analytical reactions, gave new philosophy of stills and furnace construction and operation

Gmelin Eberhard (1937-) Ger. phys., expert in low temperature calorimetry and material science

Gödel Kurt (1906–1978) Czech born US logician who gave proof of completeness of predicate logic and showed that inconsistency cannot be proved in the same system ('Consistency of Continuum Hypothesis' 1940)

Godovsky Yuli Kirillovich (1936-) Rus. chem., thermal physics of polymers (book "Thermal Methods of Investigation" 1976)

Gravelle Pierre Charles (1931-) Fr. chem , expert in adsorption calorimetry

Gray Allan P. (1931-) US calorim., early architect of DSC theory

Grolier J.-P. Etienne (1936-) Alger. born Can. chem., thermodynamics of solutions (book "Thermodynamic Properties" 2004)

460

Guggenheim Edward Armand (1901–1980)
Brit. phys. chem., known for applications
of thermodynamics and statistical
mechanics to properties of gases, mixtures,
electrolytes
Guldberg Cato Maxmilian (1836–1902),
Norw. chem., developed chemical law of
mass action, studied chemical equilibrium
and reverse reactions
Guthrie Frederick (1833–1886) Brit. phys.,
studied magnetism, electricity, vibration,
cryohydrates ('Elements Heat' 1868)
Haines Peter John (1934-), Brit. chem.,
thermomicroscopy, simultaneous TA,
polymers (book 'Thermal methods of
analysis' 1965)
*Hájek Tadeáš from Hajků (Thaddeus
Hagecius)*(1526–1600) Czech. astr.,
alchemist. and personal physician of the
Roman Emperor Rudolph II (see Rudolph)
Hakvoort Gerrit (1938-) Dutch chem.,
solid catalysts, environmental, special
instruments (book "Reactionkinetics"
1991)
Hamilton (Sir) William Rowan (1805–
1865) Irish mathematician, introducer of
quaterions, alternative formalism for tensor
and vector calculation, widely used
operation by "Hamiltonian"
Hannay James Ballantyne (1855–1931)
introductory experimentation on isothermal
mass-change curves
Harmelin Miriam (1937-), French phys.,
specialist in calorimetry and kinetics of
glassy metals
Hausdorff Felix (1868–1942) Jew math. at
the Ger.University of Bonn, known for
fractal-preceding dimension named after
him
Havel Ivan Miloš (1938-) Czech kybernet.,
expert in cognitive sciences, inventor of
spatial-scale axis
Hay James Neilson (1935-) Brit. chem.,
thermal characterization of polymers
Heide Klaus (1938-) Ger. mineralog. and
glass chem.., expert in TA (EGA DEGAS)
of natural and industrial solids (book
"Dynamische termische
Analysenmethoden" 1982)

Heisenberg Werner Carl (1901–1976) Ger.
phys., who worked on atomic structure and
founded quantum mechanics, evolved
uncertainty principle named after him,
suggested that laws of subatomic
phenomena be stated in terms of
observable properties, involved in the
unified field theory ('Das Naturbild der
heutigen Physik' 1955)
Helmholtz Hermann Ludwig Ferdinand
(1821–1894), Ger. physiol., study on
mechanism of eye and ear, theory music,
('Uber die Erhaltung der Kraft' 1847 and
'Handbuch der physiologischen
Optik'1867), credited with explication of
conservation laws, elaborated on
electrodynamics and indicated
electromagnetic theory of light, studied
vortex motion in liquids
Helmont (van) Joannes Baptist (1579-
1644) Belg. physician, confided in
alchemy but rejected 3 principles and fire
as element since it has no matter, taught
indestructibility matter, used graduated air
thermoscope to measure temperature,
thought origin of bodies are water and
ferment
Hemminger Wolfgang (1941-) Ger. phys.,
experimental and theoretical calorimetrist
(book 'Fundamentals of calorimetry' 1979)
Hench Larry (1937-), US phys.,
introducing concept of bioactivity
(inorganic glasses for implantology, book
'Bioceramics' 1993)
Herakleitos of Ephesus (540-480), natural
philosopher believing to fire as a central
principle, known for his pessimistic view
of life
*Hermes Trismegistos (Mercurius
Termaximus,* conceivably the King of
Egypt known as *Sifaos,* ~ 1996 BC, and
identified with God Hermes by Greek
literature written circa 100-300 AD*),*
legendary person accredited to the
foundation of alchemic clandestine books
Heron of Alexandria (about 1st century)
Gr. math., arithmetic solution of quadratic
equations, inventor devices operated by
steam, fire engine pumps

Hesiodos (about 8–7th century BC) Gr. poet, known for 'Theogonia' essay of myths systematic trying to give the history of world

Hess Henri Germain (1802–1850) Swiss, prof. of chem.in Rus. St. Petersburg Univ., founder of thermochemistry, formulated (Hess') Law stating that amount of heat evolved is irrespective to intermediary stages

Heyrovský Jaroslav (1890–19670 Czech phys. chem., discoverer of polarography

Hilbert David (1862–1943) Ger. math., investigated theory of numbers and relative fields, developed Hilbert space in work on integral equations

Höhne Günther W. H. (1941-) Ger. phys., expert in calorimetry (books 'Grundlagen der Kalorimetrie' 1979 and ‚DSC' 1996)

Honda Kotaro (1870–1954) Jap. phys., who pioneered thermogravimetric measurements and who introduced more strongt steel and magnets

Hooke Robert (1635–1703) Brit. naturalist, cofounder of vibration theory, who devised an equation describing elasticity that is still used today ("Hooke's Law"), worked out the correct theory of combustion, assisted Robert Boyle in studying the physics of gases and improved meteorological instruments (barometer)

Hrubý Arnošt (1919-), Czech phys., synthesis and thermal analysis of complex glassy chalcogenides, devised Hruby glassforming coefficient

Hubble Edwin Powell (1889–1953) US. astronom, initiated the study of the universe beyond our galaxy ('Observational Approach to Cosmology' 1937), classified galaxies and discovered that their radial receding velocity is proportional to their distance (Hubble's Law)

Huffman Hugh Martin (1898-1950) US phys., combustion and adiabatic calorimetry, founder of North American calorimetry conferences, known for the Huffman Memorial Award

Huygens Christian (1629–1695) invented pendulum, conservation mobility ($\approx$ mv)

Jesenak Viktor (1926-2000) Slovak chem., expert in solid-state kinetics

Jörgensen Sven Erik (1942-), Danish phys., inventor of thermodynamics for ecological systems

Joule James Prescott (1818–1889) Brit. eng. who quantified heat liberated upon the electric passage through resistance, unit joule named after him

Kamerlingh-Onnes Heike (1853–1926) Dutch phys., obtained liquid helium, found uperconductivity of mercury, studied magnetic and optical properties at low temperatures

Karhanavala Ervad M.D. (1928-1979), Indian phys., expert in nuclear energy and applied TA kinetics, humanitarian thinker and Zoroastrian priest

Kauffmann Stuart (1939-), US biol. and math., expert in biophysics aimed to the nature of life and order, proposed 4th law of thermodynamics (book "Investigations" 2002)

Kauzman Walter (1916-1980) US. chemists, studied thermodynamics and statistics, expert in theory of glasses (Kauzman point)

Keattch Cyril Jack (1928-1999) Brit. chem.., expert in thermogravinetry, author of books in same subject

Kekule Friedrich August (from Stradonitz) (1829–1896), author of carbon tetravalent chemical bonding, visualized aliphatic chain series and closed-chain structure of benzene

Kelvin, Baron of Larges (Lord *Thompson Williams*) (1824-1907) Scot. math., did important contribution in most branches of physical science, developed dynamic theory of heat, collaborated in investigating Joule-Thompson effect, propose absolute scale of temperatures, invented various electric measuring devices and even developed improved mariner's compass still used today

Kemp Richard Bernard (1941-) Brit. zoolog., biothermochemical studies and bioengineering (book "From Macromolecules to Man" 1999)

Kepler Johannes (1571–1630) most famous astronomer who laid the foundation of modern astronomy, studied nature of light and introduced concept of rays, made use of logarithm

Kettrup Antonius (1938-) Gr. chem.., expert in ecological chemistry and bioanalytical methods of TA, book "Analysis of Hazardous Substances")

Kirchhoff Gustave-Robert (1824–1887) Ger. phys., did spectrum analysis, black-body concept, formulated Kirchhof's laws of electricity

Kissinger Henry E. (1905-1979) US chem., known for kinetic evaluation method named after him,

Kolmogorow Andrej Nikolajevich (1903–1970) Rus. math., known for theory of functions, concept of probability, irrational functions ('Basic Concepts of Probability' 1936)

Komensky, see *Comenius*

Koperník viz *Copernicus*

Kubaschevsky Oswald (1912-1986) US chem., pioneering comprehensive thermodynamic databases (book 'Materials Thermochemistry' 1963)

Kurnakov Nikolaj Semenovič (1860–1941) Rus. thermoananlyst, inventor of the drum photographic recording used in thermal analysis

L'vov Boris V. (1931-) Rus. chem., spectroscopy, decomposition kinetics (book "Atomic Absorption Spectrochemical Analysis" 1970)

LaGinestra Aldo (1930-1993) Ital. phys., expert in thermal analysis of biological systems

Lagrange Joseph Louis (1736–1813) Fr. math., showed mechanics could be found on the principle of least action, studied perturbations, hydrodynamics, developed calculus of variations, partial differential equations

Landau Lev Davidovich (1908–1968) Rus. phys., developer of thermodynamic theory of second-order transformations,

superfluidity, low temperature physics ('Continuum Mechanics', 'Hydrodynamics and Theory of Elasticity' 1944)

Langier-Kužniarová Anna (1931-) Pol. geolog, mineralogy and TA of organo-clay complexes (book "Thermograms of minerals" 1967)

Laplace Pierre Simon (1749–1827) Fr. math., laid foundation of thermochemistry, theory of probability, made much use of potential partial differential equations since named after him, conducted experiments of specific heat and heat of combustion, developed ice calorimeter

LaViolette Paul A. (1938 –) US. astrophys., developer of unusual subquantum kinetics and continuous creation model of the universe, and novel approach to microphysics that accounts for forces in a unified manner

Lavoisier Antoine Laurent (1743–1794) Fr. chem., discovered relation between combustion and oxygen, divided substances into elements and compounds, explained respiration, disproved phlogiston, introduced quantitative methods to chemistry

Lazarev Vladislav Borisovich (1929-1994), Rus. thermoanal., expert in inorganic materials

LeChatelier Henry-Louis (1850–1936) Fr. metallurg., worked on chemistry of silicates and cements, physics of flames, thermodynamics, used first the dependence sample vs. environmental temperature, devised optical pyrometer

LeChatelier Luis (1815–1926) Brit. mine eng., tested products with aluminum content, patented steel production, thermometry – use of thermocouples

Legendre Adrien Marie (1752–1833) Fr. math., gave important works on elliptic integrals, laws of quadratic reciprocity, created spherical harmonics, known for Legendre transformations used in thermodynamics

Leibniz Gottfried Wilhelm (1646–1716) Ger. math., showed the lost of motion after collision called 'vis viva' (similarly to 'vis

mortua'), introduced modern mathematical notations, postulated theory of monads (building block of universe) and even designed wagon wheels
Lemery Nicolas (1645–1715) Fr. chem., founder of numerous applications of chemistry in medicine, studied theory of volcanoes, authored textbooks on chemistry ('Cours de chymie' 1675) invented workable internal combustion engine, electric motor and signal system for railroads, studied galvanoplastic reproduction
Leonardo Da Vinci (1452–1519) It. artists, one of greatest and most versatile geniuses, commencing those moving bodies can transfer motion which in total is unchanged, among other designed gliders, parachute, elevator, steam gun, studied human anatomy, etc.
Leukippos of Miletos (~ 500–440 BC), Gr. phil., cofounder of atomistic theory, bodies are in circular motion move from center, motion is made possible by positioning empty space
Lewis Gilbert Newton (1875–1946) Amer. chem., studied thermodynamics and free energy of substances, valence and structure of molecules ('Anatomy of Science' 1926)
Libavius Andreas (Liebau) (1540-1616), Ger. physic., histor. and alchemist who described dry reactions in assaying metallic ores, detailed aqueous analysis ('Alchemia' 1597 in which second edition 1606 is the section 'De Pyrotechnica')
Liesegang Raphael Eduard (1869–1947) Ger. phys. chem., expert in dye chemistry who worked systematically in periodic precipitations (known for Liesegang rings)
Lifschitz Evgenni Mikhaylovich. (1915-1985) Rus. phys., expert in statistical physics, fluid- and electro-dynamics (book 'Course of Theoretical Physics' 1979)
Linnaeus (Carl von Linne) (1707–1778), Swed. Botanists known as father of modern synthetic botany, known for reversing the Clausius scale (freezing 100–boiling 0)

Liptay Geörgy (1932-) Hung. thermoanal., expert in simultaneous TA methods (book "Atlas of TA curves" 1971-1976)
Liu Zhen-Hai (1936-) Chin. chem., thermal analysis of macromolecules (book "Calorimetric Measurements" 2002)
Lobachevski Nikolaj Ivanovich (1793–1856), Rus. math. who gave innovative curved geometry (targeted to concave surfacevof a sphere)
Logvinyenko Vladimir A. (1937-) Russ. chem.., expert in TA study of coordination compounds, author of books on same subject
Lombardi Gianni (1939-) Ital geol., expert in mineralogic characterizations, early initiator of ICTA
Lomonosov Michail Vasilyevich (1711–1765) Rus. math., founder of Moscow university, introduced comprehensive structure of non-Euclidean geometry
Lønvik Knut (1935-) Norweg. phys, inventor of thermal sonimetry
Lorentz Hendrik Antoon (1853–1928), Dutch phys., authority in quantum physics, electromagnetism, thermodynamics, radiation, behavior of light, electron theory of matter, hydrodynamics (mostly cited for Lorentz transformation)
Lulla, Lullius Raimundus (1235-1315) physician and alchemist, devised what he considered an infallible method of proving faith and reason, invented mechanical device ('ars magna') which combined subjects predicated of propositions thus producing valid conclusions
Mach Ernst (1838–1916) Aust. phil. and phys., known for his discussion of Newton's Principia and critique of conceptual monstrosity of an absolute space ('The Science of Mechanics' 1883)
Maciejewski Marek (1940-) Pol. chem., inventor of pulse (hyphenated) methods of thermal analysis, expert in heterogeneous kinetics
Mackenzie Robert Cameron (1920–2000) Scot. chem.., expert in clay chemistry, founder of International confederation on thermal analysis (ICTA)

464

Maimonides (Moses ben Maimon called Rambam) (1135–1204) Sephardic born Jewish physic. and phil., foremost intellectual figure of medieval Judaism, who tried to merge belief with learning and who recovered Aristotle's ideas ('Guide of the Perplexed' or 'Treatises on Logic' 1165)

Mandelbrot Bezat Benoit (1924–) Pol. born Fr. math., based mathematical theories for erratic chance phenomena and self-similarity, fractal dimension

Mantegna N. Rosario (1942-) Ital. phys., endeavoring to bridge physics and economy (book 'Econophysics' 2000)

Marcus Marci Jan (from Kronland) (1595-1667) unfamiliar Bohemian scientist, author 'De proportione motu', already noting principles of the light diffraction, studied impact of bouncer balls

Marti Erwin (1932-) Swiss chem., industrial thermal analysis, pharmaceutics (book "Angewande chemische Thermoanalytik" 1979)

Matějka Josef (1892-1960) Czech chem.., inventor of 'rational analysis' ofor thermal decomposition of ceramic raw materials (clays)

Maupertuis Pierre Louis Moreau (1698–1759) Fr. math., known for the principle of 'least action'

Maxwell James Clerk (1831–1879) Brit. phys., validated kinetic theory of gases, applied dynamical equations in generalized Lagrangian form and showed that electromagnetic action travels through space in transverse waves, as does light, symmetrical (Maxwell) equations of continuous nature of electric and magnetic field used today (book 'Matter and Motion' 1876)

Mayer (von) Julius Robert (1814–1878) Ger. phys., determined quantitatively equivalence of heat and work, studied principle of conservation laws even extended to living and cosmic phenomena

Mayow John (1641-1679) Brit. chem., studied similarity between chem. process of combustion and physiological function of respiration, showed that only part of air is used during burning

McAdie Henry George (1930-) Canad. chem., thermal properties of materials, standardization, environmental research, co-founder of NATAS

Mchedlov-Petrosyan Otar P. (1917-1997) Ukrain. chem., expert in concrete technology and thermochemistry

Mendeleyev Dmitryi Ivanovich (1834–1907), Rus. chem., gave periodic classification of elements, discovered periodicity in their chem. and phys. properties, investigated thermal expansion of liquids ('Principles of Chemistry' 1868)

Meyer Johann Friedrich (1705–1765) Ger. chem., known for theory of 'acidum pingui', contestant of Black's theory

Milesians School see *Thales*

Mimkes Jürgen (1939-) Ger. phys., expert in diffusion and econophysics, applied thermodynamic laws for societal behavior

Moiseev German K. (1932-), Russ. metalurg., expert on calculation of thermodynamic data (book 'Gibbs energy for some inorganic materials' 1997)

Murphy Cornelius Bernard (1918–1994) US chem., expert in TA instrumentation, cofounder of ICTA

Napier John (1550–1617) Scot. math., best known inventor of logarithm, originator of Napier's rules of circular parts for solution of spherical triangles and also Napier's bones (antecedent of a logarithmic rule)

Nernst Hermann Walther (1864–1941) Ger. phys., devised theory of electrothermic potentials, elaborated the third law of thermodynamics and made use of low temperature calorimetry

Neumann (von) Johann (Jánoš) (1903–1957) Hung. born US math., introduced premises of electronic computing devices and theory of games, math. logic and theory of continuous groups, showed math. equivalence of Schrodinger's wave mechanics and Heisenberg's matrix mechanics ('Mathematical Foundation of Quantum Mechanics' 1931)

Newcomen Thomas (1663–1729) Brit. inventor, known for atmospheric steam engine and its application for pumping water from mines

Newton (Sir) Issac (1642–1726) Brit. math. (most famous phys. and yet unfamiliar alchem.), prominent scientist who founded the discipline of mechanics, extensively described elsewhere

Niinisto Lauri (1941–) Fin. chem.., application of TA methods in thin films and optolectronic

Noether Max (1844–1921), Ger. math., theory of algebraic functions, symmetry aspects

Norton Thomas (1437–1514) Brit. alchemist who recognized importance of color, odor and taste as guides in chemical analysis, improved thermal regime of furnaces

Núñez-Regueira Lisardo (1939–) Spain. phys., energy recovery (book 'Biological calorimetry' 1982)

Odling William (1829–1921) Brit. chem., did early research on problems of valence and bounding, proposed the table of elements ('Outlines of Chemistry' 1869)

Ohm Georg Simon (1789–1854) Ger. phys., found (Ohm's) Law relating resistance to voltage and current strength, studied temperature resistance of metals, unit of electrical resistance named after him

Oinopides of Chia ($\approx$ 500 BC) Gr. math. and astr., supposable discoverer of ecliptic

Onsager Lars (1903–1976) Norw. chem., researched protonic semiconductors, helped theory of dielectrics, reciprocal transport relations in irreversible thermodynamics named after him

Opfermann Johanes (1940–2004) Ger. phys., expert in semiconductive propertics of polymers, architect of Netzsch kinetic softwares

Oppermann Heindrich (1934–), Ger.phys., thermochemical data, solid-state reactions

Ostwald Friedrich Wilhelm (1853–1932) Ger. chem., founder of modern physical chemistry, research in equilibrium and

rates of chem. Reactions, protagonists of so called 'energetism'

Otsuka Ryohei (1923-1996) Jap. phys., expert in mineralogical chemistry

Otto Nikolaus August (1832–1891) inventor of internal combustion engine

Ozawa Takeo (1932–) Jap. phys., expert in TA methodology, modulated modes and kinetics (Ozawa evaluation method)

 Paoletti Piero (1931–) Ital. chem., thermodynamics and calorimetry of complexes (book 'Ossiduriduzione' 1978)

Papin Denis (1647–1712) Fr. phys., improved air pump, discovered principle of siphon, demonstrated (and practically used) that at increased/decreased pressure the boiling point is raised/lowered, improved gunpowder and preceded steam engine

Paracelsus Theophrastus Bombastus von Hohenheim (Philippus Aureolus) (1493-1541) Ger. alchem. and phil., postulated internal 'archei' which acted as alchemists within body separating pure from impure, concept resulted in view of local centers of disease as imbalance of humors (liquid) throughout the body

Pascal Blaise (1623–1662) Fr. math., known for Pascal triangles, builder of first mechanically calculating machines as computer precursors, improved the theory of probability and combinatorial analysis, increased knowledge of atmospheric pressure (denied 'horror vacui'), founder of hydrodynamics (law bears his name)

Patzier Michal Ignac (1748–1811) Slovak metallurg., author of indirect amalgamating technology and organizer of the first internat. conf. on natural sciences (in 'Skelne Teplice' 1786)

Paulik Ferenc (1922–) Hung. thermoanal., inventor of simultaneous TA methods, known for commerce of 'Derivatograph', books on the same subjects

Pavlyuchenko Michail Michajlovič (1909-1975) Rus. chem.., theorist in heterogeneous reactions, author of books on same subjects

Peano Guisepe (1858–1932) It. math., founder of symbolic logic, non-Euclidean geometry, known for construction of space-filling curve named after him (Peano's axioms)

Pelovski Yoncho (1942-) Bulg. chem., thermal decomposition, simultaneous TA methods (book 'Technology of inorganic wastes' 1986)

Peltier Jean Charles Athanase (1785–1845) Fr. phys., watchmaker, discovered a thermo-electric reduction of temperature which effect is named after him

Penrose Roger (1931–) US. math., suggesting that all calculation about both micro- and macro- worlds should use complex numbers, (requires reformulation of major laws of physics) proposed a new model of universe whose building blocks he called 'twisters' (Penrose's tiling).

Petit Alexis-Therese (1791–1820) Fr. phys., co-developed methods of determining thermal expansion and specific heats of solid bodies

Philolaos of Tarentum ($\approx$ 500 BC) Ger. phil., held the earth is not center of universe but that is stars and planet circle about central fire

Pictet-Turrentin Marc-Auguste (1752–1825), Swiss phys., who attempted to measure velocity of heat

Piloyan Georgyi Ovanyesovich (1919-1989) Russ. phys., early treaties on theoretical thermal analysis ('Intro to Thermal Analysis" 1964)

Planck Max Carl Ernst Ludwig (1858–1947) Ger. phys., best know for Planck's constant representing quantum action, blackbody radiation, thermodynamics, physics before his quantum theory is often called classical

Platon (Plato) (427-347 BC) most famous Ger. phil., well characterized elsewhere, Platonism in science has general meaning of emphasis on a priori abstract mathematical thinking

Poincare Henri Jules (1854-1912) Fr. math., gave theory of functions, researched differential equations, theory of astronomical orbits, 3-body problem, theory of light, dimension and relativity ('Thermodynamique' 1892, 'Calculs des probabilites' 1896, 'La science et l'hypothe`se' 1906), unfamiliar herald of the theory of relativity

Poncelet Jean Victor (1788–1867) Fr. math., formulated principle of continuity, termed energy, gave practicable theory of turbines, Poncelet's overfall (meaning device) in hydrology

Popper (sir) Raimund. Karl (1902–1994) Austr. sci. and phil., known for theory of cognition and science (epistemology), defender of open thinking and society ('Logic der Forschung')

Prandtl Ludwig (1875–1953) Ger. phys., founder of modern hydrodynamics and aerodynamics, proved sound barrier, boundary layer on moving surfaces in liquids, Prandtl number named after him

Presl Jan Svatopluk (1791–1849) Czech. chem., originator of modern nomenclature in chemistry and botany

Priestley Joseph (1733–1804) Brit. chem., phlogistonist, explained some composites of air, history of electricity and light

Prigogine Ilya (1917–2003) Rus. born Belgian phys. chemist, inventor of nonequilibrium thermodynamics, propagator of the theory of chaos ("La nouvelle alliance avec la nature")

Proks Ivo (1926-) Czecho-Slovak chem.., expert in historical thermodynamics, inventor of periodic thermal analysis

Proust Louis Joseph (1754–1826) Fr. chem., established experimentally (Proust's) Law of definite portions, discovered sugar to exist in some vegetables

Ptolemaios Klaudios ($\sim$ 85–165 AD) Egyp. born, Gr. astr. who gave the first plausible explanation of (the Earth-centered) celestial motions, studied trigonometry and stereographic projections, attempted theory of refraction ('Syntaxix megale', 'Almagest' – most influential work in astronomy)

Pysiak Janusz J. (1933-) Pol. chem..,
expert in kinetics of solid-state reactions
Pythagoras of Samos (582–494 BC) Gr.
phil., best known for Pythagorean theorem,
credited with discovery of chief musical
intervals, attempted to interpret world
through numbers and classified them into
odd and even, stressed deductive method in
geometry, presupposed central fire around
which celestial bodies circle
Ramsay (Sir) William (1852–1916) Brit.
geolog., gave asymptotic continuous
record of heating hydroxides, geographical
mapping ('stratigraphy')
Rankine William John Macquorn (1820–
1872) Scot. phys., researched
thermodynamic theory of steam, engine
performance, water waves
Raoult Francois Marie (1830–1901) Fr.
phys., developed (Raoult's) Law for vapor
pressure of solvent in solution being
proportional to the number ratio of
solvent/solute molecules, demonstrated
depression of freezing points
proportionally to concentration of
dissolved substances
Rayleigh see Strutt
Reading Mike (1946-) Brit. phys., inventor
of modulated temperature techniques
Reaumur de Rene-Antoine Ferchault
(1683–1757) Fr. chem. and naturalist,
invented method of tinning iron and
porcelain, his temperature scale reading
from 0 to 80, showed impact of heat on
insect development
Redfern John P. (1933-), Brit. chem.,
promoter of TA and developer of advanced
TG systems, co-founder of ICTA in 1965
Regnault Henri Victor (1810–1878) Fr.
chem., measured specific heats, vapor
pressures of mixtures (Regnault's
hygrometer), participated in early
adjustment of phys. constants and laws
Regner Albert (1905–1970) Czech phys.
chem. who gave thermodynamic basis to
electrochemical technology, first class
educator
Reinizer Friedrich (1857–1927), Czech
born discoverer of cholesterol (including

its metamorphosis and stoichiometry
formulae $C_{27}H_{46}O$), known for introducing
the field of liquid crystals (latter
widespread by *Otto Lehmann*)
Richardson John Michael (1935-2004)
Brit. chem.., expert in quantitative
calorimetry (DSC)
Riemann George Friedrich Bernhard
(1826–1866) Ger. math., introduced idea
of finite but unbounded space (Riemann's
functions and prime numbers), devised
innovative geometry of the saddle-like
space, theory of electromagnetic action
Riga Alan T. (1937-) US chem.., polymers
and pharmacy science, lubricants and bio-
sensors (book 'Material characterization by
TA' 1991)
Roberts-Austin (Sir) William Chandler
(1843–1902) Brit. metal., designed
automatic recording pyrometer with Pt-
thermocouples for high-temperature study,
demonstrated that that diffusion can occur
between attached sheets of gold and led
*Rodovsky Bavor jun. from Bavorov (or
Hustiřan)* (1526–1592), Czech alchemists,
author of possibly the first book on
cookery
Rong-zu Hu (1938-) Chin. chem.,
thermodynamics of energetic materials,
kinetics of exothermic decompositions
Rouquerol Jean (1937-) Fr. chem.,
microcalorimetry, adsorption, co-inventor
of sample controlled thermal analysis
Rowland Henry Augustus (1848–1901)
Amer. phys., gave mechanical equivalent
of heat and of the ohm, studied magnetic
action due to electric convection (book
'Elements of Physics' 1900)
Rudberg Friedrich Emanuel Jakob (1800–
1839) Swed. phys., made refraction
measurements, inventor of heating and
cooling data for investigating alloys
Rudolph II (1552-1612), Rome Emperor
and Bohemian King, famous aesthete and
upholder of alchemy boom in Prague
Rumford (see Thompson Benjamin)
Rünge Ferdinand Fridlieb (1794–1867)
Ger. phil. And chem. who studied
processes of dyeing and first noticed

468

creation of fractal structures that
anticipated to become the archetype for
artists ('Farbenchemie' Vol. 1, 2 & 3,
Berlin 1834–1850)
Šatava Vladimír (1922-) Czech chem.,
expert in thermodynamics and chemistry of
cements, inventor of hydrothermal analysis
(book "Physical Chemistry of Silicates"
1962)
Scheele Carl Wilhelm (1742–1786) Swed.
chem., favored phlogiston theory,
demonstrated presence of calcium
phosphate in bones, discovered many new
substances (oxygen, acids, toxic gases)
Schrödinger Erwin (1887–1961) Austr.
phys., research in specific heats of solids,
statistical thermodynamics, showed that
matrix mechanics can be replaced by wave
mechanics, which pout new basis to
quantum-mechanics, known for
Schrödinger wave equation (solution for
the stationary state known as
'eigenfunction')
Schultze Dieter (1937-) Gr. chem.., expert
in simultaneous TA techniques (book
"Differentialthermoanalysen" 1969),
Sedziwoj Michael (Sendivogius) (1556–
1646) Pol. and/or Moravian alchemist who
joined the service to Rudolph II, known for
emphasizing air for life (book 'Novum
Lumen Chymicum' 1614)
Seebeck Thomas Johann (1770–1831) Brit.
phys., devised thermocouple, built a
polariscope, studied heat radiation,
thermomagnetic effect (known as
Seebeck's effect)
Segal Eugene (1933-) Rom. phys., expert
in heterogeneous kinetics of non-
isothermal processes (book
"Nonisothermal kinetics" 1983)
Seifert Hans-Joachim (1930-), Ger, chem..,
expert in construction of phase diagrams
Šesták Jaroslav (1938-), Czech.
phys.chem., expert on thermodynamics (of
glasses), known for application of fractal
kinetics (Sestak-Berggren equation),
propagator of interdisciplinarity, books on
thermophysical propertis of solids, glass

Shannon Claude Elwood (1916–1980) US.
math., research on Boolean algebra,
cryptography, pioneered information
theory - full statement of which appeared
in 'The Mathematical Theory of
Communication' (1949)
Sharp John H. (1938-) Brit. chem., expert
in cements, kinetics of solid-state reactions
Sierpinski Waclaw (1882–1969) Pol. math.
know from the construction known as
Sierpinski gasket, researched logical
foundation of mathematical and topology
('General Topology' 1952)
Šimerka Václav (1819–1887) Czech priest
and math., who introduced valuation in
psychology (logarithmic connotation of
feelings) providing basis of theory of
information
Simon Judit (1937-) Hung. phys., founder
and editor (since 1967) of the Journal of
Thermal Analysis (Academiai Kiado)
Sitter Willem de (1872–1934) Dutch astr.
who proposed that the universe is an
expanding space-time continuum with
motion and no matter ('Astronomical
Aspects of Theory of Relativity' 1933)
Škramovský Stanislav (1901–1983) Czech
phys. chem. and co-inventor of
thermogravimetry through his own-
designed 'statmograph'
Smykatz-Kloss Werner (1938-) Ger.
geolog., thermal analysis of minerals (book
"DTA in Mineralogy" 1974)
Sokrates (469-399 BC) Gr. phil., left no
writings, devoted his life to educating
youth
Sood Din Dayal (1939-) Ind. chem.,
nuclear fuel materials, thermodynamic
calculations (book 'Frontiers in nuclear
chemistry' 1997)
Sorai Michio (1939-) Jap. calor., molecular
thermodyn., thermochromic phenomena
Sørensen Toft Ole (1933-) Danish
metalurg., co-inventor of rate controlled
thermal analysis (book "Nonstoichiometric
Oxides" 1981)
Stahl Georg Ernst (1660–1734) Ger.
chem., renamed Becher's 'terra pinguis' as
phlogiston, observed acids have different

strength, propounded a view of fermentation

Stanley Eugene H. (1936-) US phys., theoretic in phase transitions and its application to economy (book 'Econophysics' 2000)

Stephenson George (1781–1848) Brit. eng., designed and build a steam locomotive, founder of Brit. railroad system

Stevens Smith Stanley (1906–1973) Amer. psych., experimental psychology where the power law is named after him and used for the measurements and psychological scaling ('Varieties of Temperament' 1942 or 'Experimental Study of Design Objectives' 1947)

Stoch Leszek (1931-) Pol. chem., expert in thermochemistry of solids (glasses)

Stokes (Sir) George Gabriel (1819–1903) Brit. math., laid foundation of scientific hydrodynamics, theory of fluid motion, Stokes' Law describes motion of small spheres in viscous fluid, established semi-convergent series used with Bessel and Furrier series, studied variation in gravity

Stolcius Daniel (1600–1660) Czech alchemist, author of 'Viridarium chimicum' and cofounder of mystic society 'Fraternitas Roseae Crucis'

Strnad Zdeněk (1939-) Czech chem., expert in glass, inventor of bioactive dental implants (book ‚Glassceramics: nucleation, phase-separation and crystalization' 1986)

Strouhal Čeněk (Vincenc) (1850–1922) Czech phys., first professor of experimental physics to the Czech Technical Univ., known for work in acoustics (Strouhal's eddy pitch) and thernodynamics (book 'Thermics' 1906)

Strutt (baron Rayleigh) John William (1842–1919) Brit. math., theory of sound, dynamics and resonance of elastic bodies, contributor to optics, acoustics and electricity, hydrodynamics, (Rayleigh number named after him)

Suga Hiroshi (1930-) Jap. phys. chem., expert in glass transition determination and definition, nonequilibrium studies, disordered solids, calorimetry

Sunner Stig (1917-1980) Swed. phys., inventor of combustion rotating bomb calorimeter, known for Sunner Memorial Award of US Calorimetry Conferences

Szakó János (1926-2001), Ruman. phys., specialist on nonisothermal chemical kinetics (book "World of Atoms and Molecules" 1963)

Szilárd Leó (1898-1964) Jewish Hung.-US phys., first to think of building atomic bomb, create chain reaction (Be+I), molecular thermodynamic concepts

Tammann Gustav Hendrich Johan Appollon (1861–1938) Ger. chem., research in inorganic chemistry ('Lehrbuch der metallkunde' 1914)

Tan Zhi-Cheng (1941-) Chin. chem., phase transition calorimetry, energetic materials (book "Chinese Chemistry Encyclopedia" 1989)

Tesla Nikola (1856–1943) Croatian electrician US resident, invented Tesla motor and system of alternating current power transmission

Thales of Miletos (624-548 BC) Gr. math., tried to find naturalistic instead of mythological interoperation of nature, invented logical proof in geometry, determined sun's course, studied static electricity

Theophrastos of Eresos in Lesbos (370–285 BC) Gr. phil., considered as a founder of botany as a systematic study, known for caricaturing various human ethical types, studied mineralogy, meteorology, physiology, physics

Thompson Benjamin (Count Rumford) (1753–1814) Brit. phys., disapproved caloric showing heat as motion, tried to calculate equivalents of heat, invented shadow photometer, water compensation calorimeter, even improved functioning of fireplaces and chimneys

Thomsen Hans Peter Julius (1826–1909) Danish phys., important thermochemist, who tried to determine the absolute values of chemical forces in order to improve yet vague concept of affinity ('Thermochemische Untersuchungen' 1906)

Thomson William (see *Kelvin* Baron of Largs)

Tian Albert (1880-1972) Fr. calorim., inventor of heat-flux microcalorimetrz and known for heat balance equation named after him

Torricelli Evangelista (1608–1647) It. math., noted the pressure of air, inventor of mercuri barometer and thermometer, improved telescope and microscope, investigate theory of projectiles

Truesdell Clifford Ambrose (1921–) US. phys., known for founding the basis of rational thermodynamics (book 'Rational thermodynamics' 1964)

Turi Edith A. (1935-) Hung. born US chem., polymer science, TA education (book 'Thermal characterization of polymeric materials' 1996)

Turing Alan Mathison (1912–1954) Brit. orig. math., US cryptography decoder, theoretical researcher of complex systems known as 'Turing machine' (framework for computing any decidable function) and the 'Turing test' (for evaluating whether machines are 'thinking')

Tyemkin Michail Isaakovič (1908–1979) Rus. math. and phys., essential contributions to the thermodynamics of solids

Tyndall John (1820–1893) Brit. chem., studied diamagnetism, absorption and radiation of heat by gases, demonstrated dispersion of light beam by suspended particles in colloids (Tyndall's effect) values calculation of fuels upon their chem. composition

Van der Waals Johannes Diderik (1837–1923) Dutch phys., research on gaseous and liquid phases, determined so-called perfect and real gases, thermodynamic theory of capillarity, know for Van der Waals forces between dielectric molecules

van't Hoff Jacobus Henricus (1852–1911) Dutch chem., father of phys. chem., relating thermodynamics to chem. reactions, laws regulating chemical equilibrium, melting points, steam pressure, introduced concept of chem. affinity

Várhelyi Czaba (1925-) Rom. chem., coordination chemistry, decomposition kinetics, book in same subjet

Varschavski Ari (1940-) Chile phys., expert in DSC analysis of elementary processes in metals

Vieille Paul (1854–1934) Fr. chem., first to measure the heat of explosion under oxygen pressure, inventor of calorimetric bomb

Vitruvius Pollio Marcus ($\approx$ 100 BC) It. arch., authority on architecture for centuries, studied hydraulics, clocks, mensuration, geometry, mech. engineering

Vold Marjorie Jean (1913–1969) US chem., colloid chemistry, theory of DTA

Vopěnka Petr (1935-) Czech math. and phil., inventor of the alternative theory of sets, books on same and various philosophy aspects

Waage Peter (1833–1900) Norw. chem., developed so called Guildberg-Waage's Law of mass action

Wadsö Ingemar (1933-) Swed. phys., cofounder of precise calorimetry

Wald František (1861-1930) Czech.chem., originator of content determination of oxygen and manganese in steel

Waterston John James (1811–1883), Brit. phys., tried to interconnect 'vis viva' with temperature, equipartition theorem

Watt James (1736–1819) Scot. engng., inventor of steam engine with a separate condenser, used a conversion for reciprocating motion to rotary by sun-and-planet gear, improved combustion furnace, unit watt is named in his honor

Weber Ernst Heinrich (1795–1878) Ger. anatom., profounder of famous (Weber-Fechner's) physiology Law, made studies of acoustic and wave motion, pioneered studies on nervous impulses

Wendlandt William Wesley (1927-2000) US chem., expert in TA techniques, founder and editor of Thermochimica

Acta, (book 'Thermal Methods of Analysis' 1962)

Wichterle Otto (1913–1998), Czech chem., inventor of widespread contact hydrogen lenses, organic polymers and plastics (kaprolaktan - silon)

Wiedemann Hans G. (1928-) Ger. born, Swiss thermoanal., inventor of progressive thermobalances designed for Mettler (book 'Chemische thermodynamik und thermoananlytik' 1979)

Wiener Norbert (1894–1964) US. math., major contributor to cybernetic concepts (book 'Cybernetics' 1948), theory of probability ('Nonlinear Problems in random Theory' 1958). He defined cybernetics as a discipline concerned with the comparative study of control mechanisms in the nervous system and computers (book "Human Use of Human Beings" 1950 and "Cybernetics of the Nervous System" 1965).

Wilburn Fred W. (1925-) Brit. thermoanal., heat transfer models, glass-making reactions

Wilcke Johan Carl (1732–1796) Ger. phys., formulated independently theory of specific heats, studied electric dispersion and Leyden jar and accepted theory of 2 fluids for electricity

Wunderlich Bernhard (1931-) Ger. born, US phys., expert in thermodynamics of polymers, propagator and theorist of modulated TA methods (books 'Macromolecular physics' 1973 and 'Thermal Analysis' 1990)

Xenophanes of Kolophon (565-470 BC) Gr. phil., solved problem of combinatorial analysis, worked on theory of primary numbers, wrote history on geometry

Yariv Shmuel (1934-) Izr. chem.., expert in clay chemistry and coupled TA techniques (book 'Organo-clay complexes" 2002)

Zadeh Lofti A. (1931–) US. math., introducer of fuzzy logic as a tool for modeling human reasoning

Zanotto Edgar (1944-) Brazil.phys., inventor of various nucleation laws in crystallization of glasses

Zelenkiewicz Wojciech Wladyslaw (1933-) Pol. chem., co-initiator of calorimetric theory, (book "Theory of Calorimetry" 2002),

Zenon of Elea (490–430) Gr. math., regarded as inventor of dialectic, used paradoxes to illustrate his philosophical arguments

Zhabotinsky Anatol Markovich (1938–) Rus. born, US phys., cofounder and explainer of oscillatory modes of chemical reactions

Živkovič Živan D. (1939-) Serb. metalurg., expert in reaction kinetics, founder and editor J. Mining Metal. (book 'Principles of metallurgical thermodynamics ' 1997)

Zu Chong-Zhi (Cu Ts'ung or Chohung Chi) (430–510 Chinese math., astron. and engineer often cited in literature

About the author

Jaroslav Šesták, was born in the village 'Držkov' (North Bohemian Mountains) where he still possesses a small farm. He obtained an MEng in silicate chemistry (Prague 1962) and PhD in solid state physics (1968), the latter was approved while he spent a year at the University of Missouri at Rolla (UMR 1970) as an assistant professor in ceramics. He got married when working in Sweden (1969, Studsvik Nuclear Research Center) with a MEng graduate from Prague (Věra) who joined him in the USA where she received her second degree in ceramics (M.S., UMR 1970). Since then Jaroslav and Věra have had two children, a daughter Elizabeth (Bětka, *1977) and a son Paul (*1980).

Jaroslav's scientific proficiency is in experimental and theoretical studies related to thermodynamics of materials, particularly glasses. After the fall of communism he received an honorary DSc in material engineering (Prague 1990) in and became a full professor in material sciences and education (1993). He has edited and authored 13 books and monographs, and published some 300 papers that have received almost 2500 citations. He was a founding member of *Thermochimica Acta* (1970) and is a member of the editorial boards of *Journal of Thermal Analysis* and *Journal of Mining and Metallurgy* as well as participating in other scientific and educational boards. He has given over 150 invited keynote lectures. He was presented with various scientific awards: NATAS (USA 1974), Kurnakov (USSR 1985), Bodenheimer (Israel 1987), ICTAC (England 1992), Hanus (Czech Chemical Society) and Heyrovsky (Czech Academy of Sciences) medals (Prague 1998 and 2000 respectively) and was appointed an honorary member of the Czech Engineering Academy (2004). He is a renowned teacher and mentor who has tried to introduce many new methods of interdisciplinary learning (endeavoring to bridge sciences and humanities) both at home (cofounder of Faculty of Humanities of the Charles University in Prague and Institute of Interdisciplinary Studies of the West Bohemian University in Pilsen, teaching at Technical Universities of Liberec and Pardubice) and abroad (founding member of the Faculty of Energy Science of Kyoto University (Japan 1996) and lecturing at various universities in the USA, Norway, Italy, Chile, Argentina, Taiwan, etc,). Beside his scientific career he was a league basketball player, mountaineer (Himalaya, Caucasus, Pamir, Andes and the Alps – earning funds as an occasional window-cleaner roping down tall buildings), ski instructor, politician (deputy and member of the Prague government 1994-1998, and a candidate for a seat in the Czech parliament) and enthusiastic globetrotter (notoriously carrying a sleeping bag in his backpack while participating at scientific conferences). He is also a recognized photographer who is famous for 'trying to capture the beauty that is incensed by his heart, which is as yet unblemished by the daily rush of an apathetic over-industrialized society'. He has held over twenty photo-exhibitions (such as Smíchov City Hall 1998, EcceTerra gallery 2000, at the occasion of 10th anniversary of the Western Bohemian University 2001, Klamovka gallery 2003, Franzensbad and Prague Academy of Sciences 2005, etc.) and as a renowned scientist, he was invited to exhibit at a number of international conferences (e.g., Tokyo 1992, Cordoba 1995, Zakopane 1997, Balatonfured 1998, Copenhagen 2000, Vancouver 2002). His address is: V stráni 3, CZ-15000 Prague,Czech Republic; Email: sestak@fzu.cz.

Printed in the United Kingdom
by Lightning Source UK Ltd.
134274UK00002BA/1/A